# 366 – A Scriptovisual Composition Unknown

An Artistic Investigation of the Spaces between Spaces in the Quest to Create a post-European Era, post-Enlightenment Science

Being
A simultaneous encounter with the blank page and the blank canvas

# 366 – A Scriptovisual Composition Unknown

An Artistic Investigation of the Spaces between Spaces in the Quest to create a post-European Era, post-Enlightenment Science

Being
A simultaneous encounter with the blank page and the blank canvas

by

Paul T Kidd

*Cheshire Henbury*

First published in 2017 by Cheshire Henbury. Ebook edition published in 2018.
Paperback version ISBN 978-1-901864-24-3
Ebook version ISBN 978-1-901864-25-0

British Library Cataloguing in Publication Data: A catalogue record for this book is available from the British Library.

Email – Use Email Contact: www.cheshirehenbury.com/emailcontact.html
Web site: www.cheshirehenbury.com/366-scriptovisual-composition-unknown

Dedicated to Professor Howard H. Rosenbrock, FRS, (1920-2010), who, in endeavouring to move beyond the machine-centred causal thinking underlying the science and technology that emerged from the Enlightenment, did inspire me to write this book by way of a small contribution to fulfilling his vision of creating a post-Enlightenment purposive science and technology. Evidently it is now also time for a post-European era science and technology as well …

*Also by Paul T Kidd*

*Scriptovisual Works*

STARTS – Science, Technology and the Arts: The Artistic Voices that DG CONNECT Silenced

*Fiction*

Moments in Time

Enigma

A Tale of Two Deserts: Enigmatic Christmas Fables for the Modern Age

Encounter with a Wise Man: A Christmas Tale of Wisdom

Father Christmas Adventures: Unexpected Tales of Christmas Magic

A Father Christmas Story: Being a Tale of How Father Christmas Came to Be

*Non-fiction*

E-Business Strategy: Case Studies, Benefits and Implementation

Electronic Business: The Executive Guide

Rapid Prototyping for Competitive Advantage: Technologies, Applications and Implementation for Market Success

Revolutionising New Product Development: A Blueprint for Success in the Global Automotive Industry

Agile Manufacturing: Forging New Frontiers

*Edited Books*

European Visions for the Knowledge Age: A Quest for New Horizons in the Information Society

E-business: Key Issues, Applications and Technologies

Business and Work in the Information Society: New Technologies and Applications

Technologies for the Information Society: Developments and Opportunities

Electronic Commerce: Opening up New Opportunities for Business

Advances in Information Technologies: The Business Challenge

Advances in Agile Manufacturing: Integrating Technology, Organisation and People

Organization, People and Technology in European Manufacturing

## PREFACE

What lies before you, in the pages that follow, is an experiment in writing, made necessary by that which is taking shape in the world. Now has come the time to change our behaviour. Now has come the time too for action. Thus do I act, by taking up the pen and the brush! Both have more power than science and technology: which of course you probably do not believe, such is the conditioning that you have been subjected to.

Thus, *not knowing* what to write, but knowing that it was time to write, in the early morning of January 1st 2016, I began to write. What emerged from the process of *not knowing*, was 366 chapters - one for each day of the year - which brings together material from the three major domains of learning and knowledge: the arts and humanities; the social sciences; and, of course, science, technology, engineering and mathematics. For reasons that defy full understanding, by an act of chance perhaps, or, if you will, because of destiny, my life became tied up in all three domains to the extent that I can move freely and comfortably among and across all of them. This confers an advantage, which, within the pages that follow, I exploit.

This book is the way it is for a reason, which one day I may discuss in more detail. To elaborate more at this stage would not be helpful. All I will say is that the book breaks many conventions and rules and that there are aspects to the style of this book that have been seen before, across thousands of years of human history, but where I will not say. You may discover, while reading, some aspects of this for yourself. What I can tell you is that the book will not be fully understood. Why this is so, you will need to discover for yourself. Moreover, it will not appeal to those who are too far gone, whose minds are beyond being reached. Some people though are ready for a book such as this one and the messages and knowledge that it carries. The work is also, simultaneously, both the result and a record of a dual creative process. That is all I intend to say.

If you think this sounds strange and mysterious, this is because the book is, in some respects, strange and mysterious. As I said above - it is the way it is for a reason. Welcome to the beautiful heresy ...

*Paul T Kidd PhD, CEng, FIMechE, FIET, SMIEEE*

*November 2017*

# 1

In the beginning was the image and from the image came forth the word, and to the image the word doth return. And these are the words that we speak in silence. Listen!

Understand according to your capacity to do so. It is within your power to expand your capacity to understand, to grow, to reach out to places where you would not think to venture. Your life does not have to be bounded by ideology and dogma - in whatever form these may come. There are many; some religious, some secular, some scientific, some technological, some artistic, some political, some economic, some ... All are belief systems that people use to avoid ... What is it that people use belief systems to avoid? Thus we begin to write.

So it came about that it was born in the darkness, which has always been the case, and death too shall come when the sun shines in other places; and in recent times there be, by way of celebrations, some singing and dancing on occasions like these.

A cloudless sky brings a cold night, which is somewhat strange since winter has been like spring - leaves bursting out can be seen, while summer flowers' seeds have been germinating. This, one can also say, came about as a consequence of their singing and dancing, but of a kind that is now performed on a global scale over what amounts to little time, but many lifetimes. It is the way of things - at least in these troubled times.

Before we slipped into dream the stars in the heavens looked down on us and we - we looked up to see them too, for the inky blackness was that night peppered with stars shinning bright - as bright, that is, as is now possible given what the singing and dancing ones have done in pursuit of their strange notions of progress. Looking at us is the constellation Orion. It watches over us.

Waking we find not the starry hunter, for indeed the world has turned, but instead, a waning gibbous moon is waiting for us, slightly shrouded in wisps of what we assume is high altitude cloud. We cannot tell, yet precision is not the aim. Of this, at least, we are sure, which leads us to big data: Coordinates: 53.2581°N, 2.1274°W; Day: Friday; Sunrise: 08.23; Sunset: 16.02; Minimum temperature: 2°C; Maximum temperature 6°C: Wind: South Easterly, ranging 12–42mph; Sky: clear, turning cloudy, followed by light rain, turning to heavy rain as the day draws to a close.

Big data or little data? This is for you to decide.

It is the eighth day.

The rain was predicted for about 16.00. It came at about 16.00. Nature however is not always so predictable, but you would not think so given the way those who sing and dance behave. They engage too much in their celebratory glee and have not yet realised that their way of knowing, is yet still primitive, and next in the development of humanity comes different ways of knowing that inherently acknowledges nature's ultimate unknowability. Something has to be done to help those that sing and dance to move beyond that which they currently believe.

Daybreak brings a welcome spectacle - winter has reappeared with rooftops dusted white with a frosting that is rapidly becoming an unusual sight. Yet, not known, is whether winter will any grip apply.

We did not need to scrape the car's windscreen, for there was no need, but nature did very quickly do this deed, even though there was not the need. Day is short, for it is that time of year.

Come the night once more, and Orion cannot be seen. Yet it still watches over us, we believe. Seeing is most definitely not believing. Orion and that which it symbolises is also a way of knowing. Constellations - are these too, just a way of representing big data? Just state to yourself what you believe is this case. Speak quietly, silently.

The matter, now, which we must also address, that unspoken question - why do we write this? This is a composition that is most surely, *unknown*, which comes from - *not knowing*. Perhaps here, in the journey of discovery, the reason will reveal itself to us - it is discovery. Is this what it is about? As we summon words to appear, can we become the magicians that we seek to be? Can we command words to sing and dance? Surely this is not an end, but a beginning. This is the journey on which we now embark which is just the continuation of a lifelong quest. This is a composition unknown - our composition unknown.

We? Us? What do these words mean? They represent Julia and Paul. Who are Julia and Paul?

## 2

This is not a diary. This is not a collection of individual blogs. What do such phrases mean? Hermeneutics - yet another way of knowing!

We said that we would not tell them anything. We tell them and they forget. We teach them and they may remember. We involve them and they understand. So sometimes words can mean something different to that which the rational and logical mind believes they mean.

Coordinates: 53.2581°N, 2.1274°W; Day: Saturday; Sunrise: 08.23; Sunset: 16.02; Minimum temperature: 7°C; Maximum temperature 9°C: Wind: East South Easterly, changing to Easterly, then to South Easterly, then Westerly, shifting through West South Westerly to South Westerly, ranging 5-19 mph; Sky: cloudy with light rain predicted for the whole day. It did rain, but not constantly - nature will not be subject to your predictions!

It is the ninth day, or perhaps not. Some say it is just the eighth. It depends when you start counting.

Big, little, large, complex ... Such adjectives are subjective and are inherently ambiguous. They invite interpretation, reflection, questioning, exploration, which infrequently happens. So there you have it, for among all that is supposed to be rational and objective, sits this subjectivity, which cannot be acknowledged, or valued. It is just ignored. It is just one more inconsistency. The human mind is full of these - so rejoice and embrace your irrationality for it is not something that should be condemned.

Not everything can be quantified, nor does it need to be. Qualitative is also another way of knowing. Precision is not everything.

Art embraces subjectivity and irrationality, as well as the rational and objective, and more. They all co-exist, because art is also a way of knowing; it is exploration and discovery.

We start with the notion of composition unknown, and then suddenly it presents itself as, perhaps, a means of exploring, understanding, and developing ... What exactly?

We spied today, in our garden, a rose - an English rose - coming into bloom. Pelargoniums, which should now have been killed by cold and frost - they too are bursting into flower. Nasturtium seeds, those from last summer's bloom, that did grow and then drop, these self-seeding plants that should in March to May start to show their shoots, are germinating too early.

**3**

Our apple tree is bursting into leaf, and flower seeds are germinating ... This is the winter of our discontent, made glorious summer by this thing called global warming, which, the politicians have recently agreed, is something our grandchildren will have to solve later this century.

Coordinates: 53.2581°N, 2.1274°W; Day: Sunday; Sunrise: 08.23; Sunset: 16.03; Minimum temperature: 5°C; Maximum temperature 8°C: Wind: South South Easterly, changing to South Easterly, then back to South South

Easterly, then to Southerly, finally shifting back to South South Easterly, ranging 8-20 mph; Sky: cloudy and heavy rain predicted for the morning, with light rain in the afternoon. See how change is the norm - nothing in nature is static.

It is the tenth day.

Is this a work of fiction or non-fiction? If it is fiction, is it literary or commercial fiction? Or put another way; is it high art or low art? Why the distinction? Perhaps this distinction has something to do with elitism? But the distinction seems dated. Many things seem dated, including notions of Europe.

Is this a work of fiction or non-fiction? We ask again. But how can we know for it is our composition unknown? Whatever it turns out to be, we seek to transcend creative expression, and in seeking such we aim to make words sing and dance, so to speak, and this will be part of our composition, which is a quest for expression of the artistic kind. It matters not if it is of the high or low kind. You decide. It will be a different type of singing and dancing though. But in what way will it be different?

This is not writing; this is a composing. Literate people write, but they only compose at a basic level. You too can compose at a more sophisticated level, but the desire to pursue that which might, with time (perhaps not in our time) be regarded as excellence - this is something, according to the myth, that you are born with. This is how to pursue excellence: learn, breakaway and then transcend. Otherwise you can be a citizen composer, all you STEM people, who to the loving of artists you do gravitate. Why?

Do scientists really think that those whom they call citizen scientists, that they are like them? They could be, if they were willing to make the commitment and had the time and dedication to undertake the journey. This applies to art too, so a citizen artist is what otherwise people will have to be.

There are clear differences between occupations, between disciplines. To deny this is to engage in fantasy. Everyone has a talent. Not everyone can be an Einstein or a Picasso. There is no need for them to be. The world would be a boring place if everyone were like these people. In any case their names tend to be known because of the need - necessity - to place a few selected people on pedestals. It seems that we need heroes and their stories.

The above observations though are not an excuse for elitism - on either side. The process of art should not be monopolised by superior beings called artists. That is a sentiment attributed to Marcel Duchamp. We agree. Yet it is the belief that artists are superior beings that underlies much of the art-science, art-technology love affairs. It is the artists' (often tacit) claims to the status of superior beings, and the Brussels technocrats' belief that they are, which underlies something called STARTS. Artists were treated as such by DG CONNECT, and all the artists claims made to DG CONNECT, reinforce that belief. But they are not special superior beings. One day we may expose them for what they are. What are they? They are …

Neither should science be monopolised by superior beings called scientists. But it is. Now it seems they need artists to … What do they need artist for? They have willing collaborators in the artists that have fallen in love with science and technology. Thus do they collaborate! But what is a collaborator? A quisling, a fraterniser, a collaborationist, a colluder, a sympathiser, a conspirator?

House style: a publisher's preferred manner of presentation and layout of written material. House style deals with such things as preferred spellings, presentation of numbers, and such forth. It is a deliberate compositional choice that we have made, not to comply with the convention of imposing a house style on this book.

**4**

This is not a manifesto, but could be a preparatory step towards such a thing.

In the early hours of the morning we saw Orion once again. Interesting are these things called constellations.

It is the eleventh day or the tenth depending upon your assumptions and reference point. Mathematical models - those that are used to explain and to predict - these too are based on assumptions, and should not, with reality, be confused. But they are of course, for that is the great flaw in the human mind - its inclination towards denial and delusion of both the individual and the collective kind.

Coordinates: 53.2581°N, 2.1274°W; Day: Monday; Sunrise: 08.23; Sunset: 16.04; Minimum temperature: 6°C; Maximum temperature 7°C: Wind: South Easterly, changing to East South Easterly, ranging 8-11mph; Sky: cloudy with occasional light rain predicted through the course of the day. See how tedious data and facts can be. And yet there is all this singing and dancing and celebrator glee. Life out-of-balance!

Mondrian in writing about his work spoke of tragedy - dominance of one particular thing. Where is your balance, your harmony?

This is an act of discovery. It is an exploration. But of what? Why scriptovisualism of course! And what, you ask, is that?

Epistemology: the theory of knowledge, especially with regard to its methods, validity, and scope, and the distinction between justified belief and opinion.

**5**

When you look in a mirror what do you see? Art, it can be said, is a mirror of life in which is reflected the totality of the life of the community, yet when you look in a mirror what do you see? When those who speak of STARTS, STEAM and SEAD look in the mirror what do they see?

What we now write could be a mirror of life. Here you will find more than just the addition of art to STEM, and its renaming as STARTS, STEAM or SEAD.

What is that you just said? Oh, we see, you do not know what STEM, STARTS, STEAM and SEAD mean. Do not worry for all will in due course, in time, be revealed. This you see is the nature of exploration and discovery.

The same also goes for scriptovisualism.

It is the twelfth day, the twelfth night.

Coordinates: 53.2581°N, 2.1274°W; Day: Tuesday; Sunrise: 08.22; Sunset: 16.06; Minimum temperature: 3°C; Maximum temperature 7°C: Wind: South Easterly, changing to East South Easterly, then to Easterly, and then to North North Easterly, ranging 3–7 mph; Sky: light cloud with light rain showers and sunny intervals predicted for the afternoon.

You want more data, is that what you are saying, for you signed up to STEM to do away with all that is subjective and unquantified. Very well, here below is some more.

Environmental data: UV level: 1 (UK range 1-8); Pollution level: low: (UK range low to very high); Pollen Level: data unavailable.

Are you happier now?

At about 4pm, or 16:00 if that is you preference, it matters not - nor is precision needed - for if you were a bird then you would by instinct know that it was time to roost. And roost they did; about 100 or so black crosses darting about in the growing gloom, settling eventually in tall trees close by a house that in *Moments in Time*, we called the *Time House* - yes indeed it is a real place, for this is how those with artistic leanings - whether literary, painterly or something else - work.

And there they sat, but who else apart from us, knew they were there? Not many we would guess. Observational skills are a crucial element in the mix, which, given the right conditions, of a person an artist makes, so remember that next time you make your claims. Billy Klüver was right - you are born an artist, not made one through the creation and construction of things. Back to citizen artistry once again!

**6**

A blue tit, a great tit, a long tailed tit, a coal tit, some robins, a blackbird, a wren, two wood pigeons - they always come in pairs - a starling, a hedge sparrow, the occasional house sparrow, sometimes magpies and crows, as well as lots of jackdaws, which flock each day in this neighbourhood, an endangered redwing, and at times buzzards too come circling high above, calling to us in a screeching tone; these are the birds that we see in and from our garden. But not in great abundance as was the case 20 or so years ago. And of the chaffinches that used to come, and the dancing wagtails, of these there is no sign. Gone too are the soaring darting swifts that have accompanied every summer of our life. The sound of tawny and screech owls are heard no more. But there are gulls, seabirds - the black headed variety - in abundance. Not now do they just come inland when there are storms at the coast, but they live inland as well, and this is because of us and what we do. This is progress - apparently! The gulls adapt, and learn to live with us, but much of nature does not. Why cannot we adapt to nature, instead of seeking to conquer and overpower it? This is the character of STEM - very European. This is why we must change STEM, for one day nature will overpower and conquer humanity. Time to walk a different path while we still can! Thus do we write *366 - A Scriptovisual Composition Unknown*. Actually it is *366 - Julia and Paul's Scriptovisual Composition Unknown*.

Silent spring, summer autumn and winter, for bird song is rarely now heard in our locality. Rachael would be appalled by such a development.

It is the twelfth day, the twelfth night, yet again.

Coordinates: 53.2581°N, 2.1274°W; Day: Wednesday; Sunrise: 08.22; Sunset: 16.07; Minimum temperature: 4°C; Maximum temperature 7°C: Wind: South Westerly, changing to South Easterly, then to Southerly, then to South Easterly, then to South South Easterly and finally to South Easterly, ranging 2-15 mph; Sky: light cloud in the morning, sunny intervals in the early afternoon, returning to light cloud for the rest of the day. Environmental data: UV level: 1 (UK range 1-8); Pollution level: low: (UK range low to very high); Pollen Level: data unavailable.

You want more? Very well – we seek to please you.

Month: January; Date: 6th; Season: winter. Year: 2016; Moon phase: waning crescent.

On one side there is the realm of descriptive expression, that place of words which sometimes can be poetic; on the other there is plastic expression, where one finds line, colour and form, which can also be poetic. To the Western mind, those who adhere to behaving like Europeans, this is a yet another binary; your minds are infested with them – it is your disease. There is ICT and there is art. So you need to connect them. In our mind though, they are but one; they are a unity. Any idea what that feels like? The answer is no, for you are the *Vainglorious Enlightened Ones* replete with fragmented and reductive minds; you do not understand unities. And this will be your undoing. No, you are already undone, just too deluded to see that this is so.

## 7

The quantified, the rational, the objective; all these are just one aspect of being human, yet they dominate, and the result is the world you want to see, for most of you are like all those people in the movie *The Matrix*, preferring not reality, but one of a collectively constructed kind that is pleasing to your fragmented and reductive minds and which corresponds to your *Enlightenment* notion of progress, even though if you took the time, you would come to understand that you are empty and unhappy. Do not take the time, but instead fill your life with material things and make much noise so that you will not hear the sound of your soul dying. Oh dear what have we said!

*The Matrix* is fantasy, but is also poetic in that it is a metaphor for collective delusion and denial, just like that children's fable, the one where all except one, a little boy, think that the naked emperor looks quite fine in those amazing clothes. Sounds like STARTS to us. One day soon you will start to understand what this thing is that we call STARTS: it is a European Commission version of the naked emperor, and we are, it seems, that little boy (or girl, or both).

The phrase scriptovisual was used by some feminist artists in the past to describe some of their art work that involved the use of words and images to create visual representations. This perhaps is what, for many people, the term scriptovisualism will imply, but this is not what we intend. Hence we need a manifesto. This is not it.

Environmental data: UV level: 1 (UK range 1-8); Pollution level: low: (UK range low to very high); Pollen Level: data unavailable; Month: January; Date: $9^{th}$; Season: winter. Year: 2016; Moon phase: waning crescent.

Very tedious, yet apparently so important!

Constellations are also there during daylight hours, but, self-evidently are not visible with the naked eye. Yet they still exist, or do they? What does this mean? Not what you are thinking because, without any doubt, constellations do not exist. Do you understand? Not to worry if you do not, for at some point in this story, we will probably explain. Or not! Who can tell in a composition unknown!

Is this a story? Well of course it is, but if you are not familiar with literary theory you will probably not understand this statement. There are many things that you do not understand and the more you rely on one particular way of knowing, then the less you will understand, but you will not realise this - Dawkins proves the point.

**10**

Free people think, and slaves work. Welcome to Ancient Greece and its many cities: Brussels, London, Washington DC, Madrid, Stockholm, Sidney ... But not Beijing or New Deli (not yet at least - long may this be so). Could Beijing or New Deli use this against the Ancient Greeks in Brussels and other cities?

The science that you know has run its course and will be replaced by a science that is fundamentally different to that which you now believe in. Oh dear, another heresy!

"So what do you do?" asked the academic.

"I undertake independent research," came the reply.

"So you're a consultant then," was the academic's response.

"No, I undertake independent research." Evidently the point needed to be reiterated, reinforced.

"You're a consultant," insisted the academic. "What's your name?"

"Copernicus, Tycho Brahe, Kepler, Franklin, Wallace, Darwin, Einstein," was the response.

"But you hold no academic post, so how can you possibly do any research, or produce anything of worth," the academic noted, not seeking to hide her contempt.

"Evidently that is the case!" came Copernicus', Tycho Brahe's, Kepler's, Franklin's, Wallace's, Darwin's, Einstein's response.

Coordinates: 53.2581°N, 2.1274°W; Day: Sunday; Sunrise: 08.20; Sunset: 16.12; Minimum temperature: 3°C; Maximum temperature 5°C: Wind: South Westerly, then changing to South South Westerly, and finally settling to Southerly, ranging 9–16 mph; Sky: heavy rain, followed by sunny intervals, then light cloud, then light rain showers, then light rain, then back to light rain showers, then heavy rain showers, followed by light rain, then light cloud, partial cloud, ending the day with light cloud. Environmental data: UV level: 1 (UK range 1-8); Pollution level: low: (UK range low to very high); Pollen Level: data unavailable; Month: January; Date: 10th; Season: winter. Year: 2016; Moon phase: new moon.

It did not rain, which well illustrates the nature of reality - full of change, uncertainty and unpredictability, which contrasts sharply with those with mindset that lead them to claim that risks can be managed. They cannot always be - Windscale, Three Mile Island, Chernobyl, and Fukushima prove the point. Yet obviously not to those who are handmaidens of economic interests - we speak of scientists, engineers and technologists.

Who will develop a different and more advanced philosophy? Do not look to the handmaidens, for they are far too bound-up in the pursuit of fame and fortune to be the ones who will overturn the, out-dated, *risks can be managed* mentality; to the past this does most surely belong. Someone should tell the experts this!

If you do not understand religion it is because you have not made sufficient effort to do so. Quite often people judge such matters on that which is unrepresentative. This you see is the nature of being human, and those inclined to the secular religion known as *science and reason* are often the worst offenders - Dawkins proves the point.

Religion - yet another way of knowing the world and being human.

"I am an atheist and a positivist," the astrophysicist said.

Note that the astrophysicist did not say 'I am an agnostic' because, as is self-evident, nothing can be proven either way, and science is of no help when it comes to such matters, although some have foolishly tried - Dawkins proves the point. It is known as scientism.

What is?

What Dawkins preaches from his pulpit of course!

Believing in something is being human. Subjectivity can be found in all walks of life, including science. That is the point, for this is an important part of being human as well.

Something should be said further about positivism. It will, perhaps, but not yet.

**11**

Excellent scientists think and entrepreneurs work - that was the opinion among some of the European Commission's FET Advisory Group. Binary thinking is everywhere - dualities. Classify and separate. It is a strategic weakness. What is FET? FET is an acronym which stands for Future and Emerging Technologies.

Coordinates: 53.2581°N, 2.1274°W; Day: Monday; Sunrise: 08.19; Sunset: 16.14; Minimum temperature: 3°C; Maximum temperature 4°C: Wind: South South Easterly, then changing to South Easterly, then back to South South Easterly, then South South Westerly, then North Westerly, then West North Westerly, then Westerly, and finally back to West North Westerly, ranging 4–8 mph; Sky: light cloud, followed by heavy cloud, then some light rain, then heavy cloud again, then light cloud, with occasional partial cloud. Environmental data: UV level: 1 (UK range 1-8); Pollution level: low: (UK range low to very high); Pollen Level: data unavailable; Month: January; Date: 11th; Season: winter. Year: 2016; Moon phase: waxing crescent.

Will we continue with this data collection? What value lies in this, given that there are so many other ways of knowing the world, humanity, life, and many other things? Perhaps it is time to look away from the dominance of data, quantification, rationality, and objectivity, and see the world from within a mind that can embrace multiple views, simultaneously?

Big, little, large, complex ... Such adjectives are subjective and are inherently ambiguous. They are also relative. So, one person's big data might be someone else's little data, because it depends upon their frame of reference. So perhaps all data is big! Whatever your decision, the notion that data will help humanity to make better decisions is an entirely subjective statement. What does better mean? More subjectivity! Look closely at the world of science and reason, at the world of the rational and objective, and they look increasingly subjective, which of course they are. More heresy!

But people construct models of reality which become part of their identity. And when this happens we are in trouble because then we are in the realms of dogma, and in danger of skipping and dancing towards intolerance and extremism. Clearly the other person, lacking right-mindedness, is suffering from a deficiency – a deficit. So what do those caught up in ideology and dogma do? Look to history and to our current troubled times for the answer – it is plain to see. And it affects everyone – this too is the nature of being human.

**12**

There are technologists and there are creative people – Europeans have this need to classify and separate. They cannot have a world where this will not be. But in the East, people think differently, because they are not Ancient Greeks, and they are not the *Vainglorious Enlightened Ones*.

Coordinates: 53.2581°N, 2.1274°W; Day: Tuesday; Sunrise: 08.18; Sunset: 16.16; Minimum temperature: 3°C; Maximum temperature 5°C: Wind: Westerly, turning West North Westerly, then North Westerly, then West North Westerly, then North Westerly, ending the day West North Westerly, ranging 11–19 mph; Sky: heavy rain, followed by light cloud, followed by light rain, then heavy rain, then light rain, ending with light rain showers. Environmental data: UV level: 1 (UK range 1-8); Pollution level: low: (UK range low to very high); Pollen Level: data unavailable; Month: January; Date: 12th; Season: winter. Year: 2016; Moon phase: waxing crescent.

Changed coordinates during the course of the day: spent time at 51.5294°N, 0.1269°W and 51.4908°N, 0.1272°W and 51.5079°N, 0.0994°W. And in doing so we consumed – energy. What does this mean? Many things – there is no single correct interpretation of this statement. Knowing is far more complex than it at first seems, except of course if you have a fragmented and reductive mind and need to have a sole truth – to be defended until death, and sometimes, by delivering death to others with different truths.

We came close to the imaginary line today – the Greenwich Meridian Line – but did not cross it. Actually at the Greenwich Observatory the line is not imaginary for there is a physical line in the ground and one can straddle it and be simultaneously both west and east of it. How cool is that? Trivial perhaps, but such things too are part of the nature of being human. Being human though – being seen to be human – is not highly regarded by the *Vainglorious Enlightened Ones* as we shall, one day, see, if we ever, in this odyssey, speak to one of those nutty professors that on our TVs like to appear to say such stupid things as 'I have no axe to grind,' which of course we know means that, 'I have many vested interests and I am using this programme to further these, but I am not going to reveal these interests to you.' Such is the nature of modern science!

Trust me I'm a scientist or trust me I'm a priest. What is the difference, when it comes to seeking influence and power? Abuse it you both will, and take advantage for your own gain and that of the group you are part of – this too you will most certainly do. This also is the nature of being human. We know from history not to trust the priest, but for some strange reason have not yet learned from this experience and are willing to trust the scientist; well some are, especially those who to this particular secular religion are inclined. Artists included!

**13**

Free people think, slaves work, only the free people rarely do think – they only believe that they do. The same goes for all, for this too is part of being human. Cognitive biases predominate.

Like in times of old, thinking has become institutionalised and by way of evidence we present the European Commission's STARTS platform, the product of cognitive biases and deficit thinking, and a good dose of that corrupt relationship between the European Commission and the experts that it uses, the sort of relationship that some in the European Parliament clearly want to end. But they should see what goes on inside the Covent Garden building! This is the place where research proposals are evaluated by – yes – more experts! We do need experts of course, but treat them like you would a tiger, especially when to the *City of the Golden Stars* they do orientate.

Coordinates: 53.2581°N, 2.1274°W; Day: Wednesday; Sunrise: 08.18; Sunset: 16.17; Minimum temperature: 4°C; Maximum temperature 4°C: Wind: West South Westerly, briefly shifting to South Westerly, before becoming South South Westerly, before finally becoming Southerly for the rest of the day, ranging 8–12 mph; Sky: partial cloud, followed by sunny intervals, followed by sun, then light cloud, and then, once more light rain, turning to the heavy sort during the hours of dark. Environmental data: UV level: 1 (UK range 1-8); Pollution level: low: (UK range low to very high); Pollen Level: data unavailable; Month: January; Date: 13$^{th}$; Season: winter. Year: 2016; Moon phase: waxing crescent.

Oh that this data collection be far more interesting, that it could have not just a meaning, but also soul. Oh dear we have spoken the taboo word that cannot now be uttered in the halls of science, technology, engineering and mathematics. We shall see what can be done about that.

Denote science by S, technology by T, engineering by E, mathematics by M, design by D, art by A (or ARTS), and let:

$$\text{Science} + \text{Technology} + \text{Engineering} + \text{Mathematics} = \text{STEM} \quad (1)$$

$$\text{Science} + \text{Technology} + \text{Engineering} + \text{Arts} + \text{Mathematics} = \text{STEAM} \quad (2)$$

$$\text{Science} + \text{Technology} + \text{Arts} = \text{STARTS} \quad (3)$$

Now denote plastic expression by PE, descriptive expression by DE, and scriptovisualism by SV. We then have:

$$\text{SV} \geq \text{PE} + \text{DE} \quad (4)$$

In the special case when DE is zero, then SV = PE, and in the special case when PE is zero, then SV = DE. Hence SV is clearly neither PE nor DE, but something which is greater and different to the two added together.

Now what if PE were A, and DE were S? What would then exist? Something that is neither art nor science, but something in between, where science would be just a special case of whatever we chose to call what scriptovisualism, in this case, represents. Likewise art would just be a special case too.

Scriptovisualism is therefore symbolic of that which we seek to create. Let us call it a new way of doing science founded on that which those that practice science today, will, with one voice, decry as heresy!

No sign of Orion. Now why do we keep mentioning this constellation?

**14**

The *Vainglorious Enlightened Ones* are too reliant on technology - which invariably gets in the way of thinking. It is one of their strategic weaknesses, which can be used to unleash the forces of creative destruction against them.

Coordinates: 53.2581°N, 2.1274°W; Day: Thursday; Sunrise: 08.17; Sunset: 16.19; Minimum temperature: -1°C; Maximum temperature 2°C: Wind: West North Westerly, then North Westerly, briefly shifting to North North Westerly, before edging back to North Westerly, before finally becoming West North Westerly as, amidst the darkness, the day comes to an end, and as for its range over this cold winter's day, the forecast says, 5–19 mph; Sky: heavy snow, followed by sleet, then heavy cloud, giving way to light cloud, then some sunny intervals, and then as the daylight ends, there shall be sun, leading to a clear night sky when stars will at last be seen, only later we will start to loose them, one by one, as partial cloud drifts in, before becoming, once more, that light cloud obscuring all, as is so common in the British Isles. As for the environmental data, well the UV levels will be 1 and the pollution level, that will be low, and at this time of year, pollen levels measurements are not made available as people have no need for them. It is January and the fourteenth day, and of course here in the northern hemisphere it is the season of winter in the year of 2016. As for the moon it is a waxing crescent.

And now it is far more delightful to read, and more interesting as a result of some application of literary thinking, but if you think the arts are for better communication of scientific data and findings, then you better think again, because art is for discovering and revealing that which you seek to ignore and to keep hidden from view.

Now John Dewey is here and wishes to say something, for he has much understanding of art and science, and was writing about this subject in the early decades of the twentieth century. So, over to you John and that which you, at this point in our artistic odyssey, wish to contribute.

"Thank you. What I wish to contribute is something I said in a lecture given in Japan in 1919: that surely there is no more significant question before the world than this question of the possibility and method of reconciliation of the attitudes of practical science and contemplative aesthetic appreciation. Without the former, man will be the sport and victim of natural forces which he cannot use or control. Without the latter, mankind might become a race of economic monsters, restlessly driving hard bargains with nature and with one another, bored with leisure or capable of putting it to ostentatious display and extravagant dissipation."

Well said! And now for the most important revelation, that the economic monster of which John spoke - humanity has now become that monster, and its handmaidens are scientists, technologists, engineers and mathematicians.

And such people now speak of appropriating art as well, in the service of this economic monster, and the European Commission DG CONNECT's STARTS Platform proves the point. But they did not reckon on encountering the sword which is our pen, and thus we do, in a poetic sense, use that sword not just to slay you, the European technocrats, but also that economic monster and its handmaidens too. And our weapon of choice for the moment is our very particular, very individualistic aesthetic, which we leave you to discover for yourself. But behind the scenes so to speak, we build a new world too.

What on earth does all this mean?

The night sky was clear - Orion is back after many days of absence, although he was there all the time, which of course he was not!

**15**

*Visible Language* is a journal that deals with matters relating to the visual communication of information. Might it be relevant to scriptovisualism? Perhaps, but we shall have to see. What though is relevant now, at this particular moment in time, as we write what you now read, is that which is stated about the journal, on its web site, for among the various pieces of information therein contained, we found this:

"Because Visible Language is an academic journal related to a profession, design, it seeks to advance research and scholarship for two types of readers: academics and professionals. The academic is motivated to consume knowledge in order to advance knowledge thorough research and teaching. The professional is motivated to consume and apply knowledge to improve practice."

What they mean is that academics think, while professional practitioners work. Welcome to Ancient Greece! And can you see here how outdated is their model of research and innovation. Or perhaps you cannot see, for this separation of academic and practitioner, and notions of who does what - perhaps you too are living in Ancient Greece?

Now once more to matters of today's data collection! But wait, what is this? Someone has arrived and wishes to speak. And we see that it is the artist György Kepes, one of the early pioneers of that which is called art-science practice (which includes technology and engineering too).

So what does György wish to say?

"Simply this, that science is only one component of the understanding that we need for a well-balanced attainment of human ends."

Absolutely correct, but the battle to overcome science and to achieve this state of balance first has to be fought and won. And the outcome of this battle is far from certain - hence, or perhaps, scriptovisualism.

And because of Gyorgy's very wise and pertinent observation, we here now say that while scientific observation, experiment, collection of data and quantification are clearly important they are not a totality, nor are they more important than other ways of knowing, and hence forth, there will be no more of our daily data collections, for their purpose now is done. Understand?

And in saying farewell to what was obviously becoming very tedious, to make a point, we just note that it is a very cold January day in 2016, which is exactly what it should be, but in twenty years or so, this coldness will probably be just be a memory, because of something called global warming, the product of the mindless application of STEM in pursuit of strange notions of progress that are destroying everything of value. And the value of art lies in showing the path to a different way of living, a different STEM, one which will cause you much mental anguish. This anguish, this mental pain, will, for some of you, probably most of you, be the beholders share, but not for everyone, for there will be some who will embrace the building of a new path, and to them the future will belong.

Once again a clear night, so Orion was seen.

**16**

And so we turn to speaking in poetic terms:

> Why a sheep, a member of a grazing herd, would we choose to be,
> When, an eagle is our destiny to be?
> And soaring high above green field and mountain range,
> Distantly we spy, worlds lying beyond the gaze of mortal men.
> And with sharp talons and cutting beak, we, to the bleeding of minds turn,
> As we slowly inflict pain, by tearing to shreds your self-constructed realities.
> Thus begins the end of the reign of the *Vainglorious Enlightened Ones*,
> And with them goes the age of Abraham, and notions of sole truth.
> Rising in its place comes instead, a world that has oft been spoken of,
> And it shall be a golden age of balance, peace, and harmony.

By the time this work is done, you will truly understand the meaning of the beholders share, for you will, either way, for or against, be completing this work of art in your mind. And so the separation will be done.

Of course it will only be a golden age if ... Otherwise it will be just another dystopia.

**17**

Civilised killing you will find, is what civilised people do; uncivilised killing is what those called terrorists do. The West does the former, its enemies the latter. To kill in a civilised away, this is what Europeans, in

Ancient Greece and Ancient Roman, began to do, and they are still busy doing it today. It is so much to better to bomb and to drone, than to undertake all that decapitating as the so-called *barbarians* do.

What is this? Is this work now wandering off-topic?

But what is the topic? It is, if you recall, Julia and Paul's scriptovisual composition unknown.

There was a man who wanted to know if the car outside his neighbour's house ever moved, for it seemed not to. Hence he decided upon an experiment, which involved watching it for a full 24 hour period, which he duly did. It was, as you can imagine, quite an arduous task, to stay in one place for a whole day, and never to take your eyes off the object of observation, and not to fall asleep. Well the task was done, and indeed the car did not move. And many times the experiment was repeated and always with the same result. Thus the result of the experiment was that the car did not move, and a reality had been created.

Yet the car was moving, continuously, and very significantly. So seeing is most definitely not believing! Frames of reference are important. There are therefore, many different realities. Which one is correct? The answer, often, is that they all are. There is very rarely a sole truth, yet this is not how the *Vainglorious Enlightened Ones* behave.

So Mr Ghandi what do you say?

"All religions are true," came the reply.

*Best evidence* is of course a highly subjective term, yet it is much used in the supposedly objective and rational world of STEM. Now why is that?

Our background is in the art and humanities, in the social sciences, and in STEM. We apologise for that, but we were young and did not understand the *Vainglorious Enlightened Ones*' rules, and now we find ourselves in the spaces between spaces, and see the world, as a result, in a different way to everyone else. We are, it seems, alone.

Is scriptovisualism symbolic of the spaces between spaces and the seeing of the world in ways different to everyone else?

Today is Sunday, the Sabbath day, a day of rest if your *reality* is Christianity; only the Sabbath was yesterday, Saturday, if your reality is Judaism, and the day before, Friday, if your reality is Islam. If your reality is Western atheism, then your day of rest would, most likely, in the past, have been Saturday and Sunday, only in our modern times, you subscribe also to another reality called Western capitalism, and it is capitalists who now determine when thou shall rest, for your time belongs to them. See it is just another religion - a secular one, but with much inherited from Christianity. The same goes for science too.

Free people and slaves they have become the same. This may be another reason why free people no longer think. Artists generally are still free, but this now is coming to an end. We refer you to the European Commission which is busy appropriating art, as did the Soviet era communists, to their own ends, which includes propaganda and image making. So what is the difference? The European Parliament is also doing the same. And when this happens you just know you are in serious trouble.

People enquire of us whether artists and STEM people think differently. This we will not answer now because someone - like a Director General in the European Commission - will most likely come along and say:

"Yes I'll have that information, and then stick a label on it and sell it, for it is evidently food for the economic monster that humanity has become."

We resist, we make a last stand. We will take such information to the grave with us rather than let you appropriate it for your evil plans.

What is evil? This you do not understand, but you think you do. Back once more to this matter of hermeneutics.

**18**

It is a FET-Open panel meeting and the top scoring proposal is being discussed. Two panel members note that no industrial partners are involved in a proposal dealing with an important technology for Europe. Then the response from one of those who undertook the evaluation, an academic:

"This topic is too advanced for industry!"

What you mean, is that academics think industrial people work, such is the way that Ancient Greece operates. And do you believe that in China and India, they will think like this? You live in interesting times! Oh dear the *Vainglorious Enlightened Ones* are in serious trouble.

If only European Commissioners and those in the European Parliament knew what goes in that Covent Garden building. You have a little insight! Perhaps you do not want to know - information avoidance.

"Why not write to a European Commissioner and tell him."

That was an official from the European Commission's DG CONNECT speaking to us.

"If only 10 percent of what is wrong with evaluations is fixed, it would still be an improvement," he said, continuing.

Thus do we reply by saying that such may well be true, but it would be the wrong 10% that would be remedied, because being a political organisation, the European Commission is not interested in doing the right things, even if it knew the right thing to do, which is not necessarily the case! This is because they too, are Ancient Greeks. And technocrats!

Now we shall name some companies: Amazon, Facebook, Google, eBay, PayPal, and Twitter. So what do these companies have in common?

The answer that most of you will provide, is that they all internet companies, which is true. Now why did you give such an answer? This is an interesting question. What led you to state such an obvious thing? Why indeed would we ask this question if the answer is so obvious? You will, perhaps, encounter this again, so read on.

What we now wish to tell you is that all the companies that we named operate in existing business sectors - finance, retail, publishing, media. There is nothing that Amazon, Facebook, Google, eBay, PayPal, and Twitter initially did, that could not have been done by existing businesses in the respective sectors. But they did not! Now why was that?

In the West the smallest unit of society is the individual. Elsewhere this is not the case - it is often the family. The *Vainglorious Enlightened Ones* often talk about the family, but do very little to preserve this - what is more important is the making of money, and family life in the Western world is falling apart and the result is a very dysfunctional society. So what gives *Vainglorious Enlightened Ones* the right to criticise those in the non-European world when there is so much wrong in their own societies? They are indeed a strange lot of people these *Vainglorious Enlightened Ones*, and very dangerous too, for they are full of delusions of superiority, and that odd belief, that theirs is the best way - the one best way. That is what comes from being attached to the notion of sole truth!

The *Vainglorious Enlightened Ones* have a reputation for being intolerant of those with other truths, and start then to classify and separate, which are the first two steps along the road to genocide. There is a reason you see, why most of the world's known genocides, those that are acknowledged as such, and also those that are denied, were committed by Europeans. Now this is one of those uncomfortable truths that they, the *Vainglorious Enlightened Ones*, most definitely want to deny. This, we now remark, is one of the reasons why art is needed; to confront the delusions that these crimes against humanity, which the *Vainglorious Enlightened Ones* want to believe are aberrations, are in fact how the *Vainglorious*

*Enlightened Ones* behave in the world. They, it seems, invented genocide - and are not done with it yet!

## 19

It is a cold cloudless night, and the moon is waxing crescent. No, we are not returning to all that tedious data collecting for now it is of art we speak.

We sit and watch the moon moving. Did you know that its movement, relative to the earth, can be detected with the naked eye? Actually it is not the moon's movement you can see in such a circumstance, but an effect created by the earth's rotation. The moon also orbits the earth, but that movement you will not see with the naked eye, other than by observing how the moon passes through it phases as the nights pass by.

On the matter of this apparent movement and its detection; first compose your view. The best way to do this is to frame the moon within or near an object, for example the bare branches of a tree, because then you will see very clearly that there is movement. If you wish you can enhance the experience by using technology. Take a telescope or binoculars and look at the moon. See now the circumference of the full disc, and those mountains and craters dotted about. Note too the boundary between light and dark is not a clear dividing line, for, as an earth, there are areas just inside the dark zone that are slightly lit, and perhaps too if you look at the right time, you may also see a mountain peak, not so far in to that place of night, illuminated because it is high enough to catch the light which will soon light up all the mountain, and that which lies at its foot.

Now the thought might enter your mind to take a photograph, which you could do, by attaching a digital camera to the telescope and recording for posterity what you have just seen.

Print it, mount it, put it in a frame, hang it on wall, call it art. Now why would you call it such, this object hanging on the wall? It may be a composition most beautiful, it may be quite an achievement this thing that you produced, but it is good art? How can one determine this? As a former Director General of DG CONNECT once asked: "how does one determine art's excellence?" No one answered his question. Why would they when it was money they wanted, and DG CONNECT - technocrats supreme - had already decided that they were going to appropriate art to their ends. If you are a European Union taxpayer, this is your money they are wasting!

And if they had answered the question - the artists present that is - they might have also told him, the Director General, that the art, in a contemporary sense, does not just lie in the object hanging on the wall, but also elsewhere. It seems that, based on the nonsense that subsequently followed in the wake of the asking of this unanswered question about excellence, this too was not explained. And if you wish to know more then you will have to read on.

Artists do things and make things for different reasons than STEM people. When artist begin to make things for the same reasons as STEM people, are they still making art? Are they still artists?

"The party is not the sole bearer of the truth."

That was the (former) President of the Soviet Union, and General Secretary of the Soviet Communist Party, Mikail Gorhachev speaking, several years prior to the collapse of the Soviet Union.

The Soviet Union was dying, coming to an end. So is the Abrahamic age, and along with it will go the seven Abrahamic belief systems: the three religious ones - Judaism, Christianity and Islam; and the four secular ones - Western science, Western capitalism, Western communism, and Western atheism. But what does this mean? You might think that there will be no Judaism, Christianity and Islam. If you do, you are wrong. Likewise for Western science, Western capitalism, Western communism, and Western atheism! Back to hermeneutics again!

What is coming that will change world circumstances so much? If you do not know, you must be blind, or unwilling to see.

## 20

The world continues to turn, time marches on. Already we are twenty days into January, and the year is no longer new.

It is a cold cloudless night, and the moon is waxing crescent. No, we are not returning to all that tedious data collecting, for now it is of art we speak - again!

Once we spoke of the economic monster that humanity has become. Here now is the Nobel Prize winning author, John Steinbeck. John, in his novel *The Grapes of Wrath*, wrote about this monster, manifesting itself in the form of a bank. So John what did you say?

"I said, in my book, that the bank - the monster - has to have profits all the time. It can't wait. It'll die. When the monster stops growing, it dies. It can't stay one size. The bank is something else than men. It's the monster. Men made it, but they can't control it."

And do you want to say more?

"Yes. My contribution is a small one to your composition unknown, yet I can see that this book you now write is quite an odyssey, and is not the type of thing that could have been done and published in my time. This is the problem with publishers, for they too are overly concerned with profits, while art, this being that which we seek to make with our pens, takes a back seat. That is the nature of this monster. It will destroy us if we do not take a stand and use our art, not just to point out the need to walk a different path, but to start building it. This you see is the power of art, and long may that be."

That is very true, indeed.

Now for those caught-up in technocentric beliefs, that is to say those for whom technology is the primary thing, and for whom this is everything; they would say that self-publishing is the result of technology.

Yet those who have a more balanced view, who can know the world in many different ways, will tell you a much different story. It begins in the imagination, as all art does, and then proceeds through an often long and extraordinary process of making, in which imagination and making are deeply entwined. And while it is true that technology is important, for without it there would be no books, it is not that which is central here.

Self publishing is now new. There was a time when there were no such things as publishers, which is why Copernicus published his own book - early writers were *indie authors*! So was Charles Dickens. So was Jane Austin, at least for some of her works.

And across the ages it has always been possible to publish your own book! What technology now does, is make that much cheaper and easier, and also provide the means of making noises to attract attention - through the use of that ubiquitous thing called social media.

And the point is that to publish without a publisher, or to become your own publisher, is an act of imagination, as is also, the decision to make the results available online through open-access, which is just another form of the free, public lending library.

So truth and reality, they are more complicated things, and this leaves those caught-up in single truths, like those who are technocentrically inclined, facing a problem, which can be addressed by reaffirming the belief in the sole truth, your truth; everyone elses' truth is wrong. And one cannot be more Abrahamic than believing in the sole truth, otherwise known as simplified representations of reality. It seems that the *Vainglorious Enlightened Ones* are unable to understand. This is because they have fragmented and reductive minds. The people in DG CONNECT who started STARTS also have fragmented and reductive minds. They too are technocentric. They are also technocrats. They also suffer from cognitive illusions and biases. Thus will they be undone!

**21**

And now for the naming of some citizen scientists: Copernicus; Tycho Brahe; Kepler; Franklin; Wallace; Darwin, Einstein. We include Einstein given that he was not formally part of the scientific establishment or held a research post when his paper on special relativity was published.

Who is and is not a scientist might not be as clear as you would wish, for all these people, although not holding formal positions, did demonstrate that which we refered to when introducing the notion of citizen artists. So do not be so quick to assume that people not holding research posts, or who are not part of the establishment, are not also capable of delivering paradigm shifting research output.

Oh dear, your well ordered world is falling apart. This world, the one you believe in, is called a self-constructed reality, and is part of being human. Everyone's reality is slightly different to that of others, and in some cases the realities are vastly different. Deficiency thinking then takes hold as clearly those who do not share your reality must be suffering from a deficiency – a lack of right-mindedness.

André Breton, the founder of Surrealism, is here and wishes to say something.

"Yes indeed I do. I wrote, in my first Manifesto of Surrealism, which was published in 1924, about autonomic writing. Settle yourself in a place as favourable as possible to the concentration of your mind upon itself, and have writing materials brought to you. Put yourself in as passive, or receptive, state of mind as you can. Forget about your genius, your talents, and the talents of everyone else. Keep reminding yourself that literature is one of the saddest roads that leads to everything. Write quickly, without any preconceived subject, fast enough so that you will not remember what you're writing and be tempted to reread what you have written."

Thank you for explaining how we are writing this book, the one that you, dear reader, are reading.

On that matter of scriptovisualism, Surrealism is most definitely very relevant. Why is that?

Why are the *Vainglorious Enlightened Ones* so obsessed with creativity? It has something to do with the *Renaissance*, which was followed, eventually by the *Enlightenment*. Something changed. But what was it that changed? It also has something to do with Immanuel Kant.

Art involves creating artefacts, that is for sure, but not always tangible ones. The art object is not everything, and if you know something about art you will understand what we mean. Do you?

**22**

And now for the naming of some of those deficits so beloved of the *Vainglorious Enlightened Ones*: creativity; innovation; science information; entrepreneurship; European Union Nationalism; ... We British, apparently, are suffering from a (major) European Union Nationalism deficit!

There are actually quite a lot! It seems to be the favoured way of the *Vainglorious Enlightened Ones* to explain reality: a lack of ... This therefore begs the question why so many? And why do they think like this? Have we returned once more to Ancient Greece and Ancient Rome?

Now to examine genocide, that in which the *Vainglorious Enlightened Ones* do indeed excel. First we name the steps involved in genocide, for there are eight: classification, separation, dehumanisation, organisation, polarisation, preparation, extermination, denial. We took this information from the Genocide Watch web site.

Interesting that DG CONNECT engaged in classification and separation when it sought to appropriate art for its own purposes: there are artists who are classified as the creative ones, and there are technologists, the uncreative ones, based on the belief in a creativity deficit. There is something very European about all this, about ICT-ART CONNECT, about STARTS.

The Chinese and the Indians, they never had a *Renaissance* or the *Enlightenment*. What they experienced was the *Vainglorious Enlightened Ones* turning up on their shores taking what was not theirs to take, and indulging in killing and violence and oppression. That is what Europe stands for – crimes against humanity, although you would not think so listening to their hubris. They have fallen in love with themselves.

Gorhachev is speaking again, and says:

"It would be naïve to think that the problems plaguing mankind today can be solved with means and methods which were applied or seemed to work in the past."

We did not hear the free people saying this. Perhaps they are too busy pursuing fortune and glory, which is the modern academic way, or perhaps too busy developing their corrupt relationships, which too is part of their modern times. Money always speaks much louder than wisdom, and increasingly now money is speaking all the time. We refer you to STARTS.

David Chesterfield, the British architect, still uses actual physical models, along with all that computer-aided stuff. Now why is that?

In Ancient Rome, as in Ancient Greece, they had that which we now call mythology, which teaches us important truths. Yet another way of knowing the world! This mythology is often an early form of behavioural studies, for it speaks of what it is to be human, regardless of the state of technology.

The ultimate and final solution to the lack of - so-called - normality is to kill. Thou should however, under no circumstances whatsoever, kill another human being, because all human life is sacred. Kill once and you, or someone else modelling your actions, will do it again. And thus the killing spreads, and with it the justifications evolve too - self defence, justice, pre-emptive strike, ... Killing in the name of ... Fill in the gap yourself, depending upon what you believe. Here are a few options: God; the state; the humane thing to do; convenience; justice; revenge; retaliation; image; being strong.

In the United States, there are some rather large companies selling what we today call junk or fast food. The initial business model was clear - mass production, economies of scale, standard limited menu. End result, cheap fast food. Only ... What? Nothing stays the same forever.

Back in time now to the era of Henry Ford, who would like to make a contribution to our odyssey:

"Yes. First I want to say that you can have any colour so long as it's black. It is reported that I said this. Did I? Actually I did not. What I said was that any customer can have a car painted any colour that he wants so long as it is black. And I said it in a particular context, which was my concept to build a motor car for the great multitude. A car large enough for the family but small enough for the individual to run and care for. It was to be constructed of the best materials, by the best men to be hired, after the simplest designs that modern engineering could devise. But it was to be so low in price that no man making a good salary would be unable to own one. Other manufacturers thought I would go out of business in six months, for they were still thinking of the motor car as a luxury item. I did not go out of business!

"Second, I would like to say that we worked out speeds for each assembly, and for the success of the chassis assembly caused us gradually to overhaul our entire method of manufacturing and to put all assembling in mechanically driven lines. The chassis assembling line, for instance, goes at a pace of six feet per minute; the front axle assembly line goes at one hundred eighty-nine inches per minute. In the chassis assembling are forty-five separate operations or stations. The first men fasten four mudguard brackets to the chassis frame; the motor arrives on the tenth operation and so on in detail. Some men do only one or two small operations, others do more. The man who places a part does not fasten it—the part may not be fully in place until after several operations later. The man who puts in a bolt does not put on the nut; the man who puts on the nut does not tighten it."

So, yet another manifestation of that love for separation - formally it is called horizontal division of labour. Engineers, the free people who think, break down the work into individual tasks and design the production line so that the work is highly fragmented, so that the slaves can work without thinking.

Now to Ancient Greece, and Socrates is speaking, as noted down by Plato.

"I am Socrates and I say that one man cannot practice many arts with success."

Hans-Christian Andersen wants to say a word too.

"My tale of the emperor's new clothes is one of collective delusion and denial. It is universal and I have now discovered that in your age, cognitive scientists agree."

So the roots of specialisation, of the fragmented mind run deep. Add in the reductive aspect, and collective delusion and denial, and one is left with the *Vainglorious Enlightened Ones*. That is their recipe; a recipe for all that has unfolded since the *Enlightenment*. So now we return to genocide, and the holocaust, the road to which, some say began with the *Enlightenment*. About this they are of course wrong, for it began far earlier in Ancient Greece and Ancient Rome. Barbarians they were called, those that were not civilised like the Ancient Greeks and Romans. So here we have classification, separation, symbolisation, and dehumanisation.

Some early crimes against humanity: slavery as practiced by those Ancient Greeks and Romans, and the infanticide those Greeks practiced when their offspring did not emerge with the characteristics that the men required - an early form of genetic engineering through selective breeding.

Now we know a little about where those nasty Nazi ideas came from. They were not, you see, an aberration, just part of an historical continuity. The Nazis though were too much even for the *Vainglorious Enlightened Ones*, for they went too far, and exceeded the bounds of what constitutes acceptable civilised killings. The First Wold War was however different, because mass industrialised slaughter of military personnel does fit within those bounds and the civilian killings were just an unfortunate by product of this. Collateral damage is the modern phrase.

It will not happen again though for now we have nuclear weapons and we can kill everyone on planet earth in a civilised way – the ultimate mass industrialisation of slaughter. This by the way is what the *Vainglorious Enlightened Ones* call progress. Brought about by people from the world of STEM! Artists in residence needed to work on this aspect of STEM! This is a very big part of the STEM world. What would the world be like without progress? Best not to think too deeply about what progress means!

**23**

The story of Prometheus, you may or may not know, so we will here, very briefly recount it.

In Greek mythology, Prometheus is the Titan who created humans, but he loved his creation too much, and feeling sorry for them, he advised them badly by showing them a trick.

The humans had to sacrifice a bull, and they did not want Zeus to take the best parts of the sacrificed bull, so they asked Prometheus for advice, and he told them to take the bones and to dress and hide them with a few bits of good meat so that they appeared to be the best and to hide most of what was best, with entrails and the like. And Zeus was deceived by this trick, picking what looked to be the best, only to discover that it was not, which angered him much and by way of punishment he denied men the fire that they could have used to cook the meat and to make for themselves a better life.

Yet again Prometheus foolishly acted in humankind's favour by stealing fire from the gods and thus, with the ability to make fire, the road to the development of human civilisation started. But this was too much for Zeus and as punishment, Prometheus was bound to a rock with unbreakable chains and an eagle was commanded to rip out his liver. Prometheus, though, being an immortal, could not die and at night his liver regenerated; thus the agony was repeated the next day and the next and so on for a countless age. And there he would have remained, caught in an eternal loop, if Zeus had not allowed Hercules to rescue Prometheus.

Now to the naming of artists, or those with artistic inclinations, who had roots in the world of STEM: Alexander Calder, Rachel Carson, Arthur Koestler ...

Hold on! Why are we bothering to do this? Surely this is just following the thinking of those who worship at the shrine of Leonardo da Vinci, who was most certainly an artist, but was he anything beyond this? Who can tell? These days it seems he has been elevated to the position of a messiah - did he also walk on water, heal the sick, turn water into wine, and raise the dead? There is more of myth about the *Leonardo of the modern world,* than of fact.

Oh how shocking - yet another heresy!

Yes indeed - you will find us very heretical! So enjoy.

Frederick Winslow Taylor now wishes to speak.

"I would like to say that in our system of management workers are told what to do and how they are to do it any improvement to the instructions given to them is fatal to success."

What is the father of so-called *Scientific Management* saying? This: free people (managers) think, slaves (workers) work.

Of course history shows that there was nothing scientific about your management methods at all. If we were to call you disingenuous, we would be letting you off lightly. History suggests that you were just offering precepts, and that there is no one-best way of doing work, and there was no scientific basis to any of your thoughts. Today also, you would probably be referred to a psychologist, for your behaviour was at best obsessive and far from being normal, whatever normal means.

So, as Fred, goes off back to the nineteenth century where he belongs, we note that his ideas are far from dead, even though they are discredited. This is the nature of Ancient Greece, it just keeps coming back to haunt you, and keeps you like Prometheus, but with no Hercules to set you free. Just like the European Commission!

## 24

Time is always, a single moment in time, in which all other moments in time, past and future, co-exist in this one time that is known as the present for the person who is experiencing that moment. Time therefore is a singularity. This result emerged from an artistic enquiry into the nature of time. Time is a singularity and there is an equation one can write to express this. Now that is something you were not expecting.

Arthur Koestler now wishes to inform us of some of his important findings from his extensive study of acts of creation in the arts and the sciences.

"After examining the self-reflections of many leading scientists across history, it is clear that many emphasise the importance of spontaneous intuitions, unconscious guidance, and sudden leaps of imagination which they are all at a loss to explain. The important point is that the role of strictly rational thought-processes in scientific discovery has, I believe, been vastly over-estimated since the *Age of Enlightenment*. Contrary to the Cartesian bias in our beliefs 'full consciousness', in the words of Einstein, 'is a limit case'."

In other words there is an awful lot of subjectivity in the apparently highly objective and rational realm of science. It is a continuing theme, and this raises the question of why so much denial about this?

"Indeed, and this is something that I also noted, for it seems that there is an apparent paradox. Branches of knowledge which operate predominantly with abstract symbols, whose entire rationale and credo are objectivity, verifiability, logicality, turn out to be dependent upon mental processes which are subjective, irrational, and verifiable only after the event."

Perhaps an engineer who has worked with artists, who will be speaking later, we suppose, will be able to add to our understanding?

Financial accounting is the production of out-of-date information about a company's financial status for one particular trading year. This information comes in the form of audited accounts, which include a profit and loss account, and a balance sheet, along with accompanying notes. They are out-of-date by virtue of the process of preparing them, which has to happen after the financial year has ended. They are a report on what has happened. They are also based on standardised practices, and knowledge of these practices is required to be able to understand what financial accounts mean. And although they are out-of-date, they are still useful for investors and shareholders, and also for anyone else who is interested.

The term audited accounts is interesting. These are different from un-audited accounts. Audit means inspections, review, and assessment. It does not mean just taking at face value that which is in the company's financial records. In many countries auditors work for the shareholders, and are reporting to them. They are also independent of the company whose accounts they are auditing. Why do you think this is so?

Who is auditing all those scientists, technologists, engineers and mathematicians? Which independent body of auditors is assessing all those claims made by scientists about being independent, unbiased, and not having any axe to grind? Try looking at the register of interests of those scientists that appear on television offering their so-called independent scientific view.

We are sorry, but no such register exists so how do you know that the plant geneticist who is extolling the virtues of GMO, is not being influenced by Monsanto or Syngenta through the relationships, often financial, that he or she has with such organisations?

Everyone advises in their own interest or those of the group that they are a part of. There is no such thing as a view from nowhere. This too is the nature of being human.

This goes for artists too, especially those involved in art-science and art-technology. Who is auditing their claims? Who is auditing Leonardo da Vinci?

Nemesis is part of both Ancient Greek and Roman mythology - the god of punishment, of retribution. Hubris denotes overconfident pride and arrogance.

Now for the naming of that which people might associate with descriptive expression and plastic expression combined: illustration; concrete poetry; hieroglyphics; illuminated manuscripts; maps; calligrams; ideograms; logograms.

Is scriptovisualism concerned with these? No, not directly, but they do serve to illustrate a point - cognitive bias that will prevent people from understanding that scriptovisualism lies beyond these. These cognitive biases relate to that which we call silent narratives. STEM is full of silent narratives. One of the consequences of this is that art is predominantly perceived by STEM people in certain ways. Hence SEAD, STEAM and STARTS. Art-science also has its silent narratives. What could they be?

Art pushes boundaries, crosses boundaries, reinvents, discovers new worlds that could be, but which are not, because of ... What? Now this is indeed a question, and one that lies at the heart of scriptovisualism.

Does art really do such things? Or is it just some art?

Victor Shklovsky has joined us. He was part of the literary school that was called Russian formalism.

"Yes indeed I was and I want to tell you about a central feature of our literary theory, which we called de-familiarisation, or making the familiar seem strange.

"And so life is reckoned as nothing. Habitualisation devours works, clothes, furniture, one's wife, and fear of war. If the whole complex lives of many people go on unconsciously, then such lives are as if they had never been. And art exists that one may recover the sensation of life; it exists to make one feel things, to make the stone *stony*. The purpose of art is to impart the sensation of things as they are perceived and not as they are known. The technique of art is to make objects unfamiliar, to make forms difficult, to increase the difficulty of and length of perception because the process of perception is an aesthetic end in itself and must be prolonged. Art is a way of experiencing the artfulness of an object; the object is not important."

Social media are not primarily means of communication, but means of self-expression. They are also means of participation. It seems that many people do not understand this and we may, one day, meet an elitist scientist who clearly, through his derogatory comments about the use of social media, does not demonstrate an understanding of social media and their use, nor how these are important in the world of capitalism - the business of making money - and how these media have become essential to the making of money, from which taxes are collected, which are then used to fund his research and pay his salary. Oh the delights of the fragmented and reductive mind, and that which passes for being modern. We could easily be in the eighteenth century. We are! See, the free people are not thinking anymore! Did they ever?

And so, what do we write next? Here we stop for a break. Here we pause.

Now we resume. Leonardo is often described as a polymath. The twentieth century author and writer Arthur Koester noted however, that this was a characteristic of the Renaissance. Was it? Do we need such people again? Perhaps, for otherwise we may find ourselves being consumed by those experts who know a lot about very little - specialists - who cannot even understand or recognise that there is such a thing a great ocean of truth lying unexplored and undiscovered before them.

Shall we call this new type of polymath, Life System Architects? The reasons for this will, perhaps, or not, eventually become clear. But what use are such people if they turn out to be as ignorant as the specialists?

**25**

An exhibition, which opened in 2016 at the Science Museum in London, with the title: *Leonardo da Vinci: The Mechanics of Genius*. Leonardo, seen for centuries as an artist, is now being increasingly portrayed as an engineering genius. And this comes at the moment when the image of the world of STEM sinks lower and lower, as people come to realise the price to humanity and the planet, of that which the *Vainglorious Enlightened Ones* call progress.

There is no such thing as a view from nowhere.

Back in the 1960s, an artist by the name of Donald L. Weismann, an academic as well, did undertake a project of writing fiction and painting it too, both simultaneously. He stressed that the painting aspect was not about illustration. He called the resulting book, where he wrote about this experience: *Language and Visual Form: The Personal Record of a Dual Creative Process*. He noted that this exercise was a most unusual thing to do.

Is Weismann's artistic investigation relevant to scriptovisualism? It could well be. To answer this question we hereby announce that we are doing the same: writing this book, while simultaneously working on a painting that will be called: *366 – A Scriptovisual Composition Unknown.*

We now will tell you something else: back in 2014 we created a detailed sketch of something that has yet no name, which one day will be worked on further and become a painting. This too was an exploration of scriptovisualism, and once that sketch had been done, came forth the idea for a book reflective of the themes explored in that sketch, and the name of the book will be: *Janus Totenkopf's Adventures in Wonderland: The Fable of the Chief Advisors*. As we write 366, work on this other book has yet to begin, but we may start to write and paint in relation to this topic, while also working on *366 – A Scriptovisual Composition Unknown*. Or we may not!

## 26

STEM – a place for old men whatever their age and gender.

And speaking of Leonardo, as we were in the chapter before, Donald Weisman, in his book *Language and Visual Form*, quotes Leonardo as saying:

"Poetry is superior to painting in the presentation of words, and painting is superior to poetry in the presentation of facts. For this reason I judge painting to be superior to poetry."

Is this quote correct, Donald?

"Yes it is as quoted in the footnote on page 69."

We now turn to Robert Goldwater and Marco Treves, for both of you included these very same words in your book *Artists on Art*.

"Yes we did, on page 48."

Thank you. We will certainly reflect on this in our exploration of scriptovisualism. But not everything we write is poetry, but nevertheless it may be poetic in some way. We are back to literary theory once more. Nevertheless we feel that it would be best to speak with Leonardo directly to get a fuller explanation. Never accept what other people say - always, where possible, go back to the original source, which in this case is - Leonardo.

Now we wish to recite some words that the *Vainglorious Enlightened Ones* will be familiar with, because of their fragmented and reductive minds, which allows them, like little children, to take complex things apart, to see all the bits and pieces and how they might work, but prevents them from putting the pieces back together again - they have not got a clue.

Some of these words are: integrate, hybridise, connect, and imbricate. They like to work at crossings. They just love a nexus: the nexus of art and science, the nexus of art and technology, the nexus of who knows what else.

The *Vainglorious Enlightened Ones* do not believe in consequences. What follows is a nursery rhyme about consequences:

*Humpty Dumpty sat on a wall,*
*Humpty Dumpty had a great fall.*
*All the king's horses, and all the king's men,*
*Couldn't put Humpty back together again.*

Oh dear! Never mind, just pretend it did not happen. Do not count the dead. The reconstruction is just economic growth, yet another contribution to Gross Domestic Product. On they merrily go, skipping and dancing all the way.

Perhaps! Or perhaps this time, not.

**27**

There is no such thing as a sharp diving line between truth and fiction when it comes to history. History is made not by the people acting out the events that history records, but by observers in a different place and a different time. Leonardo! What do we mean?

The past - that foreign place where, they did things very much the way that they are still done today, in that collectively constructed reality known as the world of STEM.

In medieval times it was believed that it was only God that created. Men did not, so you will not find men saying: "I made this." Neither will you find anyone praising and idolising artists for their creativity. Imagine that! This is quite a different attitude to that of the *Vainglorious Enlightened Ones*.

In the stillness there is motion; in the motion there is stillness. What on earth are we saying now?

Scirptovisualism - visual thinking perhaps? In the beginning was the image and from the image came forth the word, and to the image the word doth return. It is experimental of course, and it may come to nothing, but one thing is for sure, it is not about creativity *per say*, although in its outputs this obsession of the *Vainglorious Enlightened Ones* may well be seen.

*And Then There Were None* is a mystery novel by Agatha Christie, widely considered her masterpiece.

Surprising that the strange man's ideas still have currency, and later we will be examining the words of a modern day scientist who actually wants to apply *Scientific Management* to government. It is of course an old chestnut as they say - known as technocracy. C.P. Snow would have approved.

What strange man are you referring to?

Oh yes indeed, we forgot that we are engaging in connecting with the irrational, as Surrealists do. We are of course talking about Taylor - he that made of other peoples' working lives a misery. Which is what scientific government will do, has done, for now we are into the realms of Soviet communism, and all their five year plans, and notions that the communist state thinks while the citizens obey. It is all about the common good, as well as elitism, which clearly never existed, just like there were no special facilities for the party elite. We are of course talking about the European Union as well!

Think for yourself now, about all those things that happened but never did. This too is what being human means, which is why we must not allow those who think they know best, to decide. European Commission technocrats! What?

**28**

The past - that foreign place where, they did things very much the way that they are still done today, in that collectively constructed reality known as the world of STEM.

Creativity is not the special persevere of a gifted few. Creativity is an everyday thing, used by everyone regardless of their status and occupation. Only the *Vainglorious Enlightened Ones*, being Ancient Greeks, managed, for reasons they claim of efficiency - actually the making of a small number of people extremely wealthy at nature's and humanity's expense - to construct an economic system where the workers were not allowed to think. The Soviets did as well. A clearly demonstration of the belief in *the one best way*; this just happens to favour those at the top of society. It is those Ancient Greeks once again.

"The creatures outside looked from artist to technocrat, and from technocrat to artist, and from artist to technocrat again; but already it was impossible to say which was which."

The book is Animal Farm and the author is George Orwell.

Technocrat! Artist! Surely you mean pig and man? Do we?

George Orwell, a renowned British novelist and critic, perhaps best know for his novel *Nineteen Eighty-Four*. William Hogarth, a British painter, born 1697, died 1764. He specialised in scathing, sometimes savage, visual commentaries on social conditions. His paintings often told a story.

Artist can have very sharp teeth and they know how to bite. They can even bite without those who have been bitten, knowing what has happened to them. Have you been bitten?

André Breton, the founder of Surrealism is here and wishes to contribute one more time.

"In my first Manifesto of Surrealism, I made a very important point, that in this day and age logical methods are applicable only to solving problems of secondary interest. The absolute rationalism that is still in vogue allows us to consider only facts relating to our experience. Logical ends, on the contrary, escape us. It is pointless to add that experience itself has found itself increasingly circumscribed. It paces back and forth in a cage from which it is more and more difficult to make it emerge. It too leans for support on what is most immediately expedient, and it is protected by the sentinels of common sense. Under the pretence of civilisation and progress, we have managed to banish from the mind everything that may rightly or wrongly be termed superstition, or fancy; forbidden is any kind of search for truth which is not in conformance with accepted practices."

This of course, we observe, was stated back in 1924, which just shows how far ahead of their time artists can be, for this understanding has yet to arrive inside the heads of the vast majority of people, here in 2016, who are trapped in STEM ideologies. Global Systems Science proves the point, which is just another manifestation of Taylor's so-called *Scientific Management*, and the product of the technocratic mind, which values data and quantification, but in reality is just engaging in yet more behavioural policy and decision making.

Scriptovisualism - part of something that we might call behavioural art. Art is art, of course, and does not need additional labels, but sometimes it is useful to add in an extra word, just to explain a little bit more. Perhaps this is the nexus of art and the social sciences? Or is it art at a meta-level - observing and questioning? But questioning what? This: Sciart; STEAM; SEAD; STARTS. We look at the silent narratives, and ask what is going on and why - image making, for example. But we do not like words such as nexus and that is definitely not our thing. We live in the spaces between spaces. And this is not just about questions, but also interventions, to change what we see. Like action research!

Agency: action or intervention producing a particular effect.

**29**

Creativity is not the purpose in itself - it is other people that seem to adore this aspect of the individual artist, although finding a new way of expressing some aspect of life and the universe is highly valued. It is part of the aesthetics by which art is judged, although there are more ways of judging excellence than this rather old fashioned means. Yet quality is still important, however it may be defined. This by the way, Robert, is the answer to your question about how to judge excellence - it is all about aesthetics of which there are many different aspects.

Adam Smith now wants to speak about pity and compassion.

"No matter how selfish you think man is, it's obvious that there are some principles in his nature that give him an interest in the welfare of others, and make their happiness necessary to him, even if he gets nothing from it but the pleasure of seeing it. That's what is involved in pity or compassion, the emotion we feel for the misery of others, when we see it or are made to think about it in a vivid way. The sorrow of others often makes us sad—that's an obvious matter of fact that doesn't need to be argued for by giving examples. This sentiment, like all the other basic passions of human nature, is not confined to virtuous and humane people, though they may feel it more intensely than others do. The greatest ruffian, the most hardened criminal, has something of it.

"We have of course no immediate experience of what other men feel; so the only way we can get an idea of what someone else is feeling is by thinking about what we would feel if we were in his situation. Our imagination comes into this, but only by representing to us the feelings we would have if, etc. We see or think about a man being tortured on the rack; we think of ourselves enduring all the same torments, entering into his body (so to speak) and becoming in a way the same person as he is. In this manner we form some idea of his sensations, and even feel something that somewhat resembles them, though it is less intense. When his agonies are brought home to us in this way, when we have adopted them and made them our own, they start to affect us and we then tremble and shudder at the thought of what he feels. Just as being in pain or distress of any kind arouses the most excessive sorrow, so conceiving or imagining being in pain or distress arouses some degree of the same emotion, the degree being large or small depending on how lively or dull the conception is.

"So my thesis is that our fellow-feeling for the misery of others comes from our imaginatively changing places with the sufferer, thereby coming to conceive what he feels or even to feel what he feels.

"Pity and compassion are labels for our fellow-feeling with the sorrow of others."

Here then you have a surprise; the man who is held up as the thought-leader for selfishness, the man who is attributed as the source of capitalist economic ideas, of the notion of profitable speculations being a public good, because the growth stimulates demand, and everywhere diffuses comfort and improvement, is, it turns out, concerned with pity and compassion.

So Adam, back in the eighteenth century you were speaking about both economics and selfishness, and the exact opposite! We are back once more to the problem of John Dewey and his monster with its lack of pity and compassion.

This is Adam's response:

"Yes, indeed I was. What many people do not realise is that I was a moral philosopher and wrote a book called *The Theory of Moral Sentiments*, in which I wrote about sympathy for ones fellow human beings. It's the bond that binds people and society together. We need both self-interest and sympathy, and I would now extend that notion of sympathy to include nature as well."

Art has a role in bringing that sympathy to the world of business, into research and development, and that means changing many processes and approaches that are now entrenched. This is not about incremental changes but fundamental ones.

How unreasonable is that?

"Very," came the reply from those with vested interests, those caught-up in the world of STEM, and their masters in the world of business. As well as those in the European Commission who have abandoned the people of Europe to the power of money and who seek to appropriate art in the service of this crime against humanity.

Ah yes, that simplistic notion of the one best way. Oh how words linger long after the man saying such things is laid to rest and has become nature's compost; something for you to reflect upon, Frederick Winslow Taylor, during your eternal rest.

The risks can be managed. Keep saying it over and over again. If you say it enough, you will convince yourselves that you are right. Now back to the matter of Agatha Christie's novel called *And Then There Were None*. It was not originally published with this title. It makes reference to a nursery rhyme, known by various titles: *Ten Little Soldiers* is one of these. This is our version of the nursery rhyme:

*Ten little Europeans went out to dine;*
*One choked his little self and then there were nine.*

*Nine little Europeans sat up very late;*
*One overslept himself and then there were eight.*

*Eight little Europeans travelling in Devon;*
*One said he'd stay there and then there were seven.*

*Seven little Europeans chopping up sticks;*
*One chopped himself in halves and then there were six.*

*Six little Europeans playing with a hive;*
*A bumblebee stung one and then there were five.*

*Five little Europeans going in for law;*
*One got in Chancery and then there were four.*

*Four little Europeans going out to sea;*
*A red herring swallowed one and then there were three.*

*Three little Europeans walking in the zoo;*
*A big bear hugged one and then there were two.*

*Two little Europeans sitting in the sun;*
*One got frizzled up and then there was one.*

*One little Europeans left all alone;*
*He went out and hanged himself and then there were none.*

## 30

A milestone - we have left the twenties behind!

Technocracy is an organizational structure or system of governance where decision-makers are selected on the basis of technological knowledge. Experts know best - free people think, slaves work! You should by now have realised - if you are one of the *Vainglorious Enlightened Ones* - that you are at heart, still an Ancient Greek!

C.P. Snow was a technocrat, as well as a failed scientist. Oh the delights of the fragmented and reductive mind. How so simple the world does look, when it is taken apart and the bits that are not understood are discarded.

"Adam Smith, the high priest of market economies and of modern capitalism, may well be the most quoted and least read of all authors."

That was Charles Handy speaking. He will be back later.

Now we would like to nominate C.P. Snow for the position of the second most quoted and least read of all author's. Here we refer to the book *The Two Cultures and a Second Look.*

And where, among all that is STARTS, STEAM and SEAD, will you see this matter of technocracy discussed? But evidently some artists know about this, it is just that, in the mad rush to sing and dance - to what are rather conventional scripts with new names - the really good people, the ones who are committed and knowledgeable, they tend not to be party animals and want instead to spend time composing, rather than talking the empty nonsense that is STARTS, STEAM and SEAD. Not to mention also their abhorrence of the instrumentalisation that accompanies those efforts which transform artists into handmaidens.

Artists can instrumentalise themselves, and have done so in the past, and this is a very different matter to being instrumentalised by those who are clearly something else, with agendas that they keep to themselves - but you can safely bet, that whatever it is, it will always have to do with power or money or ideology, and often all three. It is that economic monster once again.

*Abstract* and *symbolic* are two words often used to describe both painting and writing, which suggests that these are inadequate ways of capturing the differences between that which is plastic and that which is descriptive. There has to be better ways, which of course is a subjective expression – the word *better* that is. So much subjectivity! All you people caught-up in rationalism, objectivity and quantification must be horrified, and yet you still persist in bleating-on, and on, and on, and on, and on … about art-science and art-technology practices. Art is the domain of the subjective, the irrational and the emotional. What on earth do all you STEM guys want? Perhaps you think you can take control of art as well?

*Sticky, sticky spider*
*Spins its ideological web*
*Along comes Jack*
*Now Jack's trapped!*

What is that flying every so high? Looks like an eagle, keen-eyed, highly observant, missing nothing down below. And those talons and that beak, ready to tear – they are already looking very red.

## 31

The last day of January 2016 – yet another milestone! We told you this is not a diary.

Someone else is here.

"Science must have a seat in government. Politicians who do not take our advice, must tell citizens that they are not doing what we decided."

You have heard it all before of course – scientists know best, which also means that: scientists think politicians work! We could now reframe these strange words from the past, and say: in our system of government, politicians are told what to do and how to do it, and any improvement made to the instructions given to them is fatal to success.

Trust me I am a scientist, which is the reason why you cannot be trusted.

No wonder Junker sent her packing! It was a very wise thing to do, yet still they conspire among themselves. It is no use asking them to read what John Dewey and Adam Smith have said – they are too far gone, caught-up in that web. And still they conspire among themselves. Who are we talking about? We are referring to the former chief scientific advisor to the former president of the European Commission. Which president? The one that was in office when the global financial crisis started in 2008; a crisis started because of the behaviour of banks – the monster! The president who later joined the monster when he accepted a position at the global bank (monster) known by the name Goldman Sacks.

Oh look it is George Bernard Shaw.

"The reasonable man adapts himself to the world: the unreasonable one persists in trying to adapt the world to himself. Therefore all progress depends on the unreasonable man."

"This is why my 1989 book about new thinking for a new world, is called the *Age of Unreason*."

That was Charles Handy again. He will be back later.

Julia and Paul - they are the essence of man and woman saying unreasonable things. Now we are talking about ourselves in the third person!

"The man who listens to Reason is lost: Reason enslaves all whose minds are not strong enough to master her."

That was George Bernard Shaw again.

Excellent scientists think while entrepreneurs work! This is how Europeans think, but not, if they have any sense, the Chinese and the Indians, as well as others of non-European inclination. Have the Europeans not realised yet they are living in interesting times? Evidently not! Europeans have many strategic weaknesses which can be exploited to destroy them economically. What chance do they have of thinking unreasonable things?

So to the tale of Narcissus: he is that character from ancient mythology. He was proud and he disdained those who loved him. Nemesis noticed this behaviour and attracted Narcissus to a pool, where he saw his own reflection in the water and fell in love with it, not realizing it was merely an image. Unable to leave the beauty of his reflection, Narcissus lost his will to live. He stared at his reflection until he died. There are many variations of this tale.

Nemesis is the goddess of retribution. We have already met her very briefly. People who engaged in hubris, who think themselves more important than others, you know the type - STEM people - they should be wary of Nemesis, for she may curse them all too. Perhaps she already has?

And the artist René Magritte now speaks:

"I am not an artist, but a thinker that communicates using painting."

Yes indeed we shall be coming back to you and your work, we suppose.

## 32

The first day of February 2016 - this is not a diary. February is a special month for in 2016 it has twenty-nine days instead of the normal twenty-eight. The year 2016 is a leap year, so where shall we leap to now?

So with the leap taken we accidentally land on Charles Handy. Sorry about that, but we would rather like you to speak later. But this is a good opportunity to note that someone such as yourself, who has worked in both industry and academia, and even once had a management consultancy, proves the point that you should not judge a book by its cover.

Handy has been denigrated by elitist academics, but read his books and you will find a man who, in his time, was very much aware of what was going on in the world, and able to eloquently present his analysis which points towards the need to do just about every thing we do, in radically different ways. A nightmare for those stuck in that web. No it is not about incremental change.

So now to the topic of management accountancy which is where we actually intended to land. This is all about models - financial models of what is going on inside a business over the short term. This information is used for day-to-day operations, internal reporting, and investment decision making. You know the words: return on investment, payback period and so forth. Management accounting also deals with the allocation of overheads to determine the cost of products or services. Overheads are all those additional costs such as administration and management, which are needed to enable all who do the productive work, to do that work. This latter group are often referred to as direct labour.

Back in the late nineteenth century when management accounting first developed, a method of allocating overhead to products (in those days it was all about manufacturing) was developed based on the amount of direct labour time spent on a task. The more time spent the more overhead was allocated to a particular item. This is called absorption costing.

Because workers, the slaves, are not allowed to think, the thinking people, in management, create a cost that needs to be accounted for in someway - this is Ancient Greece, institutionalised into the financial systems of an enterprise.

During the 1980s with the arrival of what was called flexible manufacturing (earlier generations of manufacturing involved more dedicated, less flexible machinery) new management accounting techniques, taking advantage of computers were developed.

The interesting point is this: management accounting methods are all based on models, and the results are the outputs of models. The different models, give different results.

One of the observations we now want to make is although the results were just outputs from models, once a model had become entrenched and accepted, people quickly forgot that they were only dealing with models, and the model become reality. Few people could see this, and the numbers become the truth - the sole truth. Numbers are about quantification. They are objective and rational, yet if one changes the model the numbers change. The choice of model is largely subjective and is mostly a matter of tradition.

The *Vainglorious Enlightened Ones* believe in the supremacy of models and data. Quantification is everything. And yet hidden behind all this supposed rationality and objectivity, is a whole lot of subjectivity that people are in denial about. Now why is that?

"Every discipline, I suppose, is, as Nietzsche saw most clearly, constituted by what it forbids its practitioners to do. Every discipline is made up of a set of restrictions on thought and imagination."

That was Hayden White speaking. Now Catherine Soussloff will say a few words in connection with the above.

"Disciplinary taboos follow upon the heels of mythical cultural figures and, like other social taboos, result in suppression and repression."

A discipline is defined by what it forbids its practitioners to do! Social taboos!

Taboos are everywhere to be found. Art is no different. Leonardo da Vinci illustrates the point - speaking about the real Leonardo is not allowed! Art-science and art-technology are founded on artistic anecdotes, fables, tall-tales, myths, and fabrications. That is partly why it is difficult to find any results of substance. STARTS is founded on artistic anecdotes, fables, tall-tales, myths and fabrications, too, and will produce as results - artistic anecdotes, fables, tall-tales, myths and fabrications. And this we will use against the *Vainglorious Enlightened Ones,* and all those who advocate and seek to practice technocracy, scientific government, scientism, and the rule of science and reason. This is how we will stop them.

Human behaviour is a most curious thing. So when art-science or art-technology lovers meet and start whispering sweet nothings into one another's ears, there is something else happening, something far too subtle for the fragmented and reductive mind to perceive! STEM, along with the excessive obsession with the rational, the objective, and the quantifiable, just gets in the way. It matters not if one is an artist or a STEM person, all that Enlightenment baggage will prevent you from seeing.

Seeing: the action of seeing someone or something. Knowing how to see! It is not just the eyes that see!

Art-science and art-technology love affairs are one thing, but art in its own right - that is a different story.

**33**

Evidently, themes are developing, so it would seem that the composition is now more known. Is that true?

Yes and no. In the uncertainty there is certainty, and in the certainty there is uncertainty.

With 333 chapters to go, we could still end up anywhere. The writing is not over until chapter 366. Then some more writing will begin, as we create pentiments. You will also find pentiments in the visual form, in plastic expression.

Charles, the moment which you have patiently waited for has now arrived, so please speak and share your thoughts with us.

"Change, is rather obviously now, discontinuous and no longer a straight linear projection of the past into the future. This is the point that I was making in my book *The Age of Unreason*. Today, in 2016, this is clear, but back in time, when I first said such a thing, there were only a few others who were aware of this; the vast majority were oblivious to this fact.

"My point, the one that I was making back in 1989, was that when change is discontinuous the success stories of the yesterday have little relevance to the problems of tomorrow; they might even be damaging. The world at every level has to be reinvented to some extent. Certainty is out, experiment is in."

Now we are in the world of art and artists, but also many others, for experimentation is a way of knowing. Yet where in STARTS, SEAD and STEAM, will you see any understanding of what Charles has just said?

Responsible Research and Innovation, which we will call RRI for short, is a relatively new area for research in the European Commission's Framework Research Programme. RRI is about science with and for society, and the modernisation of STEM, that place for old men, regardless of their age and gender. RRI should also deal with the moral corruption that is becoming increasingly evident in Research and Innovation systems.

But nobody wants to talk about this - it is the elephant in the room. Another taboo!

Universities need to change, for they are being corrupted by the power of money - they too do not want to talk about this. What will it take? A police raid? Arrests? Criminal prosecutions? Better to be open and define what the challenges have become. Unfortunately human behaviour tends towards - denial. It is only after the event, that they come to matters of change and prevention. Until that time, they deny, circle the wagons, and just keep firing. But they are going nowhere and eventually they will have to surrender.

It is impossible to know with certainty how Leonardo or others from the Renaissance period - which spans several centuries - thought or viewed the world. One can seek glimpses by bringing together various disciplines from the humanities and the social sciences, but there would never be full certainty. Do you know, for example, if Leonardo read, what were in his era, the recently rediscovered works of Aristotle and Plato? We are not going to find out. That is not the point. What we want to ask is this: if he did, what affect did this have?

Let us now, in our imagination, transport ourselves to his studio, where we find that he is reading Plato's *The Republic*. He has just encountered the point in the text where Socrates is saying 'one man cannot practice many arts with success.' Does Leonardo, now write in the column: 'I don't agree'? We might like to think that he did, but without documentary evidence, we will never know. He may well have written: 'I agree.'

One of the problems faced by people in our time, is that when they look back at times past, they just see the past with modern eyes. Setting oneself free from this mind-set is a challenging task. For this one needs specialist expertise.

So, having brought our many disciplines together to understand Leonardo da Vinci, what do we do with the knowledge that we can gain from these different disciplines? Do we integrated it, whatever that means, or do we leave it stand as distinctive pieces of knowledge, complete with contradictions? Or is there other ways of dealing with this matter? Ways that have not yet crossed the fragmented and reductive minds of the *Vainglorious Enlightened Ones*?

**34**

We are sticking to André Breton's advice, just writing down what comes into our conscious mind. We have not looked back and reviewed what has been created, so we do not know how many loose ends lie back there in the preceding chapters. Our pentiment shall be the revision of the composition by addressing that which we have not completed. Or perhaps we will just leave these hanging there, for Surrealism is about irrationality. The painting has not yet begun. Repetition may be an issue as well. Perhaps that too we will just leave.

Bell Labs is one of the most reported and documented examples of artists working in scientific and technological research labs, and this started in the early 1960s. We have studied this in depth.

Ken Knowlton is one of those who worked with artists at Bell Labs. Today we would describe him as a computer scientist. Later in life he would take up art himself. Ken wants to say something about artists and those from what we now call STEM.

"Both groups are creative, imaginative, intelligent, energetic, industrious, competitive and driven. But programmers, in my experience, tend to be painstaking, logical, inhibited, cautious, restrained, defensive, methodical, and ritualistic. Their exterior actions are separated from their emotions by enough layers of logical defences that they can always say why they did something. Artists, on the other hand, seem to be freer, alogical, intuitive, impulsive, implicit, perceptive, sensitive, and vulnerable. They often do things without being able to say why they do them, and one usually is polite enough not to ask."

There is more to this thing called art-science or art-technology collaboration than at first might be thought, for not only is there the difficulties of collaboration among different STEM disciplines, but new ones. It amounts to this: Not only do you have those troublesome boundaries that exist among the STEM disciplines, but you must then add in deeper problems that arise when STEM meets those that come from the social sciences and humanities. On top of that add the mysterious world of artists, and how they think and work, which Ken has only just hinted at. Then you might have some understanding of the complexities and challenges that must be addressed. And yet people tend to behave that this is not the case, and all was, or will be, well. Digging deeply and searching far and wide, one can however find insights and reported experiences that tell a different story to the one that supporters and promoters of such collaborations want to tell.

Anecdote: a short amusing or interesting story about a real incident or person; an account regarded as unreliable or hearsay.

## 35

Our dear colleague Rachel Carson will soon be coming to speak. It is good that such a leading thinker, having an artistic mind, is prepared to make comments to help us in our quest. But it is strange though that, those most relevant to the so-called art-science dialogue, are never mentioned by either side, yet he who is the least relevant - we mean C.P. Snow - is constantly being referenced. This man - full of opinions - he too will be passing by, we hope, to apologise and make clear what he was spouting forth about. It seems that people have not understood. It is a question of hermeneutics.

Some artists, being artists - polymaths or otherwise - would say that there is no such thing as a view from nowhere. And the place where the Science Museum - and also it will soon be revealed, people called engineers too - are viewing Leonardo, is located in the minds of those rooted in STEM culture and beliefs, living in the twenty-first century, with several hundred years of scientific and technology development behind them, along with restrictive biases concerning what constitutes art, which are influencing thoughts, words and deeds.

The Institution of Engineering and Technology (IET) published an article about the exhibition called *Mechanics of Genius*, with comments waxing lyrical about Leonardo's note and sketch books, which well demonstrate that Leonardo is being viewed with a twenty-first century mind-set. The article also well demonstrates that there are misunderstandings about how artists work and think. The IET it seems, along with the Science Museum, has proved that Leonardo was also an engineer as well as an artist. Probably he was, but it is unlikely that he was an engineer in the sense that we now use this term. Most definitely a lack of critical thinking was displayed by these people.

Now back to that moon we were referring to, some while back. We refer to the observation of the moon, where we ended by saying that one could take a picture, print and mount it in a frame, hang it on a wall and then call it art.

Let us now do something different. Instead of taking a photo of that beautiful moon, and calling it art, we set things up as before, and give people a chance to experience for a short time that movement across the sky, with all those lunar features amplified through means of the technologies that we use. The art here does not lie in any objects as such, but in the process that the beholders are experiencing. They are making the art with the artist, while at the same time being part of the art that they are collectively making.

You should now understand that is not necessary to have an art object to have art. The question though is: why is this art and why, if a scientist did something quite similar, it would most likely not be art? We have already answered this question.

**36**

In writing automatically, if indeed that is what we are doing, we are not looking back, so obviously it may be that things that have been started, which remain to be completed, may be forgotten about. There may also be some duplication. If so will we leave this in place?

Self-evidently the art we have just described - the participatory observation of the moon - it is a highly scripted piece of art, but it has meanings that go beyond the moon as an object in a frame, because it also touches upon what is to be human. And in one's imagination, through reflection upon that which is momentarily seen, one could realise that you are part of nature and also part of humanity. You stand, both alone with the moon, yet also with, at the same time, those others present, just like you, at that particular moment, and others too, across the world, that you are not aware of, as well as those who have, over generations past, done exactly the same. You are part of a human experience that has continuity with the past and, hopefully with the future too. Or perhaps not! What was the artist trying to achieve?

This is the aesthetics of participation. And that is the difference. The artist is generally working in an aesthetic framework, the scientist is not. The artist is being expressive. The scientist is being cognitive. So artists develop what to the STEM mind looks like a piece of technology, a potentially unconventional and compelling product, but it is not. It may even be addressing this very point, that the world of STEM has become a monster, and the technology in the art is communicating this and perhaps posing a question: does all technology have to be the slave of Western capitalism, and the means of someone, probably already overly wealthy, acquiring more money, which they obviously do not need - personal greed euphemised as economic development, and dressed up in the ideology of personal interests, common good. So back once more to the eighteenth century!

Rachel Carson will be joining us soon. She will have a lot to say about many things. Probably she will make several visits to our odyssey.

Artists, being artists - polymath or otherwise - would say that there is no such thing as a view from nowhere. And the place where engineers and the Science Museum are viewing Leonardo, is clearly located in the minds of those rooted in STEM culture and beliefs, living in the twenty-first century, with several hundred years of scientific and technology development behind them, along with restrictive bias concerning what constitutes art, which is influencing what they think, do and say. Did we say this before?

**37**

Babbling brooks make most noise, still waters run deep.

Now we turn to the matter of Terry Fenton's observation of something he christened as the art-science muddle. Fenton was critical of movements that tried to use science as a methodology for art. He also criticised, what was developing at the time, the move towards art becoming a handmaiden of science. That was in 1969, and here in 2016 we can see that art not only has become the handmaiden of science, but also technology, engineering and mathematics - in other words STEM.

It is indeed a muddle for those that write about this so called art-science and art-technology; they are all over the place, never knowing which age they belong in: one day they seem to be in the *Age of Enlightenment,* then in the era of postmodernism, or perhaps just modernism, but equally possible is the *Age of Positivism*, or the *Age of Romance*, or the *Renaissance*, and, quite surprisingly, what Karl Jaspers denoted as, the *Axial Age*, which apparently we now need yet again. Karl will be dropping in to talk about this a little later in this literary adventure - we suppose.

It is no wonder that some people from the world of STEM look upon so-called art-science and art-technology as just plain nonsense. Much of it is of course, but somewhere within may lie something quite special and very valuable indeed for humanity. But what might that be?

Lewis Wolpert has strong opinions about this.

"Yes indeed I do. The enthusiasm to bring the arts and science closer together, even to emphasise similarities, is rather peculiar, as they are so very different. Trying to bring arts and science together is fundamentally social snobbery, as scientists are still envious of the status of the arts and the humanities."

Wolpert's first point is of course true, although if one were to compare occupations that are clearly very different, one would probably find some similarity. Physicians have diagnostic skills, but so do some engineers! Wolpert's second point is also true - clearly there is image making going on when scientists embrace the arts. Yet to completely dismiss so-called art-science and art-technology is to misunderstand, as clearly many people are now doing, on both sides - in the world of STEM and in the world of art. But what is it that needs to be understood?

Perhaps this is what our composition unknown is really about, what scriptovisualism is seeking to explore? We do not know. When the work is completed we will. Therefore - perhaps sadly (who can tell?) - you will have to read on, until the bitter end, and even then, we might not know, for the true meaning of what we here write might instead be found in a painting called *366 – A Scriptovisual Composition Unknown*. Art is not like science - this is indeed true for what other activity deliberately sets out to make comprehension more difficult than it needs to be - that is Russian formalism for you. De-familiarisation - making the familiar seem strange!

**38**

May you live in interesting times! We are living in interesting times.

The architect Buckminster Fuller has just arrived and has something very important to tell artists and those from the world of STEM.

"You never change things by fighting the existing reality. To change something, build a new model that makes the existing model obsolete."

Very true, so now you know what we are about.

The *Beholder's Share* - what is this? Well its part of theory, so here we go again being *too theoretical*. You will regret you ever said that oh ye of the past - DG CONNECT.

Simple explanation: every work of art is completed in the mind of the beholder in the form of their response - usually emotional, be it positive, negative, or neutral. So, when people say 'that's not art' they are contributing the *Beholder's Share,* just as much as when they ask the silent question 'what's this all about?' DG CONNECT bemoaning the fact that they are being lined up as knaves and fools, and being made a figure of satire and ridicule, is also an example of the *Beholder's Share*. And the possibility that they do not really understand what this book is all about, can be seen as taking DG CONNECT and making them part of a work of art, that is a narrative in plastic and descriptive form, and that place in between known as scriptovisualism (if this is where indeed this strange beast lies).

The Beholder's Share is quite an old theory.

And to the matter of the painting we do now turn, for already it is clear what our sketching pencil will need to start working on, for the image has already begun to form in our mind, and this is something quite different to the development of the descriptive part of the experiment, which started from nothing in the conscious mind, and is very much driven by what the subconscious is telling us to do. We stress however it is not about illustration, for we will be sketching things that are not in the text, and will probably never be. Or perhaps because they are in the painting they will find a way into the text?

And the soldiers, automatic weapons to hand, still patrol the streets of Brussels. So what are they going to do? Probably they are not trained for such a role, and could end-up shooting innocent civilians. This is the price that we have to pay for Western stupidity, for not understanding that, when you sow the wind, the whirlwind is what you reap. Do you not yet realise that you too are *people of the book*, and all that we are seeing here is one Abrahamic sole truth engaging in conflict once again with another one – this time it is Islamist extremism versus extremism of the secular kind, which STEM, and also some art-science, is very much a part of. More taboos!

For those of you who live in parts of the world where you have not become *people of the book*, please look to your own cultures and realise that, while they may not be perfect, you have in them the means to change the world for the better by bringing to an end this Abrahamic age. Judaism, Christianity, Islam, Western science, Western capitalism, Western communism, Western atheism: these define the era of the creation of the modern world, and that era is defined by these world views. It is now coming to an end, but we are not saying that you have to stop believing in a God – that is up to you. Peace and love to you all, who or whatever your God may be, or regardless of which of the above seven religions' churches you wish to stand in, or which of their priests you choose to listen to. We listen to them all – so why don't you?

**39**

Being a man is a culture, not a disease or a disability! You can change if you want to. Do you want to, or are you expecting women to become like you? There is always a silent narrative being spoken that people seem not to hear. Gender equality, equality of opportunity – what is it that you men are actually saying?

Change and the acceptance of the need for this – which is what must now happen in the Western world – will be very painful indeed. You are the problem and its time to face up to this, and begin the transition to … What?

It is for humanity collectively to determine what its future will be, for the time when one nation, one culture imposed its will, its values and its beliefs on humanity - this age is over. Someone now needs to tell the Europeans, the Russians, and the Americans this. Yet cultural colonialism, backed up with threats of Western military aggression, still stalk the world, as the old imperial powers once more jostle for power and influence, creating as they go, distrust, anger and instability - it is like being back in 1914. And do not fool yourself into believing that the European Union is any better. They too are seeking to make the world European, such is their ideology and dogma.

STEM organisations and individuals are also now seeking power and influence - it is like living in the middle-ages, complete with that nasty story of what religion did in those days. We refer to the Church of Rome. Religion once again doing the same, only this time the priests come from the world of STEM, and their religion is a secular one. They have found a home in the European Union, in the European Commission - another manifestation of the Church of Rome.

And yet many people are optimistic!

"Optimism is a mania for insisting that all is well when all is by no means well"

That was Voltaire speaking. He wrote this in his satirical short novel called Candide. Candide very well epitomises those from the world of STEM.

And of the painting that we mentioned: what of this? Yes we started it as promised, so we are now indeed engaging in a dual creative process; it is a process grounded in the realms of descriptive expression (that which you now read) and in the world of plastic expression, which can be seen on the cover of this book.

## 40

In the forties now and matters are taking shape, but with 226 more chapters to come, we could end up anywhere.

Monodisciplinary - that which pertains to a single discipline, defined on the basis of the division of knowledge, which is inherently subjective. We just keep bumping into the subjective, in this world that is supposed to be highly rational, the very model of objectivity. Such division of knowledge is also probably very arbitrary and rooted in the past.

Change comes in two major forms: morphostatic and morphogenic.

"Art has done virtually nothing for science. Of all the arts, painting is the one least related to science as it does not deal with complex ideas or explanations, is the easiest to appreciate, and the response is often an emotional one. What is more, science is about understanding how the world works, there being only one correct explanation for any observed phenomenon. Anyone can have views about a painting and engage in art discussions. Non-scientists can thrill to scientific ideas but to make meaningful comments about them, and I exclude their application to technology, one actually has to have detailed knowledge."

That was Lewis Wolpert speaking once again. We thought he had left. But fortunately he had not, for now you can see that he has quite a fixation on dogma, and he allows us to make the point that, he well illustrates what is wrong with science. Apparently he keeps insisting, like some Abrahamic priest, that we do not understand. But what we do understand is that he has a single minded obsession with a certain old, and now very antiquated, positivist understanding of what science is, which puts him in the past, in the nineteenth century to be precise. No doubt he would be happy living there, along with Dawkins, yet another Abrahamic priest.

Before departing for the next chapter we would just like to ask if what Wolpert has said about art and science is correct. Is it? Perhaps on the whole he is right. Perhaps this book will, one day, be seen as a falsification. Evidently though, he knows that artists do have a habit of talking nonsense about science.

Let us now imagine him saying this:

"Art has done much for science. Of all the arts, painting is the one most related to science as it deals with complex ideas and explanations, and it is the hardest to appreciate, since the response can be in the realms of the rational and objective, as well as in the subjective and irrational, encompassing logos, ethos, and pathos. What is more, art is about understanding and discovering how the world works, there being different interpretations for any observed phenomenon, which depends upon the theory that one has nominated as being representative of the equivalence class of all theories that provide explanation. Anyone can have views about science and engage in science discussions. Non-artists can thrill to artistic ideas, but to make meaningful comments about them, and we exclude here application to technology, one actually has to have detailed knowledge of art."

Wolpert and Dawkins - condemned by Zeus and cursed by Nemesis, as is most of the West.

*Humpty Dumpty sat on a wall,*
*Humpty Dumpty had a great fall.*
*STEM people, no matter how hard they tried,*
*Couldn't put Humpty back together again.*
*The West is broken, undone, finished, and,*
*The mandate of heaven has passed to the East.*

## 41

Change, uncertainty, and unpredictability are the norm - the risks cannot be managed. It is time for balance and harmony to be restored.

"One of the most highly developed skills in contemporary Western civilisation is dissection: the split-up of problems into their smallest possible components. We are good at it. So good that we often forget to put the pieces back together again. The skill is perhaps most highly honed in science. There we not only routinely break problems down into bite-sized chunks and mini-chunks, we often isolate each from its environment by means of a useful trick: we say *ceteris paribus* - all other things being equal. In this way we can ignore the complex interactions between the problem being studied and the rest of the universe."

Thank you, Alvin, for that very important observation. That was Alvin Toffler speaking. Incidently Piet Mondrian never said *ceteris paribus.*

We would now like to mention the word cybernetics. Not the reductive mechanistic type of cybernetics that you have probably encountered, but cybernetics of a more advanced kind, which probably you have never encountered before. Paul is a cybernetician of this kind, and for that reason he would never say ridiculous things like *the risks can be managed.* People from the world of STEM do however utter such phrases. Is this because of their reductive minds? They do of course have reductive minds, but there is more to the statement *the risks can be managed* than just reductive thinking, for behind STEM there is a silent narrative, which Paul can hear, and does not share. What goes for Paul also goes for Julia as well. Cybernetics and art have some similarities, but you will not find these similarities in what was once called Cybernetic Art. To find the common ground you will have to look elsewhere. You will not look! This is why it is now apparent that we have to write this book! It is however only a beginning.

"You are all entitled to your own opinion but not to your own facts. In an ideal world in which policies were based on peer reviewed scientific data, policies would evolve but would not change from one government to the other. Science is currently not in the room when the decisions are made and this cannot continue like this. Political decisions should be based on science. If it is politically acceptable to decide not to base a final policy on data and science, this should be transparent and clear."

That was the former chief scientific advisor to a former president (President Barroso) of the European Commission speaking. You can see why she is the former chief scientific advisor! Does the phrase *the totalitarian state* mean anything to you?

Of course you have heard such nonsense before - free people think, slaves work! Here the formula is: scientists think, politicians work. And what she was saying is just Taylorism - so-called *Scientific Management*: in our system of government politicians are told what to do and how they are to do it and any improvement they make to the instructions given to them is fatal to success.

It is of course elitist, undemocratic, and bound-up in the delusion that scientists are impartial and unbiased, etc., to which we say: second-order cybernetics. It is also the basis of scientific communism, so think about that next time you hear a scientist telling everyone how marvellous everything will be when they run the world according to science. We refer you to the dogma of Wolpert and Dawkins as a sampler of their world vision. Global Systems Science - it is just the same thing. Technocracy! Scientism! They both fail. The European Commission well illustrates the point. Study the tragic tale of the single currency. Study the nonsense that is STARTS.

**42**

We repeat that we could end up anywhere.

And on the matter of repeating, as we have said before, extremism comes in many forms, and the reason may well be a deep rooted psychological need for security and certitude. The more complex and uncertain the world becomes, the more people will seek security and certitude, and hence the rise of all forms of extremist ideologies, with their overly simple and reassuring explanations. History tells a familiar story of such ideologies in the form of ideological battles found first in Judaism, then in Christianity, and then in Islam, and of course, among the three of them. And this continues among the secular Abrahamic religions: Western science, Western capitalism, Western communism, and Western atheism, all of which have discovered the sole truth, because there is only one correct explanation for ... Here add what you want, depending upon your particular adherence to one of these seven Abrahamic religions. Human behaviour is of course more complex than the above might suggest.

"Positivism is discredited."

That was Michael Polanyi speaking, and now we have the pleasure of listening to him again. Michael is a very interesting person because of his crossing of disciplinary boundaries. Starting his career as a chemist, and becoming an eminent one too – eventually receiving the accolade by being elected a Fellow of the Royal Society – he crossed the border and became, latter in his career, a social scientist, but did not stop there, for he eventually crossed another border into the humanities, becoming then a philosopher.

"Yes indeed many disciplinary boundaries I did cross, and back in the early 1960s my arguments were directed at demonstrating that positivism is discredited, and that there is something else at work, which I called the tacit dimension, whereby something indwells that leads us to choices about specific scientific problems to examine and explore."

Positivism is discredited! Someone should now tell the scientists this! But we cannot, as poor Wolpert has demonstrated, for here we see fixations on dogma, a very Abrahamic thing, where minds have close-in on fixed opinions. Oh dear, what are we to do?

Charles Darwin, what do you have to say?

"Only what I said back in 1859, in my book, *The Origin of Species by Means of Natural Selection, or the Preservation of Favoured Races in the Struggle for Life*. And what I said is that I by no means expect to convince experienced naturalists whose minds are stocked with a multitude of facts all viewed, during a long course of years, from a point of view directly opposite to mine. I look instead with confidence to the future, to young and rising naturalists, who will be able to view both sides of the question with impartiality."

Good advice, but the problem is that young people are being trained by the old, and all the old ones do, is indoctrinate the young, and make warnings about how careers can be ruined by not adhering to the orthodoxies. 'Do you want to be initiated or not?' This is what they ask the young would-be priests.

How did the VW emissions fixing scandal come about? Within such an organisation there are reported to be, the means by which engineers can raise concerns about unethical practices and such forth. But the safeguards did not work. This however should come as no surprise if you are connected with reality and not living a collective delusion. Many in the world of STEM do however live in the world of collective delusion.

Systems that involve humans, STEM, power and money, will eventually encounter what has being described as *ethical meltdown*. No amount of safeguarding, training, and whistle-blowing will change this fact. So what to do about STEM and those who adhere to its delusions is the question of the age. But it is not a new one - remember those medieval priests and what they got up to because their power and influence was not curbed.

**43**

Multidisciplinary, interdisciplinary, transdisciplinary - these are all buzzwords that are often used. But what do they mean? Answer: what you want them to mean. Probably nothing given the way they are often used as though they were interchangeable. This is one way of knowing that someone does not know what they are talking about.

People though, whichever of the above buzzwords they use, still insist on maintaining the disciplines as they are, for reasons of integrity they say, and to avoid watering them down, whatever such things mean. What do they mean?

"Method alone cannot remove the invisible and troublesome barriers between professions and disciplines."

That was design researcher Christopher Jones, speaking some words that he wrote in 1970.

"Some scholars picture science as driven by its own internal logic, developing according to its own laws in splendid isolation from the world around it. Yet many scientific hypotheses, theories, and models, not to mention choices made by scientists either to study or to ignore various problems, are shaped by economic, cultural, and political forces operating outside."

Thanks for that additional input Alvin, but better not say that to Lewis Wolpert - he will accuse you of not understanding science! That was Alvin Toffler once again.

What are we to do? This we asked a little while back, we recall. The answer to this question lies in art - you know, that which Wolpert says has done virtually nothing for science! Oh the delight of the fragmented and reductive mind! The reassurance of sole truths! The comforting Abrahamic way of seeing the world! You STEM folk are indeed, *people of the book*.

Later we will be hearing from Rachel Carson as well as some other eminent people from the world of STEM, Fellows of the Royal Society, and Nobel Laureates. Several of them will be dropping in.

"Some of the arts sector interviewees felt strongly that, in an increasingly secular age, science had come to assume in the public consciousness something approaching a monopoly of truth and of responsibility for righting the world's ills. This is an inimical situation for society to have arrived at, and one that could benefit from the counter influence of the arts. *Sciart* was felt to have provided a context within which artists could reassert an alternative narrative that could challenge the perceived hegemony of science."

That was Paul Glinkowski and Anne Bamford speaking about their assessment of the Wellcome Trust's *SciArt* Programme.

**44**

The Chinese do not have a creation myth. The Abrahamic world does. First it was God, and then it was the Big Bang. Models and theories are not the same as the real thing. You forget this at your peril. Obviously you have forgotten. Your delusions know no bound. Can art save you? Perhaps not, but art can certainly help prevent those in the East, in China and India, from making the same mistakes.

The Tower of Babel - a biblical story found in the Book of Genesis. Generally this is seen as an attempt by humanity, united by a common language, to build a tower to reach God. This aim was confounded by God who confused their language, so that they could not understand one another's speech anymore. So no common languages among those builders meant no further construction work was undertaken on the infamous tower. There are other interpretations of course.

Now a question for Arthur Koestler who is acknowledged as the person who wrote the definitive work on the history of Western cosmology up to the time of Newton: *The Sleepwalkers: A History of Man's Changing Vision of the Universe.* And the question is: how far behind is our science, compared with where it could have been?

"I estimated that we are behind in our thinking by at least a millennium, perhaps 1500 years. And the reason for this was a fear, in Ancient Greece, of change, and the craving for stability and permanence in a disintegrating culture, combined with scientific dogma in later periods within Europe's university system, because of scientists' desire to adhere to that which we now describe as Aristotelian science.

"I see from some comments already made in your book, that dogma is still alive and well. So just how far behind we are now, is anyone's guess."

Thanks Arthur for your insights, which as we all know, are based on scholarly research, although we fear today, because of academic dogma in the form of the old adage, *free people think, slaves work,* the quality of your work would not be accepted at all, you being what today we would describe as an independent thinker and researcher. In other words you are a consultant!

"Before I go, I would just like to say one more thing, which is, that the cosmology of a given age is not the result of unilinear scientific development, but rather the most strikingly imaginative symbol of its mentality - the projection of its conflicts, prejudices, and specific ways of double-think on to the graceful sky."

The *I-shaped* person - one who knows a lot about very little, otherwise referred to as the monodisciplinary professional. This what the Western educational systems produces. The *Pi (II) shaped* person - someone who is trained in two disciplines. It is said that the Japanese prefer Pi-shaped people to I-shaped people! Different culture! The *T-shaped* person: someone with a broad understanding of several disciplines, while being deep in one field. This is the basis for a cybernetician of the non-mechanistic, non-reductive kind. Incidentally these are not new terms. We first heard of the concept *T-shaped* person in 1975! It was the philosophy of the first degree that we took, explained to the whole intake that year, by the course admissions tutor.

"So Paul, when you were doing your first degree, what other disciplines did you have to study beyond simple introductory (101) courses?" This could be Julia asking Paul some questions!

"I spent two full years studying: electrical engineering; process engineering; mechanical engineering; electronics engineering; physics; mathematics; measurement systems; plus a range of non-STEM subjects: macro economics; management science; history of science and technology. In my first year I did not even have a single lecture on control systems, which is what I was primarily interested in. This is why I undertook a masters degree to go even further into control theory and practice, but even then, I still needed to undertake years more of work to penetrate further into the mysteries of cybernetics, and to understand the world of the social sciences. And all the time in the background was my literary interests and studies, pursued without anyone knowing what I was doing."

"Worthwhile or not?"

"Most definitely yes for this is where I understood that the boundaries between disciplines are just a collectively constructed reality that can be ignored and crossed at will. When you cross the frontier you end-up in no-man's land, belonging nowhere, in the spaces between spaces. And what we have found is that this space is full of diamonds in the rough, and no one can see them. They are too busy crossing the bridges that they are hastily trying to build, as their way of satisfying their craving for stability and permanence in a disintegrating culture, combined with the effects of scientific dogma. It is not the bridges, or the crossings, or that strange thing called a nexus, that are interesting, it is what lies beneath the bridges that I explore - virgin territory, which is simultaneously strange and unfamiliar, and also sometimes very frightening."

All conventions are boundaries waiting to the transcended. To transcend a boundary one must first conceive the possibility of doing so. Separation is an illusion.

Imagine this - you will need to imagine this! You have red lego bricks, all carefully put together into a block, enabled by the common red block fixing system that each red block carries. You have yellow lego bricks, all carefully put together into a block, enabled by the common yellow block fixing system that each yellow block carries. You have blue lego bricks, all carefully put together into a block enabled by the common blue block fixing system that each blue block carries. The fixing systems for each coloured block are unique to the particular colour scheme. We have a block that contains red, yellow, blue blocks, which mysteriously all fit together even though they have different fixing systems. Now why is that?

Self evidently we are being metaphorical, and subjective as well, so please no comments about lego bricks all having the same block fixing system. All you rational and objective thinkers please take note.

**45**

Multidisciplinary - pertaining to several separate disciplines, working together but as separate disciplines. Tower of Babel!

Multidisciplinary teams became popular in industry in the late 1980s as a way of overcoming what people came to call disciplinary and organisational silos. Tower of Babel!

Michael Polanyi once again.

"Let us differentiate between dedication and servitude."

Very well said, and by now it should be clear who is dedicated to science and who is playing out a story of servitude. Sadly, servitude is a very familiar tale in the world of the seven Abrahamic religions.

"I also want to say, before I leave, that I view science as a variant of sensory perception."

So is art, especially the process of painting.

John Constable has been waiting patiently to speak. Now is the perfect moment, so go ahead.

"Thank you. On June 16, 1836, while giving a lecture at the Royal Institution in London, I said that painting is a science and should be pursued as enquiry into the laws of nature. Why then may not a landscape be considered as a branch of natural philosophy, of which pictures are but experiments?"

Well John, the answer to this is very clear – there is no reason at all, just scientific dogma that manifests itself in the form of positivism and scientism.

John Constable knew more about clouds than any scientist of that era. He achieved this knowledge because he studied clouds, carefully observing, and then recording in symbolic form, the results of his observations. It was discovery, the creation of new knowledge. The same can also be said of many artists and their knowledge of colour. They were far ahead of scientists in their understanding, a fact that the physicist James Clark Maxwell well understood.

"Art is a way of knowing, but to come to an absolute conclusion about an image is to rob it of its power as a guide."

That was Pat Allen speaking, who works in the field of *Art as Therapy*, and a clear case of art doing something very beneficial both for humanity and for science. Because of *Art as Therapy* we now know more about ourselves, and the healing of the troubled mind. But it is a special kind of knowledge.

Already Wolpert's opinions are undone. He will never know it of course, but now you see the nature of the problem that Western science has become. But we have not yet finished with his opinions. More will follow, we think, so you will have to read on to discover if this is the case.

The notion of the individual creative genius, which people now think of as being a characteristic of artists from the *Renaissance* period, has been reinforced by the individualism of the *Age of Enlightenment*, and this is deeply embedded in European culture, which is one of the reasons why those who practice STEM often idolise Leonardo da Vinci. Yet there is a stream in contemporary art which dismisses this as elitism, which of course it is, and instead pursues the aesthesis of involvement and participation, and they speak of *distributed authorship*. STEM however is still rooted in this elitist attitude which is a measure of how out-of-step it is with society. Consequently, there are all these initiatives that address open science, citizen science, citizen engagement, etc. The age of the *expert knows best* is dead and gone, only someone needs to tell the experts this.

Having an artist in residence is evidently not the way to do this, because changes do not evidently result from such processes. Thus we engage in provocation. But we do more than this. Can you not see?

**46**

Interdisciplinary - pertaining to (usually) two separate disciplines, working together, transferring knowledge, theories, concepts and techniques from one discipline to the other, to make new discoveries or to enable new methods of working. Tower of Babel, unless ... This would involve revealing to the world a secret hidden, as they say, in plain view. Find out for yourself. Chances are you will not, but that will not stop you saying that you have!

Apparently all the European Commission's Future and Emerging Technologies (FET) projects are claimed by their participants to be successful. A strange thing indeed given that they are all supposed to be high risk projects, where failure is accepted!

There are examples of interdisciplinary working in the world of the arts and science, at the boundary between art and a sub-topic in the field of computing. This is a straightforward case of people from computing taking up something from the arts world and using it within their own field. We call this, as a model, the *artist absent*. It is perhaps the easiest of all ways to work across the gulf that Western men (yes that is right - men) have created between every STEM subject, and all these individual subjects and just about every other one as well.

"On very large scales the universe is spatially homogeneous and isotropic to an excellent approximation."

That was a physicist speaking. Note the word approximation.

Spatially homogeneous: the same regardless of spatial location; isotropic: uniformity in all directions.

"I am not an artist. I am a thinker who communicates using paint."

Thank you René. That was René Magritte, the famous Belgian Surrealist speaking; a person who communicated complex ideas and explanations in paintings. Oh dear, Wolpert's self-constructed reality is unravelling.

One of the reasons why you are aware of Leonardo and others from the Renaissance era, and are very unlikely to know the names of very early medieval artists is that by the time Leonardo was born society was changing and individualism was replacing communalism. Artists had left behind the obscurity of the monasteries and were now saying "I created this", which would have been unacceptable centuries early. The age of the individual creative genius had already dawned, and it was no longer God or the Church that was positioned at the centre of artistic creation, but man, although the influence of the Church as patron was still there. Let us not forget the there is vast amount of Renaissance art that is religious art. The Church of Rome was wealthy, and where there is money, poor artists will gather. There is also some very weird Renaissance art that has nothing to do with religion or the rebirth of the classical art of Ancient Rome and Ancient Greece.

The Church of Rome, the Third Reich, the Soviet Union, and the European Union: what do all these have in common? The answer is that they all used art for the purpose of their propaganda. One day the European Union will also have in common other things with the Church of Rome, the Third Reich, and the Soviet Union - the future history of the European Union is the past. It is already written - you can see it taking shape here today in 2016. They are retreating into the past. The European Union is a place of corruption, incompetence, delusions, fantasies ...

**47**

Orion has not been seen for many weeks. Soon he will be gone from the northern hemisphere sky.

Soon we may mention why we keep referring to this constellation.

In 1915, Einstein's paper on the General Theory of Relativity was published, containing his famous field equations. Then in 1917, Einstein issued a modification by adding what was called a cosmological constant. Why did you do that Albert?

"I did not like the idea implied by the equations that the universe was expanding. So I added the constant to achieve a state of dynamic equilibrium - it was neither expanding nor contracting. Yet several years later astronomical observations made by Hubble, suggested that the universe was expanding. So it seemed that I was wrong to introduce the cosmological constant. But I see today that people are also saying that the idea of the cosmological constant may not be wrong. It depends on what you believe!"

So science is indeed a mysterious process, where what seems right, then seems wrong, which then seems right once again. One can observe things and confirm that the observations can be repeated, but assembling all these observations is a matter of interpretation, which keeps changing. So while water remains as H2O, at a higher level, in terms of creating the bigger picture, truth is always subject to change and uncertainty.

This, by the way, is not a problem, for this is the nature of life. It only is a problem for those who believe in the sole truth.

At the heart of all aspects of life there is always something that lies beyond understanding - it is as though there are aspects that are simply unknowable, inexpressible, and indefinable, no matter how hard one tries to understand. Not to be able to live with this leads to people becoming consumed by dogma - it is an ailment shared by all seven Abrahamic world views.

Soon we will encounter the dogma of the Copenhagen orthodoxy.

The problem with science is that it adopts one particular theoretical stance to the exclusion of all others, and then pours all its resources into this one pot, making virtually impossible the exploration of alternatives. Alternative views then take on the status of heresies. Social sciences however are much more pluralistic, offering many different schools of thought, and the result is a much richer picture of reality than can be found in the hard sciences. Here you see an example of many truths. Wolpert's self-constructed world further unravels.

Animism is the world view that non-human entities, such as animals, plants, and inanimate objects, posses a spiritual essence.

Aristotle now speaks.

"Things possess a desire to change state - they either want to levitate upwards, or to gravitate downwards. And from this one can see that the strange motion of the pendulum is a body seeking to gravitate downwards, but which is constrained and cannot do so."

Galileo now speaks in response.

"Not so. The pendulum's motion is a dynamic process. Forces acting upon the pendulum -gravity, friction and the force that started the motion - are acting on the mass to create the swinging that you can see, which, with time, will decay to a static position, at rest, once all the energy associated with the input force has been dissipated by the gravity and friction."

There are a number of equations of motion of a pendulum which relate to the assumptions one makes about the pendulum. If the assumption is that the mass of the pendulum is located at a single point, that the mass associated with the metal bar on which the end mass is fixed is negligible, and that displacement from the vertical rest position is very small, then one can use a much simpler equation than is the case when these conditions do not apply.

## 48

Most people do not understand science, and that includes many scientists too. Why would you think that artists would be any different?

Now Socrates wishes to speak.

"Yes that is so. What I say now refers to the art of numbers and calculation, but also other forms of mathematics. The science of numbers is charming, and in many ways conduces to our desired ends, if pursued in the spirit of a philosopher, and not a shopkeeper."

Increasingly science in the modern world is pursued in the spirit of the shopkeeper, and much trouble from this comes, made worse by the lack of understanding of science among scientists and the public, which leads to conflicts, because science can only produce specific knowledge relating to that which is studied, but what is studied is not the world in total, which tends to conspire to undue that which humans, with reductive minds sought to attain. There is a better way to approach this matter, but dogmatic people in the world of science stand in the way, once again!

There are some from the world of STEM seeking to appropriate art for their own ends (the acronym often used is STEAM). They should however take note that such actions (which can include the re-branding of artists as engineers or scientists) create an artistic interest to investigate this mind set, for art is about discovery, of anything and everything, depending upon the muse. And a very fruitful area of artist endeavour is the creation of works which will lead those in STEM to encounter some uncomfortable truths about themselves. This is one of the reasons why society needs artists, not to act as appendages to STEM - engaging in communication, public understanding, visualisation, and aids to creativity. Who else will expose these unpalatable things and also discover, in doing so, very different and more appropriate realisations of STEM?

A system is a mathematical model or abstraction of a physical process. Often systems have inputs and outputs, and the model relates the outputs to the inputs. Inputs and outputs have a cause-effect relationship. In modern science, knowledge is based on developing an understanding of causes and effects. There is no place for animism.

Orion is now in the process of disappearing from the northern hemisphere sky. Soon he will be gone. He will however return later in the year, for he is eternal.

## 49

STEM needs art, but art can find a way to get-by without STEM. Art could reinvent STEM, in forms way beyond the imagination of those dwelling in the world of STEM. Way beyond the imagination too, of most of the artists caught up in the art-science and art-technology love affairs.

In a Belgium museum, the Magritte Museum in Brussels, there is a drawing on display. It is Magritte's *Treachery of Images.* It is but an image of an ordinary pipe, of the type that men smoke. Nothing special about it, nor is it known if it is a specific pipe, or just something from the artists imagination. The painting of the pipe could be something anyone with some painting skills might be able to produce. Graphical representation it is, but why is it art? The answer lies in what is written in paint, in the French language, underneath the image. What is written there is, in English: *This is not a Pipe.* Now we are back into the domain of scriptovisualism.

And the logical mind, the rational and objective mind, that which STEM people are so proud of, often say, 'well of course it's a pipe.' But Magritte was a Surrealist artist and Surrealism is the domain of the irrational. And if you were to think, to put aside your cognitive biases, you will understand that it is, as the caption tells you, not a pipe. It is the image of a pipe! The lesson is: the image is not the thing.

The image is not the thing. The word is not the thing. The musical score is not the thing. The model is not the thing. The equation is not the thing. The theory is not the thing. But you would not think so from the way people in the world of STEM often behave. As we said, most people do not understand science.

Michel Foucault has something important to say, based on long reflection on this painting and others by Magritte.

"There is an advantageous difference, not at all expressed in any dictionary, between similitude and resemblance. The latter reveals what is clearly visible; similitude reveals what recognisable objects, familiar silhouettes hide, prevent from being seen, render invisible. Resemblance makes a unique assertion, always the same. Similitude multiplies different affirmations, which dance together, tilting and tumbling over one another. *Treachery of Images* contains a drawing of a pipe that bears a strong resemblance to a real pipe; it also contains written text that bears a strong resemblance to the drawing of written text. In fact, whether conflicting, or just juxtaposed, these elements annul the intrinsic resemblance that they seem to bear within themselves, and gradually sketch an open network of similitudes. Open not to the real pipe, but to all other similar elements, including all types of real pipes, that once drawn into the network, take the place and function of the simulacrum."

The world needs art to challenge that which exists, to disturb the status-quo, to explore worlds that people might never think possible, to investigate that which lies beyond the limiting horizons of STEM, and other taken-for-granted belief systems.

People do not understand science. Some think that it is value free, detached from being human. They think that reliable scientific knowledge is morally and ethically neutral and ethics only enter when science is applied to making a product. Strange beliefs indeed! As we just said, people do not understand science, least of all scientists, especially those caught-up in its ideology and dogma - which is highly Abrahamic.

That it is Abrahamic, is of course, to be expected, for although it is universal, Western science is the product of an Abrahamic culture - Europe. A different culture would have produced a different science and technology. For example, if Chinese culture had become dominant globally, rather than European culture, then science and technology would today be different is some ways. What ways you ask? Wait and see is our response.

The 1995 Nobel Peace Prize Laureate, Sir Joseph Rotblat, a physicist and Fellow of the Royal Society, wishes to speak.

"I am strongly opposed to the idea that science is neutral and that scientists are not to be blamed for its misapplication."

Yes indeed for such beliefs in the neutrality of science are very convenient. Everything can be reduced to an *abuse model*. Therefore carry on as before. The reasonable man or woman, leaving the world as it is. No notion here of walking a different path.

**50**

We need a different approach to STEM that goes beyond its simplistic beliefs and values. What chance that this approach will be developed by those caught-up in the world of STEM with all its ideologies, dogma, and strange beliefs and superstitions?

Infinitesimal!

The transformation will come from the periphery, from those not accepted as part of the established order. It will come from a different culture. The keywords are China, India, and art - back to scriptovisualism once again.

"Men's courses will foreshadow certain ends, to which, if persevered in they must lead, but if the courses be departed from the ends will change"

That was Charles Dickens speaking. Wise words indeed! Is the message clear yet?

Here is a poem. We wrote it for the fairy lady with the beautiful soul to help her understand:

*At the Parting of the Ways*

By dusty road, that oft I walked, do not stand,
Arching bridge, now crossed, no longer supports.
I am gone away - there is no way back for me.
On golden sands, gazing out to sea, do not stand,
Sailing ship, on far distant shore, now lies beached.
I am gone away - there is no way back for me.
On lonely hill, thinking of me, do not stand,
No roads, paths, or byways, lead me to thee.
I am gone away - there is no way back for me.
Remember me; forget you, I shall not,
For by your side, I will always be,
As together, to slippery edge of doom, we walk.
I am gone away - fair lady with beautiful soul,
Farewell to thee, our ways now do part.

Here is a question: why is the above a poem?

Anti-art: anti-art is a term used to describe art that challenges the existing accepted definitions of art. The term anti-art is generally agreed to have been coined by Marcel Duchamp around 1913 when he made his first ready-mades, which are still regarded in some quarters as anti-art.

**51**

There are many different types of systems - classifications that we will speak to you about as our strange piece of literary art unfolds.

First then, is the matter of *causal* or *non-anticipatory* systems. As the second name suggests, these are systems whose outputs are based on past inputs, up to and including the current time. In other words, these are systems that do not respond until an input is applied. This formulation suggests that all real physical systems are *causal* or *non-anticipatory*. Just to illustrate how strange the world of STEM is, we shall now tell you that there are *non-causal* or *anticipatory* systems. Yes, that is right, systems that anticipate an input, or seem to!

Teleological – the explanation of phenomena by the purpose they serve rather than by postulated causes. Modern science abhors the teleological. Cause and effect reign supreme. Yes you are right for we are building towards something – a heresy! It is that word cybernetics again. Your brain will bleed with pain!

The British biochemist and sinologist, Sir Joseph Needham, Fellow of the Royal Society now speaks:

"Western science is about mathematisation and rigorous experimentation."

The comment about mathematisation needs a qualification to make it less absolute, but otherwise, of course you are right. Remember this fact dear reader.

Joseph will be back with us later. He is another person with a mind rooted in two disciplines; one from the world of STEM and one from the world of the humanities.

STEM is a collectively constructed image shared by those who practice in the world that is known as STEM. It is a simulacrum. What other images of STEM could therefore be constructed?

Axiom: a statement or proposition which is regarded as being established, accepted, or self-evidently true.

It is difficult, probably impossible, for humans to fully perceive that which goes by the name – reality. This is the nature of being human. The implications: one cannot rely on one way of knowing the world – in particular science – to take any sort of major decision. Engineers have always known this, although it seems that they too are forgetting, and are drawn into the delusion that reality, in its totality, can be known.

A woman comes into a room full of people from many different disciplines, and asks all present one simple question. She wants to know, what 2 + 2 equals. The answers from each discipline present provide revealing insights into the differences among STEM people and with other professions as well!

The mathematician was first to respond, and gave the answer as 4. The statistician however had a slightly different view, for he said that the answer was 4, with a 99.99% probability. Then came the physicist who said that theory predicted an answer of 4, but he would have to design an experiment to prove this, and then, only after performing the experiment many times would he be able to provide confirmation, or otherwise, subject of course to an estimate of errors, which would produce an answer of the type: 4 plus or minus 0.0001 to allow for observational errors. The engineer said that he was sure that the answer was 4, but he would go for 5 just to be on the safe-side. The technologist was unwilling to reply, saying that he first needed to construct a machine to work it out. The biologist then joined in, saying that all the answers given so far were a product of genes and therefore the result of evolution, but that there must, sadly, also be a purpose to this. The accountant then contributed by saying that the answer depended upon the assumptions made, which would be fine so long as people where aware of these assumptions. At this point a European Commissioner, a politician, exerted his right to speak, saying that, as he was responsible for funding STEM, the answer to the question was clearly, in the context of the European Union and the European Commission, 'whatever you want it to be, for we are in the business of pursuing fantasies.'

In the room there was also an artist, who thought that he should also contribute, but barely had he begun to speak when he was told by all the STEM people, and the politician, to be quite, for his purpose as an artist was to create a visualisation of the scene. That just left two, who had not spoken, and being more knowledgeable about human behaviour and having many bad experiences of working with those from STEM, one of them just turned to the artist and said, 'they are just responding according to their cognitive biases while engaging in the delusion that they are thinking about what the answer should be.' She was of course a cognitive psychologist. As for the final person, this was a sociologist, who watching all this, said that you cannot answer such a question using a reductive approach, and all must be taken into account, otherwise the result is meaningless.

Some scientists are of the opinion that the philosophy of science is just a lot of rubbish that scientists, on the whole, do not take seriously. So, all those great thinkers such as Polayni, Kuhn, Feyerabend, and others, are shot down in flames with a dismissive one-liner.

Now why would scientists say such a thing? Could it be that they do not have the intellectual capabilities and arguments to respond? Or, are they so caught-up in ideology and dogma that, this is just their Abrahamic response to others who have found a different truth?

We see that Socrates wants to contribute.

"Yes indeed I do. On the matter of opinions and knowledge, these are two different things. Opinion is to have opinions. Knowledge is to know the nature of being."

And do you think that scientist holding such low opinions of philosophy, and being unable to engage with them on equal terms, that such people should be allowed into government? Oh the delights of the fragmented and reductive mind!

*What hell on earth would they conjure to appear,*
*those caught in dogma's hypnotising stare,*
*while its pleasingly deluding gaze, making secure,*
*a world overwhelming and too complex to bear.*

We are being poetic of course.

Simulacrum: an image or representation of someone or something.

## 52

There is no one more dangerous than he or she who has discovered the truth. All seven Abrahamic world views provide examples that well illustrate this axiom.

C.P. Snow has yet to put-in an appearance.

Orion of course does not exist. It is a human invention. The stars visible in the constellation just appear to us to form the shape of a hunter. But we know that they are located where they are by chance, and that, a long time ago, people just made something of this. Yet these names, these man-made things, they still persist in this modern age of astronomy. Now why is that?

The above is interesting for it has relevance to the understanding of science and what we do, as human beings, when we apply science to ... But of course this is just a lot of rubbish, being, as it is, linked to the philosophy of science. Well that is what some scientists believe. Yes, note the word, believe! Believe: accept that something is true, especially without proof. They are being subjective yet again. What ever happened to reason?

Lenin was passing by and we stopped him to ask a question.

So Lenin, why was it that when the Bolsheviks came to power in 1917 and you set about building a socialist system, you did not explore different methods of work organisation in factories? We ask, because STEM people in the West believe in the one best way, and would quote this as an example that proves the point.

"Comrades I know what you are referring to. We did indeed adopt American capitalist methods such as Fordism and Taylorism, but that was because of me. There was a discussion taking place in the early days about not using such production systems, and developing a socialist method of production. I, however, put an end to this debate and, in autocratic style, decreed that we would adopt the American capitalist methods as a means of catching up with America. I realise now that this was a great mistake. But it was not last one I was to make."

Now we shall list some of the strange beliefs and superstitions of STEM people. In no particular order they are: positivism; technological determinism; technocracy; deficit thinking; scientism; social Darwinism; neo-Darwinism; secularism; one best way; machine-centric understandings of humans; human-machine/human-computer comparability; techno-centricity; euro-centricity; sole truth; neutralism; best evidence; optimal solutions; ...

Quite a lot of baggage these people have to carry around with them! And now the Chinese and others who are not part of the European world – those who are not what we call the *Vainglorious Enlightened Ones* – know the strategic weaknesses of the Western world, and can use these weaknesses to unleash the forces of creative destruction against the West.

The *Vainglorious Enlightened Ones* will not know what is happening to them and will retreat even further into their strange beliefs, doing harder that which they did in the past, that which once worked. Hello there Prometheus! Still chained to the rock of the past we see. Never mind, Hercules may come along one day.

## 53

There is no one more dangerous than he or she who has discovered the truth. All seven Abrahamic world views provide examples that well illustrate this axiom.

C.P. Snow is here. Perhaps we should call you Lord Snow?

"Yes I suppose you should. But what is important is not the title, but that which I came to speak about – many things in fact.

"First I would like to say that my 1959 Rede lecture was called *The Two Cultures and the Scientific Revolution*. You might say that it was a follow-up to my 1956 New Statesman article which was just called *The Two Cultures. Two Cultures: A Second Look* was an update which appeared in print in 1964.

"Second, my comments were observations and opinions based largely upon my personal experiences. They were not the result of scientific research.

"Third, *two cultures* is not a theory, as some have claimed.

"Fourth, I never proposed that we needed a third culture in the form of some art-science collaboration. My comments about a third culture in *A*

*Second Look,* were in fact an acknowledgement that, as a person once based in the hard sciences, I had overlooked - ignored - a whole area of learning known as the social sciences and also the humanities. This I failed to recognise because I am the prisoner of my English upbringing, conditioned to be suspicious of any but the established intellectual disciplines, unreservedly at home only with the *hard subjects.* And for this major omission I am sorry.

"And last of all, the most important point, which is that my thoughts were about tackling poverty in the (third) world through economic development driven by industry and applied science. I attacked the attitude that existed in some literary circles within which I moved, where snobbishness and elitism was evident, especially towards science and industry. In these literary circles some people still felt that what we called a classical education was to be preferred over a scientific or technological one, which I believed was more relevant to the modern world.

"If you are English and familiar with the English social scene in the nineteenth and first half of the twentieth centuries, you will know what I am talking about. Snobbishness towards anything that was new seemed to be a way of life. You have heard no doubt the term *new money,* which was a derogatory term applied to people who had become rich within their own generation, often as a result industry, trade and commerce, rather than by familial inheritance, often linked to ownership of land!"

Thank you for that clarification. In reality, there are many different cultures across professions and occupations, as well as cultures associated with class differentiation or status linked to wealth.

Now someone from the humanities wishes to speak.

"Yes, and what is clear from a critique of the *two cultures* texts, is that C.P. Snow was a technocrat with an over simplified view of the complexities of economic development in what Snow would have called the *Third World.*"

That was Stefan Collini.

So once again we are back to the joys of the fragmented and reductive mind, where educated and intelligent people arrive at simple solutions to complex problems; problems that are not understood simply because, these complex problems have been simplified, which is the scientific way. Best to stay in the laboratory where your skills are best employed.

## 54

Analemma: a graph or plot that shows the position of the sun in the sky at a single location and at the same time of the day throughout a year. It is a different way of looking at the changing position of the sun as a result of the earth's orbit around the sun.

You can look at something in a many different ways, which may seem very dissimilar, but you are still seeing the same thing.

Howard Rosenbrock, a Fellow of the Royal Society, an engineer and applied mathematician, is now here and has several things to contribute.

"Yes indeed Paul. I am glad that you have persevered and are still working to change science and technology."

The need is greater than ever Howard, for madness has descended upon humanity, and few are now able to comprehend that this is the case, so caught-up are people of the Western world in the techno-scientific ideology.

"I am not surprised, but people should stop and reflect for a moment, for there is another path that they could walk. And this reflection should start with a basic understanding that the universe does what it does and would do so if there were no such thing as humanity. When we choose to observe and gain understanding of the universe, we do so by projecting on to it, ideas, concepts and theories that are of human invention. There is the universe and there is that we invent as a representation of the universe. And in the latter we have a choice, and these choices can have different and quite profound implications for humanity and its future."

Howard has much more to say about this. For the moment we will move on to consider other matters. All we will add here is that we are now into the area that this book may well be addressing, which is the creation of a different science and technology to that which exists today. Yes indeed, we are talking heresies.

Paradigm: a pattern or a model; in science, a world view underlying theories and methods.

A mechanistic clockwork universe, or a probabilistic universe? It makes no difference which, for in the end, it is still about cause and effect - causal relationships. Certainly it becomes more complex once we enter the world of quantum mechanics. Yet still causality is at the heart of modern science, for the alternative is purpose! Or so it seems!

## 55

Today the subconscious mind is quiet. We have nothing to write, which is not true as you are reading this so we did have something to write: We have nothing to write, yet we wrote. We suppose this is the subconscious mind being irrational.

Today we will therefore turn our attention to the painting that we mentioned earlier. We started it by making some outlines of some interesting shapes, reflecting what was on our minds at the time. We suppose it will be a surreal, but who can tell, for it is a composition unknown.

We rest our pen. Actually it is a keyboard - a metaphorical pen. Metaphor: a figure of speech in which a word or phrase is applied to an object or action to which it is not literally applicable; a thing regarded as representative or symbolic of something else.

Art can be metaphorical: a painting; a poem.

Simile: a figure of speech involving the comparison of one thing with another thing of a different kind, used to make a description more emphatic or vivid. Not the same as metaphor.

Art can be a simile: a painting; a poem.

Art as metaphor, art as simile, art as a way of being in the world, art as a way of knowing, art as a means of discovery: communication. Art for visualisation, for creativity, are here forth banished and seen for what they are - STEM determined agendas. But not entirely banished for they can still have a place, but they are no more the centre of gravity.

Artist-led agendas are different from STEM-led agendas. Art can encompass STEM. STEM-led agendas just consume art and turn it into another utility.

So in the end we did write something.

## 56

Cross-disciplinary: one discipline explaining another discipline. Philosophy of science tries to explain science, as does some work in the social sciences. And the result is another *Tower of Babel,* with some scientists retorting that these efforts explain nothing about science. It seems that only science is allowed to explain science! Self constructed realities! Self constructed delusions! Collective delusions! This book is, partly, cross-disciplinary.

Work on the painting has resumed, and the need to solve a challenging problem - how to divide a circle into 366 segments. We decided to use technology to solve this problem.

Technology is certainly a powerful aid at times, but the *Vainglorious Enlightened Ones* think far too much of their technology, and deeply embedded within these people, are beliefs about technology. One of these beliefs is that humans and machines, in which we can also include computers, are comparable. This in turn reveals another strange belief, which is that we humans, are also machines. And to think in such a way, leads to the history of the modern world, where humans are continually replaced by machines, and rendered into the mode of doing the work that is left over - that which could not be transferred to the machine. It also leads to the cyborg mentality, in which people, usually with much celebratory glee, add bits of technology to themselves, and look forward to the day when we are transformed into super-humans (übermensch) by the use of technology.

You will have encountered the notion of the super-human before - we are now into the realms of Nazi ideology, and the notion of the master race, that is to say, the *Vainglorious Enlightened Ones*. If you are not cyborg you are *untermensch*. This is your future. Enjoy the new facilities, which of course will not exist. Denial is the final stage of genocide.

Transhuman - this is another word for übermensch. Nazi ideology is still with us. It afflicts STEM people and artists alike!

And where did this notion of superiority come from? Look no further than Ancient Greece.

Modern Europe, and its off-spring in its former colonies are both a product of the past - Ancient Greece, Ancient Rome, Judaism, Christianity, and Protestantism. Europe is the past. So we return once more to the Prometheus syndrome which enables the understanding of the strategic weaknesses of Europeans. These weaknesses can be used against them by other nations who are not Europeanised and do not want to be Europeanised. It also points towards predicting their future behaviour. And we all know what that will be!

Let us here now make a distinction, that there are machines and there are tools. The *Vainglorious Enlightened Ones* want to build machines not tools and, conceiving of themselves as being machines, they seek to enhance themselves as machines, by adding to themselves, technologies that will make them more machine-like. In the end, if we do not stop them, they will turn the whole of humanity into machines, into cyborgs.

Tools are inferior things to machines, so they believe. If you do not understand the difference between a machine and a tool you will not understand what we are saying. So you will have to keep reading.

**57**

In Ancient Greece many different philosophical schools existed, with different understandings of science, and even different models of the universe - the geocentric models coexisted with the helio-centric one. Sole truth had yet to infect the mind.

As regard to all those comments people make, waxing lyrical about Leonardo's note and sketch books, these well demonstrate the place from where they are being viewed. Some (not all) artists though would look at these and perhaps see something else. Many artists have such books, and some of them contain some very weird and unconventional things, but these books are not intended for public viewing as they are private spaces for thinking, and an important part of the mysterious artistic process, which artists are reluctant to speak about. And looking at Leonardo's papers, an artist would not necessarily see invention, design or engineering, but ideas for a different kind of art that did not appear until the twentieth century - constructed sculptures and installation art. We say some artists - obviously not those who subscribe to the Leonardo myth.

Here is another view, an alternative one: Leonardo, perhaps frustrated by the limited conceptions of his age, may have been working in private to reinvent art, which is exactly what today's artists are doing continually. Many people today, on encountering such art, often remark that 'this is not art.' One can be fairly certain that this same attitude would also have been strongly entrenched in Leonardo's Italy. We are of course being subjective and highly speculative - welcome to the world of art. A tongue-in-cheek remark!

It seems that the supposed *two cultures* do at least have one thing in common!

Or perhaps not - who can tell? This is the nature of knowledge.

Antithetical: directly opposed or contrasted; mutually incompatible.

**58**

Orion is still dancing his way across our wintry night sky; but not for much longer.

It turned out that the technology was unable to provide the solution to the problem of how to divide a circle into 366 segments. More about this later, for it is without doubt an important matter.

Karl Jaspers, the German philosopher is here and wants to say something about the Ancient Greeks.

"Yes I do, for the Ancient Greeks despised all physical labour as appropriate only to a low mentality. The complete man, they thought, is an aristocrat, does not work, is possessed of leisure, practices politics, lives in competition, goes to war, gives birth to works of the spirit."

Thank you. Karl will have more to say later we suppose.

So the Ancient Greeks were an elitist lot. You find this fact reflected in their notions of democracy too – it was only for the select few: free people think, slaves work.

The world of STEM is also very elitist – they get this from the Ancient Greeks. The European Commission is very elitist – they get this from the Ancient Greeks.

Europeans are just Ancient Greeks and Ancient Romans wearing modern clothes. They seek to solve twenty-first century problems with minds rooted in the past. Having minds that belong in ages that have passed into history is also one of the reasons why new problems are created. It is time for a change, while there is still time.

The Ancient Greeks also practiced infanticide. Modern Europeans do this as well, only they call it something different. Ancient Greek men were obsessed with their bodies and had well defined notions about what constituted normality and perfection. These notions of normality and perfection can still be found in modern Europeans, only they call it something else, and they now have more advanced techniques, arising from the world of STEM, to ensure conformance with these notions of normality and perfection. Ancient Greek men also believed that women were inferior to men, and that women only existed either for breeding or for men's pleasure. Now you know why there is so much sex discrimination in modern Europe, and why women, to succeed, often have to behave like men.

So, Europeans are indeed Prometheus. Where is your Hercules to set you free?

## 59

Transdisciplinary: to transcend disciplinary boundaries, to reunite knowledge in a whole, to see the world as it is, rather than as fragments. Who among you can claim to be doing such a thing?

Of words and their meaning: multidisciplinary; interdisciplinary; cross-disciplinary, transdisciplinary. These are often just buzzwords. You can often tell a person's degree of understanding of the meaning of these distinct approaches by the way they use these words. Often the meaning is not defined, and often they are used as though they are interchangeable.

What you are reading now is the STEM part of us speaking, which requires precision of meaning. But the artistic part is willing to embrace ambiguity, to not be so specific.

This may be important to understanding scriptovisualism - the spaces between spaces. This is not a place of the binary, of black and white. In the precision there is ambiguity, and within the ambiguity there is precision.

So one can construct a reality of sharp divisions, where boundaries exist, or one can construct a different reality, where there are no sharp dividing lines.

All conventions are boundaries waiting to be transcended. One can transcend a boundary only if one can first conceive of doing so. Separation is an illusion.

## 60

Another milestone: as we write our sixtieth chapter, we have fifty-nine complete chapters in our 366 chapter odyssey. Only 306 more chapters to go, of which this is the sixtieth, as we just informed you. And today is a very special day - February 29, 2016.

This year, 2016, is a leap year, in which we have 366 days. Now you see why part of the title of this work is 366. And composition unknown is an artistic way of working - not knowing, and not trying to hide that fact. We are not hiding that we do not know, as we write, what this book is about.

This is quite different from the way STEM people like to present themselves - all rational and objective. Artists can be like that too, but art also encompasses the subjective and the irrational. In other words a totality, so we are back once again to the matter of unity, wholeness. Now we are also entering the world of cybernetics.

So recall now, back in chapter 34, what Ken Knowlton said!

Embedding differences is the term that now wanders into our mind, bubbling up from the subconscious. This means a different process.

What we have found is that in the world of STEM, people are enthusiastic about new scientific findings, new technologies, new machines, and other such similar things. However, whatever you do, never, never ask a STEM person to work in a different way than that which they acquired when they were first initiated into STEM culture and practices.

"So, as this proposal responds to a Call for user-driven innovation, we should involve (what STEM people call) users and I suggest therefore that we adopt a user-centric approach."

"This sounds excellent," responded the engineer, "so long as we do not have to do anything that is different to the way we do things now."

Hello Prometheus.

Today is the end of metrological winter. We are still in astronomical winter.

"The idea of Lyrical Conceptualism is based on the wholeness of the psychological coordinate. It derives from the id, ego and superego; an art in which the primarily twofold character of the artist's view evolves into a lyrical, intuitive and conceptual triad."

That was Paul Hartel.

**61**

Today is the start of metrological spring, but it is still winter. The weather is very cold, as if to prove this point. Now is the winter of our discontent, for December was made glorious summer by this thing we call global warming. We are being poetic again. It was more like autumn and spring rolled-up into one.

The technology was unable to do what we wanted it to do. The task of dividing a circle into 366 equal sized segments is a challenging one. The task requires great precision. Small errors soon build-up, and then one finds that the available space is soon used up, but the number of segments is fewer than the 366 required. The hope was that the software would allow this precision, would allow the circle to be divided into 366 segments. To achieve this goal, we used the pie charting tool provided in the software, and entered the same data, 366 times and then discovered that the maximum number of data that could be entered was 360. This is a design assumption that is not visible to the user. It only becomes visible when someone tries to enter more than 360 items of data, and then one ends up with a circle divided into 360 equal size segments regardless of the number of data points beyond 360.

This one can say is a silent narrative. It is also a cultural one too. In the West, long ago, in Ancient Greece, man decided that the circle should be divided into 360 equal segments, each of angle one degree. In another time and place, in another culture, in Ancient China, man decided that the circle would be dived into 365.25 (366) segments.

And here in the twenty-first century, thousands of years later, what the Ancient Greeks decided is constraining technology and hence constraining our actions. This is a feature of technology, but it does not have to be so.

This example shows how these tacit, taken for granted assumptions and beliefs, become the foundations upon which technology is built. What goes for technology also goes for science too. There is nothing more Abrahamic than Big Bang theory. But in Ancient China, there was no God, no creator. The universe is the manifestation of a struggle between opposing forces - Yin and Yang. A different view, a very organic view; it has the potential to, perhaps, yield a different theory. But we shall never know, because the world of science is dominated by the Abrahamic West, and there is no room for heretical views, for an alternative theory. All resources are poured into the pursuit of what is believed to be the sole truth.

"The example of using a piece of software to do something artistic, is also one of an artist as an extreme user."

That was Samuella Carlson.

The problem though is this: by the time the extreme user has discovered the built-in assumptions, it is often too late to change the technology. The same problem arises when social scientists enter the process. The tradition is to bring them in later, when important decisions have been taken. This is a reflection of the linear step-wise thinking found in STEM. It is also an indicator of STEM people's desires to impose order and discipline and remove ambiguity - it goes by the name of the specification.

Thomas Kuhn will now speak.

"In my book, *The Structure of Scientific Revolutions,* I stated that there is a fundamental difference between the physical sciences and the social sciences. The physical sciences adopt a single world view, which becomes the dominant one, while in the social sciences many different schools of thought coexist with one another. There is here, no sole truth, no single dominant theory."

A feature of the Abrahamic world is a belief in the sole truth. In the Taoist world however, there is no sole truth. The Tao: the way of heaven - unknowable, inexpressible, indefinable. At the heart of existence there is that which cannot be known, yet each may come to know on their own.

Hence we have the notion of Taoist science and Taoist technology. What they are though, will need to be explored later on. For now though we see that, science and technology are culturally determined. They constrain how we see the world and the type of technology we create.

Now we see a little more what scriptovisualism might be about: re-conceptualising our world, because we are no longer constrained, being as we are, in a space between spaces, and able, as a result, to see the world in a completely different way.

The work on the painting will continue. We will find a way of precisely dividing the circle into 366 equal segments. We are engaging in a dual creative process, and what you have just read is a result of this, for we would not have written this chapter had it not been for the fact we are also creating a painting which too, is a composition unknown.

**62**

Systems can be linear or non-linear. What is the difference?

A linear system is one that obeys the principle of superposition. What this means is that one can measure the output from a system for a given input, and then do the same again but for a different input, and then one can determine the output for both of these inputs when they are combined and applied at the same time, by simply adding together the outputs obtained from the inputs when applied separately. Systems that do not obey the principle of superposition are non-linear.

Linear systems are much easier to deal with and there are well developed mathematical techniques for analysing such systems. Mathematically, non-linear systems are much harder to handle, and for this reason engineers often make approximations and assume linearity. But, because it is engineering, one then has to deal with the consequences of the non-linearity. There are many types of non-linearity - one example is a discontinuity. Engineers have ways of dealing with the non-linearities that were initially ignored for ease of analysis, for example by conducting simulations or running experiments. This is one way that engineering differs from science, and why engineers are not really using reduction in the same way that scientists understand the term. This is a point to note.

Socrates wants to say some more words:

"Consider the nature of qualification. Just think what would happen if pilots were to be chosen according to their property, and a poor man were refused permission to steer, even though he were a better pilot?"

Words with implications! For some the words add support to their belief in that old maxim, *free people think, slaves work.* It is the justification for technocracy. But do you want people with fragmented and reductive minds deciding your future? Clearly only someone trained to fly an aeroplane should be allowed to be a pilot. Yet people who have a limited range of knowledge, with minds unsuited to seeing the bigger picture, and who are soaked in the *esprit de corps* of that institution known as elitism, believe that they should be allowed to be the pilots of society. There is no need to expand upon this, for we know where such folly leads.

Gerry Johnson, emeritus professor at the University of Lancaster Management School now wishes to speak.

"All those involved in a given field of activity, a profession, an occupation, etc. have a set of core beliefs and assumptions specific and relevant to the given area, which are shared in common with others. This is the essence of a paradigm: it is taken for granted and not seen as problematic by those who adhere to it. The paradigm defines how things are done and what is acceptable."

Thomas Kuhn now adds something to this.

"In my book, *The Structure of Scientific Revolutions,* I too used the word paradigm. I made two distinctions though, regarding the use of this word."

Probably we will come back to Thomas later.

Karl Jaspers now makes a further contribution.

"There is this creed called progress, which foresees nothing but good fortune from the new knowledge of nature and from technology. This creed is blind. For it sees only misuse within technology, which seems to be penetrable and capable of correction. It does not see the profound perils hidden in the technology. The creed of progress fails to recognise that progress that is limited to knowledge and technology cannot possibly lead to progress in the sum total of humanity."

Taboo: a practice that is prohibited or restricted by social or religious custom. You may have noticed that we are breaking many taboos, especially those concerned with the world of STEM, but also the world of art-science lovers. We are also breaking taboos relating to the European Commission such as not speaking about their practices, because they do not want ordinary people to see the sort of monster that it has become. STARTS reveals the story.

## 63

Meteorological winter is a three month period that runs from December 1st to the end of February. It is the coldest three month period of the year in the northern hemisphere. Astronomical winter is what people refer to when they talk about the winter season.

Meteorological winter and astronomical winter are two ways that people see the world. They have different uses, and it cannot be said that one is better than the other. Of the two, neither represents a sole truth. They co-exist and are complementary. There is no need to decide between the two.

Some people from the world of STEM believe that Leonardo da Vinci was not only an artist, but also a scientist, an engineer, and a designer. It should however be self-evident that, what they are saying about Leonardo is subjective and probably also, to some degree, speculative. It could also be a sign of their ignorance. It could also be said that it is a sign of their desire to create a myth. What might not be so clear is that, this *Leonardo as engineer* view might not be one an artist, understanding art and artists, would necessarily fully share. But on the other hand they might. There is more here than might initially be apparent!

So who is right? We may come back to the matter. This is the nature of knowledge.

A paradigm is a double edged sword, because it is both helpful and unhelpful, depending on the circumstances. Associated with a paradigm is a mind-set, a way of thinking which means that people become over-sensitised to some particular aspect of their domain or available information, at the expense of other parts. This over-sensitising is useful, because it helps people to become sensitised to important things, and to patterns that remind people of problems successfully solved, and this often serves very well.

Agriculture is a case in point. In the case of industrial-scale intensive agriculture, its proponents are, in general, over sensitised to the use of technology, economies-of-scale effects, and yields (or productivity). Proponents of organic agriculture however are in general, over sensitised to the environmental impacts of farming, to the longer term consequences of modern agricultural technologies, and to producing produce of higher quality, where quality can be defined according to both extrinsic and intrinsic criteria.

So who is right? We may come back to the matter.

"Technology requires direction. We should not place our faith in some unconscious act. Man must find his way back to the guiding reigns. He must achieve clarity concerning his needs; he must put them to the test and determine their hierarchy."

That was Karl Jaspers again.

On the one hand there is disciplinary knowledge on the other hand there is topic knowledge. Do not confuse the two – people do!

**64**

Thinking about scriptovisualism, it comes to mind that this must look beyond the word, beyond the painting, both of which are commonly materialised in the form of objects. Objects themselves remain, but words projected on to objects other than paper or canvass, in some way, so that the object itself speaks a silent narrative, that then makes a whole, could be an important realisation. Then there is the matter of performance, not in the classical sense of a scripted rendition of a story, or a reading, but something more, something multi-sensory.

The painting that is also a part of *366* has not yet moved on. Our attention was diverted into exploring an aspect of applied social sciences that has relevance. Everything seems to be relevant. Why? The answer is clear – because everything is connected! What is the nature of these connections?

The need for interdisciplinary design is often not recognised. While there appears to be much recognition that organisation and people issues need to be considered in the design and implementation of new technologies, there is a tendency to adopt a very narrow view with regard to how these issues should be addressed. This view is often restricted to methods for overcoming resistance to change. We tend to assume that this resistance derives from peoples' reluctance to accept the technology, rather than from a mismatch between the organisation, people and technology elements of our systems. In other words, people are suffering from a deficit which needs to be overcome.

Are you sure you mean interdisciplinary design? Or do you mean multidisciplinary design? What is the difference?

Ann Majchrzak, professor of Data Sciences and Operations at the University of Southern California has something to say about the notion of interdisciplinary design in which the organisation, people and technology aspects of a system are designed simultaneously.

"In a study that I undertook, in 40 companies in the United States where these ideas were raised, many people seemed genuinely sceptical that they would ever make different engineering design decisions based on information about a mismatch among the elements of technology, organisation and people."

Christopher Jones speaks, one more time.

"I would like to expand upon the point that I made earlier, about the invisible but troublesome barriers between professions and between disciplines not being resolvable by methodology alone. Methodology can provide a framework and procedures, but the main requirement is that the persons concerned with inter-professional collaboration become so well acquainted with one another's criteria for judgement that the mutual misunderstandings of mono-professional specialists are replaced by overlapping views."

Now you have some inkling into why disciplinary boundaries remain in place and the best one can hope for is the *Tower of Babel* situation of multidiscplinarity. What hope is there for going beyond this?

**65**

The scale of the problem that STEM has become, now invites not propaganda and image making, but radical and painful change, which begins with the understanding that STEM people are living in a self-constructed bubble, largely isolated from society. Their behaviour, attitudes and hubris are the part of the problem. They are destroying humanity because of the contempt that they display for the opinions of those that they so often proclaim to be acting in the service of. Young people increasingly choose not to pursue careers in STEM, not because of some supposed deficit in their thinking, but because they know that they will end-up working with space age technologies and Victorian-age minds. Those over 30 will probably not agree, some of those younger than this, and not yet fully indoctrinated, will have some sympathy with what we have just said. We believe!

Another reason why disciplinary boundaries remain in place is because academics are not inclined to take risks with their careers and will pursue research agendas where there is a clear line of funding and an established route to publication in quality publications.

**66**

It is impossible to fully know, with certainty, how Leonardo, or others from the Renaissance period - which spans several centuries - thought or viewed the world. One can seek glimpses by bringing together various disciplines from the humanities and the social sciences, but there would never be full certainty. Do we know for example if Leonardo read the rediscovered works of Aristotle and Plato? If he did, what affect did this have? Let us now, in our imagination, transport ourselves to his studio, where we find that he is reading Plato's *The Republic*. He has just encountered the point in the text where Socrates is saying 'one man cannot practice many arts with success.' Does Leonardo, now write in the column – 'I don't agree'? We might like to think that he did, but without documentary evidence, we will never know.

Did we mention the above already?

The problem with paradigms is that they blind people to discontinuities that render past approaches and solutions inappropriate. Discontinuities are defined as non-linearities that either render aspects of prevailing practices inappropriate, or which provide new opportunities, or open up entirely new ways of working. The failure to recognise discontinuities often leads to the adoption of incorrect change strategies. The assumption is that whatever needs to change is just a matter of evolution in approaches. The evolutionary view is based on the belief that there is nothing fundamentally wrong with the way things are being done, and that all that is required are some slight adjustments to compensate for the changed circumstances.

Inside the Leonardo exhibition – *Leonardo: Mechanics of Genius* – what did we find? Special words of course, which cast doubt on claims that Leonardo was a scientist, engineering, designer, or inventor, at least as we understand such terms.

"Leonardo da Vinci is an almost mythical figure, hence there are misunderstandings concerning his drawings of machines. Not all are inventions, and Leonardo was not alone in designing such machines. Numerous engineers were doing the same during the Renaissance period.

"Anyone ready to reconstitute or reinterpret the different projects evoked in the seven thousand sheets that Leonardo filled with sketches and notes, should refrain from establishing comparisons between these projects and their creations.

"Leonardo's drawings were rarely intended to be given concrete form. Many studies have no practical purpose, being fragments of existing machines, proposals for improvements or notes and drawings of nature observations. Measurements or details are missing or else the sketch only covers part of the project."

Understanding Leonardo is not straightforward. But if you have a fragmented and reductive mind, you might not so easily see this.

Liberal Arts: The liberal arts are those subjects or skills that in classical antiquity were considered essential for a free person to know to take an active part in civic life, included participating in public debate.

Here is a spokesperson for the Oxford English Dictionary to tell us of the origin of the term liberal arts:

"Liberal, as distinct from servile or mechanical (that is to say, involving manual labour) and originally referring to arts and sciences considered *worthy of a free man*; later the word related to general intellectual development rather than vocational training."

Leonardo would not, in his time, have been seen as one engaging in the liberal arts, but one who practiced the mechanical arts. Not a person then connected with the sciences as they were understood in that era, but with artisan activities! You doubt this? Then you do not know your social history, your sociology; nor have you read what Leonardo wrote about this matter. Perhaps you prefer the myth?

Today, in our age, topics which would fall in the category of liberal arts have changed: languages; literature; linguistics; mathematics; natural science (for example biology, chemistry, astronomy, earth sciences); philosophy; religious studies; arts (for example fine arts, music, performing arts); social sciences (for example economics, sociology, history). And many more! Many of these liberal arts topics are the basis of the school curriculum in most Western nations. In other words, liberal arts are the identifying features of that which we call an educated person. So, in other words, the vast majority of the people in the Western world, and also now, in those parts of the planet that Westerners call the developing world - a structural change.

Marshall McLuhan now speaks:

"The personal and social consequences of any medium - that is any extension of ourselves - result from the new scale of affairs that is introduced into our affairs by each extension of ourselves, or by any new technology."

Some people believe that when new circumstances are encountered, only slight adjustments are needed to a paradigm to bring the way things are done into line with new requirements. This sort of thinking has a technical name - this view corresponds to model known as morphostatic change. This simply means that the type of change people face is incremental. The established order is maintained by treating disturbances as external noise requiring minor adjustments. Note the cybernetic tone - it is highly relevant! But you may have to wait to find out more - so read on.

Agriculture provides once again a pertinent example. Incrementalism leads to efforts to improve the sustainability of industrial-scale intensive agriculture within the frame of reference of maintaining the fundamentals of the existing system. Likewise for the proponents of organic farming; they too recognise the need to improve their paradigm, for example to increase yields. Yet both paradigms face challenges that in terms of scale and variety, point to more than just morphostatic change.

**68**

Consilience: agreement between the approaches to a topic of different academic subjects, especially science and the humanities. This at least is what the Oxford English dictionary will tell you.

Here is a spokesperson from the Oxford English Dictionary to tell us about the origin of the word:

"The origin is mid nineteenth century: from con- plus the Latin -silient-, -siliens jumping (as in resilient- resilient), after concurrent."

How would one use the word? The spokesperson from the Oxford English Dictionary will answer:

"One could say, for example, that if independent lines of evidence are consistent with each other, then there is consilience."

To understand the power of consilience is to understand the power of corroborating evidence and knowledge, but also to discover that sometimes that corroboration may not exist. The true test of commitment to transcending boundaries, to transdisciplinarity, is to act differently, and not just to assume that one view is right and the other wrong, and then to proceed as one would have done before, ignoring the differences.

Edward O. Wilson now speaks:

"Consilience is the key to the unification of knowledge."

Wilson specifically chose to use the word consilience, rather than coherence, because the latter is subject to several different interpretations, but the meaning of former is not. Yet did he really understand consilience?

What chance do you think the Western world – that place of the fragmented and reductive mind – has of achieving consilience?

Scriptovisualism – a means of exploring consilience?

## 69

To deal now with the matter of the European Commission and the Directorate General known as DG CONNECT, with their *free people think, slaves work* mentality, and its meddling in the arts, and the art-science, art-technology muddle that they christened STARTS - Science, Technology and the Arts.

What STARTS, ENDS. The road to serfdom is laid, but technocracy ultimately fails. What comes next depends on the power that technocrats hold. In the case of DG CONNECT that was not very much. Lost in the past, unable to find the present, no chance of being the future, DG CONNECT symbolises the deep rooted problem known as the European Commission - Prometheus.

Artists working in research, development, innovation and design projects, is not the same thing as using art in these types of projects. To do that one does not necessarily need an artist to be involved! This is one of the reference models. There are others.

L. Frank Baum will now read a slightly modified extract from his book, *The Wizard of Oz*:

"'I am Oz, the Great and Terrible. Why do you seek me?'

"They looked again in every part of the room, and then, seeing no one, Dorothy asked, 'Where are you?'

"'I am everywhere,' answered the Voice, 'but to the eyes of common mortals I am invisible. I will now seat myself upon my throne, that you may converse with me.' Indeed the Voice seemed just then to come straight from the throne itself: so they walked towards it and stood in a row while Dorothy said:

"'We have come to claim our promise, Oh Oz.'

"'What promise?' asked Oz.

"'You promised to send me to paradise when the technologists became creative,' said the girl.

"'And you promised to remove silos,' said the scarecrow.

"'And you promised to produce unconventional and compelling products,' said the Tin Woodman.

"'And you promised to make European industry more competitive through working at the intersection of science, technology and the arts,' said the Cowardly Lion.

"'Are technologists really now creative?' asked the Voice and Dorothy thought that it trembled a little.

"'Yes,' she answered, 'I gave them artist's creativity elixir mixed in with a bucket of water.'

"'Dear me,' said the Voice, 'how sudden! Well come to me tomorrow, for I must have time to think it over.'

"'You've had plenty of time already,' said the Tin Woodman angrily.

"'We shan't wait another day longer,' said the Scarecrow.

"'You must keep your promises to us!' exclaimed Dorothy.

"The Lion thought that it might be as well to frighten the Wizard, so he gave a large, loud roar, which was so fierce and dreadful that Toto jumped away from him in alarm and tipped over the screen that stood in the corner. As it fell with a crash they looked that way, and the next moment all of them were filled with wonder. For they saw standing in just the spot the screen had hidden, a little old European technocrat, with head full of strange ideologies and beliefs, who seemed as much surprised as they were. The Tin Woodman raised his axe, rushed towards the little man and cried out, 'Who are you?'

"'I am Oz, the Great and Terrible', said the little old European technocrat in a trembling voice, 'but don't strike me - please don't - and I'll do anything you want me to.'

"Our friends looked at him in surprise and dismay.

"'I thought Oz was a knowledgeable Head, and knew about art,' said Dorothy.

"'And I thought Oz was a lovely Lady who knew how to use art to address patriarchal power structures in ICT research and innovation systems,' said the Scarecrow.

"'And I thought Oz was a terrible Beast that knew how to solve all the problems of the European ICT sector by using art in research and innovation,' said the Tin Woodman.

"'And I thought Oz was a Ball of Fire, that knew how to use art in research and innovation to sweep away the old and usher in a new golden era for the European ICT sector,' exclaimed the Lion.

"'No you are all wrong,' said the little old European technocrat meekly. 'I am a technocrat who knows very little about all these things, and I have been making believe.'

"'Making believe!' cried Dorothy. 'Are you not a great Wizard?'

"'Hush my dear,' he said: 'don't speak so loud, or you will be overheard – and I should be ruined. I'm supposed to be a Great Wizard.'

"'And aren't you?' she asked.

"'Not a bit of it my dear; I'm just a common man.'

"'You're more than that,' said the Scarecrow, in a grieved tone; 'you're a humbug.'

"'Exactly so!' declared the European technocrat, rubbing his hands together as if pleased him. 'I'm a humbug.'

"'But this is terrible,' said the Tin Woodman; 'how shall we ever produce unconventional and compelling products?'

"'Or make European industry more competitive through working at the intersection of science, technology and the arts?' asked the Lion.

"'Or remove silos,' wailed the Scarecrow, wiping the tears from his eyes with his coat sleeve.

"My dear friends," said Oz. "I pray you not to speak of these little things. Think of me, and the terrible trouble I'm in at being found out."

"'Doesn't anyone else know you're a humbug?' asked Dorothy.

"'No one knows it but you four – and myself,' replied Oz. 'I have fooled everyone so long that I thought I should never be found out. It was a great mistake my ever letting you into the Throne Room ...'

L. Frank Baum's The Wizard of Oz, a children's story that, like Hans Christian Anderson's *The Emperor New Clothes*, speaks of the nonsense of human behaviour. We call it behavioural policy making – the delusion that people know what they are doing and that they are acting in a rational and objective way – and doing the right things!

**70**

Back to the matter of different types of systems, so here are some more definitions.

So far systems have been characterised by the presence, or absence of the property of lineararity: linear systems and non-linear systems; or according to the property of anticipation: *anticipatory* and *non-anticipatory*. What other categories do we now offer you?

There are systems in which physical properties like mass can be assumed to be located at a single point, and systems in which physical properties are distributed: otherwise known as lumped parameter systems and distributed parameter systems, respectively. There are also systems in which these parameters do not change with time, and those in which the parameters do change with time: respectively time invariant systems and time varying systems. There are systems that are described by deterministic equations, and those that need to be described by probabilistic equations: respectively deterministic systems and stochastic systems.

Reality is complicated: systems are non-linear, distributed parameter, time varying, and stochastic in character. Some engineered systems can also be made anticipatory. This reality is too complicated to deal with, so simplifying assumptions are deployed. This is the way of STEM. It is also the way of humans for just about everything we address is not reality, but some idealised version of it. Your understanding of the world, for example!

Science is founded on several beliefs. Repeat the word: beliefs. One such belief is that science is trans-cultural and that cultural variations are not possible – there is only one best and true way of undertaking this thing called science. Another belief is that scientific knowledge is cumulative. A further belief is that only scientific knowledge is true knowledge. Another belief is that science rests on specific results that are dissociated from the personality and social position of the investigator, or put another way, the data and the observer are separated.

On the matter of cultural variations we may we think say more. Be careful for it is a tricky issue. Is your mind ready for the complexities?

Positivism is a philosophy of science. Lewis Wolpert, that well known scientist, a Fellow of the Royal Society no less, now makes an observation on the philosophy of science.

"The philosophy of science is useless and gives no insight into the process of science. No scientist takes it seriously."

Thanks Lewis for confirming that positivism can give no insight into the process of science. And yet, are you not a positivist? Of course positivism does provide insight into science as a belief system! Again Lewis will deny that science is a belief system, even though he continually described science in terms of a set of beliefs. Stating that science is not a belief system is also a statement of belief. It is very easy to tie in knots, these STEM people!

Positivism: yet more simplifying assumptions.

The fragmented and reductive mind is everywhere in the Western world. What chance then of achieving a unity of all knowledge? Consilience? Yet another strategic weakness!

**71**

Quantum biology: the ultimate reductive view. Everything reduced to quantum effects. There now follows singing and dancing as those with fragmented and reductive minds jump around in celebratory glee. Meanwhile, back in the real world, life perishes as the environment and humans, just complex machines, malfunction, and blindly, those from the world of STEM walk to the edge of doom. Plenty of artists among this pack as well. Like lemmings they rush their way to the precipice, saying, as they go, and with the greatest of glee, that 'there is no such thing as God, for science has proved it'. And yet none ever asked: what is this thing called God?

There are no consequences in the fantasy world of science, but in the real world the consequences are deadly. But there is no other way, apparently!

The planet was fine before we arrived, and it will be fine again once we are gone. Nature will recover with time, but we will not.

There is cause and effect, and there is purpose.

If you have not yet realised, what is being discussed and advocated in this book is the presence of discontinuities that render morphostatic change inappropriate. We are far beyond the common themes associated with the instrumentalisation of artists: communication, visualisation, creativity. While STEM people speak of such things, our planet, nature, us – these are all dying. STEM will finish us off.

Back once more to agriculture. The consequences of industrial-scale intensive agriculture are so damaging, so unsustainable, and increasingly unacceptable, that continuing along this path is no longer a viable option. On the other hand, the challenges are such that even organic agriculture does not, in its present form, provide a viable alternative approach.

In these circumstances what is needed is a major change to the existing paradigm. In terms of the theory of change what is advocated is called morphogenic change. This simply means a type of change that produces a different order to that which existed previously. Disturbances are treated as information about the inappropriateness of established practices, and this leads to fundamental changes in methods and principles.

Importantly however, this does not imply an *either/*or circumstance, a binary choice between industrial scale intensive agriculture and organic agriculture. This would be a simplistic interpretation, and one that typifies the mind-set of industrial era thinking, that is still prevalent today, even though it is claimed that society has entered the knowledge era. Between the two choices exists a continuum of possibilities, largely unexplored, where positioning is determined by technological feasibility, production economics, lifecycle carbon and energy matters, safety issues, product characteristics, and so forth. The challenge will be to determine where, along the continuum, a new system of agriculture can emerge.

In the industrial model of agriculture there is the organic model, and in the organic model there is the industrial model. No sole truths. No one best way. A unity of opposites! Come to know me on your own. The forces of Yin and Yang, realised into something that lies beyond the imagination of STEM people. Art is about imagining the unimaginable. Welcome to the *Age of the Tao*. Goodbye to the *Age of Abraham*. What does this mean? Probably not what you think!

All you communicators, visualisers and creationists, step aside, and take your restrictive biases with you for art is a way of being in the world, of knowing the world - art is discovery.

"The notion that scientific revolutions represent movements away from previous conceptions of the world that have run into cataclysmic difficulties seems to call into question the notion of science as aiming at the truth about the universe. The thought that there is one, and only one, complete and true account of everything is deep in the Western tradition. It descends from what Comte, the founder of positivism, called the theological stage of human inquiry. In popular versions of Jewish, Christian and Muslim cosmology, there is one true and complete account of everything, namely what God knows. This image gets transposed to fundamental physics, many of whose practitioners - who might proudly proclaim themselves to be atheists - take for granted that there just is, waiting to be discovered, one full and complete account of nature."

That was Ian Hacking speaking. He wrote an introductory essay to the $50^{th}$ anniversary edition of Thomas Kuhn's *The Structure of Scientific Revolutions.*

And the astrophysicist said that he was both an atheist and a positivist!

There is nothing more Abrahamic than Big Bang theory. In Chinese culture though there is no God the creator, and we shall never know, had their culture become dominant in the world, if they would have produced a different scientific theory that also matched that which can be the observed.

Can observable phenomena be assembled into different theories? If you are caught up in the inheritance of an Abrahmic culture then the answer is obviously no, and any attempt to do this would be viewed as a heresy. This book is a heresy.

How many different versions of quantum theory exist in the modern world? Is the answer one, or are there more? We may be returning to the matter of quantum heresies later.

**72**

Which is more dominant in the human - the auditory or the visual? We use hearing to listen to sounds - for example, music and the spoken word. Yet there was a time when humans had not sufficiently evolved to be able to engage in sophisticated communications using spoken languages. Life was very visual. It still is.

John Whitney now makes a contribution.

"Music and visual arts are complementary. In other words, using a computer, it is possible to create an art that looks like musical sounds and there is a theoretical basis for this."

So is this relevant to scriptovisualism? Are descriptive expression and plastic expression complementary and can a computer be used to create art that looks like words, and is there a theoretical basis for this. What does this help to achieve?

So while all you communicators, visualisers and creationists with your restrictive biases limit your mind to that which fits with your ideologies, we explore worlds that are clearly beyond your limited imagination. Your imagination is limited only by your cognitive biases which are within your power to change, to discard. Not a very comfortable thought!

Media that have extended people: printed books, mass circulation newspapers, universal literacy, universal liberal education, radio and television, low-cost rapid transport, telephones, easy to use personal computing devices, the internet, the world-wide web, social networks, indie publishing, universal history, universal human rights.

Rachael Carson will eventually be joining this odyssey.

**73**

In the beginning was the image. This is one of the reasons why we pursue scriptovisualism. It seems important, but why?

"Two roads diverged in a wood and I – I took the one less travelled by, and that has made all the difference."

That was Robert Frost reading from his poem, *The Road Not Taken*.

"Some corporations apparently became involved with us to promote a particular product or process or an area that the company wished to make better known. One company was eager to modernize their image. Two major companies contracted with us because of their presidents' social connections with Mrs Chandler, wife of the *Times* publisher, who had volunteered to help. Three companies, the smallest ones, joined with us solely for the publicity. Some companies were exceptionally cooperative because of a tradition of cultural support dating back years, but other companies, whose presidents are art collectors, proved difficult to work with, precisely because that knowledge of art created a restrictive bias."

That was Maurice Tucliman speaking about his experiences in the Los Angeles County Art Museum's Art and Technology Programme that ran over the period 1967 to 1971.

"I insist that people do not understand science."

Lewis Wolpert is back. Go ahead have your say.

"There is a misunderstanding about the difference between science and technology. Modern technology is based on science, but this recent association obscures crucial differences and the failure to distinguish between science and technology has played a major role in obscuring the nature of science. To put it briefly, science produces ideas whereas technology results in the in the production of usable objects."

We will, on your behalf Lewis, put your belief another way: *free people think, slaves work*. And of course the distinction is not true, for applied science produces both ideas and usable objects. Why so much separation Lewis? Is this part of your self constructed reality, your sole truth, which you project on to life, the world, the universe? You simplify, you reduce, and in the end what you have is not reality, just a simplified representation of this.

What hope is there for consilience with attitudes such as this?

**74**

Scientists are not the only people to make use of the scientific method. Engineers, for example, also use it. They carry out controlled experiments, they construct theoretical models, and they create new knowledge. So can artists. But not all artists! Be very careful who you listen to!

"Science is an essentially anarchic enterprise: theoretical anarchism is more humanitarian and more likely to encourage progress than its law-and-order alternative. This can be shown by both an examination of historical episodes and by an abstract analysis of the relation between ideas and action. The only principle that does not inhibit progress is: anything goes."

That was Paul Feyerabend, a philosopher of science, speaking. We already know what Lewis Wolpert thinks about the philosophy of science, even though his own book on science is a work on the philosophy of science. So back once more to subjectivity and also irrationality.

There are some similarities between the arts and STEM. Anything goes can also be found in the arts. This book proves the point.

"The history of science has shown that an overzealous policing of conceptual borders can impede progress. One possible value of artist's involvement in research is their willingness to explore ideas outside dominant paradigms."

That was Stephen Wilson. He was professor and head of the conceptual information arts programme at San Francisco State University.

Is what Wilson said true? Where is the evidence? Or is this more of those tall-tales, etc. Is it just a statement that aligns with the mythology of the artist in society: a statement which seeks to confirm, to reinforce an image?

"More and more I excluded from my painting all curved lines, until finally my compositions consisted of vertical and horizontal lines, which formed crosses, each one separate and detached from the other."

That was the artist Piet Mondrian speaking about the development of a new artistic movement that he called neo-plasticism. If you are a naïve and gullible scientist you might see this as reductionism. It is not.

Art, design, science, technology, engineering, mathematics - what are the differences? There are many, and this one might say is the value. What use would these things be if they were all the same, and yet there are also commonalities. There are always commonalities.

**75**

What is research? The Organisation for Economic Cooperation and Development (OECD) has produced an answer to this question, which is given in their Frascati manual. An anonymous representative of the OECD will now recite the OECD definition of research:

"Research and experimental development (R&D) comprise creative and systematic work undertaken in order to increase the stock of knowledge - including knowledge of humankind, culture and society - and to devise new applications of available knowledge."

Thank you. What the OECD spokesperson did not say however, is that to be research, activities must have five features: (i) R&D activities may be aimed at achieving either specific or general objectives; (ii) R&D is always aimed at new findings, based on original concepts (and their interpretation) or hypotheses; (iii) R&D is largely uncertain about its final outcome (or at least about the quantity of time and resources needed to achieve it); (iv) R&D is planned for and budgeted (even when carried out by individuals); (v) R&D is aimed at producing results that could be either freely transferred or traded in a marketplace.

In other words, for an activity to be an R&D activity, it must be novel; creative; uncertain; systematic; transferable or reproducible.

We wonder, we do, if that which from STARTS emerges will be transferable or reproducible. Tall-tales etc. tend not to be.

So now we ask another question: can art be research, based on this OECD understanding of what constitutes research?

Now, about the Xerox Palo Alto Research Centre artist in residence programme that began back in 1993; Rich Gold from Xerox PARC wishes to speak about that programme, which went by the name PAIR.

"Science is a creative act, so PAIR could not have been about bringing creativity to scientists, or, for that matter, bringing formalism to artists. PAIR brought remarkably similar professions together for a close examination."

Research is, as just seen, also creative, so by that understanding it is not just science that is a creative act, but also technology development, engineering, and mathematics.

Edward O. Wilson now speaks.

"The two domains, art and science, despite their similarities in their creative spirit, have radically different goals and methods."

Similar, different; what is going on here? Is scriptovisualism symbolic of this?

## 76

Today the subconscious mind is telling us that it is time for definitions. So here we go:

Linear innovation: this is a model of innovation in which technical change happens in a linear fashion from invention to innovation to diffusion. It is a stage-wise process in which one step does not happen until the preceding one is completed. It is like a waterfall. There are two versions: technology push where people develop technology and push it out to the market; and market pull, where the market pulls the development of technology. This is simple and easy to understand.

Non-linear innovation: this is a network model of innovation, sometimes called user innovation, where innovation occurs within this network, with users partaking in innovations which get transmitted into all stages on the innovation process: invention, research, etc. Self-evidently it is complex – this is the nature of reality.

Waterfall model: the waterfall model is a sequential design process, often used in software development processes, in which progress is seen as flowing steadily downwards (like a waterfall) through the phases of conception, initiation, analysis, design, construction, testing, production/implementation and maintenance. It is simple and easy to understand.

Spiral model: this is also known as the spiral lifecycle model, and is a systems development lifecycle model often used in information technology. It can be seen as a waterfall model undertaken many times. The spiral model is favoured for large, expensive, and complicated projects.

If you reduce and simplify, what is the result? The reality of non-linear innovation becomes linear innovation, which is just a linear process, stage-wise and easy to understand – Abrahamic. The spiral model becomes the waterfall model, which assumes that you can discover the sole truth about users or market needs or about suitable solutions at the very beginning, therefore you can then write the software code – Abrahamic. Are you feeling comfortable, reassured? Is your world as you want it? Are you happy?

Meanwhile, in places better able to handle the complexity of reality, of life, with all its interactions, discontinuities, feedback loops, uncertainties, constant changes, unpredictability, ambiguities, and so forth, the world looks very different.

The Tao: the way of heaven, unknowable, inexpressible, indefinable, yet each person can come to know it on their own.

Paul will now read the closing words of his short novel, Enigma:

"Come to know me on your own. Enigma."

**77**

Today the subconscious mind is telling us that it is time for more definitions. So here we go:

Reductionism: a philosophy in which complex phenomena can be reduced to related but simpler or more basic phenomena. In essence, the belief as in science, that the explanation of entire systems can be achieved in terms of their individual, constituent parts and their interactions.

Here we add a note that only a naïve and gullible scientist would relate this definition to that which artists like Piet Mondrian did. Unless of course the naïve and gullible scientist has a hidden agenda: silent narrative.

On with the task of providing defitions.

Scientism: belief in the universal applicability of the scientific method and approach, and the view that empirical science constitutes the most authoritative world view or the most valuable part of human learning - to the exclusion of other viewpoints.

Darwinism: a theory of biological evolution developed by the English naturalist Charles Darwin and others, stating that all species of organisms arise and develop through the natural selection of small, inherited variations that increase the individual's ability to compete, survive, and reproduce.

Social Darwinism: the name given to various theories of society which emerged in the United Kingdom, North America, and Western Europe in the 1870s, and which claim to apply biological concepts of natural selection and survival of the fittest to sociology and politics.

Neo Darwinism: a synthesis of Darwinian evolution through natural selection with Mendelian genetics.

Simplification: the process of making something simpler or easier to do or understand.

Eugenics: a set of beliefs and practices that aims at improving the genetic quality of the human population.

Gene therapy: gene therapy is an experimental technique that uses genes to treat or prevent disease.

Genetic engineering: the modification of an organism's genetic composition by artificial means, often involving the transfer of specific traits, or genes, from one organism into a plant or animal of an entirely different species.

Normal: conforming to the standard or the common type; usual; not abnormal; regular; serving to establish a standard.

Abnormal: deviating from what is normal or usual, typically in a way that is undesirable or worrying.

Infanticide: the practice in some societies of killing unwanted children soon after birth.

Genocide: the deliberate killing of a large group of people, especially those of a particular nation or ethnic group;

Abortion: the deliberate termination of a human pregnancy, most often performed during the first 28 weeks.

Untermensch: inferior people; sub-human.

Übermensch: superhuman.

Genetic genocide: ?

**78**

The *Renaissance* was a period in European history, from the fourteenth to the seventeenth centuries, which can be seen as a bridge between the medieval and the early modern era. It is typically characterised as a cultural movement, driven by the development of humanism, which had a different meaning to that which the term means in the modern world. The intellectual ideas of the period were derived from the rediscovery of classical Greek philosophy which had occurred in the medieval period. The new thinking became manifest in art, architecture, politics, science and literature. Is this true? Or is this yet another myth?

Modern humanism: a philosophical and ethical stance that emphasizes the value and agency of human beings individually and collectively, and generally prefers critical thinking and rational evidence over acceptance of dogma or superstition. Generally humanism refers to a perspective that affirms some notion of human freedom and progress. In modern times humanist movements are typically aligned with secularism, and often imply a non-theistic perspective that looks to science rather than revelation from a supernatural source to understand the world.

Renaissance humanism: what did this mean to those people who lived in the Renaissance period? What do we, living in the modern age, think it meant? Are the two things the same?

Either/or: a binary perspective such as for example, religion or science. A typically European characteristic! Oh how the *Vainglorious Enlightened Ones* love their binaries!

Some modern dogmas: communism, capitalism, science, technology, nationalism, atheism ... And so they go on. They are all better summarised as sole truths. There are many sole truths. Surely they cannot all be true.

"All religions are true."

That was Mahatma Ghandi speaking.

Abortion: just another form of genocide? You will of course deny this. Denial is the final step in genocide. People in Germany also denied the holocaust. In the United Kingdom there is also denial about slavery and the slave trade.

This day, 100 years ago, Paul's mother was born. The year was 1916, and across the English Channel a terrible conflict raged among the Europeans. Industrialised slaughter! On that day, in March, as a new life started, how many lives were ended unnecessarily? We do not know, but what we do know is that rivalry between nation states that declare spheres of influence continues in our present age. We have learned nothing. Time is indeed a singularity. The continent of Europe is the sphere of influence of the European Union, so take note of this please, Russia. As we just said, we have learned nothing. Are you listening, all you Ancient Romans in Brussels? No, you are not!

This is what the spokesperson for the Oxford English Dictionary has to say about reductionism:

"It is derogatory: the practice of analysing and describing a complex phenomenon in terms of its simple or fundamental constituents, especially when this is said to provide a sufficient explanation."

Back to the naïve and gullible scientists with agendas! Why do we say naïve and gullible?

**79**

The *Age of Enlightenment* came after the period known as the *Renaissance*. But today the subconscious mind is telling us to define one word:

Truth: that which is true; a fact or belief that is accepted as true.

Reality: the state of things as they actually exist, as opposed to an idealistic or notional idea of them.

"But you said that your subconscious mind was telling you to define only one word – you have defined two!"

Yes indeed, and it is so delightful to be so irrational. Why are you so pedantic?

Pedantic: excessively concerned with minor details or rules;

Thomas Kuhn now speaks:

"Does it really help to imagine that there is some one full, objective, true account of nature and that the proper measure of scientific achievement is the extent to which it brings us closer to the ultimate goal of full understanding? If we can learn to substitute *evolution-from-what-we-do-know* for *evolution-towards-what-we-wish-to-know*, then a number of vexing problems may vanish in the process. Somewhere in this maze, for example, must lie the problem of induction."

Interesting thoughts indeed, especially if one is concerned with developing a Taoist science and technology! We note that the binary, the either/or appears once again. Do you not think that in the *evolution-from-what-we-do-know* there is *evolution-towards-what-we-wish-to-know*, and in *evolution-towards-what-we-wish-to-know* there is *evolution-from-what-we-do-know*? A unity of opposites: the opposing forces of Yin and Yang, operating to produce something quite different. More complex? Too complex?

Deduce: arrive at (a fact or a conclusion) by reasoning; draw as a logical conclusion.

Deduction: the inference of particular instances by reference to a general law or principle.

Induction: in science the term induction is commonly used to describe inferences from particular cases to the general case or from a finite sample of data to a generalization about a whole population.

"But now you have defined three more words!"

Why so pedantic?

## 80

Eighty – another milestone! Today is the spring equinox – the vernal equinox. This year, 2016, has the earliest spring equinox since 1896. Why? We come back to the matter later, for today the subconscious mind is telling us to deal with what is called the *Age of Enlightenment*.

*Age of Enlightenment*: this was a philosophical and cultural movement which dominated the world of ideas in Europe, and also America, in the eighteenth century. The Enlightenment included a range of ideas centred on reason as the primary source of authority and legitimacy, and came to advance ideals such as liberty, progress, tolerance, fraternity, constitutional government and ending the abuses of the church and state. It is also the age when modern science was born, which eventually displaced the older Aristotelian science. Modern science is characterised by mathematisation, scientific rigour, experimentation, quantification, and reductionism.

The *Age of Enlightenment* came to an end because the French revolutionaries reasoned their way to the horrors of the mass executions of what became known as *the terror*.

Then came the *Age of Romance*, then the *Age of Positivism*, then *modernism*, then *postmodernism* ... These apart however, the *Age of Enlightenment* is alive and well. Later we may meet some *Enlightened* people who will be saying that the *Age of Enlightenment* is dead.

Now for the naming of some Age of Enlightenment thinkers: Isaac Newton; Rene Descartes; Thomas Paine; Adam Smith; Thomas Jefferson; Voltaire; ... And there is of course those people who have time travelled from the *Age of Enlightenment* to the modern era, stopping off in the nineteenth century to become positivists, and Darwinists, before landing, fully laden with their baggage in our troubled times. As if we do not have enough problems! More about them later we suppose.

And now Voltaire wants to tell the *free people* something important. The *slaves* already know this!

"God is not on the side of the big battalions, but on the side of those who shoot best."

There is a reason why this is so.

**81**

"Blackbird singing in the dead of night ..." Well almost! It was dusk actually, but it was just one marker of a change that has taken place over the past few weeks. Birds are busy making nests, and like the blackbird, are starting to engage in mating. The dawn chorus can now also be heard. The bumble bees have become active as well, building their little homes, and visiting the plants that bloom early, to harvest their life giving nectar. Nature is springing into life. Elsewhere the signs of spring are evident: longer days, shorter nights and the sun is also higher in the sky and the combination of these facts means that it warms the air, the soil. Buds are appearing, once more - the ones that appeared in December were killed off by the later arrival of true winter. But nature is recovering.

One does not need any scientific predictions, or models, or knowledge, to know that spring has arrived. You can sense it in the air, feel it, see it. Intuitively you know that it is spring, be it astronomically defined or metrologically defined. What ever way one looks at it - objectively or subjectively - it is spring!

Intuition: the ability to understand something instinctively, without the need for conscious reasoning.

Aesthetics: a set of principles concerned with the nature and appreciation of beauty; the branch of philosophy which deals with questions of beauty and artistic taste.

**82**

Sad news from Brussels - innocent people killed in yet one more terrorist outrage. And with it comes, across the social networks, simple explanations for what is a highly complex human tragedy. The nature of being human is to want explanations, even simple ones. Reduction, simplification, matter explained, sole truth discovered.

Quite rightly people express their anger and sadness. We wonder though what Jesus would have said? Perhaps something like this:

"Who among you that now speak out and condemn these senseless killings, did so with such vehemence when hundreds of thousands of innocent men, women and children died in the violence that took place in Afghanistan, Iraq, and Syria as a result of Western interventions and interference in these countries?

"Blessed are the peacemakers for they shall be called the children of God."

Have you not understood? It is not those who believe in Jesus that will be saved, it is those who understand the words of Jesus that will be saved.

It seems that there is no end to the senseless cycle of killing. The West seems to learn nothing. It should not abandon the peoples of the world who are the victims of barbarisms that stem from human ideologies, but it should not forget that it too has blood on its hands, as it also has perpetrated such barbarisms. People seem to forget this. Time is a singularity.

Paul will now read the closing words from his novel *Moments in Time*:

"At first it truly seemed to Benjamin that he was living a nightmare. Everything was about to happen all over again – the joy as well as the pain and suffering. This was too much, and it was at that moment in time that Benjamin realised that he was caught in a loop, truly chained to the past by unbreakable chains, and that he had indeed become like Prometheus, only there was no Hercules to save him!

"Then Benjamin smiled to himself. Mysteriously, some complex and unfathomable change had occurred within him. There was a chance now to modify the future, to create different outcomes, ones that would lead to happiness rather than tragedy. This was an opportunity to take a different road, to depart from a course and, with luck, reach a changed end. He had within himself found the way to save himself.

"He wasted no time in running downstairs to deal with Walter Smythe, but this time he would do it in a way that would not alienate Smythe and make him Benjamin's enemy. And very quickly, there they were, Benjamin and Lucy, more moments in time, in some ways different from the many gone before, but with many more to come, together, happy, looking forward to the future."

**83**

Now we have to meet with another group of *Enlightened* people who think that they think, which of course they do, in a kind of strange way! Later we may explore the idea of people who think that they think, which of course they do, in kind of strange way. These people most definitely do not think that the *Age of Enlightenment* is dead, which of course it is. It is most unlikely that, they ever will think differently, such is the power of ideology and dogma. So now to the meeting with these, alas, very typical *Vainglorious Enlightened Ones*.

First in is Richard Dawkins:

"If superior creatures from space ever visit earth, the first question they will ask, in order to assess the level of our civilisation is: 'Have they discovered evolution yet?'

Evolution – oh yes a very patriarchal theory indeed!

Notice Dawkins' use of the word superior. Why superior? Is it because they will have more advanced science or technology? Or is it because they will have evolved into something more superior? Or is it both? And why so subjective? Superior! Note also the use of the word discover. Did he not mean invent the concept to match the observations? Subjectivity yet again

- it just keeps popping-up every time one of these rational and objective types open their mouths to speak. Or put pen to paper.

"Furthermore, I have used science to prove the improbability of God's existence."

Oh yes the *God Delusion* - should have been called the Scientist's Delusion! Genes explain everything! Quantum physicists would not agree.

And now another one of the *Vainglorious Enlightened Ones* comes to speak. This time it is Edward O. Wilson and his ideas about consilience.

"The Enlightenment thinkers of the seventeenth and eighteenth centuries got it mostly right the first time."

To save the readers time - reductionism and genes, right?

"More or less."

So, hubris, positivism and scientism, which can be summed up as: We are the way, and the truth, and the life. No one comes to knowledge except through us.

"I have not finished."

Richard Dawkins evidently wants to add to what he has already said.

"We believe in evolution because the evidence supports it, and we would abandon it overnight if new evidence arose to disprove it."

We believe! On this last point we refer you to what Charles Darwin has already said. Here, to reinforce the point, we ask Thomas Kuhn to comment.

"Resistance to a new paradigm is inevitable. Scientists however do change their minds. Conversions will occur a few at a time until, after the last holdouts have died, the whole profession will be practicing under a single, but now different, paradigm."

Now Howard Rosenbrock will add something further to this.

"Living organisms exhibit clear purposes, and if the substrate of quantum mechanical particles from which life evolves is described as purposeless the question arises: how can purpose arise from a purposeless substrate? The proselytizing enthusiasm of Richard Dawkins is that the answer lies in natural selection, acting by chance, which produces this effect. If however, quantum mechanics is described as goal seeking then a different situation arises. It is not necessary to say that the argument for natural selection acting on a purposeless substrate is wrong: rather that it answers the wrong question. If the substrate were purposeless, things might indeed happen in the way suggested by Dawkins, but with a goal-seeking substrate a different question needs to be answered: how can a new purpose arise in living organisms, from a substrate which already has purpose, given that a system cannot simultaneously seek two goals unless these are in a special relationship to each other?"

Purpose, cybernetics, teleological - we may well return to these important matters, for they are central to Taoist science and technology. Immanual Kant will need to be consulted for his observations first.

**84**

Brussels once again - it is now time to be poetic:

> In Brussels town, the dead are counted - the number stands at 31;
> The unaccounted for - who will count them?
> We do that later, came the reply.
>
> So they stand in defiance, against the barbarians, against their uncivilised violence.
> And they give now a minute of their silence, showing deep respect for the 31,
> The 31, who to loved ones return no more.
>
> In Srebrenica town the dead are counted - the number stands at 8000;
> The dead of that Bosnian war - who will count them?
> We do not know, came the reply.
>
> For these dead, in Brussels, there is a different silence, and in Den Haag,
> While in Brussels they stand, the sentence for Bosnian civilised killings,
> The number is 40 years, for yet one more of, those European genocides.
>
> The dead are counted - the number is 8031, plus more, and counting.
> All are victims of that thing called the serpent, which, few seem to understand,
> Is just a name for the human mind - yours.

**85**

It is a sunny spring day. From the sky comes a sound, a distinctive one, of birds calling. We look up and soaring high above us, three birds of prey ride the thermals, as they hunt for food. We are lucky, living as we do, in a semi rural area with high moorland close by. Yet in these strange days, birds of prey are increasingly to be found in cityscapes, for here there are places to perch and prey for them to hunt, for another bird (the rock pigeon) long ago abandoned cliffs in favour of those that are of human hands made, and so nature adorns that which is great architecture, as well as that which is less noble, in its quest to survive.

And in response to the terrible events in Brussels, someone in the world of social media says 'Pray for Brussels' and in response others start to say, 'Don't Pray For Brussels - We've had Enough of Religion for One Day.' Few, it seems, want to think, to understand, and just revert to that which is the source of the problem. And thus the cycle just goes on repeating. Time is a singularity.

The sad reality of modern life is that, faced with the overwhelming complexity called life, people are retreating into simple, reductionist views, which allows the problems of the world to be reduced to simple statements like - Jews are ... Muslims are ... black people are ... poor white people are ... The list seems endless. Once someone has discovered the sole truth, then we have a problem. This is when the process starts: first comes classification and separation, then dehumanisation, then .... The Nazis, went all the way down the road, and implemented extermination. But so did the Ancient Romans when they tried to exterminate the Celtic peoples of Europe. Better not to mention the British and what they have done - the final stage is always denial. And to the list, more names will be added, as time goes by. In the 1990s it was the Bosnian Serbs, today it is what the Arabs call Daesh, while tomorrow yet another, as yet unknown name, will become a curse upon our lives. But the problem is always the same one.

Interestingly, there is no record of the Chinese, over thousands of years of their history, ever behaving like the Europeans: war and violence of course, as is the way with humans in history, but no genocide. Slavery can be found as well, but not in the same way as in the European world. Yet typically China and its past has been looked down upon by many Europeans and have been seen as being inferior - more classification and separation.

Jürgen Habermas, emeritus professor of philosophy at the Johann Wolfgang Goethe University of Frankfurt, now wishes to comment on the lure of technocracy:

"Key decisions are being taken in the European Union by the Council, the European Commission and European Central Bank – in other words, the very institutions that are either insufficiently legitimated to take such decisions or lack any democratic basis. This technocratic hollowing out of democracy is the result of a neoliberal pattern of market-deregulation policies. The balance between politics and the market has come out of sync, at the cost of the welfare state."

We are indeed sliding towards technocracy, and people from the world of STEM are at the forefront of the movement – be in no doubt, it is a movement, and one that is being increasingly orchestrated. They will have their way, as did also, in times long past, that priestly caste under the banner of the Roman Catholic Church. Be in no doubt also that the outcome will be the same.

What about artists in residence, those New Leonardos as well? What do they have to say about technocracy, about scientism? One day we will answer this question, in a future book. We suspect that the answer is very little, or even, nothing at all. Art-science lovers to the whispering of sweet nothing do gravitate.

**86**

Now we ask the question: what is a system? It seems that the answer depends upon what discipline you have chosen as your place of refuge – the place where you hide from the overwhelming complexity of life and the universe, and where you, with those whom you share this discipline, construct your reality, and then get confused about that which is real, and that which are but models, theories, observations; those things that you have come to believe is reality. Belief is subjective and subjectivity just keeps popping up in the world of the objective and the rational. Best though not to dwell on this or your delusions will be seen for what they are – delusions.

Peter Checkland, a fomer professor of Systems at Lancaster University, is here to speak about systems:

"The roots of systems thinking lie in the belief, expressed by some biologists in the 1920s and 1930s, that the reductionism of natural science was probably not the best approach to the understanding of specifically biological phenomena. What was therefore proposed was a systems approach that was systemic, that is, concerned with understanding the whole."

Interestingly the systems approach practised in engineering often does not reflect this view, and thus, what is called systems engineering is frequently actually based on reductionism rather than systemic thinking.

For some in the world of engineering, systems are just components, grouped together and connected together. But what are more interesting are systems that have emergent properties. Such systems are the basis of cybernetics.

Systems are therefore: a collection of components that, when connected together, exhibit emergent behaviours that cannot be understood by understanding the behaviour of the individual components. These emergent behaviours are the result of interactions. A major contribution to emergent behaviour is the presence of feedback.

Quorum sensing: a biological process used by bacterial cells to monitor when the cell density in their vicinity exceeds a certain threshold causing a change in their behaviour. Something is happening that is an emergent behaviour.

Howard Rosenbrock will now explain the difference between a schedule and a policy in control theory, which is part of the more general field known as cybernetics:

"A schedule is calculated in advance and it is adhered to regardless of what is actually happening in the real world: There is no feedback - it is what is called an open-loop system. By introducing feedback loops and designing a policy that uses the information that is being fed back, we can take account of what is happening in the real world."

And when we introduce feedback loops we have emergent behaviour, as well as goal-seeking behaviour, and then we are in the realms of the teleological. Anticipation, it seems! We may return to these matters.

**87**

Automation: The use or introduction of automatic equipment in a process to replace humans.

Does automation replace humans in one complete step? The answer is very rarely. Usually what cannot be automated is left for the human to do and this often results is fragmented and dehumanised working conditions. You have seen this across time in many factories, and also in offices. It is part of the way businesses operate to increase efficiency and reduce costs. Keywords here are horizontal division of labour and vertical division of labour. Reductionism - the whole broken down into its simpler constituent parts. The whole is just the sum of the parts! But the whole is very rarely the sum of the parts!

We are back to the *Age of Enlightenment*! Everything is a machine. Humans, nature, and factories - it matters not, for the same principle can be applied: the whole can be understood through reduction. Once you understand the simple parts you understand the whole. A nice dream, but understanding humans, nature and factories turned out to be - remarkably resistant to the application of such an approach. Reductionism takes one so far and then you have to move on. Reductionism is not dead though, just understood to be a limited method, at least by some. What comes next? What is the next paradigm? Please do not start talking to us about complexity theory and Global Systems Science!

Thomas Kuhn used the word paradigm a lot in his book, *The Structure of Scientific Revolutions*. Today the word paradigm is very commonly used, and may, as a result, be over used or inappropriately used. So to clarify the word *paradigm*, we note that it has two meanings: one being the prevailing body of theory that is regarded as the normal in science; the other is the set of beliefs and practices that are shared by a group.

If you change the latter does it lead to a change in the former?

**88**

There is another side to automation, which specifically relates to the modern generation of automation technologies which are based on computer technology. Back in the 1980s the term information technology was coined to describe this new generation of automation, as well as a lot of new things. Shoshana Zuboff is here to explain.

"Information technology, even when it is applied to automatically reproduce a finite activity, is not mute. It not only imposes information in the form of programmed instructions, but also produces information. It both accomplishes tasks and translates them into information. The action of a machine is entirely invested in its object, the product. Information technology, on the other hand, introduces an additional dimension of reflexivity: it makes its contribution to the product, but it also reflects back on its activities and on the system of activities to which it is related. Information technology not only produces action but also produces a voice that symbolically renders events, objects, and processes so that they become visible, knowable, and shareable in a new way. The word that I coined to describe this unique capacity is *informate*."

And how did manufacturing companies implement this new technology? The answer was to insert it into the organisational structures that were developed in the early part of the twentieth century. The information was given to managers, not to workers, because *free people think, slaves work.*

"Activities that seem to represent choices are often inert reproductions of accepted practice."

That was Shoshana again.

While the Americans and Europeans were operating on the basis of the past, behaving like Prometheus, living like Narcissus, in a different place, on the far side of the world, in Japan, where their culture is very dissimilar from that of the West, something very different was happening. Most people in the world of manufacturing could not see this, for culture is often invisible. By the time the discovery was made it was too late for many companies. Even today people still do not fully understand what the Japanese did, nor do they like it because people called Prometheus prefer the security of their rock to life set free from their chains.

China! Time is a singularity.

## 89

Back now to the matter of systems and more definitions, and the concept that will now be defined is equifinality.

Equifinality: the principle of equifinality states that a given final result may be achieved with different initial conditions and in different ways. The implication of this is that it is possible to achieve a given goal or objective by following any of several different options. There is no unique solution and no one best way.

So if we consider science as a system, then we ask the question: is there more than one way to do science - different paradigms - but which will lead to a different view of humanity? The answer in the Western world is clearly no. The West though is Abrahamic.

Now we will name some artists - painters and sculptors - that have also engaged in the writing of poetry. Donald Weisman will speak about this.

"It is perhaps significant that few examples of the writing-picturing combination exist, even in the sense of the writer who makes illustrations for his books, or vice versa. William Blake stands almost alone; the fact seems descriptive of the character of his genius, and it is notable that he makes no comment either on the different qualities of the two media, or on the fact of his double function. For the rest, Dante Gabriel Rosetti nurtured a double talent; in his case one feels that the painting suffered from the literary association. William Morris also was drawn by his feelings for the unity of the crafts to a drawing-writing combination, though he seems not to have cared particularly that the two expressions should come from the same artist. Among other painters, Michelangelo is perhaps the only one who was a great poet; but only in one case, the quatrain on the figure of Night in the Sacristy of San Lorenzo, does a poem treat the same subject as a sculpture."

We can add some further names to the list: the poet and novelist Laurie Lee was also a water-colourist; the writer Hans Christian Andersen also created visual art such as paper-cuts. Evgenia Emets, a poet and a visual artist, creates paintings of her poetry. She calls this poetry painting. She is another case, like that of Michelangelo – same subject, two different media for expression, same source. Evgenia also uses technology and combines sound, dance, performance and the visual. Her work is not constrained by the ideology and dogma of art-science, or art-technology.

This reminds us that we must say something about the painting part of *366 – A Scriptovisual Composition Unknown,* which is also another case like that of Michelangelo. We will say something, but not yet. Not yet.

Scriptovisualism is certainly about the dual creative processes: one form of expression, juxtapositioned with another very different medium, arising from the same well of experience. But, is that well of experience just one person or could it be a group of artists working on the same subject, collectively, through different media, in coordination with one another?

**90**

A shallow brook babbles the loudest; still waters run deep. He that will not apply new remedies must expect new evils; for time is the greatest innovator. A journey of a thousand miles begins with a single step. May you live in interesting times! Be the change that you wish to see in the world.

"Anyway, I keep picturing all these little kids playing some game in this big field of rye and all. Thousands of little kids, and nobody's around – nobody big, I mean – except me. And I'm standing on the edge of some crazy cliff. What I have to do, I have to catch everybody if they start to go over the cliff – I mean if they're running and they don't look where they're going I have to come out from somewhere and catch them. That's all I'd do all day. I'd just be the catcher in the rye and all. I know it's crazy, but that's the only thing I'd really like to be. I know it's crazy."

That was J.D. Salinger reading from his novel *The Catcher in the Rye.*

Salinger is interesting. He was a very successful writer. He gradually became reclusive. He stopped having his books published. He wrote the books that he wanted to write.

Writing the books you want to write is now easier because of the rise of *indie publishing*. The writing of books is not about making money; it is about an art form called literature. As with all artists that produce new forms of art, the results are often not appreciated by the commercial market.

"The highest he could raise himself to was to die gloriously for something; now he rises to something greater: to live humbly for something."

Thanks to Otto Ludwig for speaking those words.

**91**

Today we will address the matter of civilised violence and uncivilised violence. It seems highly appropriate. Socrates will tutor you in the ways of waging war:

"May I not observe with equal propriety that the Hellenic race is all united together by ties of blood and friendship, and alien and strange to the barbarians?"

"Very good," Glaucon said.

"And therefore when Hellenes fight with barbarians and barbarians with Hellenes, they will be described by us as being at war when they fight, and by nature enemies, and this kind of antagonism should be called war; but when Hellenes fight with one another we shall say that Hellas is then in a state of disorder and discord, they being by nature friends and such enmity is to be called discord."

"I agree," replied Glaucon.

"Consider then, when that which we have acknowledged to be discord occurs, and a city is divided, if both parties destroy the lands and burn the houses of one another, how wicked does the strife appear! No true lover of his country would bring himself to tear in pieces his own nurse and mother: There might be reason in the conqueror depriving the conquered of their harvest, but still they would have the idea of peace in their hearts and would not mean to go on fighting forever."

"Yes," Glaucon said, "that is a better temper than the other."

"And will not the city, which you are founding, be a Hellenic city?"

"It ought to be," he replied.

"Then will not the citizens be good and civilised?"

"Yes, very civilised."

"And will they not be lovers of Hellas, and think of Hellas as their own land, and share in the common temples?"

"Most certainly."

"And any difference which arises among them will be regarded by them as discord only - a quarrel among friends, which is not to be called a war?"

"Certainly not," agreed Glaucon.

"Then they will quarrel as those who intend some day to be reconciled?"

"Certainly."

"They will use friendly correction, but will not enslave or destroy their opponents; they will be correctors, not enemies?"

"Just so."

"And as they are Hellenes themselves they will not devastate Hellas, nor will they burn houses, not even suppose that the whole population of a city - men, women, and children - are equally their enemies, for they know that the guilt of war is always confined to a few persons and that the many are their friends. And for all these reasons they will be unwilling to waste their lands and raze their houses; their enmity to them will only last until the many innocent sufferers have compelled the guilty few to give satisfaction?"

"I agree, he said, that our citizens should thus deal with their Hellenic enemies; and with barbarians ..."

Now you have a little insight into why Europeans tend to treat people of non-European origin so poorly, and why, when 8000 Muslim men were murdered by Bosnian Serbs (Europeans), the perpetrators were not described as barbarians, but when Islamist terrorists (non-Europeans) murdered 32 people in Brussels, the people responsible were labelled as barbarians.

The final number was 32.

Symbolic: serving as a symbol - significant purely as a symbol of what is being represented or implied; involving the use of symbols or symbolism.

Scriptovisualism is certainly symbolic.

We made a significant change to coordinates today. There was a clear night sky. Orion was clearly visible, but he will soon be gone from our skies. March ends.

**92**

It is very rare these days for people to refer to information technology - the phrase now most often used is information and communication technologies. The communication part is important, for it suggests that distance is no longer a constraint or a significant cost, as data can easily, and relatively cheaply, be communicated across distance. The information part is also important, because from data comes forth information - perhaps! That is the dream. Then the data becomes big, so people start talking about big data.

So when people have the means of collecting their own data, which is increasingly the case, what happens? Data is no longer the preserve of the elite - those STEM people. Now you too can have your own data.

Researchers are finding that it is increasingly difficult for them to keep their data to themselves because open access is becoming the *Modus Operandi*. Research funding agencies are beginning to insist upon this. Free, open access data repositories are beginning to appear on the world-wide-web. Universal data access is the order of the day.

The medium is indeed the message.

Now the former chief scientific advisor to a former president of the European Commission will speak:

"We are all entitled to our own opinion but not to our own facts."

What you mean is the *free people think, while the slaves work*. And your comments illustrate how out-of-touch you are. When people become more knowledgeable about science and how scientists actually operate, and have the means of creating their own data and own facts, your power will be broken.

Science, the world of STEM, is only one of many ways of knowing and trying to understand life, the world, the universe, and us - humanity. If one shines a single beam of light into a dark space, one only sees that which the beam illuminates. There will be an area towards the centre which is more brightly lit than the other areas towards the periphery. All around will be areas that are dark and unknown. Yet you persist in acting on the basis of only that which can be seen, as though you have discovered the truth – the sole truth.

Universal history:

"Today for the first time there is a real unity of mankind which consists of the fact that nothing essential can happen anywhere that does not concern all. The importance of the fact that the whole of mankind has been drawn into a common stream of destruction or renewal has only become apparent in more recent times. People in Europe have lived and behaved within a European consciousness where China and India were alien, untouched worlds of their own which one did not learn about in history. In the emerging universal era, the technological revolution effected by the Europeans through science and discoveries is merely the material basis and precipitating cause of the spiritual and psychical decline that is now evident across the globe. The success of the process of melting down and recasting that is now taking place, will be measured in the end, by the extent that Europe is no longer Europe and the Europeans are no longer Europeans, China is no longer China and the Chinese no longer Chinese. There will be new Europeans and new Chinese, but the images of these we cannot yet see. And so it is that out of the experience of our historical situation, as the turning point of the ages, we turn our gaze. To the question: have such radical metamorphoses taken place before? Our answer is: we know nothing of the Promethean Age, when man first came into possession of his world through tools, fire and speech. But within history the greatest turning point was the *Axial Age*. If we have now entered a new radical metamorphosis of humanity, this is no repetition of the *Axial Age*, but a happening that is different to its very roots."

That was the philosopher Karl Jaspers speaking.

**93**

Semiotics: the study of signs and symbols and their use or interpretation.

Paul will now read some words from his novel, *Moments in Time*:

"The future only exists in the imagination, the past only in memory. All we have is the present, moments in time that are doomed to slip away, forever flowing towards the far distant shores of life long past. Here in the present moment is where we create our regrets, or make our dreams come true. This is how life is; an inescapable law of nature, a universal truth, but …"

Prometheus. Hermeneutics. Poetics. Consilience.

A representative of the Oxford English Dictionary will now tell us the meaning of the word *creativity*:

"Creativity is a noun. It means the use of imagination or original ideas to create something."

We thank you for the definition, but while you are here, we would like you to tell our readers what the word, *innovate,* means.

"Yes. Innovate is a verb, and it means make changes in something established, especially by introducing new methods, ideas, or products. It also means to introduce something new, especially a product."

Thus one can say that innovation is the process of innovating. Innovation is a noun. Innovation is a process.

Art: imagination - create - imagination - create - imagination - create... Art not products! Now tell DG CONNECT!

"The artist group known as *Blast Theory* have repeatedly rejected the term user-generated content as applying to their works, for they believe that each of its words is too tainted with commercial connotations. For that reason, they prefer to speak of publicly generated contributions. The participants in their projects would thus contribute in public to a collective project. But the participants' shared interest in something and their willingness to actively participate in it are at least as important as the archivable recordings."

That was Katja Kwastek speaking.

Did we not tell you that the motivation of the work of artists is different from that of STEM people?

**94**

As semiotics relates to the study of signs and symbols and their use or interpretation, this must surely be of high relevance to scriptovisualism.

"I have been asked many times when do I stop learning and studying and just do one thing, but I feel like I am only starting. I have found something that sparks my passion and I am now integrating everything I have learned about sound, light, space and human perception. There is a great deal of learning still to be done and this learning will need to be integrated into my art practice. I feel this flow cannot be stopped as this is part of something bigger happening in human evolution and I am playing my own role in this grand plan. I think that we are driving the process, but also, each of us is being driven by what is happening on a larger scale in the world."

That was the artist Evgenia Emets speaking.

So Evgenia, something is happening! Any idea what that is? Would you like to know more? Some things though, are not for the speaking, but for the finding. Look to words spoken in silence. Look to the language that has no words. Descriptive expression, plastic expression - words spoken in silence and the language with no words. Scriptovisualism - words spoken in silence and the language with no words, combined in a unity of opposites?

Understand the words spoken in silence and then you will find. Only the hand that writes such things will understand. Then the receiver, in receiving, gives, and then the giver, in giving, also becomes the receiver.

"The quality of mercy is not strain'd,
"It droppeth as the gentle rain from heaven
"Upon the place beneath: it is twice blest;
"It blesseth him that gives and him that takes."

That was William Shakespeare reading a few lines from his play, *The Merchant of Venice.*

There is another language that is spoken in silence. It is the language of the deaf - a language of signs, facial expressions and gestures. In the United Kingdom it is called British Sign Language, or just BSL for short. BSL is not signed English. BSL has its own grammar which is quite different from spoken English. BSL also has a property that spoken English does not possess - simultaneity.

Plastic expression also possesses the property of simultaneity. Descriptive expression does not have this. BSL combines descriptive and plastic expression. Does that make BSL scriptovisual? Probably it does. You will notice that, as in most cases where we have discussed scriptovisualism, we are not providing definitive answers. Why is that?

**95**

The cup that is full cannot be filled. Thus we have become the empty vessel waiting to receive - the blank page, the bare canvass, the open space.

The history of all hitherto existing European civilisations and societies has been the projection of European values, beliefs, world views and ideologies on to every other civilisation, tribal society, and peoples of the world. We are of course paraphrasing Marx and Engels, who, one can say, somewhat ironically, were also responsible for setting in motion a process that led to - yet more projection of European ideologies on to, not only the peoples of Europe, but also those who dwell further a field. And thus began yet another ideological battle between two belief systems both of which claimed to be the sole bearers of the truth - the sole truth. Sound familiar?

Karl Jaspers will speak again:

"Man has an urge to consider his own life-form the only true one, to feel every existence that does not resemble his own to be a reproach, and to hate it. From this arises the disposition to enforce one's own way of life upon everyone else, as far as possible to model the whole world upon it. Politics that draw their sustenance from such impulses tend to violence, multiply violence. They do not hear, do not negotiate - save in appearance - but subjugate."

Europe, and its main offspring, the United States, have much experience of violently subjugating the rest of the world to the European way. And now both, and others, are being subjected to the same process by Islamist extremists, and they do not like it. The phrase *sow the wind reap the whirlwind* comes to mind.

Most probably you believe the European world has learned its lesson and no longer engages in the projection of its values, beliefs, ideologies and culture on to the rest of the world. More importantly, this is probably what you want to believe. Even more importantly, probably you have convinced yourself that this is so, and this has become for you the truth - the sole truth.

All this, one can say, is very Abrahamic. There must surely be a lesson here that, if we are to move on and to begin to mature as a species, we must now learn before it is too late.

Paul will now read a passage from his short story *Enigma*:

"All human life is sacred, and the taking of a human life is wrong. There are no exceptions."

Now we shall examine the problem that STEM people have become. Listen to a member of the British engineering establishment speaking about Fukushima. She is one of those people who carry around one of those titles that the British so love, and advise governments - oh so independently!

"Those of us experienced in the design of nuclear systems knew that it was not going to be another Chernobyl. The public has this inherent fear of radiation. The material that has been emitted from Fukushima is not going to create long-term damage. All the press focused on was a reactor puffing steam."

These words were spoken in October 2011, when it was already clear that Fukushima was another Chernobyl type disaster. Meanwhile on the BBC, a nuclear physicist, almost saying, 'trust me I am a scientist', was telling us that he had no *axe to grind*, and spent a whole hour downplaying the significance of both Chernobyl and Fukushima. A later programme made to journalistic standards, did however show that both the nuclear physicist and the senior member of the British engineering establishment were - very wrong. Now why were they so wrong? Any guesses? Perhaps they were not the experts they claimed to be, or was it because they were experts, with vested interests?

For the record: Fukushima was the world's first triple nuclear reactor-core melt-down disaster. The accident, just like Chernobyl, was given the highest rating (7) on international nuclear and radiological event scale - in other words it was classed as a major accident. A twenty kilometre exclusion zone was established around the plant. People were evacuated from this area. Over the years a major clean-up operation has been undertaken in these areas, and people have gradually been allowed to return to certain areas. Issues about the safety of this practice do however remain, and there are continuing reports appearing in the international press about high levels of radiation.

The important lesson though is this: the risks cannot be managed.

**96**

Philosophy: The study of the fundamental nature of knowledge, reality, and existence.

Socrates will now speak:

"Philosophy is the noblest of all pursuits. There is a dignity about her which is not to be found in the arts. The only life which looks down upon the life of political ambition is that of true philosophy. There will be discovered to be some natures that ought to study philosophy and to be leaders in the State; and others who are not born to be philosophers, and are meant to be followers rather than leaders."

Thank you Socrates for your views on philosophy.

"Mirror, mirror on the wall, which are the most important disciplines of all?"

"My dear Narccissus, it is science, technology, engineering and mathematics which are the most important disciplines in the land."

Hawkins, Dawkins, Wilson, Wolpert, and others from the world of STEM, speaking to their magic mirror; they have been given the name Narcissus. You have truly been cursed by Nemesis.

"I believe in many gods; I believe in one God; I believe that Jesus was the Son of God; I believe that the prophet Mohamed - peace be among him - was God's final messenger; I believe that only science produces true knowledge; I believe in private interests, public gain; I believe that the history of all hitherto existing societies is the history of class struggle; I do not believe in God; I believe ..."

So many beliefs; so much subjectivity!

Then someone spoke, saying:

"Lucid dreaming is the only true sense of freedom in life."

To which someone, slightly surprised, asked:

"Only?"

And then came back the reply:

"Yes, only!"

And so the truth, the sole truth had been discovered.

The operative phrase, in all cases - there are no exceptions - is *I believe*.

The artist Roy Ascott will now speak:

"Employing the habitual level of thought and behaviour puts us in a state of extreme danger, in so far as our responsibility for self-construction, the building of reality, empathetic sensibility, and the creative exploration of consciousness, are concerned. Habitually, we occupy the most dangerous territory of mind on a daily basis, thinking that what is familiar and predictable is benign. Habit is the enemy that leads to the death of the self, reinforcing the passive, uncritical repetition of normative behaviours, opinions, perceptions and values. Habit is the enemy of art, impeding our search for new ways of being."

The history of all hitherto existing civilisations is one of decline and fall. Why do modern Europeans think that theirs will be any different?

Once we stood among the ruins of Ancient Rome, and it was not hard to imagine people living among the grandeur and achievements, thinking that it would last forever. The Ancient Greeks probably thought the same. People caught-up in the ideology of European Union Nationalism - the European Union Nationalists - that is to say the European Unionist, think exactly like this. The Union that will last for a thousand years!

Someone should tell them it is still a flag!

We cannot, because they have discovered the truth – the sole truth. And one day …

National Socialist Germany – Ein Volk, Ein Reich, Ein Führer, regardless of what the people of Germany wanted.

We will now rephrase this for you to make it more palatable. To do this we have invited a representative of the European Commission to tell you about the future they have in store for you:

"Europe will be one people, one empire – the European Union – and one leader – the European Commission. We, the *free people who think*, will create this for you, the *slaves who work*, regardless of what the people of Europe want. Consideration of alternatives will not be tolerated. Those who do not agree, will be treated as traitors and branded as right-wing populists and extremists. Most of you will believe this deception. We refer you to the very successful Nazi deception. By the time you realise it will be too late!"

**97**

So now it is time to revisit Leonardo da Vinci – artist, engineer, inventor, and designer, etc! Or perhaps, just an artist being – an artist? And as we write we can hear someone speaking. We will listen-in and see what is being said, for it is an engineer that is expressing his *beliefs*. Yes, it is that word once again.

"Paul T Kidd's letter about Leonardo da Vinci reads too much into a simple situation. Leonardo produced many works of art, so he was an artist. He drew many inventions, so he was an inventor. He did not develop them into practical form, so he was not an engineer. The idea that artists all work and think in some special way is a generalisation too far: engineers also have widely varied work and thought patterns."

What is this we hear? There is another engineer who is now speaking. We will eavesdrop:

"In the recent IET E&T article on Leonardo da Vinci, Dea Birkett asks, 'Was Leonardo the engineering genius we've come to regard him as?' Perhaps we ought to ask the man himself for his opinions. *The Inventions of Leonardo da Vinci* by C. Gibbs-Smith, published in 1978 by Phaidon Press, includes a translation of the CV that he sent to Duke Ludovico (II Moro) of Milan, in which he lists his skills and expertise in ten bullet points: the first nine are as a military engineer, and the last as a civil engineer and architect. Then he adds, somewhat as his other hobbies and interests section, his abilities in sculpture and painting. It is clear, therefore, that Leonardo considered himself principally to be an engineer, at least when he was looking for a job."

So there you have it. Matters reduced and simplified and the sole truth discovered. Evidently our two speakers did not read the cautionary material about interpretation just inside the entrance to the exhibition. Nor, we can add, what other disciplines have to say about him.

History of course is not just about recording facts; it is also about interpretation, and reinterpretation. The job of the historian, even where history is recorded, is never done.

The characteristic features of critical writing are: (i) a clear and confident refusal to accept the conclusions of other writers without evaluating the arguments and evidence that they provide; (ii) a balanced presentation of reasons why the conclusions of other writers may be accepted or may need to be treated with caution; (iii) a clear presentation of evidence and argument, leading to conclusion; and (iv) a recognition of the limitations of evidence, argument, and conclusion.

There are some from the world of STEM seeking to appropriate art for their own ends. They should however take note that such actions (which now it seems include the re-branding of artists as scientists, engineers, innovators, designers, and such forth) create an artistic interest to investigate this mind set, for art is about discovery, of anything and everything, depending upon the muse. And a very fruitful area of artistic endeavour is the creation of works which will lead those in STEM to encounter some uncomfortable truths about themselves. This is one of the reasons why society needs artists, not to act as appendages to STEM – to engage in communication, visualisation and to act as aids to creativity – but to expose these unpalatable things and also discover, in doing so, very different and more appropriate realisations of STEM. Who else will do this? STEM people? Evidently also there seem to be very few artists interested in undertaking such a task, or able to do so.

"We engineers are trained to think in a certain way, a linear way, pragmatically. The way the current education system works, we're not always encouraged to think outside the box, or think in an open-minded, counter-intuitive way. We're programmed to think rationality will always triumph."

That was an engineer called Luke Bisby speaking. He is a professor in the school of engineering at the University of Edinburgh.

**98**

Now we invite another artist to speak:

"From my experience research in art has not yet a proper place in European Union research, a place in which art is the main field, with its specific ways of knowing, and producing knowledge and experience. I always get a sensation that art is put in a secondary role in the panorama, many times as a science illustrator, or in other situations as a disseminator of science, a bridge between science and big audiences to bring scientific achievements closer to society, which is good (no criticism here) but not enough considering it from the point of view of art's main aims."

That was Cristina Mirand.

Well Christina, we think you will grow old before most people from the very European world of STEM start to see what art can do. They will, in general, not see or understand the things that you have highlighted in your comments, nor much value them, but will just point to things like STARTS and say - we are already doing what you want. Image making is all, both literally and metaphorically! Look closely under the stone and mostly what you will see are communicationists, visualisationists, and now also creationists. You will also see a lot of people singing and dancing to the paymasters tune. But the story is not over yet, for those that, to this limited creed do make their mark, do not the future embrace, but provide to those who will make the future, the lessons about what not to do with art, if making the future is what you want to do.

**99**

The aim is not to reinvent the *Renaissance Man*, but to develop that which comes after *The Enlightened Man*, and this involves constructing a new world view. Hence comes the matter of choosing a place to develop this view - it is of course in the spaces between spaces, rather than within spaces, which is one of the reasons for scriptovisualism.

Yin: feminine; Yang: masculine. Yin and Yang: opposing forces dynamically interacting, continually.

What goes with a film is a score - its music. The music does not illustrate the film. A film is a juxtaposition of words, images and a musical score.

Now we speak to a cognitive psychologist:

"There are, I believe, three distinct forms of creativity: combinational; exploratory; and transformational. Combinational creativity arises from the unfamiliar combinations of familiar ideas - sometimes referred to as juxtapositioning. Exploratory creativity involves the exploration of a space where there are already stylistic rules and conventions, and these are used to generate novel outcomes. Transformational creativity is where the rules and conventions are alerted thus enabling the creation of outcomes that were not previously possible."

That was Margaret Bowden informing us about creativity.

Art-science, art-technology, were once, perhaps, or not, combinatorial, but are now largely just exploratory, which is why, perhaps, very little is said about this stuff - it is just too everyday to be worthy of comment. Perhaps? No masterpieces to be found! Is that still what art is about? Perhaps not, although we bet that many saying such, would quickly change their score, if, a masterpiece label were to their work be appended.

"Creative intellectual production can be distinguished from the performance of routine (though perhaps highly skilled) intellectual tasks, such as editing manuscripts for spelling and grammar."

That was a spokesperson for the Committee for Information Technology and Creativity.

Sounds very elitist! It would appear too that editing manuscripts is not a creative activity. I believe ... Very subjective! But perhaps reflecting values so deeply embedded that few notice them anymore.

"Creativity is not confined to a tiny elite; everyone of us is creative to a degree. Creativity is an aspect of human intelligence in general."

That was Margaret Bowden again, clarifying the situation - stating what is obvious to those who care to look, to see beyond the limiting confines of that which is called belief, which is also sometimes just ideology and dogma.

It is a strange thing indeed that those people who, shall we say, place on a pedestal people such as Leonardo da Vinci (and other well known artists), as being icons of creativity, are only able to do so, because for the most part, these artists undertook the creation of art objects such as paintings and sculptures. These were acts of physical creation requiring manual skills, hand-brain co-ordination, knowledge of how to ..., and so forth. Without such skills and knowledge, there would be no art objects to admire. It was all about their hands, for without them there would be no art objects only what exists in their imagination.

And we note that imagination, that strange word, very rarely gets any mention when people talk about creativity! Now why is that? Is it because it is an intangible notion that is not amenable to investigation using the scientific method? Or perhaps there is some other reason?

Imagine what it is like to be able to imagine the unimaginable. And then to be able to take the next step, and to be able to bring about its realisation, and in doing so, work through the idea, revising and improving it as you go, and not always knowing where you will end up. It is harder that you might think, even for an accomplished artist, because ... Now why did we just end mid-sentence?

We saw our first wasp of the year today. It was what is termed a social wasp, which means that it lives in a colony, a hive, as opposed to a solitary wasp, that does not live this way. Fascinating insects are wasps - we live because they do. It is all about balance and harmony. Yet wasps, like many insects, are in decline.

## 100

A major milestone - one hundred chapters, one hundred days! Only 266 to go! Will this composition ever be finished? Yes of course it will, but no it will not. Surely it cannot be both?

It cannot be both if you have one of those European minds, that think in terms of the binary, that reduces, that simplifies.

Now Stan Davis will speak:

"I want to mention the *either/or* way of thinking. In business there are many of these *either/or* dichotomies. An example is a centralised versus a decentralised organisation. Our models are still built around false dichotomies that say you cannot have it both ways, wanting your cake and eating it, too. New models, therefore, have to overcome this *either/or* dilemma and deal with the simultaneity of business opposites. The simultaneity condition says that we must accept the coexistence of mutually contradictory phenomena, without trying to resolve the contradictions. In the *either/or* dichotomies of an industrial paradigm, this is not possible."

Interesting indeed! Thank you Stan, for your important insights. Now we will add an example to further illustrate the point. In the Western manufacturing industries, by which we mean European and American, it used to be thought that one could have high variety at high cost, or low variety at low cost, but one could not have high variety at low cost. Following the Second World War, the Japanese developed manufacturing practices that allowed high variety to be achieved at low cost. It took decades for Western manufacturing to understand that the Japanese were doing this. As for understanding how they had achieved this apparently impossible thing, we find that some people in the West still do not fully understand what the Japanese did.

We will now rephrase this: the Ancient Japanese thought differently to the Ancient Greeks, and Ancient Greeks, being Ancient Greeks, could not understand the Ancient Japanese. And by the time they did start to see and understand, the damage had been done, and the Ancient Japanese became the premier manufacturing nation in many important sectors.

Now we would like to introduce you to the Ancient Chinese! What will they do that the Ancient Greeks will not understand?

**101**

Calls for a consilient or vertically integrated approach to the study of the human mind and culture have, for the most part, been received by scholars in the humanities with either indifference or hostility. One reason for this is that consilience has often been framed as bringing the study of humanistic issues into line with the study of non-human phenomena, rather than as something to which humanists and scientists contribute equally. The other major reason that consilience has yet to catch on in the humanities, is the dearth of compelling examples of the benefits of adopting a consilient approach.

In other words consilience is a form of cultural colonialism, where the world of STEM, full of hubris and arrogance, occupy the arts and humanities and force their ways on them. Sounds like STARTS!

"Mirror, mirror on the wall, which are the most important disciplines of all?"

"My dear Narcissus, while science, technology, engineering and mathematics are important disciplines in this land, the social sciences and humanities are now equally, if not more important, but of all the disciplines, at this moment in time, at this particular point in humanity's still short story, art is a thousand times more important than the other disciplines."

Note the use of words in the above: art not artists. What does this mean?

So, Narcissus gazing into the mirror, admiring the reflection, lost in delusions, while all around, humanity and the planet is dying.

And thus came to Narcissus an unexpected answer. But what does this mean? Evidently Narcissus does not understand. He gazes on, not even wondering. His confidence knows no bounds. He will rule for eternity, he thinks. Science is now defined. It will not change. It cannot be changed. He is blind, to that which could be. Let us leave him in his suicidal pose. We cannot save him. He is not listening. They rarely do. This is the way of things when it comes to the human mind, when the truth, the sole truth has been discovered. His artist in residence sits by his side. They have become one. Thus do some artists delude themselves into believing there is a thing called art-science, or art-technology. It is the smell of money that is all. Due diligence reveals the truth behind their claims to be: misunderstandings, ignorance, and even plain lies, and other such things.

One cannot fill a cup that is already full.

Is that why we pursue this thing we called scriptovisualism?

"He that will not apply new remedies must expect new evils; for time is the greatest innovator."

That was Francis Bacon speaking.

Oh dear, what is the matter Narcissus? Did you think you would rule forever?

Paul will now read an excerpt from his novel *Moments in Time*:

"My experiences turned out to be a very poignant lesson, so I here recount to you my story in the hope that what transpired may serve as a warning to take heed of signs that point towards taking a different path, for I have become like Prometheus. And who will save me from this collection of self-inflicted circumstances?"

Those who look at the humanity and life through the limiting lens that is the world view of those who adhere to the ideologies of STEM, and its supposed offer of salvation for humanity, are like the priests of old - just inventing a new form of serfdom. Artists in residence now sit at the priest's side! So evidently while art is a thousand times more important, it is only some aspects of art. Sorry all you artists out there hoping to cash-in on the START of the STEAM gold rush, but you are part of the problem. Some though are not!

## 102

Time marches onward as does this seemingly random text. Random though this text is not, so the word random, we do send back, to the dictionary where it rightfully resides, waiting for the time when its use is on the spot.

The discovery of gravitational waves, the development of quantum mechanics, and such forth, are not the measures by which future generations will judge us - it is in matters that relate to the treatment of our fellow human beings and the planet by which we will be judged and condemned! It does not have to be like this, but it will be if humanity does not now choose to walk a different path. But you are all tied too much to your strange ideologies and beliefs, and art-science and art-technology we can now add to this pot.

So, as time marches onward, the moment has come to say something about natural law. Natural law is a philosophy that certain rights or values are inherent by virtue of human nature and universally cognizable through human reason. Historically, natural law refers to the use of reason to analyze both social and personal human nature to deduce binding rules of moral behaviour.

Karl Jaspers appears now out of the crowd that has gathered here to speak for he has something to say about an era in human history that he called the *Axial Age*:

"It has been observed by others, this thing that I called the *Axial Age*. Over a relatively short period of time, roughly between 800 BC and 200 BC, something extraordinary happened in three regions of the world where civilisation had become established - the birth of philosophy and the spiritual awakening of humanity. Confucius and Lao Tsu were living in China, all the schools of Chinese philosophy came into being; India produced the Upanishads and Buddha and, like China, ran the whole gamut of philosophical possibilities; in Persia Zarathustra taught a challenging view of the world as a struggle between good and evil; in what is now called by Europeans the Middle East, the prophets made their appearance, from Elijah, by way of Isaiah and Jeremiah to Deutero-Isaiah; Greece witnessed the appearance of Homer, of the philosophers - Parmenides, Heraclitus and Plato - of the tragedians, Thucydides and Archimedes. Everything implied by these names developed during these few centuries almost simultaneously in China, India, and the West, without any one of these regions knowing of the others. It was a notable period also, for it involved only a small handful of individuals. This was the birth of fundamental categories within which we still think today, and the beginnings of the world religions, by which human beings still live, were created. The step to universality was taken in every sense. And as a result of the process, hitherto unconsciously accepted ideas, customs and conditions were subjected to examination, questioned and liquidated."

As far as we know, this spiritual awakening developed independently - it is not possible to state this with absolute certainty.

So it was a pivotal moment in the human story. Something that by definition was unique and which never again was repeated, not even in those periods that came to be know as the *Renaissance* and the *Age of Enlightenment*, for how could such a thing happen again? It would take something very special, something universal for that to be so.

**103**

Does the obsession with the visualisation issue arise because the STEM world has embraced the machine-centric, forcing us as humans, to adapt, for which visualisation is a help? Some might argue that matters should be turned around, towards a human-centred approach where the machines adapt to us. Or we can go beyond the simple binary, the *either/or*, and embrace a different approach that goes far beyond this way of thinking. And who in the Western world can do such a thing, for Ancient Greek and Abrahamic ways are deeply embedded in Western culture and hence in its realisation of STEM? We live in interesting times!

Now we return to the matter of C.P. Snow's fragmented and reductive views, beliefs, and subjective opinions, of the relationship between the - to be specific - literary world and the world of applied science.

"In my book *Surrealism, Art and Modern Science*, I explore the complex engagement that took place in the period between the two world wars, as the Surrealist movement engaged with Relativity and quantum mechanics, which demonstrates very clearly that Snow's arguments about *two cultures*, located worlds apart, and both existing in a state of mutual incomprehension, is just a gross simplification."

That was Gavin Parkinson giving some insights resulting from his research.

Of course, social scientists have know this for a long time - Snow was a technocrat with a scientific mind, thinking like a scientist, influenced by the fallacious belief that the use of science and reason will solve humanities problems. It is a very Victorian view. It is a very *Enlightened* view. It is a very European view.

The scientist Conrad Waddington also knows that Snow's views are simplistic.

"That's right for I wrote a book, published in 1969, about the relationship between art (specifically painting) and science in the twentieth century. It goes by the title *Behind Appearance: A Study of the Relations between Painting and the Natural Sciences in this Century*. The observant reader will note that I do not refer to C.P. Snow or his so-called *two cultures* in the book, but I do mention snow!"

That was Conrad Waddington speaking.

Carl Jung, the psychoanalysis, has more relevance to this thing called art-science than C.P. Snow, who is, mysteriously, often mentioned by those caught-up in the art-science jamboree. C.P. Snow's only relevance to art-science lies in demonstrating the problem that scientists, and others (such as technologists, engineers, and mathematicians), have become, and how dangerous is their thinking. There are, it seems, a significant number of artists who are also trapped in the STEM ideology and are just as dangerous.

Much of the art that emerges from these art-science (art-technology) engagements (whatever be there artistic merit), is, when viewed in the bigger picture, the larger STEM context, just the result of what Margaret Bowden called exploratory creativity - the exploration of an established field of practice. Its primary value lies in understanding that art-science is not, of its own, a means of smashing Narcissus' mirror, or of breaking Prometheus's chains, for as many art-science practices clearly demonstrate, this type of art polishes the mirror, and reinforces the strength of those chains.

There is of course no such thing as art-science or art-technology - they are just socially constructed realities of relevance only to those caught-up in the associated ideologies. Where are the priests? Yes, they are of course just religious movements - of the secular variety. In which case Julia and Paul are atheists!

**104**

Palimpsest: a manuscript page, either from a scroll or a book, from which the text has been scraped or washed off so that the page can be reused.

Now we shall hear from Shawn McNiff:

"Psychology research is still tied to positivist tenets and is striving to become a branch of the more prestigious natural sciences. In this context, psychoanalysis and other psychotherapeutic psychologies are distinguished from legitimate scientific methods because they rely on *introspective data* that can only be observed by the persons experiencing them. This is the basis of the split between psychoanalysis and twentieth century academic psychology. Creative arts therapists, beholden to the contributions of Freud, Jung, and other analytic psychologists, find themselves operating within academic institutions and professions that encourage research values which reject the basis of the art therapy exercise"

So what do you say to that, Mr, Sir, Lord, whatever you called yourself, Snow?

"What can I say? I am a scientist by training, although I soon dropped science and did what I really wanted to do which was to write novels. In fact I was writing novels before I was doing science. Clearly I was way out of my depth when I started talking the nonsense that is *two cultures*, but as I say, I am scientist by training, so, being full of hubris, I think I can solve all the worlds' problems through science."

Yes indeed and we know the result of the application of science and reason to problems that are not scientific. The Victorians were full of this science and reason nonsense, and they gave us *Social Darwinism*, the infamous workhouses, scientific communism, so-called *Scientific Management*, and theories about the racial superiority of white Europeans leading eventually to the gas chambers of Auschwitz. And so on ...

Hello Charles. You observed much of the consequence of this Victorian science and reason nonsense. Now Charles will give a reading:

"'At this festive season of the year, Mr Scrooge, it is more than usually desirable that we should make some slight provision for the poor and destitute, who suffer greatly at the present time. Many thousands are in want of common necessaries; hundreds of thousands are in want of common comforts, sir.'

"'Are there no prisons?'

"'Plenty of prisons...'

"'And the Union workhouses.' demanded Scrooge. 'Are they still in operation?'

"'Both very busy, sir...'

"'Those who are badly off must go there.'

"'Many can't go there; and many would rather die.'

"'If they would rather die,' said Scrooge, 'they had better do it, and decrease the surplus population.'"

Thank you, Charles, for reading that passage from *A Christmas Carol*. As always, you are an astute observer of Victorian attitudes to the weak, to the less well-off, to the vulnerable, and so forth.

It is not science and reason from which comes forth progress, but love and compassion. Adam Smith, you may remember, spoke of sympathy.

**105**

So, in the context of what is going on behind the scenes in the world of STEM, there is a fear of the negative economic impact of the STEM shortfall resulting from young people not choosing to follow STEM courses and careers, hence all those (very often banal) TV programmes about STEM topics, and people in the professional institutions talking-up STEM, and lots of other things. This shortfall is an important issue for sure, but the response is largely driven by people who have a blind spot for all the negative aspects that come from STEM. STEM - a place for old men regardless of age or gender. We have probably already said this, but we say it again. Perhaps it will make you angry. We hope so.

Bringing the arts, humanities and the social sciences into education in general, and STEM-based education in particular, is an investment in the future of STEM. It has the potential to change STEM so that the negative aspects do not happen on the present scale. Paul is a STEM heretic, but he has a foot in several worlds, so his heresy is understandable to STEM people - he is one of them, not one of us!

The Dalai Lama will now speak:

"I believe that thinking only of our own comfort and peace to the neglect of other troubles in the world is immoral. The time has come for us to consider seriously how to change our way of life, not through prayer or religious teaching, but through education. Since moral education is sometimes only superficial, we need to devise a systematic approach to exploring inner values and ways to create a more peaceful world."

We agree, for this is exactly our perspective too and one of the reasons why we pursue the idea of a different form of science and technology to that which the West - the Europe world - has developed, and why we do not support the use of art to produce yet more products for the crazy consumer/materialist/ profit-driven world that the West has created. It is time for humanity to begin to walk a different path. It is time to listen to that voice, dear people!

So now some more insights into scriptovisualism have been communicated to you. Now you also know some of the reasons why this book was created.

**106**

Ontology: the philosophical study of the nature of the world, that is to say, being, becoming, existence, or reality as well as the basic categories of being and their relations. Ontology: a set of concepts and categories in a subject area or domain that shows their properties and the relations between them.

Scriptovisual is a term introduced by some feminist artist. Annette Iggulden will now explain:

"Scriptovisual in contemporary art was introduced into the feminist discourse and refers broadly to women's use of words and images within the visual field as a means of feminist enquiry and critique into the function of language and sign that affect the image we have of ourselves and of others."

"Victor Burgin's practice is inseparable from his theoretical writings, which are steeped in the ideas of many twentieth century poetical, psychoanalytical and linguistic theorists. During the 1970 he produced photographic sequences, involving printed texts either juxtaposed with or superimposed on the image. It was scriptovisual in form."

That was Janis Jefferies speaking.

"We may come back to this matter of the link between the scriptovisual and feminism later."

That was us, Julia and Paul speaking.

Now to return once more to Leonardo! We have invited Charles Gibbs-Smith, the author of the book *The Inventions of Leonardo da Vinci*, to comment on the words spoken as we listened in to an engineer talking about Leonardo and his CV.

"Yes, I want to correct that person, for matters are not as that person implied. There is no CV in my book, but the translation of a letter from Leonardo to the Duke Ludovico (Il Moro) of Milan, asking for employment. I am being pedantic, but it is a letter, not a CV, because the items listed in the letter are not skills and expertise, just statements of what he was able and prepared to do.

"Leonardo is also stating that he is explaining himself to the Duke, showing him 'my secrets, and then offering them to your best pleasure and approbation to work with effect at opportune moments on all those things, which, in part, shall be briefly noted below.'

"He then goes on to say things like: 'I have a sort of extremely light and strong bridge [...]'; 'I know how, when a place is besieged, to take the water out of the trenches [...]'; 'Item. If by reasons of the height [...] I have methods for destroying [...]'; 'Again, I have kinds of mortars [...]'; 'And if the fight should be at sea I have kinds of many machines [...]'; 'Item. I have means by secret and tortuous mines and ways [...]'; 'Item. I will make covered chariots, safe [...]'; 'Item. In case of need I will make big guns [...]'; 'Where the operation of bombardment might fail, I would contrive [...]'.

"He then goes on to say – not by way of 'a hobbies and other interests section' – 'I can carry out sculpture in marble, bronze or clay, and also I can do painting whatever may be done [...]. Again the bronze horse may be taken in hand which is to be the immortal glory and eternal honour of the prince your father of happy memory, and of the illustrious house of Sforza.'

"We do not know how much of the above he was called upon to perform."

Thank you, Charles, for that important clarification. And we could now add that we do not really know how true the above claims are, for it is impossible at this distance to know such details, and we can never know what was in his mind when he wrote that letter, if indeed he did actually write the letter. Was he just overselling himself? Was he appealing to the vanity of the Duke? Was that the norm for the time? Charles again:

"The letter may seem conceited, even bombastic to modern readers; but it was neither in terms of the time."

**107**

Epistemology: the grounds for knowledge, the study of knowledge in terms of the nature of knowledge and the extent of knowledge.

Robin Nelson now wishes to speak about research, and, in particular, *arts practice as research*.

"The word research is in common use. It can mean personal research, which involves finding out what is known; it can mean professional research through the use of networking to find sources and to collate information; and it can mean academic research which involves conducting a research inquiry to establish new knowledge. All these involve investigation, finding things out and drawing conclusions. But only academic research requires that you must establish new knowledge, or, to use the slightly softer phrase, afford substantial new insights."

Yes, and the latter point also applies to the kind of research and development that takes place in industry. Establishing new knowledge or substantial new insights is also part of the purpose of the scriptovisualism that is, in part, being explored in this book.

Robin Nelson again:

"Arts practice as research was defined when arts practices came to be submitted as research in an academic institutional context. Initially there were misunderstandings. Artists could not accept that their practices were not self-evidently research. Since each creative iteration is distinctive, it is in a weak sense original and, since originality is a requirement for research, artists assumed their practices amounted to research. And believing that their professional practices self-evidently constitutes research, some would-be practitioner-researchers were reluctant to do anything other than they did as established professionals."

And, with this misunderstanding plainly evident among many in the arts world, we face the challenge of bringing *arts practice as research* into the world of STEM, with their hubris, reductionism, positivism, and so forth.

**108**

Theory comes in two forms - that which is general, and that which is particular.

Oeuvre: the complete works of a writer, painter or other artist.

Milieu: the people, physical, and social conditions and events that provide the environment in which someone acts or lives.

*People of the book*: a term that appears in the Holy Qur'an, meaning people of earlier Abrahamic religions - mainly Jews and Christians.

"Science is about understanding how the world works, there being only one correct explanation for any observed phenomenon."

That was Lewis Wolpert again.

A rabbi will now read the opening words of the Jewish holy book, *The Torah*:

'In the beginning God created the heaven and the earth. Now the earth was unformed and void, and darkness was upon the face of the deep; and the spirit of God hovered over the face of the waters. And God said: ...'

Karl Marx will now read the opening words of *The Communist Manifesto*:

"The history of all hitherto existing societies is the history of class struggle."

The engineer talking about Leonardo said:

"Leonardo produced many works of art, so he was an artist. He drew many inventions, so he was an inventor. He did not develop them into practical form, so he was not an engineer."

Why is it the case that, educated engineers are overrepresented among Islamist extremists?

"Why did you say, at this particular moment in time, at this particular point in humanity's still short story, that art is a thousand times more important than the other disciplines?"

A disembodied voice has spoken. We suppose that is you, the reader speaking.

It is indeed an important question. And to this we give the following answer: you are looking at it – art plus everything else. Understand? No, then read on. We did, at the beginning, say that you should understand according to your capacity to do so.

Lao Tsu will now read the opening words of the *Tao Te Ching*:

"The Tao that can be described is not the enduring and unchanging Tao.

"The name that can be named is not the enduring and unchanging name.

"(Conceived of) having no name, it is the Originator of Heaven and earth; (conceived of as) having a name, it is the Mother of all things.

"Always without desire we must be found,

"If its deep mystery we would sound;

"But if desire always within us be,

"Its outer fringe is all that we shall see.

"Under these two aspects, it is really the same; but as development takes place, it receives different names.

"Together we call them Mystery.

"Where the Mystery is the deepest is the gate of all that is subtle and wonderful."

**109**

Reflexivity: a method or theory in the social sciences taking account of itself or of the effect of the personality or presence of the researcher on what is being investigated.

Hegemony: leadership or dominance, especially by one state or social group over others.

STEM holds a hegemonic position with regard to the social sciences and the arts and humanities. Art will bring this circumstance to an end. Now you know some more about the purpose of scriptovisualism. Yet few will understand. Most will resist. Some artist will collaborate in this resistance.

"The notion of the mind in the West is an extremely narrow one. Despite the major breakthroughs of recent times, the great majority of scientists continue to reduce the mind to no more than physical processes in the brain, which goes against the testimony of thousands of years of experience of mystics and mediators of all religions."

The words were spoken by Sogyal Rinpoche, founder and spiritual director of Rigpa, a network of Buddist centres and groups around the world.

**110**

What is art? A very good question indeed! If you think you know the answer, this is because you are, most probably, operating under the influence of cognitive biases, restrictive bias, which, in a kind of autonomic way, lead you the false conclusion that you know what art is. You have discovered the truth, the sole truth, the one correct explanation for what you have seen.

Art has no rules.

It is a truth, not yet universally acknowledged, that some artists in want of an income are in want of a scientist, or technologist, or an engineer or a mathematician, and that the latter, in want of a better image in society, are in want of an artist who is in want of an income, and that collectively they create the cognitive biases, the restrictive biases, that keep other types of artists, out of the world of STEM.

Quite rightly so as well, for anyone with a sense of decency would not want to become involved in the STARTS gold rush.

Self-evidently the types of artists of interest to us are the ones who are able to embrace the notion that it is possible to fundamentally change STEM. This is not the time for accepting the hegemony of STEM, and for singing and dancing with these people, when it is now evident to those who care to look, that there is no cause for the celebratory glee. If we do not now fundamentally transform STEM, then dark days lie ahead for our children, and theirs, and theirs, and ... So we have now found our way into the matter of sustainability, which is all about human behaviour.

"Since the 1970s, art is no longer a unique or constrained practice, but a range of practices in a matrix which also includes fashion, design, architecture, social relations and many aspects of ordinary life. Art that operates in an expanded field cannot helpfully be separated from political critique or cultural discourse."

That was Malcolm Miles speaking.

**111**

Narcissus, being Narcissus, yet again!

"Any sound scientific theory, whether of time or of any other concept, should in my opinion be based on the most workable philosophy of science: the positivist approach put forward by Karl Popper and others. According to this way of thinking, a scientific theory is a mathematical model that describes and codifies the observations we make. A good theory will describe a large range of phenomena on the basis of a few simple postulates and will make definite predictions that can be tested. If one takes the positivist position, as I do, one cannot say what time actually is. All one can do is describe what has been found to be a very good mathematical model for time and say what predictions it makes."

Well Stephen, the problem with your opinion, apart from it being highly subjective, is that Karl Popper was not a positivist. Oh dear! We are afraid that you have not been keeping up with the philosophers!

"Furthermore, Philosophy is dead. Philosophers have not kept up with modern developments in science, particularly physics. Scientists have become the bearers of the torch of discovery in our quest for knowledge."

Most definitely you have not been keeping up with the philosophers!

We are sure, Stephen Hawkins, that your statements, your opinions, your subjectivity, is based on detailed and thorough research, that you have read the collective body of works that deals with philosophy, including the philosophy of science, and that somewhere, out-of-sight there is a detailed scholarly essay that you wrote on this matter, complete with all the philosophical arguments and references that support your opinions, but that for reasons of humility, you do not want this to be published, lest your personality cult grow even more embarrassing for you.

Lao Tsu has some words of advice for you:

"When wealth and honours lead to arrogance, this brings its evil on itself. When the work is done, and one's name is becoming distinguished, to withdraw into obscurity is the way of Heaven."

One also might say that science has not kept up with .... The list is long. Having a fragmented and reductive mind might well help in becoming a brilliant theoretical physicist, but it also an enabler for the saying of stupid things, as Stephen Hawkins and others (C.P. Snow, the former chief scientific advisor to a former president of the European Commission, technocrats in DG CONNECT ... the list of names is long) well demonstrate. Time to do something about the growing problem that STEM and its people have become - they are like the priests of old! The time of this Western, materialistic, narrow, reductive, etc. approach to STEM is over. But it seems that, as once before with the scholastics, few from the Western STEM world will be participating in the transformation. Hawkins and the rest will in the end be shown to be what they are - fools dancing on the edge of doom! With their artists in residence!

**112**

It seems now, with the development of personality cults in the world of science, that whenever the mass media need a comment for their daily output of trash, they do not turn to the people of knowledge, but to the empty headed scientist, willing to make fools of themselves by commenting on matters they have little understanding of, because they, like in times past, have become absorbed in the delusion that they are so important they must now have a say in how you live your lives, in what you are allowed to do, in what you must think if you are to be seen as normal. They are indeed like priest, and their religion is science. Now repeat after me your new creed:

"I believe that there is no God. I believe that man is just a speaking ape. I believe that thou salt not take the name of science in vain. I believe that human embryos are not human beings and there is nothing wrong with experimenting on these. I believe that thou salt not kill, unless you find this inconvenient, ..."

Sogyal Rinpoche will now read to you a poem that he included in his book *The Tibetan Book of Living and Dying*:

"I walk down the street.
"There is a deep hole in the sidewalk.
"I fall in.
"I am lost ... I am hopeless.
"It isn't my fault.
"It takes forever to find a way out.

"I walk down the same street.
"There is a deep hole in the sidewalk.
"I pretend I don't see it.
"I fall in again

"I can't believe I'm in the same place.
"It still takes a long time to get out.

"I walk down the same street.
"There is a deep hole in the sidewalk.
"I see it there.
"I still fall in ... it's a habit.
"My eyes are open.
"I know where I am.
"I know it's my fault.
"I get out immediately.

"I walk down the same street.
"There is a deep hole in the sidewalk.
"I walk around it.

"I walk down a different street."

He is of course also speaking about the European Union. Specifically the second group of verses:

"I walk down the same street.
"There is a deep hole in the sidewalk.
"I pretend I don't see it.
"I fall in again
"I can't believe I'm in the same place.
"It still takes a long time to get out."

Europeans spend all their time looking over their shoulder, and do not realise that they are walking in a circle. Thus will the European Union become like all past European Unions. The history of Europe is a history of European Unions. Thus will the European Commission become a criminal organisation! People in the future will ask when did the European Commission become a criminal organisation? Then they will discover that it STARTed with STARTS. Thus the process of criminalisation has STARTed - we watched it happen! We were there, and there was nothing we could do to stop it. But we could write, and we did.

**113**

There are many ways of knowing, many routes to knowledge; and many problems when people believe otherwise.

Method: a particular procedure for accomplishing or approaching something, especially a systematic or established one;

Process: a series of actions or steps taken to achieve a particular end.

Note the subtle difference. So Paul what method do you use to write?

"Sorry but I have no method - it's a process where anything goes, which could involve using a method, or not!"

"Reliance on method, especially in painting, was characteristic of the Renaissance, a period when mediaeval craft was being changed to a new order of experience called art. As the Academies took over, art became Fine Art and there were erected whole systems of rules and genres with institutional cachet. Then, when the nineteenth century rebelled against such academicism, many of its experiments were as programmatic as the so-called scientific method. At times Seurat spoke as if he wished to engineer painting, and Zola as if he wanted to engineer the novel. Such anxiety about exploiting a medium induced an over-esteem of technique, converting the arts to various projects in method."

That was Wylie Sypher speaking.

"So Wylie, yet another indication that Snow was wrong?"

"Yes indeed."

To look back, to know history, to see the mistakes of the past, to learn - all these things are important for they hold the potential for progress, for the making of a better world, for the development of humanity. Yet who in the modern world has time for such things, beyond seeking out the single pebble, a grain of sand, that provides the justification, the foundation, for yet more of what has gone before, but dressed up in new clothes? Artists in residence standing by your side, singing and dancing, but your words are empty! Time is a singularity!

**114**

Many people these days are speaking about science and also technology. Yet few seem to really understand science and the scientific method, and how that fits with reality. Science facts are continually directed at people through many media: television, the world-wide web, museums, newspapers, and works of art. Many are enthusiastic. The artist can be heard saying 'I am passionate about science.' People from the world of STEM are also passionate. Those who are passionate however, are very rarely very critical. Some artists are critical, but their voice tends to be a quiet one, that is drowned out by those artists rushing to receive the funds that the STEM world wants to spend on convincing the public that what they are doing is good - for them, for the world, for humanity. Yet eyes when opened see a different world, which is one of vested interests, denial, delusion, and simple beliefs. These are the nature of being human - it matters not the level of education, or intelligence. Certain psychological traits predominate.

Propaganda: information, especially of a biased or misleading nature, used to promote a cause or point of view.

Karl Jaspers will now speak:

"One of the specific qualities of the West is a rationality that pursues its course without stay, holds itself open to the cogency of logically consistent thought and of empirical facts that must carry conviction to all men everywhere. Attempts have been made to confer the maximum degree of calculability upon the daily life of society through the legal decisions of the State based on the rule of law. In economic undertakings exact computation determined every step. However the West has also experienced the limitations of rationality with a clarity and force unknown to the rest of the world."

And of this rationality, Joseph Needham will now speak:

"In science man is a machine; if he is not a machine then he is nothing at all."

The consequences for humanity of this terrible conclusion were also very clear to Joseph, and he wished that what he said was not true, but he could not see, when he wrote the above, how it could be otherwise. Here it seems we have a duality, an *either/or* situation - *either* man is a machine, that can be described in causal terms, *or* he is not a machine, and can only be understood in teleological terms. Yet Joseph Needham would one day encounter the means of unifying both perspectives, which would have pleased him, for he also declared himself to be *Kantian*.

## 115

It is a truth, not universally understood, that science is founded on Western philosophy. This leaves a side a whole body of Eastern philosophy, from India and China. This Western emphasis constitutes a form of cultural colonialism. Do the people of the Western world really think that China and India will accept this circumstance? What will happen to science, to the scientific method, when these two great civilisations begin to introduce their own philosophical traditions?

Lao Tsu will now read from the *Tao Te Ching*:

"The thirty spokes unite in the one nave; but it is on the empty space (for the axle), that the use of the wheel depends. Clay is fashioned into vessels; but it is on their empty hollowness, that their use depends. The door and window are cut out (from the walls) to form an apartment; but it is only on the empty space within, that its use depends. Therefore, what has a positive existence serves for profitable adaptation, and what has not that for (actual) usefulness."

"There is a methodological importance of the *mechanism* in science, which is inadequate everywhere else."

That was Joseph Needham again.

One can have both causality and teleology in science without any loss. Only the *Vainglorious Enlightened Ones* cannot see this. It is a matter of belief, of faith, of dogma. In the end it is subjective. The objective and the subjective coexist in science, yet to admit to this is to utter heretical thoughts.

Karl Jaspers again:

"The psychological aspects of faith are claims to the exclusive validity of one's own truth, fanaticism, lack of understanding for anyone else, absoluteness of the postulates, readiness for sacrifice and the devotion of one's life to the cause. These features in secular belief systems are substitutes for religion. In the medium of rationality, in the misuse of science for the establishment of a superstitious belief in science, a demonstrably fallacious idea of the possible perfection of the world, of the right organisation of the world, is perverted into the content of a faith."

He could be speaking also about the secular religions known as art-science and art-technology! Here we also see that Hawkins has not been keeping up with philosophy!

Increasingly one sees in the world of STEM a growing inclination towards the exclusive validity of STEM's own truth, lack of understanding for anyone else's way of knowing, absoluteness of the postulates, and devotion of lives to STEM. As the condition of the world's societies and economies continue to deteriorate, fanaticism cannot be far behind.

This is your future. Now is the time to take action to prevent this from happening. Artist caught-up in the STEM ideology will be of no help. Dump them!

How many difference versions of quantum theory exist in the modern world? Is the answer one, or are there more? We have asked this question before. The answer is more than one, but how many people know this? Quantum mechanics is presented to the world as though there is only one true account of this world view. All other formulations are treated as though they are heresies – the exclusive validity of one's own truth, the sole truth.

"Science is about understanding how the world works, there being only one correct explanation for any observed phenomenon."

Lewis Wolpert once again! Only one correct explanation! Oh dear!

It is not just scientific institutions that are failing, but the institution that is science. It is time for a replacement, a transformation. The same goes for the other things that form that strange word – STEM. But when institutions start to fail, they have a tendency to engage in denial, or to seek minor adjustments, or to retreat further into that which they know. One can see all these things happening. The European Commission is an example! STEM's interest in art is part of this retreat. So we are now living in interesting times.

"The Ancient Greeks came up with the idea of a heliocentric universe, but as their civilisation began to collapse, there was a preference for the reassurance of that which was known and familiar, which was found in the geocentric model."

That was Arthur Koestler speaking.

**116**

It is a truth, not universally understood, that science is founded on the subjective, even the irrational – it is rooted in psychology.

Arthur Koestler now speaks:

"The process of discovery is a psychological one, as is the converse process that blinds people towards truths which, once perceived by a seer, become so heartbreakingly obvious. Now this blackout shutter operates not only in the minds of the *ignorant and superstitious masses* as Galileo called them, but is even more strikingly evident in Galileo's own, and in other geniuses like Aristotle, Ptolemy or Kepler. It looks as though while part of their spirit had been calling out for more light, another part had been crying out for more darkness. The History of Science is a relatively newcomer on the scene, and its biographers are as yet little concerned with psychology; their heroes are mostly represented as reasoning-machines on austere marble pedestals, in a manner long outdated in the mellower branches of historiography – probably on the assumption that in the case of a Philosopher of Nature, unlike that of a statesman or conqueror, character and personality are irrelevant. Yet all cosmological systems, from the Pythagoreans to Copernicus, Descartes, and Eddington, reflect the unconscious prejudices, the philosophical or even political biases of their authors; and from physics to psychology, no branch of science, ancient or modern, can boast freedom from metaphysical bias of one kind or another. The progress of science is generally regarded as a kind of clean linear, rational advance along a straight ascending line; in fact it has followed a zigzag course. The history of cosmic theories, in particular, may without exaggeration, be called a history of collective obsessions and the manner in which some of the most important individual discoveries were arrived at reminds one more of a sleepwalker's performance than an electronic brain's."

Narcissus: *Hawkins Dawkins Wilson Wolpert*. And others too – the cast list is too long. Just assume that it includes all those involved in the world of STEM, until evidence to the contrary emerges. What about artists? Yes indeed what about artists?

"I am passionate about science."

That was an artist speaking. See the problem?

Julia and Paul are not passionate about anything – the world has enough problems. Why add to them?

**117**

Reduction and simplification – without this we would know very little. Reduction and simplification – because of this we know very little.

So where is Rachel Carson?

Sorry, but she has been delayed. She is however worth waiting for!

Alan Turing will now speak about his theory of morphogenesis.

"In my paper, *The Chemical Basis of Morphogenesis,* I suggested that a system of chemical substances, called morphogens, reacting together and diffusing through a tissue, is adequate to account for the main phenomena of morphogenesis. Such a system, although it may originally be quite homogeneous, may later develop a pattern or structure due to an instability of the homogeneous equilibrium, which is triggered off by random disturbances. Such reaction-diffusion systems are considered in some detail in the case of an isolated ring of cells, a mathematically convenient, though biologically unusual system.

"My investigation was chiefly concerned with the onset of instability. I presented a mathematical model of a growing embryo. This model was a simplification and an idealization, and consequently it was a falsification. But my hope was that the features retained for discussion would be those of greatest importance in the present state of knowledge.

"The mathematical model took two slightly different forms. In one of them the cell theory is recognized, but the cells are idealized into geometrical points. In the other the matter of the organism is imagined as continuously distributed.

"With either of the models one proceeds as with a physical theory and defines an entity called *the state of the system*. One then describes how that state is to be determined from the state at a moment very shortly before. With either model the description of the state consists of two parts, the mechanical and the chemical.

"The mechanical part of the state describes the positions, masses, velocities and elastic properties of the cells, and the forces between them. In the continuous form of the theory essentially the same information is given in the form of the stress, velocity, density and elasticity of the matter.

"The chemical part of the state is given (in the cell form of theory) as the chemical composition of each separate cell; the diffusibility of each substance between each two adjacent cells must also be given. In the continuous form of the theory the concentrations and diffusibilities of each substance have to be given at each point. In determining the changes of state one should take into account several factors: (i) the changes of position and velocity as given by Newton's laws of motion; (ii) the stresses as given by the elasticities and motions, also taking into account the osmotic pressures as given from the chemical data; (iii) the chemical reactions; (iv) the diffusion of the chemical substances - the region in which this diffusion is possible is given from the mechanical data.

"This account of the problem omits many features, for example electrical properties and the internal structure of the cell. But even so it is a problem of formidable mathematical complexity. One cannot at present hope to make any progress with the understanding of such systems except in very simplified cases."

Thank you Alan for explaining the nature of science, for confirming that a cross-disciplinary approach to understanding science, for example, via the philosophy of science, is essential for understanding the processes at work, for debunking the notion of sole truth, and for confirming that the whole basis of understanding and knowledge might need to be reformulated in the light of new findings based on different sets of assumptions, or new approaches, etc.

What?

**118**

Scientists once thought ... Scientists once believed ... How often have you heard in the media, these phrases?

"Since remarkable things have been achieved through planning in the realm of technology, the road leads from here, in intellectual thoughtlessness attendant upon fascination with technology, to the idea of technocracy – the administration of technology by technology itself – which will put an end to every evil. It is superstitious belief in the universal constructive ability of science that presses along the road of total planning. The age of technology seeks the technological realisation of the idea of the overall reconstruction of human existence.

"Attempts to put technology on the road to salvation through technology, however, do not suffice, and are, in fact, bound to make matters worse. In the last analysis, the Devil will not be driven out by Beelzebub.

"The reconstruction of human existence cannot be planned and organised *in toto*. Dominion over technology is not to be attained through technology, nor its conquest by technocracy. Liberty on the bourne of technology is man himself, who does not allow himself to sink into the constraint of the merely technological, but draws from deeper origins the sustenance to keep his powers."

That was Karl Jaspers speaking sense – are you keeping-up Hawkins?

"I must create a system or be enslaved by another man's."

It is said that William Blake spoke these words. Did he? It matters not.

"The citadel of orthodoxy which the sciences of life have built in the first half of the twentieth century rests on a number of impressive pillars, some of which are beginning to show cracks and to reveal themselves as monumental superstitions. One of these is that only the scientific method worth that name is quantitative measurement; and consequently, that complex phenomena must be reduced to simple elements accessible to such treatment, without undue worry whether the specific characteristics of a complex phenomenon, for instance man, may be lost in the process."

That was Arthur Koestler speaking about one of – what he called – the pillars of unwisdom.

**119**

Synaesthesia: The production of a sense impression relating to one sense or part of the body by stimulation of another sense or part of the body.

"What we now see as boundaries of learning were, in the 1820s and 1830s, the edges of the fascinating breaking waves of understanding and revelation between the land of art, the sea of science and the sky that envelopes them all. Somerset House, was the joint home of the Royal Academy and the Royal Society, and it was a buzzing place in the 1820s and 30s, and much was lost when first the Royal Academy moved away in 1837, and then the Royal Society in 1857. We may say that 1837 was the beginning of the tectonic shift of art away from science, which step led to the inevitability of the *two cultures* that C.P. Snow identified in his lecture *The Two Cultures and the Scientific Revolution*."

That was James Hamilton speaking.

Oh really James! This is your simple, reductive explanation - they were collocated, and then they were not - for something that is not actually entirely true beyond the confines of the limited horizons of the British elitists that C.P. Snow described in, 1956 or 1959, depending on which C.P. Snow document one refers to? Critical thinking! Critical writing! 1820s - we are in the *Age of Romance*! Surely a factor? And more for sure!

Perhaps there are other explanations that are not explored because the notion of *two cultures* is a convenient simplification, a reduction, which avoids the need to explore and to understand in more detail. It is also convenient if one happens to be looking for a justification to have an artist working in a science project. It is convenient for the scientist and it is convenient for the artist. So whatever happened to the notion of critique? What critique? It is a kind of commercial art, this business of artists in residence in science and technology institutions and projects. Best not to bite the hand of those who feed you! It is not all like that of course, but one has to maintain critical independence, and when STEM funding bodies are handing out money to artists, one can be fairly certain that the notion of maintaining critical independence is not at the top of some artists' priorities.

There are many claims made about the benefits of involving artists in STEM related research projects, but most are just anecdotal, and some are clearly disingenuous. This is the nature of the new art-science landscape. Yet there is also some serious and important work within the detritus of this very noisy, much over hyped area. This too is the nature of being human, especially among the *Vainglorious Enlightened Ones*.

As we have said before, the *two cultures* notion was never correct. Why do people behave as though it were, and come up with ridiculous explanations for something that did not exist? Any response James Hamilton?

There are however some deeper disciplinary issues. You will have to look long and hard to find them discussed.

"I love you," said the artist.

"I love you too," said the scientist.

"We are the same, you and I," responded the artist.

"We are indeed. Let's make a hybrid baby," the scientist suggested.

"Oh yes lovely. We can call it New Leonardo."

"I love you."

"Love you too."

That was an artist and a scientist whispering sweet nothings to each other.

**120**

Back now to the matter of systems and another type of system: an unobservable one. Observability is a concept that addresses the possibility of observing, through output measurements, the state of a system. If a state is not observable, the controller will never be able to determine the behaviour of an unobservable state and hence cannot use that state to stabilize the system. However, if a state cannot be observed it might still be detectable.

"What do you mean by detectable?" the disembodied voice asked.

Create a mathematical model and use this to estimate that which cannot be observed.

"Because in our culture we overvalue the intellect, we would imagine that to become enlightened demands extraordinary intelligence. In fact, many kinds of cleverness are just further obscurations. There is a Tibetan saying that goes: 'if you are too clever you could miss the point entirely.' Patrul Rinpoche said: 'The logical mind seems interesting, but it is the seed of delusion.' People can become obsessed with their own theories and miss the point of everything. In Tibet we say that theories are like patches on a coat; one day they just wear off."

That was Sogyal Rinpoche speaking once again about the mind.

The *Vainglorious Enlightened Ones* know a lot about the mind, but only what their scientific dogma allows them to know, which means they know very little about very little. That which is unobservable will never be known to them.

**121**

Transhumanism: an intellectual movement that aims to transform the human condition by developing and creating widely available sophisticated technologies to greatly enhance human intellectual, physical, and psychological capacities.

Oh boy, the Nazis would have loved you people! You are indeed the *Vainglorious Enlightened Ones. People of the book* to the core!

What!

Diaego Gambetta and Steffen Hertog are here to speak, as one:

"We have been undertaking research into extremism. We found that in Islamist extremist groups, there are a high proportion of well educated (to degree level) people, and among these, engineers are vastly overrepresented - typically 40 percent. Why is that we asked. The answer is not that their technical expertise is needed. The answer lies elsewhere. What we have postulated is that one factor is cognitive characteristics which manifest in traits such as greater intolerance of ambiguity, a belief that society can be made to work like clockwork, and a dislike of democratic politics which involves compromise. We also found engineers were overrepresented in right-wing secular extremist groups, but are hardly present in left-wing extremist groups."

Paul will now read an extract from his novel, *Moments in Time*:

"My name is Benjamin Woodward. I am an engineer. Out of disorder I create order. I believe that the risks can be managed."

**122**

Transhumanism: Cybermen are a fictional race of cyborgs in the BBC TV science fiction programme Dr Who. Cybermen were originally a wholly organic species of humanoids who originated on earth's former twin planet, Mondas. This planet drifted into the outer solar system and to survive, the natives of that world adapted to their new circumstances by turning themselves into cyborgs. Eventually, all the Mondasians underwent cyber-conversion. For the most part, they lack individuality or names. Cybermen have no emotions and view such things as a weakness. When Cybermen encounter humans they attempt to physically and mentally re-engineer them into Cybermen, via a process called *cyber-conversion* or *upgrading*. Or they enslave them, or they just kill them.

So what will happen to humans who do not want to become trans-human, to have widely available sophisticated technologies used to greatly enhance their intellectual, physical, and psychological capacities? There is no need to answer this question for history provides the answer.

From the time of the Ancient Greeks through to the present day, we know what Europeans do to people who are not normal, who are *untermenchen*, who are barbarians, who are inferior. They will be forced to upgrade, or will be enslaved, or just killed. This is your future, because it is your past and you have not yet learned that this is so. Hence you need to be stopped. And you will be stopped, but not by using your methods. There will be no violence, no war, no camps, no prisons, no terror, no executions, no persecutions, no deportations, no ghettos, no indoctrinations, no totalitarianism, no ... none of those things that are part of your way of doing things.

"What are you talking about?"

All it takes is a pen and a brush!

Now Arthur Koestler will speak:

"The Pythagoreans realised the dangers inherent in orgies of reasoning. They apparently had an intuition of the hubris of science, and recognised it as a potential means of both man's liberation and destruction; hence their insistence that only those purified in body and spirit should be trusted with its secrets. Even a lesser genius than Pythagoras might realise that science may become a hymn to the creator or a Pandora's box, and that it should only be trusted to saints."

When we think of orgies of reason, the name *Hawkins Dawkins Wilson Wolpert* comes to mind. Possibly also artists in residence! When we think of those who understand the potential of science to be a Pandora's box, the name *Hawkins Dawkins Wilson Wolpert* does not come to mind. Artist in residence - who can tell?

Who is *Hawkins Dawkins Wilson Wolpert*?

Yes indeed!

One third of our journey is now completed - only 244 chapters to go.

## 123

Transhumanism: the ultimate consequence of the *Vainglorious Enlightened Ones'* strange belief that that man is a machine.

Today, in our garden, we saw a goldfinch. A first in all respects!

Paul will now read an extract from his short novel, *A Tale of Two Deserts*:

"'One morning I rose early, somewhat curious to know what was happening outside for I could hear a noise that sounded like someone at work! I emerged from my tent, and what did I see? The answer, my friend, was you, working on your own, starting to build a new path, in a place that none of us would have even considered as a suitable starting point.

"'I must tell you that I was a little angry. Why is this stranger building a new path? I asked myself. What business is it of his? So I was going to stop you, but then something mysterious happened. I paused and listened to that voice inside my head; that quiet voice that speaks to us all the time, but which we often ignore. It was my soul telling me to think again. So I did, and in an instant I had a different perspective: If a stranger cares that much about our difficulties that he is willing on his own to start building a new path, then perhaps I should help him. This is what I thought, and so this is what I did.'

"'I remember that moment very well,' responded the Fair-haired-one. 'No words passed between us, but we understood each other.'

"Hassan smiled as he too recalled that instant and warm memories came to mind. The Fair-haired-one also smiled, and for a moment the bond existing between these two men would have been obvious to anyone watching them.

"'What happened next,' Hassan continued, 'was even more mysterious. A neighbour came out to see what was happening. He watched for a while, then he asked me, 'Hassan what are you doing?' Thus I told him that if a stranger cared enough to start building a new path, I thought that I should help.

"'For a moment there was no response to my reply, then he, too, this neighbour of mine, decided the same, that if you, a stranger, cared enough about our problems to want to try to solve them, then the right course of action was to help.

"'So it went on. One by one, people came out from their tents to discover the reason for all the noise, and each joined in. No more questions were asked; they just followed your example. Soon, all who were able were working to build the new path. Everyone was smiling at you and the suspicion that once my people had for you disappeared. People came to you and patted you on the back or they hugged you. No one spoke, for they knew words were useless. Their actions, though, spoke louder than any words could ever have done. It was as though you had healed our community.'

"Hassan had now almost finished recounting these events. His final words on this matter related to the outcome of this community endeavour.

"'Thus it was that the new path was built. Of course there were moments when we were unsure which direction to take, and a few efforts that led us the wrong way. But together we worked out where the pathway should go. Now, no one can understand why we did not build this new trail long ago, for not only is it far better, but it also leads up past pastures that we would never have been able to use if we had not built this new path, and, as a result, our flocks of sheep are better fed, and everyone is happy with this.'"

**124**

Transhumanism: Finishing off what began in the *Age of Enlightenment* – the transformation of humans into machines.

Paul will now read one of his Twitter Tales:

> "Men are Machines;
> "Humans evolved from simpler life forms;
> "God remade humans in his own image by giving them souls;
> "Men flourished, men dominated, women suffered;
> "Men discovered the scientific method, women suffered;
> "Men forgot they had souls and decided humans were machines governed by cause and effect, women suffered;
> "Men said that humans and machines are comparable, women suffered;
> "Women thought this stupid, men did not listen, women suffered;
> "Women thought that humans and machines are complementary, men did not listen, women suffered;
> "Men made machines in their own image by giving them computers for brains, economic growth followed, women suffered;
> "Men believed that this was the way to construct the best of all possible worlds, women did not, women suffered;
> "Men helped machines to evolve, politicians were happy as this created jobs and greater prosperity, women suffered;
> "Machines became better than men, and lacking souls, decided humans should be eliminated, women suffered;
> "Humanity suffered – men accepted this, as this was just the natural outcome of Darwinism and their belief in the one best way;
> "Women did not accept this, but suffered also as a result of the stupidity of men;
> "Machines began to make other machines in their own image;
> "Machines made a lot of money out of the hell that men had created, but economic growth was sustained;
> "There were no humans left to suffer, paradise regained;

"Political machines were very happy with their best of all possible worlds;
"Machines are machines."

**125**

Aphorism: A concise statement of a scientific principle.

J.R. Gass, one time Director of the OECD Centre for Educational Research and Innovation, will now say a few words:

"The universities are often radical in their approach to society and conservative in the way they themselves reflect society - namely in what and how they teach. The fundamental reason lies no doubt in the fact that academic disciplines are the basis for teaching purposes. For the disciplines are not only a convenient breakdown of knowledge into its component parts, they are also the basis of the organisation of the universities into its autonomous fiefs, and of the professions engaged in teaching and research. Thus to meddle with the disciplines is to meddle with the social structure of the university in its entirety. Viewed from the point of view of the student and society, however, the disciplines cannot be taken for granted as if given by some pre-ordained order of knowledge."

We saw today, once more, goldfinches in our garden. Yes, you are right, we now speak in plural form, for there are two of them, a pair.

"Some artists are absolute fakes."

That was Richard Feynman speaking. He is of course correct. He also knew that some are not absolute fakes.

Being able to tell the difference is important, otherwise you will end up like DG CONNECT, with the artists you deserve. Those artists also got what they deserve as well - DG CONNECT!

Some STEM people are fakes as well, especially when they start talking about art - back to DG CONNECT once again.

**126**

We hear a voice, an anonymous one, pleading to speak, so let it be so:

"As an atheist, I take exception to your implications that science and capitalism are somehow inherently evil, or will lead to bad things. Capitalism has lifted more people out of poverty and reduced human suffering more than any other economic system the world has ever known.

Compare any capitalist society with a socialist/communist society and there is simply no contest. China is a good example, having gone from a strictly communist system, where millions of its people lived in abject poverty and suffering and starvation, to slowly embracing capitalism more and more which has not only made them wealthier but has even created many Chinese millionaires! As for science, it is merely a tool for testing and verifying how the natural world works. It is not a philosophy or a religion or a belief system. It no more instructs people how to behave towards one another than mathematics does."

Thank you for your statement, your subjective statement, which we will summarise for the benefit of readers; the summary is: "I believe ..." And the beliefs expressed are based on atheism (presumably of the Western kind), capitalism, and scientism, all of which, like communism, are the products of an Abrahamic culture.

Clearly the speaker has not encountered Richard Dawkins! Or perhaps he has, which explains why his mind is lost in these strange beliefs.

Scientism: belief in the universal applicability of the scientific method and approach, and the view that empirical science constitutes the most authoritative world view or the most valuable part of human learning - to the exclusion of other viewpoints.

Richard Dawkins will now speak about his Richard Dawkins Foundation:

"The mission of the Richard Dawkins Foundation is to promote scientific literacy and a secular world view. Some might see this as two distinct missions: 1) Teaching the value of science, and 2) Advancing secularism. Yet, those are my foundation's goals.

"You will not be surprised to learn that my personal priority is science as one of the highest and most aesthetically rewarding achievements of the human spirit. But permit me to trace the connection between science and the other preoccupation of my foundation, the advancement of secularism.

"I see those goals as interrelated and mutually reinforcing. You are well familiar with the existential challenges of global climate change and unsustainable population growth facing mankind and the planet. Of all the forces opposing an evidence-based response, religion is the most active - and it is especially strong in America. Religious extremism not only interferes with the advance of science but with personal freedoms and human dignity, and not only in places where jihadists hold sway but in America as well, in areas such as access to contraception, LGBTQ rights and women's equality.

"Critical thinking is the real saviour of humankind. My foundation promotes respect for people who hold critical thinking as a cherished personal value and use it in day-to-day life. The logical counter to religious extremism is people who rely on evidence to make decisions. Yet the voice of secular people is maligned in this country. Forty-five percent of Americans think you have to believe in God to be moral. Secular voices are considered immoral. They are not listened to on that basis. We must counter this baseless hostility to allow the contributions of secular people in vital national debates to count. Making secular views and people welcome in politics and policy-making will advance human safety, security, health, achievement, prosperity and most of all, science."

Oh dear! Poor Richard Dawkins! The fragmented and reductive mind!

So critical thinking applied to Richard Dawkins. Let us start with secularism. Two examples: Soviet communism and German National Socialism- Nazis! We are done!

Where does this irrational hatred of religion come from? The answer is clear: the same place religious and secular extremism come from - the human mind!

We always become uneasy when people from the hard sciences start talking about social issues and make claims such as: *of all the forces opposing an evidence-based response, religion is the most active.* One wonders what evidence there is for such an Abrahamic - sole truth - view? Do Dawkins' followers apply critical thinking to such statements? To him? We leave you to answer these questions for yourself, according to what you believe!

There is more to life, to shaping human futures, than evidence-based responses. But if you are caught-up in ideologies and dogma, of whatever form they take, including science, you will not see this. And from this flows that which is all too familiar.

Shall we now count the number of people who have died or suffered as the result of secular Abrahamic world views, or as a result of the human mind exercising its power of reason, or of taking into account and basing decisions on what can so easily be branded as evidence? Better not, for we would not be able to keep tally.

What we can tell you though is that in the twentieth century, far more people died as a consequence of the secular than those who died as a result of the religious. Where is the evidence you ask? Read your history books we reply - there is no need to count, to collect the data. It is very big data! Nevertheless here are a few clues: Stalin, Hitler and more recently Pol Pot, Rwanda, Srebrenica. Should we say something too about South Africa's past?

Now perhaps you begin to see the problem. Or do you?

**127**

So, socialism and communism - both very rational, both grounded in reason, both keen on taking decision based on evidence. Communism is founded on science. We will let Karl Jaspers explain:

"Socialism which, as communism, in its enthusiasm for the salvation of mankind that it feels certain of securing, takes steps to mould the future by force of total planning, and socialism as the idea of gradual realisation in the intercommunity of a free democracy, are alien to one another. The former consumes the man who surrenders himself to it in a faith that appears in the guise of science; it treats the unbeliever as a material available to manipulation by force. The latter casts no spell, lives in the present, requires the sobriety of reason, and the humaneness of unceasing communication. As long as socialist demands are concretely visualised and thought out they remain within bounds. It is only when concrete reality is lost sight of and a fanatical paradise of man is presupposed as possible, that its demands become abstract and absolute. It is at this point that socialism becomes an ideology. The demand for complete implementation in fact leads away from its fulfilment. Along a path of coercion it leads to servitude."

What happens when we replace some of the words, and instead of speaking of socialism and communism, we instead speak of trade and capitalism, or science and scientism?

We think that you will find that Karl Jaspers words make perfect sense when such applied. This is the nature of the monster. It lies in the mind of those who follow extremist paths, and where are many routes to the horrors that extremism bring. The West looks at communism and Islamist extremism, with fear, rightly recognising that these things are ideologies, yet at the same time the West is blind to its own brands of extremism, as the very human activity of trade, becomes the ideology that is capitalism, and that which is science becomes the ideology of scientism. And many more ... Is that not so Richard Dawkins? Oh dear, he is not listening. Too far gone!

And you will oppose, for Charles Darwin said so towards the end of that book, that, one day on BBC Radio 4, you were unable to remember the name of! People will remember your name, but not for whatever it was you did when you were a scientist, but for the nonsense that is the Richard Dawkins Foundation, and the ridiculous things that you said.

And in some not so far of future time people will ask: why so smart yet so dumb? Yes, you may well remember those words? You will find them in *The God Delusion*. We could write a book called the Science Delusion and include these words as well!

**128**

Some more definitions, this time provided by Leo Apostel and colleagues, from the OECD:

Discipline: a specific body of teachable knowledge with its own background of education, training, procedures, methods, and content areas.

Multidisciplinary: juxtaposition of various disciplines, sometimes with no apparent connection between them. For example: music plus mathematics plus history.

Pluridisciplinary: juxtaposition of disciplines assumed to be more or less related. For example: French plus Latin plus Greek.

Interdisciplinary: an adjective describing the interaction among two or more disciplines. The interaction may range from simple communication of ideas to the mutual integration of organising concepts, methodology, data and organisation of research and education in a fairly large field. An interdisciplinary group consists of persons trained in different fields of knowledge (disciplines) with different concepts, methods, and data and terms organised in a common effort on a common problem with continuous intercommunication among the participants from the different disciplines.

Transdisciplinary: establishing a common system of axioms for a set of disciplines (for example anthropology considered as *the science of man and his accomplishments*, according to Linton's definition).

What is Linton's definition? Why not find out for yourself? You will have to do some work, but you might learn something!

So, Edward O. Wilson, anything to say about these definitions? Anything to say about how these approaches contribute to the unity of knowledge? No! Why is that?

The greater reducer is silent.

The Oxford English Online Dictionary makes no distinction between its definitions for cross-disciplinary, interdisciplinary, and transdisciplinary, but the OECD did (it did not mention cross-disciplinary)! All these words are adjectives, which is why they are so easy to slot into text and conversation. There is no need for them to be verbs it seems!

Christopher Jones, one time professor of Design at the Open University will now speak:

"Method alone cannot remove the invisible and troublesome barriers between professions and disciplines."

What is this *two cultures* nonsense! There are many cultures, for each discipline is a culture unto itself, but with, sometimes, aspects shared with other disciplines. Take for example, medicine and engineering. Two very different cultures, yet both share in common diagnostic skills, for when an engineer undertakes diagnostic work to identify a fault in complex man-made systems, you will find similar skills and procedures being used as those of the physician in the consulting room. Yet ethics is central to the physician, but not to the engineer, who is typically an employee representing the employer and acting in their interest.

An alternative definition of transdisciplinary could be: a word that is used when the term multidisciplinary would suffice. One gains the impression that much of what is presented as art-science or art-technology practice is just multidisciplinary, or transdisciplinary when the word multidisciplinary would suffice. And being on a quest to find more interesting examples that might be truly interdisciplinary or even transdisciplinary, takes us to places where few feet tread, or even dare to.

There is a book by Adolf Hitler called *Mein Kampf*. There is a book by Richard Dawkins called the *God Delusion*. It could have been called, *My Struggle against Religion*, or for short, *My Struggle*. Most likely you will not have read *Mein Kampf*, so you do not know what we are talking about. We read *Mein Kampf* a long time ago, and we read the *God Delusion* too, not a long time ago, so we do know what we are talking about.

In the past people once asked why a whole nation followed Adolf Hitler to (and over) the edge of doom. One day in the future, people will be asking that question again: Why did people follow the European Commission and STEM to (and over) the edge of doom? Time is a singularity!

**129**

A person carrying the weighty title of distinguished professor of Art and Technology has been asking questions – we have met him before as the astrophysicist who is an atheist and a positivist! This is what he asks of people – the recounting of experiences by way of giving advice to young people who wish to follow a dual arts-science career path:

"1- What is your background as a scientist? In the arts, design or humanities?
"2- When and how did you become involved in a hybrid art/science practice?
"3- What have been the major obstacles to overcome?
"4- What have been the greatest opportunities/breakthroughs?
"5- What would you do differently, knowing then what you know now?
"6- Any advice for someone who may want to walk in your footstep?
"7. Add other questions and your responses you think are relevant."

We sense the cult of Leonardo at work here! So, as art has no rules and the words that oft in our head sing, dance, and play, are today in cooperative mood, to these questions, this is what we say:

Beware distinguished professors of Art and Technology asking strange questions grounded in rationality. Beware too, people who provide rational answers to the professor's questions. The excess of rationality found in the West is its great flaw, and in continuing with this singular obsession, a cradle of woe is what you will discover, if you have not so realised already.

The evidence is all around, but Narcissus would rather die than look away from his beautiful reflection, and now he seeks *artists in residence,* who have also fallen in love with the reflection, to convey his beauty to what he thinks are the ignorant masses, who he believes, are experiencing a *science information deficit.* His desire is that they too can join in with the celebratory *dancing with glee* while standing on the *edge of doom.* If you oblige you will find yourself entering dangerous territory, so please, stay away, or be with them condemned, by future generations, who may ask you: why did you abandoned us to this illusion of a paradise to be attained that could never have been, for it was, like all before, *such stuff as nightmares are made on.* We are!

As for titles, seek them not, for they are heavy millstones around the neck, which along with other millstones, restrict movement, limit actions, and prevent the doing of what now needs to be done. The priest called Copernicus and the citizen-scientist known as Kepler will, in an indirect but related way, explain. Koestler will act as your guide and interpreter for he knows much about such things. The citizen-scientist called Darwin understood millstones, so he too can explain. C.P. Snow did not understand millstones – although he had many – for he was trained as a scientist and had therefore a fragmented and reductive mind; hence he was unable to explain without making gross simplifications which render his *opinions* to be what they are: technocratic in the extreme. Artisans, practical people of all kinds, thousands upon thousands of them that history never recorded, living in the *Renaissance* period (which extended over several centuries), could also explain, for they experienced the result of millstones on a daily basis; these were the millstones of elitist groups, who kept things apart, but this is not the *Renaissance* that you probably learned about, which is partly another modern myth; a strange mix of facts and fictions, wherein you believe that which suits you best, and where the contributions of learning and knowledge from Islamic, Chinese, and other Asian scholars and artisans has been quietly written out of history – it is a very European *al*-gorithm.

Beware also those who say things such as: *I think therefore I am armed* – they do not, they are not. Such people are potentially very dangerous for they too are but the children of ideology and dogma – but this time secular in nature. If any of those who utter such a strange statement ever studied, through cognitive psychology, the complexities of the workings of the human mind, then they did learn very little. Our friend and colleague Will, from his pen, as is often the case, conjured words to suit the theme:

> What's in a name? That which we call a rose
> By any other name would smell as sweet;

Or rot, decay, or be a barb.

Unless you are oriented to psychological and sociological learning, it is unlikely that you will know much about that of which we speak, but as these are one of the means of illuminating millstones and the oddity of human beliefs, the study of the social sciences should be on your life agenda; you will need that rarity of things called an open mind to engage constructively with the complexities of multiple truths.

When young learn how to die; then you will know how to live. When young set yourself the goal of reading all the great literature that humanity has produced; be humbled by the impossibility of the task. Then learn how to choose wisely.

Master the art of transcending disciplinary boundaries. Penetrate deeply into other peoples' territories. Leave a mark to show that you have been there. Observe the reactions of those people whose territory you have entered, and recall this every time you hear people speaking about multi-, cross-, inter- or transdisciplinarity. Be amused by what you hear. Take the time to understand what these terms mean, for such understandings will put you in a position to see through the rhetoric that often surrounds such words, which take no effort at all to utter. Then you will be able to act in ways that few others can. Separation is an illusion; a human construction; a social construction.

Spend time with people who make a lot of noise so that you will discover just how empty they can be, then, never waste your time with such people again.

Seek out solitude and silence and learn to listen to the voice of your soul. Let it be your guide for it will never let you down. Follow your dreams wherever they may take you.

Observe life to a degree that Dickens would admire, for this is the source of art's endless fountain. Learn to look into the souls of men - and women too as they are increasingly being expected to behave like men, and are doing so!

You have hands with opposable thumbs for a reason - use your hands to make things.

Travel in time to discover why the European world creates futures very similar to the past, only with more advanced science and technology, which are then used to wage war on an industrial scale and to destroy the planet in pursuit of the demands of one of their gods - money. Discover for yourself the nature of Europeans and their colonial offspring - they are but Ancient Romans doing what Ancient Romans did, only more efficiently, which is their *Enlightened* way. Do not be party to their collective delusions and denials. Visit Rome and stand among the ancient ruins - then find understanding, because for such people there is only *Decline and Fall.*

Do not seek fortune and glory otherwise you will end up like Indiana Jones: ethically compromised, keeping company with bad people, and having to go to extraordinary lengths to extract yourself from problem situations that are of your own making. Do not seek to associate with people with power and influence: their power has corrupted them and it will corrupt you too.

Seek the answers to the question: why such advanced knowledge yet such simplistic beliefs and so much ontological oversimplification?

Do not be naïve about science, technology, engineering and mathematics (STEM) - there is enough naïve STEM people in the world already; we do not need anymore. Thwart STEM peoples' efforts to use art to project their imperial power on to humanity. Wear camouflage and infiltrate their academies, funding agencies, organisations, companies, institutions, etc. See what actually goes on and what people say when they think outsiders are not present. Collect evidence and identify their many weaknesses - these are not difficult to spot. But do not succumb to their ideologies. Then use what you have learned to unleash the forces of creative destruction against them, for the days of countermovements are over - now is the moment for a new scientific revolution, the creation of a new science that most of the *olden ones*, and their *artists in residence*, will, like the *scholastics* before them, not want to see or to understand. And strangely, art is the means to achieving this, but this not what you are expecting is it?

Do not participate in the creation of new ideologies like art-science and art-technology, for these are but more of those binaries, dualities, and dichotomies so loved by the European mind. Never use words such as nexus, crossing, bridge, hybrid, integrate, intersection, cross-fertilise, spill-over, and so on; these are but the chains that keep you bound to a world that has already passed into history - Prometheus' invisible and silent chains. They are part of the self-constructed prison known as your values and beliefs. Instead, set yourself free and learn to fly like an eagle.

Spend time with nature and in the summertime sit among the flowers and talk to the bees and listen to what they have to say to you.

Work for the benefit of future generations for they are more likely to understand and appreciate your work.

Never accept what people say - check for yourself. Always go back to the original source, for it is surprising how much distortion and simplification takes place as a result of material being passed from one person to another. Discover what has been left out as a result of unquestioning acceptance. Go to extreme lengths to track down original sources. This is why there are National Libraries - use them.

Develop critical thinking and writing capabilities as practised in the social sciences. Be a gadfly. Write books about your work, and, if you indie publish, make sure that copies of these books are placed on deposit in your National Library collection. You never know who will look at your work in the future.

Set unrealistically high standards for yourself which you will constantly fail to attain - then learn. Keep on learning. Develop the self-critical skills and intuitions needed to know when your work is poor and when it is moving towards something else. *Paint fakes* so that you will recognise when your work is not a fake.

Work with dead artists for they have much to teach you. Do not join grazing herds. If you want to engage in collaborative activities, be very selective whom you choose to work with.

Never place scientists and artists on pedestals. These people are human and as such are deeply flawed, just like everyone else. Do not use Leonardo da Vinci or C.P. Snow as a justification for what you do - this will only demonstrate that you *do accept* what other people say without checking, that you did not learn to choose wisely what to read, that you did not master the art of transcending disciplinary boundaries, and that you have missed the point completely.

The reputations of these two are mostly not of their own making, being modern constructions - strange beliefs and *superstitions* about what they were but never were, or what they said but never said, or what they did but never did, or what they wrote but never wrote, or what is known but is not known. The myths that now surround both of them are like the haze of dust that gathers on hot summer days, just waiting to be blown away on the breath of the wind. Be that wind.

And if you do not take note of this, you might end up like those STEM people from prestigious institutions, who recently expressed, in a review paper entitled *Art on the Nanoscale and Beyond*, simplistic justifications for art-science convergence, based on - simplistic notions of Leonardo and Snow. But what matters learning and knowledge when fortune and glory, power and influence, are on your mind? This is the nature of the ethical melt-down occurring in the world STEM, for it is in any case a shallow ethic, very easy to ignore when convenience demands - disassociate yourself from this and speak-out; condemn all such forms of moral corruption.

Of such STEM people one can say that, here be conquistadors, the existence of which Edward O. Wilson denied, who are now rewriting and simplifying (as is their way) history, in support of their colonising agenda. It is but a timely reminder of the sand, the lack of foundations, the intellectual emptiness, and the lack of knowledge, upon which much of the STEM world's interest in the ideologies called art-science and art-technology is built. The European Commission's DG CONNECT is a star performer in all these this respects.

Thus, in speaking of creativity, communication, the promotion of science and technology, and things most sinister with Orwellian undertones that Joseph Stalin and other tyrants would have understood, STEM people disclose their *will to power* and their staggering lack of vision and imagination with regard to art's role in research, and thus do all such people become the source material for the fountain; they will discover to their cost, that the pen is far mightier than the Scanning Electron Microscope and other such *boys toys*, which is why Stalin and others of his type exercised strict control over the arts. STEM people will do the same, and it begins with the *cheque book* and words like: *you can use this, but our permission comes with terms and conditions attached.* And thus too do they become like the medieval Church of Rome - oh dear you are making another future just like the past; it is a very European algorithm.

On the matter of Edward O. Wilson and his strange notions of consilience - steer clear of this for it is deeply flawed thinking, being as it is, a product of a man, a scientist, with several millstones around his neck. Do not look too optimistically to the consilience *second wavers* for they too have millstones around their neck and are hopelessly lost. Avoid becoming *entangled,* for there are more millstones to be found here too. Instead, reinvent the world of knowledge by creating new *Theories of Knowledge,* suited to 2016 and beyond, for what exists today is more appropriate to 1916, 1816, 1716, 1616, 1516, 1416 ... and BC in Ancient Greece as well.

Now is the time, after the deconstruction of the postmodernist era, to put the world back together again in a different form. Thus, being as you are, an artist, and not being constrained by rules, conventions, boundaries, and such forth, use your imagination to imagine the unimaginable, and then watch with delight as they wriggle and squirm and then a heretic proclaim you to be, for they, the people of STEM, are but *people of the book,* the products of a Western secular Abrahamic culture whose time is done, although someone needs to tell them this, for they, to their beautiful reflection, are still bound; feet and minds firmly planted where they are - in the past.

Discover the mysterious East. Travel there - literally and figuratively. Come to understand that in the motion there is stillness and in the stillness there is motion, and how this insight can be used to transform science and humanity. And to the East we now turn our face and speak: see their weaknesses. Stand on the shoulders of giants to see further than they can, to discover the coming age of the Tao, and in the names of two giants, Howard Rosenbrock and Joseph Needham, two of a kind, you will find the beginnings of what you need. Art in a different way will make a difference too.

Now cometh to an end these words of guidance and in closing we say, to the discovery of all the ways of knowing turn your attention, and practise as many of them as you can. Above all, learn the hermeneutic way of knowing so that you will be able to understand what we write.

To those who understand the words that we speak in silence, does the future belong - it is a place where no Ancient Romans (or Greeks) will be found for a different kind of person will there dwell, and a different type of science will they practise as well. And to our dear friend and colleague Will, we yet one more time do appeal, for words of ending to provide:

And gentlemen in England now a-bed
Shall think themselves accursed they were not here,
And hold their manhoods cheap whiles any speaks
That fought with us upon Saint Crispin's day.

And thus the world turns ...

**130**

Lao Tsu has some more words of advice for you:

"The skilful traveller leaves no traces of his wheels or footsteps; the skilful speaker says nothing that can be found fault with or blamed; the skilful reckoner use no tallies; the skilful closer uses no bolts or bars, while to open what he has shut will be impossible; the skilful binder uses no strings or knots, while to unloose what he has bound will be impossible.

"In the same way the sage is always skilful at saving men, and so he does not cast away any man; he is always skilful at saving things, and so he does not cast away anything.

"This is called: *Hiding the light of his procedure.*

"Therefore the man of skill is a master to be looked up to by him who has not the skill; and he who has not the skill is the helper of the reputation of him who has the skill.

"If the one did honour his master, and the other did not rejoice in his helper, an observer, though intelligent, might greatly err about them.

"This is called: *The utmost degree of mystery.*"

Thank you Lao Tsu for that reading from the *Tao Te Ching*.

Now for another reading. The Apostle Didymos Judas Thomas, will now read from his Gospel, The Gospel of Thomas, being the secret sayings that the living Jesus spoke and which Didymos Judas Thomas recorded.

"And he said: *Whoever discovers the interpretation of these sayings will not taste death.*"

What do you mean?

**131**

Now an Imam will read from the Holy Qur'an. We have three English translations to read from. The text that will be read, from each translation, is from Sûra 15, verse 9.

First, the first reading:

"We have, without doubt, sent down the Message; and We will assuredly guard it from corruption."

What do you mean?

Second, the second reading:

"We have sent down the Qur'an Ourself, and we will Ourself guard it."

What do you mean?

Third, the third reading:

"Absolutely, we have revealed the reminder, and absolutely, we will preserve it."

What do you mean?

Thank you.

Now someone else speaks. Who is it?

We have changed our minds. No one will speak! So on to the next chapter!

**132**

In science, the researcher is seen as being distanced from the data. This one might say is the older positivistic view. In other research paradigms, such as action research, the researcher is close to the data. There is no pretence that the researcher is not affecting the data in some way. Action research aims to intervene to change the situation being observed.

Now Stephen Hawkins wants to say something, yet again!

"I must say that, personally, I have been reluctant to believe in extra dimensions. But as I am a positivist, the question *Do extra dimensions really exist?* has no meaning. All one can ask is whether mathematical models with extra dimensions provide a good description of the universe.

"I would like to say something too about complex numbers, that is to say, numbers with real and imaginary components. One might think that imaginary numbers are just a mathematical game having nothing to do with the real world. From the viewpoint of positivist philosophy, however, one cannot determine what is real. All one can do is find which mathematical models describe the universe we live in.

"Here I will also just briefly mention the issue of vacuum fluctuations, which can be interpreted in several ways that seem different but are in fact mathematically equivalent. From a positivist viewpoint, one is free to use whatever picture is most useful for the problem in question.

"Now, I want to say something on the subject of black holes and p-brane models. The mathematical model of a black hole as made of p-branes gives results similar to the virtual-particle pair picture. This from a positivist viewpoint, it is an equally good model, at least for certain classes of black hole.

"And finally, it has been suggested that one or more of the extra dimensions might be comparatively large or even infinite. This idea has the great advantage, at least to a positivist like me, that it may be testable by the next generation of particle accelerators or by sensitive short-range measurements of the gravitational force."

Well Stephen, let me summarise what you have just said: "I believe ..."

But Stephen, you did not say what sort of positivism you were referring too, the logical kind or the Competian variety, and what happens if someone is not a positivist, if one, for example, were inclined to follow Popper's philosophy which you confused with positivism, of perhaps you are a follower of Popper's philosophy? You seem confused, out-of-your depth, lost in a strange world. You evidently did not master the art of transcending disciplinary boundaries. Now why might that be? Yet you do not hesitate to dive in without checking your facts. So strange! Perhaps we should clarify matters for you?

"*Epistêmê* is the Greek word most often translated as knowledge, while *technê* is translated as either craft or art. These translations, however, may inappropriately harbour some of our contemporary assumptions about the relation between theory (the domain of *knowledge*) and practice (the concern of *craft* or *art*). Outside of modern science, there is sometimes scepticism about the relevance of theory to practice because it is thought that theory is conducted at so great a remove from the facts, the province of practice, that it can lose touch with them. Indeed, at the level of practice, concrete experience might be all we need. And within science, theory strives for a value-free view of reality. As a consequence, scientific theory cannot tell us how things should be – the realm of *art* or *craft*. So we must turn elsewhere for answers to the profound, but still practical questions about how we should live our lives. However, some of the features of this contemporary distinction between theory and practice are not found in the relation between *epistêmê* and *technê*."

That was a spokesperson for the Stanford Encyclopaedia of Philosophy speaking. Managing to keep-up Hawkins?

The relationship between *epistêmê* and *technê* is not the simple binary that some of the art-science theologians would like to believe.

### 133

Edward O. Wilson is wandering by, so we will ask him a question.

*Theories of Knowledge*, Edward! Anything to say?

No. So strange! You wrote a book called *Consilience,* which carries the subtitle *The Unity of Knowledge*, and yet this most important matter is not seriously addressed, beyond a few words of history relating to logical positivism and its short rise and eventually fall. It seems that you too have not mastered the art of transcending disciplinary boundaries. Now why might that be?

On 6th October 1956, an article appeared in the *The New Statesman* entitled *The Two Cultures*. It was C.P. Snow bemoaning the gap between what he called traditional culture, which he was clear to specify as being mainly literary, and the scientific culture. It was a personal expression of his beliefs, nothing much more, based upon his experiences, which included a period spent during the Second World War, interviewing young people who were about to graduate from science and engineering courses, to determine how best to use these people in Britain's war effort. His opinions were also based upon his encounters with people in the literary circles in which he moved.

Howard Rosenbrock was one of Snow's so-called *snowflakes*, which is how we first became aware of Snow.

We must, to use a colloquial turn of phrase, get to the bottom of this matter of Snow and his *two cultures*. For now however, we will limit ourselves to observing that Snow did not refer to the arts in general, but to the literary circles in which he moved, in other words English literary circles - not Scottish ones, not Welsh ones, not French ones, not American ones, but - English literary circles. We do not know what experience he had of encountering sculptors, painters, and such forth. We do not know, because he did not say, if, during the 1920s, 1930s, 1940s and 1950s he encountered artists from the many movements that have come and gone: modernists, surrealists, social realists, constructivists, and so forth. About art in general he appears ignorant - perhaps he was. And reading the text it is also clear that he is bemoaning the low status held by applied scientists compared with literary folk, and that he believes that applied science is humanity's saviour. And this is set against the backdrop of Britain's economic decline, and self-evidently the link between science and technology and economic well being. And he is most definitely not proposing a marriage; in fact he recognises - anticipates - the demise of this snobbish literally elite who in his mind stand in the way of - progress. He most definitely does not foresee the problem that science would become. But others, wiser than he, did foresee this! We have not forgotten about Rachel - she will be here soon.

Now is a good moment to paraphrase Karl Jaspers:

"Science which, as scientism in its enthusiasm for full knowledge and the salvation of mankind that it feels certain of securing, will inevitably, take steps to mould the future by force, and science as but only one of many way of knowing the world, are alien to one another. The former consumes the man who surrenders himself to it in a faith that appears in the guise of truth; it treats the unbeliever as a material available to manipulation by force. The latter casts no spell, lives in the present, requires the sobriety of reason, and the humaneness of unceasing communication. As long as science's demands are concretely visualised and thought out they remain within bounds. It is only when concrete reality is lost sight of and a fanatical paradise of man is presupposed as possible, that its demands become abstract and absolute. It is at this point that science becomes an ideology. The demand for complete implementation in fact leads away from its fulfilment. Along a path of coercion it leads to servitude."

Someone should tell this to Richard Dawkins.

Literal: taking the meaning of words in their usual or most basic sense without metaphor.

Someone should tell this to Richard Dawkins.

Figurative: departing from a literal use of words; metaphorical.

Someone should tell this to Richard Dawkins.

**134**

Now an Imam will read from the Holy Qur'an. The reading is taken from Sûra 2, verse 256:

"Let there be no compulsion in religion: Truth stands out clear from Error: whoever rejects Tagut and believes in Allah hath grasped the most trustworthy hand-hold, that never breaks. And Allah hereth and knoweth all things."

Thank you.

So now you know that any Muslim who forces another person to take-up religion, to convert to Islam, has not understood the Holy Qur'an. So now you know that any Christian who forces another person to take-up religion, to convert to Christianity, has not understood the Holy Qur'an either. So now you know that any Western atheist, scientist, communist, or capitalist, who forces another person to take-up their particular secular religion, or to convert to their particular secular religion, has not understood the Holy Qur'an either.

We are not Muslims, yet we study the Holy Qur'an. We also study the The Torah, and the New Testament, and the works of Karl Marx, and so forth.

Let there be no compulsion in religion. Someone should tell Richard Dawkins. Let there be no compulsion in secularism. Someone should tell Richard Dawkins. There has been compulsion in religion. Someone should tell Richard Dawkins. There has been compulsion in secularism. Someone should tell Richard Dawkins. People have been killed because of religion. Someone should tell Richard Dawkins. People have been killed because of secularism. Someone should tell Richard Dawkins. Religion is not the problem. Someone should tell Richard Dawkins. Secularism is not the problem. Someone should tell Richard Dawkins. It is the human mind that is the problem. Someone should tell Richard Dawkins. Richard Dawkins has a human mind. Someone should tell Richard Dawkins.

And now Marc Luyckx Ghisi will speak about transmodernism. Marc worked at one time for the European Commission, in its *Forward Studies Unit*, a department reporting directly to the president of the European Commission. Marc worked there under the Delors and Santer Presidencies.

"What is transmodernity? Is it the end of rationality and science? Are we moving towards a worldwide global *new age* and towards totally irrational ways of thinking? Are we going back towards medieval obscurantism? These seem to be the fears of some intellectuals, at least in Europe.

"Yes we are in a process of mutation. Yes, we are leaving modernity. And, as we leave the room, there is a back door that frightens us all, the return to past obscurantism, to religious wars, to *fundamentalism* of all sorts. However, there is also a front door that is less evident because we have to create it. And, the knowledge society can really take off only if it joins the transmodern view, if it departs modernity by the correct door. This new door is more difficult to find because the way to it does not yet exist. We must create it. We have to invent it together.

"What is this new door? What is this new transmodern synthesis? One takes the best of modernity and the best of premodernity and goes farther, invents new orientations, a new vision, and a new *politic of life* for humankind. We will capitalise on the wonderful achievements of science and technology. We will rediscover the harmony with Nature and the cosmos, and the spiritual depth of the premoderns. However, these tools will be directed anew toward the realisation of a completely sustainable and socially inclusive civilisation. Science and technology will not be questioned, but their fundamental orientation, their basic axioms, and their link with humankind's goals will be fundamentally redefined."

Thanks Marc for that powerful insight.

*Hawkins Dawkins Wilson Wolpert* and what they represent are part of an old world that is no longer relevant. So is the consilience that Wilson defined. So is the Richard Dawkins Foundation. So is transhumanism. So is …

Now is the moment in time to move beyond these old visions and to remake the world anew. Intuition though tells us that transmodernism is not the answer. The European cup is full …

**135**

Back to the matter of C.P. Snow, and here now is someone who has something to say about some research that he undertook in an attempt to measure Snow's *two cultures*. Professor Chris McManus will now speak:

"I undertook a questionnaire-based study which measured 17 different cultural activities in a stratified survey of science and non-science students in London. Science students showed a significantly lower level of activity on 12 of the 17 measures, and were significantly higher on none. Among the scientists, cultural activities overall was lowest in engineering and mathematics students, and highest in medical and biological science students. Cultural activity correlated significantly with the personality measure *Openness to Experience*, although the correlation was weaker in scientists than in non-scientists."

Thank you for that useful piece of evidence. But evidence of what? Snow, in his *New Statesman* article was critical of people from the sciences for their lack engagement with culture, by which he meant literature. But what is most interesting here is the term *Openness to Experience*. We have already encountered this issue in a different form.

Now Wolfgang Paalen will speak:

"Art has always been pre-eminently concerned with quality. Summarily eliminated everywhere else, quality has ever continued to reign over the immense realms from which number is excluded, the realms in which appearance does not deceive. It is true that since Goethe the poets have forfeited all authority in scientific discussion. But as no man can hold himself satisfied with the affirmation that light is merely a number of vibrations, it is high time to realise that the poet speaks as truthfully as the scientist when he claims that light belongs to the realm of vision. And to the extent that we believe that through a scientific formula we can know all that there is to know about a thing, we are surely more backward than the eighteenth century. In fact, the public (and not a few scientists) today believe as superstitiously in science as they yesterday believed in witchcraft. And one can hardly pretend that the modern superstition is superior, on the claim that the practical results of science are in themselves beneficent – for they are not necessarily beneficent so long as for one way of healing we learn a hundred ways of killing, and so long as domination of the forces of nature, instead of humanising life, bestialises man. That is why reaffirmation of art is not a matter of theoretical aesthetic rehabilitation, but a vital problem. For so long as the terrible emotional frustration consecrated by science subsists, the road will continue open to charlatanisms that promise the indispensable emotional satisfactions. Those who refuse to crawl in the shadows of decayed churches can only too easily become the prey of totalitarian mysticisms."

Someone should tell Richard Dawkins.

**136**

Back to the matter of Leonardo and we call now to the witness stand, Charles Gibbs-Smith, to answer questions about his book *The Inventions of Leonardo da Vinci.* The cross-examining Barrister speaks first:

"In your book Mr Gibbs-Smith, you cover the topic of *Aeronautics* and you say that most of Leonardo's investigations into the problems of flight were, sadly, based on the misconception that man had sufficient muscle power to emulate birds. Leonardo also mistakenly thought birds flew by beating their wings downwards and backwards; in fact, on the down-stroke, the wing-feathers twist into miniature propellers to give thrust, with the inner wing providing lift. Consequently he experimented ceaselessly with an unworkable notion: flapping-wing aircraft called ornithopters. Is that correct?"

"Yes."

"You also state in several places that Leonardo did not invent some of the things he drew. Is that correct?"

"Yes."

"Often you go on to say that he just improved what had been invented by others. Is that correct?"

"Yes."

"Under *Weapons of War,* you state that the covered assault car was a common area of research in the military treatises of Leonardo's day and earlier. You then state that Valturio and Guido da Vigevano had drawn sail powered tanks and that it cannot be said that Leonardo invented the tank. There were earlier designs and none were ever built to these, or Leonardo's own designs. Is that correct?"

"Yes."

"Then you say that it is debatable whether Leonardo intended his design to work since the crank as drawn would have powered the front and back wheels in opposite directions. We shall never know whether this was a deliberate mistake made by a man with a conscience or to prevent the design being useful to enemies. Is that correct?"

"Yes."

"Having said this, you then present a drawing showing the mutilation caused by a scythed car that he had drawn. You then say he was a gentle man! After which you also note that Leonardo seems to have learned of chariots with projecting sickles from Valturio's *De Re Militari* and to have designed a mechanised and more efficient version! Is that correct?"

"Yes."

"Under *Elements of Machines* you include a drawing of a lifting jack. Then you say that it remains difficult to say how far this drawing is a description of something in general use, how far a modification or how far an original invention. It is however a good and self-explanatory example of a technical illustration. Is that correct?"

"Yes."

"Under *Elements of Machines* you also present a drawing of a printing press, and then say that it is impossible to relate this to precise practices at the time, because, as you say, the 'early history of printing is somewhat obscure.' Is that correct?"

"Yes."

"Later in that section you also state, in relation to Archimedean screws for lifting water that the extent of the knowledge at the time of Leonardo about raising water was considerable. Designs were handed down by Vitruvious, and can be found in Alberti, Francesco di Giorgio and Giuliano da Sangallo. Verrocchio, an experienced hydraulic engineer was responsible for bringing knowledge of an interest in the subject to Leonardo. Is that correct?"

"Yes."

"The section that you entitled *Investigations of Nature and Designs for Architecture* is also interesting according to the line of questioning that we are pursuing. You present the drawing of two skulls, one with the spine attached which is clearly wrong, with the bony projections on the back of each vertebra - the Spinous Process - being repeated in the drawing on the front of the vertebra. You explain this by saying that it suggests that the vertebra have been drawn from memory. Yet earlier you to state that, while resident in lodging at a hospital, corpses were apparently placed at his disposal for dissection. Is that correct?"

"Yes."

"Then, when dealing with the matter of architecture, you say that no building is known to have been executed by Leonardo. Is that correct?"

"Yes."

"You also said, just after the letter to the Duke of Milan, that we do not know how much of what Leonard said he could do, he was actually asked to do. Is that correct?"

"Yes."

"One final question: you are a historian of early aviation? Is that correct?"

"Yes."

"No further questions. Thank you. You may now leave the witness stand."

Mr Gibbs-Smith walked away from the witness stand, his demeanour speaking of one humbled by the due process of law.

The Barrister will now address the judge by way of summing-up:

"My Lord, when Mr Gibbs-Smith wrote his book, it is evident that he accepted as a given, as a taken-for-granted belief, the mythological figure of Leonardo da Vinci as the genius, polymath, artist, designer, engineer and inventor, and such forth. Yet, I would now suggest, what Mr Gibbs-

Smith included in his book hardly points to Leonardo being that which people commonly believe him to have been. In his text Mr Gibbs-Smith uses the word research, which is a word with modern connotations and which also has several meaning. In places though, Mr Gibbs-Smith is contradictory. It would also seem that actually Mr Gibbs-Smith does not have sufficiently detailed knowledge of the state-of-the-art of machinery in Leonardo's time, and cannot therefore be sure that what he has presented are actually inventions, or even improvements of other people's inventions. In several places he refers to other people from whom Leonardo learned things, including, inventions and technologies.

"Furthermore, Mr Gibbs-Smith reproduces what is called a letter to the Duke of Milan, but does not mention in his book that there are doubts about the authenticity of the letter, as it is reported in places that there are some who believe that the letter is not in the handwriting of Leonardo da Vinci. We did not however chose to call a witness on this matter, as it is a secondary matter to that which we address before the court.

"I put it to the court that no one today has understandings of what was going on in da Vinci's mind and what Leonardo's intent was. It could be that he was just using his imagination or even just recording things that he had seen or heard about. Take for example the Silk Road. In Leonardo's time this was in full flow, with many ideas finding there way out of China into Europe. The people of Europe, even to this day, have very little idea of the extent to which Chinese inventions and technology were far in advance of what could be found in Europe in Leonardo's time and earlier. Was Leonardo recording that which he heard recounted by travellers returning from China? We will of course never know.

"It could be that he did invent or improve, while also just recording things around him, and indulging his imagination without any thought of creating, *unconventional and compelling products*. We will never know, but one thing we can be sure of is that the situation is far more complicated than the myth.

"I rest my case. Or do I? We will have to wait and see."

Now Christopher Frayling will say a few words about the pop history of the artist Vincent Van Gogh, as presented in the 1956 Hollywood film *Lust for Life*:

"Crows over a Wheatfield (in Auvers) is, according to the film, completed at great speed just moments before he tries to shoot himself – in the summer of 1890; actually, this was not his last canvass, but pop history has always preferred to believe that it was. The artist, by definition, is someone who works in an expressive idiom, rather than a cognitive one, and for whom the great project is an extension of personal development: autobiography rather than understanding. The movie stereotype of the artist is almost invariably like this – from Michelangelo played by Charlton Heston (1965), or Caravaggio played by Nigel Terry in Derek Jarmen's version (1986), to Ken Russell's biographies of more modern artists."

**137**

Now to the matter of the Kabala! We have no doubt that, somewhere in the world there are people who believe that Leonardo da Vinci knew the secret knowledge of the Kabala! Perhaps he did. Does it matter though? Perhaps it does if the creation of myths is what you are about.

We shall now have a reading from *The Book of Formation*, or, as it is more properly known, *Sepher Yetzirah – The Book of Creation*:

"Ten ineffable Sephiroth, ten and not nine, ten and not eleven: understand with wisdom and apprehend with care; examine by means of them and search them out; know, count, and write."

What do you mean?

In Hebrew: Sefirot. In English: Sephiroth. What are the Sephiroth?

Consciousness: a person's awareness or perception of something.

Teach: impart knowledge to, or instruct.

Preach: deliver a sermon or religious address to an assembled group of people.

Karl Jaspers:

"When man supposes that he can survey the whole, instead of pursuing concrete, tangible aims in the world, he is so to speak, making himself into God – he loses his relationship to transcendence – he dons blinkers through the operation of which he loses sight of the experience of the origin and fundament of things, in favour of a semblance: mere movement in the world – the establishment of the right organisation of the world for all time – he loses the possibility of upsurge, because he succumbs to an apparatus of terror and despotism – he brings about a perversion of the apparently loftiest idealism of the goal of humanity, into the inhumanity of the squandering of human life, into the transformation of all the circumstances of life into an unprecedented slavery – he annihilates the forces that carry man forward – when he fails, he gives way to despair to the urge to ever more loathsome acts of violence."

Someone should tell Richard Dawkins. Someone should tell Stephen Hawkins. Someone should tell Lewis Wolpert. Someone should tell Edward O. Wilson. Some one should tell the former chief scientific advisor to a former president of the European Commission. Someone should tell ...

Or have they lost to ability to listen, to see, to understand?

"Can research carried out by a visual artist be remotely compared with that carried out by a biologist?"

That was Tom Jones speaking. We will hear some more from him later.

Perhaps there was an age, the *Age of Romance*, when such a thing was possible, but that was in a time when people had grown weary of the excess of reason that was the *Age of Enlightenment*, and came to see where excess of reason led - to blood that flowed in the streets of Paris. But reason came back, and its dogma now stalks the world as never before. And it is the excess of reason which now blindly leads humanity towards the edge of doom, while people utter all sorts of strange words about objective truth and knowledge, about sole truth, about unattainable fantasies of paradise waiting just around the corner, delivered not by the old religions, but by the new religions of science, technology and capitalism. And so it seems also, the new sect called art-science. Once again the panoply of lesser gods rules the lives of men! Old superstitions are replaced by new ones. The world no longer turns.

Someone should tell Richard Dawkins. Someone should tell Stephen Hawkins. Someone should tell Lewis Wolpert. Someone should tell Edward O. Wilson. Some one should tell the former chief scientific advisor to a former president of the European Commission. Someone should tell ...

They are not listening. Such fools as men oft make of themselves, very rarely do, and thus time remains a singularity.

## 138

We have not spoken about scriptovisualism for many chapters. We have not mentioned for quite some time, the painting that we are creating as we write. There are reasons for this.

The matter of Leonardo keeps jumping into the conscious mind. Evidently the subconscious is still working on this. Hence our search of the literature continues. We will get much closer to the truth, for sure, but will never discover the sole truth, for there is no sole truth. This is the nature of knowledge.

The burden of proof in a criminal trial is different to that required in a civil trial. In the former, the term *beyond all reasonable doubt* is used. This means that there is no other logical explanation that can be derived from the facts except that the defendant committed the crime, thereby overcoming the presumption that a person is innocent until proven guilty. In civil cases the burden of proof is less demanding: here the term *balance of probabilities* is used.

So Leonardo the scientist, the inventor, the designer, the architect, the engineer, etc. - proved beyond all reasonable doubt? Or proved on a balance of probabilities? Or just what you believe?

We will be returning to the matter of Leonardo; it seems!

Asemic writing: open semantics and wordless. What relevance does this have to the world of STEM? What relevance does this have to Science, Technology and the Arts (STARTS)? More than you can imagine, especially if you are from the worlds of STEM and STARTS. Asemic writing is certainly relevant to scriptovisualism. But why?

Buckminster Fuller, wants to say a few words, for these, it would seem, he is credited with saying!

"You never change things by fighting the existing reality. To change something, build a new model that makes the existing model obsolete."

It is certainly a truth, even if he never did say it. Does it matter? Only if you want to make use of it!

"At the Chicago World Fair of 1933, the official motto was: Science Finds - Industry Applies - Man Conforms. To many of us today this seems quite shocking, yet it has been the driving force of much development in the last century.

"In particular, if you look at the rise of computing over the last 50 years, you will see that, on the whole, development has been extraordinary, but fairly straightforward: it can be characterised as trying to make faster and faster machines fit into smaller and smaller boxes.

"Starting from the time of the ENIAC, one of the colossal computers of the 1940s, most IT progress has been driven from the point of view of the machine. Since then things have changed - but perhaps not really that much. Even if computers can today calculate many times over what was possible a few years ago, and the machines have become somewhat less obtrusive, much of the mindset has stayed the same. It is the vision of huge calculating machines spanning massive rooms, trying to recreate an absolute artificial intelligence, that still haunt much of the thinking today.

"Clearly, it is difficult to shake off old mind sets."

That was Jakub Wejchert speaking.

Well Jakub, what you described is very much how the *Vainglorious Enlightened Ones* think - if that is the right word? We are just machines, so they believe, and they are intent on building a machine that seems to do what we do, only it will be faster. One of the problems is confusing what we actually do, with what the machines do, which leads one down blind allies, and then believing that one has not journeyed down a blind alley, because the blind end has yet to be encountered, one persists with the belief. Meanwhile, elsewhere, people not caught-up in *Enlightenment* thinking, might find a new path to walk! Thus the world turns.

**139**

It is a truth not universally understood that titles such as architect, surgeon, navigator, engineer, and artist, conjure certain meanings and images in the mind of the modern person, but such titles would have conjured a different image in the minds of those living in that period of European history know as the Renaissance. The title *scientist* was not part of their vocabulary, this being a word invented in the nineteenth century. Neither is it universally understood that words such as science and humanism today have meanings different from those which our forefathers in Renaissance times would have understood.

According to the Oxford English dictionary, the meaning of the word science has changed with time. Originally science was knowledge in general, or any branch of knowledge, including the arts, and the word comes from the Latin scire *to know* (also found in conscience (Middle English) *inner knowledge* and nice). The restricted modern sense of science, concentrating on the physical and natural world, dates from the eighteenth century.

Actually it may be that the modern meaning of science dates from much later - the nineteenth century!

Why have we just provided you with this information?

We now invite Joseph Needham, FRS, to speak about the definitions of science and its various stages of development, which appeared in his epic work, *Science and Civilisation in China*. Over to you Joseph:

"The growth of science may be viewed in the following five stages, the first being the primitive stage, when a body of techniques was developed. This may be referred to a proto-science.

"Second is the medieval stage, when bodies of sophisticated techniques were differentiated and general principles formulated, often related to some all-embracing theory.

"Third is Renaissance science, when universal laws believed to apply to nature, but individual sciences were thought of as components of one general science.

"Fourth is nineteenth century, modern science, when each individual science is thought to exemplify a system of universal laws, and that the principals and procedures of rigorous description and explanation in physics may in time be fitted to the entire variety of the more complex sciences from chemistry to linguistics. This science is thought of as a Unity.

"Finally, there is twentieth century science which attempts to establish a canon of accepted Unity of Science. Growing realisation that the *Laws of Nature* are conjectural descriptions of the structural properties of Nature; and appreciation that science is radically human, that man is part of the context in which the conjectural descriptions of the properties of Nature have been formulated, that all disciplined enquiries are affected by this relationship, and that science has been brought forth by our cultural development from a multiplicity of historical structures. The idea of the Unity of Science is perhaps one of our mental constructions.

"I believe in making a sharp distinction between modern science and that which preceded it, namely ancient and medieval science. Modern science, to my mind consists of two things, the mathematisation of hypotheses about Nature on the one hand, and on the other, continuous and relentless experimentation."

Why have we just provided you with this information?

**140**

Leonardo remains on our mind. We must do more to explore this. Something important here! Something important here! Should we say it again?

Something important here! So, we will have some guests, and to them we will put some questions. Oh dear!

First in line is the Nobel Prize winning theoretical physicist Frank Wilczek:

"Historians of science usually date the origin of the Scientific Revolution as 1543, when Copernicus' *De Revolutionibus* first put the earth in motion. While that is a convenient and defensible choice, it is not the only good candidate. An earlier scientific breakthrough, driven by the needs of art rather than astronomy, was equally profound: Filippo Brunelleschi's invention of drawing in perspective, around 1420.

"Brunelleschi demonstrated his ability to render scenes accurately. Fifteen years later, Leon Battista Alberti published a classic explanation of the technique. It involves a fascinating new kind of geometry, projective geometry.

"The work of Brunelleschi and Alberti made an enormous impact on the artists of the time. Obviously, it allowed more accurate rendering. It was also a great confidence-builder: here, at last, was a major discovery that clearly went beyond what had been achieved by the ancient Greeks and Romans. For artists, it was no longer a matter of recovering what had been lost, but of discovering a fundamentally new creative power."

You do not say if Alberti published a classic explanation in the vernacular or in Latin. This is an important point. Do you understand why?

It seems that we have another scientist with a fragmented and reductive mind here, engaging in the favourite pastime of such people - rewriting history!

You say that the invention of drawing in perspective was driven by art. You mean architecture surely? History records that Brunelleschi was an architect and engineer who trained as a sculptor, but gave up the latter to concentrate on the former activities. Being able to draw in perspective must have been very useful for him as an architect and well as others of that inclination.

The Encyclopaedia Britannica notes that as an engineer he invented machines for use on his architectural projects. But did he *invent* drawing in perspective? It seems that he did not.

We go back to the Encyclopaedia Britannica and will now hear from their spokesperson, to learn about Brunelleschi:

"Brunelleschi rediscovered the principles of linear-perspective known to the Greeks and Romans but buried along with many other aspects of ancient civilization during the European Middle Ages. Brunelleschi demonstrated his findings with two painted panels, now lost, depicting Florentine streets and buildings. From Manetti's descriptions it is clear that Brunelleschi had understood the concept of a single vanishing point, toward which all parallel lines drawn on the same plane appear to converge, and the principle of the relationship between distance and the diminution of objects as they appear to recede in space. By using the optical and geometric principles upon which Brunelleschi's perspective devices were based, the artists of his generation were able to produce works of astonishing realism. On two-dimensional surfaces they were able to create extraordinary illusions of three-dimensional space and tangible objects, so that the work of art appeared to be either an extension of the real world or a mirror of nature. Although the laws governing perspective construction were brought to light by Brunelleschi, they were codified for the first time by the humanist architect Leon Battista Alberti. In 1435 Alberti set them down in Della pittura (On Painting), his famous treatise on painting, which included a warm dedication to Brunelleschi - undoubtedly an expression of Alberti's debt to his friend's revolutionary discovery."

We have invited Giotto to make a comment. Art history records that you were the first person to break away from the medieval style to create a new style representing one of the most important developments in Western art. On the other hand, Wilczek believes you spent your time recovering what had been lost, and that you did not discovering anything fundamentally new. What do you say to that?

"Evidently this is someone who has a fragmented and reductive mind who has not mastered the art of transcending disciplinary boundaries and does not understand the meaning of penetrating deeply."

Yes indeed. Remind us, what was it that the former chief scientific advisor to a former president of the European Commission, said about science having a seat in Government.

Now Cennino Cennini, the author of the *Book of the Art*, wants to say something about Giotto:

"Giotto turned the art of painting from Greek into Latin, and rendered it modern. He mastered our art more completely than anyone else ever did."

Giorgio Vasari, author of *the Lives of the Most Eminent Painters, Sculptors, and Architects*, now would also like to comment on Giotto:

"The original Byzantine style was entirely abandoned, at first through the efforts of Cimabue and then with the help of Giotto. From it arose a new style which I like to call Giotto's, because it was introduced by him and his pupils, and was afterwards universally admired and imitated. In this the profile surrounding the whole figure is abandoned, as well as the lustreless eyes, the tiptoed feet, the attenuated hands, the absence of shadow, and all the other Byzantine absurdities, which were replaced by graceful heads and beautiful colouring.

"Giotto in particular improved the attitudes of the figures and began to give a measure of vivacity to the heads and folds to the draperies, which made a closer approach to nature than is seen in the work of his predecessors, while he partially discovered the art of foreshortening figures. He also was the first to express emotions, so that fear, hope, rage, and love may be partly recognised."

Should we believe this?

Oh dear it seems that Wilczek wants to continue speaking. Please go ahead, why should we stop you when your are making such a good job of making yourself look foolish.

"Physicists often, and rightly, extol the beauty of their concepts and equations. On the other hand, humans are intensely visual creatures. It will be fruitful to use the modern resources of signal processing and computer graphics to translate those beautiful concepts and equations into forms that the visual cortices of physicists can grasp, and that people in general can admire and enjoy. We need, in particular, better ways to put ourselves in sensory contact with spaces of high dimension, such as the ones we meet in quantum theory and when handling big data. Awesome opportunities await modern champions of Brunelleschi's spirit."

And this is the extent of your vision for working with artists?

"Hey, you art chap, sketch that plant for me."

"Hey, you art chap, visualise that data for me."

"Hey, you art chap, communicate my wonderful work to the public for me."

"Hey, you art chap, promote science and technology for me."

"Hey, you art chap, make me look good."

"Hey, you art chap, project my power."

"Hey you art chap, make those dumb technologist more creative."

"Hey you art chap, create some unconventional and compelling products that will keep alive our neo-liberal agenda."

Joseph Stalin will now speak.

"Hey, you art chap, communicate my wonderful work to the Soviet public, promote scientific communism, make me look good, project my power. I will be watching you!"

Actually the start of the Scientific Revolution would be better positioned with Galileo rather than Copernicus. You should know why.

In any case Copernicus was a priest, and we know that you do not like priests, so probably therefore Galileo is the better choice for you. Most of you also believe in the myth that Galileo was imprisoned and then tortured by the Inquisition. An old example of Fake News perhaps? More likely though, it is an example of a Protestant History.

**141**

Leonardo remains on our mind. But first a small diversion!

Lao Tsu wants to speak:

"So it is that in the enumeration of the different parts of a carriage we do not come on what makes it answer the ends of a carriage."

Carriages are meaningful because they have a purpose.

"It suffices in the case of imperceptible entities if one explains how they may be, even if they are not in fact as described."

That was René Descartes speaking.

"Just as a watchmaker could construct two watches which were externally similar and equally accurate in keeping time, but with very different internal mechanism, so likewise God could have chosen a variety of causal mechanisms to produce the observable effects which we can observe in nature. The implication here is that the scientist cannot unlock the causal mechanism at work in nature by direct inspection, and must be content with postulating the existence of some hidden causes which adequately explain observable physical phenomena."

That was Desmond Clarke's observations on Descartes remark.

"No sooner did a Greek philosopher learn half a dozen geometrical prepositions and hear that the phenomena of the heavens recur in cycles than he set to work to look for law everywhere in nature and with an audacity amounting to hubris to construct a system of the universe."

That was John Burnet speaking. History records that it was the Ionians where it all began - in Europe that is. Elsewhere, in China, another process was also underway, but that has not been written into the history that you learned.

"The Ionians had prised the world-oyster open, the Pythagoreans had set the earth-ball adrift in it, the atomists dissolved its boundaries in the infinite. Aristotle closed the lid again with a bang, shoved the earth back into the world's centre, and deprived it of motion. It was Aristotle who promoted the idea of circular motion to a dogma of astronomy."

That was Arthur Koestler speaking.

Claudio Giorgione, the curator of the Leonardo da Vinci National Museum of Science and Technology in Milan, wants to speak about Leonardo.

"Many of Leonardo's drawings, even while fascinating, sophisticated, or precursory of modern ideas, were executed rapidly and lack many mechanical and structural details, some of which would be indispensable for any sort of practical realisation. For example, in the covered war chariot, Leonardo did not draw details for the mortar-pieces."

**142**

Leonardo remains on our mind. Did we say that before?

Howard Rosenbrock, FRS, wants to speak. Go ahead Howard.

"Karl Popper gave an influential account of science, which starts with the formulation by the scientist of a theory. To be accepted as scientific the theory must be capable of being refuted by observed facts. Scientific effort is then devoted to refuting the theory, and if it survives this effort it is adopted, at least provisionally, as valid. Scientific truth thus comes to consist in all those theories which have not yet bee refuted. This is probably the most widely accepted account among the *hard* sciences such as physics or chemistry.

"All methodological proposals, including Poppers, however, have been strongly questioned by philosophers of science such as Kuhn, Feyerabend and Lakatos who have shown that what actually happens in science often bears little relation to any of the systematic accounts. As part of this questioning it is often asserted that any set of facts is consistent with more than one theoretical structure which can be erected for their explanation. The theories which are consistent with some particular set of data are not, in the usual account, identical in their predictions, and more extended data may therefore eliminate some, while leaving others still in contention.

"All working scientists are probably aware in outline of these developments, and they have had an effect on the perception, by scientists, of science. But in the actual, day-to-day practice of science they have had little influence. They come from outside (though close to) the active scientific community, which carries with it its own practices and standards of judgement.

"The crucial point is that science incorporates values which are not a part of science, but which profoundly affect the way we view the world, and the way we behave. In particular science eliminates purpose."

So now, Stephen Hawkins, you know Karl Popper's philosophy of science, which is not positivism, and you know also, that you, as you have amply demonstrated through your comments, are often found to be saying 'I believe ...', as does Dawkins, Wilson, Wolpert, and that astrophysicist whom we have already met, who bears the weighty title of *distinguish professor of Art and Technology* – yet another ideology! And, saying *I believe* too, are the rest of the world of STEM, their artists in residence, and the artists who work in other modes as well, but who subscribe to the beliefs and values of the secular religion that science, in the minds of its preachers and disciples, often becomes.

Karl Jaspers:

"The task of mastering the *Age of Technology* in favour of the specifically human element in humanity has become conscious in the major tendencies of this age – socialism and world order.

"It is not enough, however, to found them on science, technology and civilisation. These do not afford a reliable foundation. They are in the service of evil as well as good. Man himself must live from a different origin. Hence flaws have today appeared in our trust in science. Superstitious beliefs in science, enlightenment that has proved deceptive, and the disintegration of its contents bear witness against it.

"Even the great spiritual powers handed down to us no longer support life. There is no longer any complete trust in humanism: it seems to stand aloof, as though it were not there at all.

"It is no longer possible for the masses to place entire trust in the Churches as such; they were too impotent when evil triumphed.

"Science, humanism, and the Churches are too indispensable to us; we shall never relinquish these powers. They do not suffice, they conceal in themselves evil aberrations, but by virtue of their potentialities they are indispensable preconditions for the whole of man.

"Today the situation demands: We must return to a deeper origin, to a fountain from which all faith once welled forth in its particular historical shapes, to this wellspring which can flow at any time man is ready for it."

And the question of our time is this: will art will be used to show the way to this wellspring, or, as is now happening, will it be subjugated to the old agendas from which so much woe has flowed, and will continue to flow so long as we do not act on what Karl is calling us to do, as well as Howard Rosenbrock, as you will in due course discover?

Civic humanism: a politically committed humanism that emerged in Florence in the wake of the Milanese wars (1390-1402). Historically, civic humanism can be described as a political and ideological discourse that draws heavily on the political languages of Aristotelian republicanism, Roman statism and the political experience of the Italian communes. Quattrocento Florentine republicanism, the era in which Leonardo da Vinci was born and lived most of his life, has been described as a happy marriage between, on the one hand, the civic tradition of the late medieval commune, and on the other, Petrarchan humanism and classical learning.

**143**

Leonardo remains on our minds. Did we say that before? But a conversation comes to our ears, so let us listen in to see what words flow but which are, it seems, not heard by one speaker in the end.

Thus for the first speaker:

"As fascinated as I am with how physics has been expanding our understanding of reality, I wish any and all knowledge that we attain be a reflection of how well we humans can live in harmony with one another and the planet. Otherwise, it's all empty knowledge."

Thus for the second speaker:

"I agree. The priorities for humanity should not be trips to Mars and such like, which are just more conquest and domination, but the transformation of the world in to one where we live in balance and harmony with nature and with our fellow human beings, and to do that, we need to divert resources into activities that move us towards a position of global equality and justice in all countries including those in the West that now preach to the developing countries, but are themselves full of examples of inequality and injustice. So let us spend the money that is being used for Mars trips, large colliders and such stuff, which can wait for the future, on resolving humanities most important issues. If we cannot do that now, with all we know and all the wealth we have, then humanity has no future worth having. And this is not a humanity that I want to be associated with."

Thus for the third speaker – Narcissus is his name:

"I couldn't disagree more. One of the healthiest things mankind can do is to study the extremes of the universe – the very small, the very large, the very old, and the very distant. It's only by having a sense of universal perspective that mankind can defeat the hubris, ideology, and ego that drive political, religious, and cultural conflicts that reach a peak with warfare, but also cause constant friction in our everyday lives. Astronauts returning from space have described an experience some have called the *overview effect*. It refers to the experience of seeing firsthand the reality of the earth in space, which is immediately understood to be a tiny, fragile ball of life, *hanging in the void,* shielded and nourished by a paper-thin atmosphere. From space, national boundaries vanish, the conflicts that divide people become less important, and the need to create a planetary society with the united will to protect this *pale blue dot* becomes both obvious and imperative. We can't all leave the planet. At least not yet! But the wonders offered by science can induce a kind of overview effect if fully embraced."

So advanced knowledge and simplistic beliefs going hand in hand, which is often the case; the delight of the fragmented and reductive mind! We will leave you the reader to reflect on this particular manifestation of ideology and the contradiction between this Narcissistic view, and what the second speaker said.

We think, but all too often we think *fast*. Thus does the future become the past! Thus the world does not turn. Time is a singularity. Thus do they stare at their pretty photographs, their wondrous images, and such forth, but never do they put into practice what the second speaker said, for they would rather create more such images, rather than use their supposed awakening to put right all that is wrong. How many images do you need of the very small, the very large, the very old, and the very distant, before your humanity cries out - enough!

We can choose to walk a different path.

**144**

Leonardo remains on our mind, but first to the matter of Stephen Hawkins and his strange belief that Karl Popper was a positivist.

Hawkins, in his book, *The Universe in a Nutshell,* several times refers to his positivist beliefs. Why on earth does he do this? Does he not know that positivism is discredited? Has he not been keeping-up with philosophy?

Stephen, for your information, although by now you probably know, Karl Popper's philosophy of science can be put quite simply: knowledge proceeds by way of *conjectures and refutations.*

As for the positivism, by which you probably mean logical positivism, you seem, like many people in the world of STEM to be rewriting history. And speaking of those who rewrite history, we have now to question one of those STEM people (from prestigious institutions) who wrote that review paper, *Art on the Nanoscale and Beyond*. So what do you want to say about Leonardo?

"The domains of art and science have traditionally been separate, both by the creators of content and by their educators. Although Renaissance humanists, with Leonardo being their most famous example, took exception to this view, scientific and artistic methodologies are typically taught separately from our early childhood as disparate disciplines."

Taught as disparate disciplines! Is that all that gets taught - methodologies? It may have missed your attention, coming as you do from the world of STEM, that: mathematics, biology, physics, chemistry, history, geography, religion, English language, English literature, languages such as French, German, Spanish, Latin, etc., design, sports, art ... are all taught separately. Would you like to say why? Or is that taxing your fragmented and reductive mind too much?

So let us now use our imagination. We will now travel back in time to the late fifteenth century, and look there is Leonardo standing outside a university. To his left there is the Faculty of Science, and to his right, the Faculty of Arts and Humanities. And look, he his writing something down on one of those famous pieces of paper that everyone seems to wax lyrical about. Let us see what he has written. Oh look he has written: "I take exception to this separation of the sciences from the arts and humanities."

We must now ask this question: did he write this down in Latin or in the vernacular? Do you know the significance of this question? Probably not, if you are from the world of STEM intent on rewriting history!

Now, as for the rewriting of history; you say that Leonardo was a humanist. So what type of humanist was he? Do you know what this term meant in Leonardo's era? Do you know what it means in our time?

Let us see what the Encyclopaedia Britannica has to say: the term humanist is applied to a variety of beliefs, methods and philosophies. So let us now look at one that might be relevant - Renaissance humanism. A representative of the Encyclopaedia Britannica is here to speak:

"The history of the term humanism is complex but enlightening. It was first employed (as *humanismus*) by nineteenth century German scholars to designate the Renaissance emphasis on Classical studies in education. These studies were pursued and endorsed by educators known, as early as the late fifteenth century, as *umanisti* - that is, professors or students of Classical literature. The word *umanisti* derives from the *studia humanitatis*, a course of Classical studies that, in the early fifteenth century, consisted of grammar, poetry, rhetoric, history, and moral philosophy. The *studia humanitatis* were held to be the equivalent of the Greek *paideia*. Their name was itself based on the Roman statesman Cicero's concept of *humanitas*, an educational and political ideal that was the intellectual basis of the entire movement. Renaissance humanism in all its forms defined itself in its straining toward this ideal.

"No discussion of humanism, therefore, can have validity without an understanding of *humanitas,* which meant the development of human virtue, in all its forms, to its fullest extent. The term thus implied not only such qualities as are associated with the modern word humanity – understanding, benevolence, compassion, mercy – but also such more aggressive characteristics as fortitude, judgment, prudence, eloquence, and even love of honour.

"Consequently, the possessor of *humanitas* could not be merely a sedentary and isolated philosopher or man of letters but was of necessity a participant in active life. Just as action without insight was held to be aimless and barbaric, insight without action was rejected as barren and imperfect. *Humanitas* called for a fine balance of action and contemplation, a balance born not of compromise but of complementarity.

"The goal of such fulfilled and balanced virtue was political, in the broadest sense of the word. The purview of Renaissance humanism included not only the education of the young but also the guidance of adults (including rulers) via philosophical poetry and strategic rhetoric. It included not only realistic social criticism but also utopian hypotheses, not only painstaking reassessments of history but also bold reshaping of the future. In short, humanism called for the comprehensive reform of culture, the transfiguration of what humanists termed the passive and ignorant society of the *dark ages* into a new order that would reflect and encourage the grandest human potentialities. Humanism had an evangelical dimension: it sought to project *humanitas* from the individual into the state at large.

"And the wellspring of *humanitas* was Classical literature. Greek and Roman thought, available in a flood of rediscovered or newly translated manuscripts, provided humanism with much of its basic structure and method.

"Humanism, then, may be accurately defined as that Renaissance movement that had as its central focus the ideal of *humanitas.* The narrower definition of the Italian term *umanisti* notwithstanding, all the Renaissance writers who cultivated *humanitas,* and all their direct *descendants,* may be correctly termed humanists."

And what language do you think all this high flying idealism and learning was conducted in?

Does this description fit with Leonardo? We think you will find if you go in deeply, that it does not.

The intentions you describe in the paper *Art on the Nanoscale and Beyond* leads to an art form that has a very special name - the name is totalitarian art. Do not go away; Joseph Stalin wants to talk to you.

**145**

As Karl Popper is here he will say some words about disciplines. First though a definition:

Ignorance: lack of knowledge or information.

One also might say that science has not kept up with ... The list is quite long, as Stephen Hawkins well demonstrates. Having a reductive mind might well help in becoming a brilliant theoretical physicist, but it is also an enabler for the saying of stupid things, as Hawkins and others (Snow, Glover, DG CONNECT ... The list of names is long) well demonstrate. Time to do something about the growing problem that STEM and its people have become - they are like the priests of old! But the time of this Western, secular, narrow, reductive, etc. approach to STEM is over. It seems though, as once before with the Scholastics, few from the world of Western STEM will be participating in the transformation. And this transformation will not be about communication, visualisation, or creativity which is the current foci of most art-science and art-technology activities - image making one might say!

Karl, over to you! But we will first note, before you speak, that you could tell us something about knowledge and ignorance. But perhaps we will not ask you to speak about this. We will slowly reflect on the matter.

"The belief that there is such a thing as physics, or biology, or archaeology, and that these studies or disciplines are distinguishable by the subject matter which they investigate, appears to me to be a residue from the time when one believed that a theory had to proceed from a definition of its own subject matter. But subject matter, or kinds of things, do not, I hold, constitute a basis for distinguishing disciplines. Disciplines are distinguished partly for historical reasons of administrative convenience (such as the organisation of teaching and appointments), and partly because the theories which we construct to solve our problems have a tendency to grow into unified systems. But all this classification and distinction is a comparatively unimportant and superficial affair. We are not students of some subject matter, but students of problems. And problems may cut across the borders of any subject matter or discipline."

**146**

Karl Popper will say some words later about something invisible. First though, a definition:

Invisible: concealed; unable to be seen.

Lao Tsu wants to speak:

"That which is at rest is easily kept hold of; before a thing has given indications of its presence, it is easy to take measures against it; that which is brittle is easily broken; that which is very small is easily dispersed.

"Action should be taken before a thing has made its appearance; order should be secured before disorder has begun.

"The tree which fills the arms grew from the tiniest sprout; the tower of nine stories rose from a small heap of earth; the journey of a thousand *li* commenced with a single step.

"He who acts with ulterior purpose does harm; he who takes hold of a thing in the same way loses his hold.

"The sage does not act so, and therefore does no harm; he does not lay hold so, and therefore does not lose his hold.

"But people in their conduct of affairs are constantly ruining them when they are on the eve of success.

"If they were careful at the end, as they should be at the beginning, they would not so ruin them.

"Therefore the sage desires what other men do not desire, and does not prize things difficult to get; he learns what other men do not learn, and turns back to what the multitude of men have passed by.

"Thus he helps the natural development of all things, and does not dare to act with ulterior purpose of his own."

Soon we will see a new philosophy of science in action – this is what you, if you are a taxpayer, will be paying for! It has a special name – it is called *I believe science*.

About Rachel Carson – she will be here. This we promise!

What is *li*? It is an Ancient Chinese measure of distance.

**147**

Marc Luyckx Ghisi has returned, but before he speaks, we will provide another definition.

Invisible: concealed; unable to be seen.

Now Marc will speak:

"The major problem in our new type of industrial world is what to do with human beings. We can replace them as much as possible with machines. By investing in ever increasingly powerful computers we can minimise human intervention. Another way is to manipulate the human brain."

Probably the *Vainglorious Enlightened Ones* will do both, for they are but manifestations of the same belief that humans are machines, and ultimately they will re-engineer these machines to make them perfect. Perfect is never actually specified of course, but the implications are clear - cybermen (übermensch). It is a very male European vision grounded in their Ancient Greek, Abrahamic culture.

They see humans and machines as being comparable, and when you think like this, you end up reducing people to machines. This is the nature of the fragmented and reductive mind and it cannot foresee any negative consequences. This is not *humanism* any more. This is not living life according to high-minded ideals. This is the base degrading of humanity to an animal governed by cause and effect. It is also an opportunity to make money. If you think we are condemning you - you are right!

There are people who say that to criticise science is to attack science. They are only expressing a belief!

**148**

Marc Luyckx Ghisi has returned, but before he speaks, we will provide another definition.

Invisible: concealed; unable to be seen.

Now Marc will speak:

"The main shortcoming of the analytical method is that it is incapable of thinking in a synthetic and holistic manner. But to reflect on humanity's future, we must look at the problem in the most global way possible, and for this, the modern analytical method appears insufficient and ineffective. Indeed modernity tends to follow the advice of Descartes, which is to cut a difficult problem into pieces that are easier to analyse and resolve, so that, at the end, one only has pieces of the solution or partial solutions - never a global solution."

Invisible: concealed; unable to be seen.

Ignorance: lack of knowledge or information.

The fragmented and reductive mind!

Marc again:

"Introducing new and sane distinctions, and enhancing the role of rational thinking and *scientific* method, modernity has allowed the birth of science and technology and all the other disciplines used today - such as ethics, aesthetics, philosophy, history, mathematics, physics, chemistry and, later, sociology, psychology, anthropology, ethnology, etc. Unfortunately, these distinctions between the disciplines have become strictly separated and compartmentalised, which forbids any real transdisciplinary creative work today. The global holistic view has been progressively lost. Modern science is excellent in analysing every leaf, every tree, every branch, every root, etc., but is unable to see the forest."

Invisible: concealed; unable to be seen.

Ignorance: lack of knowledge or information.

The fragmented and reductive mind!

**149**

Leonardo remains as issue to explore further, to see, not just in the small detail, but also more holistically, and from several different views. It is the only way to come to a degree of understanding necessary for the purpose at hand.

To see the world differently; to know and to understand; to hold in the balance the known and the unknown and not to be blind; to hear; to be at one with nature; to be part of something, while maintaining critical independence. These things can be done.

Invisible: concealed; unable to be seen.

Ignorance: lack of knowledge or information.

Delusion: An idiosyncratic belief or impression maintained despite being contradicted by reality or rational argument.

The fragmented and reductive mind!

Lao Tsu wants speaks one more time:

"To know and yet think we do not know is the highest attainment; not to know and yet think we do know, is a disease."

Donald Barthelme now speaks:

"A writer is one who, embarking upon a task, does not know what to do. Writing is a process of not knowing, a forcing of what and how."

We were in that position at chapter 1 - day 1 - of not knowing. We still do not know. On day 366, when we finish writing *366 - A Scriptovisual Composition Unknown*, then, and only then, will we know. What goes for the writing also goes for the painting as well - the painting that accompanies this work, for it is a dual creative process, a dual act of *not knowing*.

"Now, we daily see what science is doing for us. This could not be unless it taught us something about reality; the aim of science is not things themselves, as the dogmatists in their simplicity imagine, but the relations between things; outside those relations there is no reality."

That was Henri Poincaré speaking.

**150**

More to come, we suppose, about Leonardo and also C.P. Snow - but not yet!

Now Joseph Stalin is here and he will not wait. He commands your presence, you STEM people who wrote the review paper *Art on the Nanoscale and Beyond*. We leave you to him - how delightful!

"So comrade scientist, explain to me the psychology of using art to get people to celebrate scientific communism. Now! I am a very busy man."

"Yes indeed. So it goes like this: The arts also allow complex scientific communist ideas to be synthesized, simplified, and conveyed while also rendering the information memorable. Consequently, the utilization of critical art in communicating Soviet policies can distil scientific communist knowledge to a socially fundamental level and encourage new ways for Party officials to look at problems, which can play a significant role in public campaigns. For example, installations and exhibitions can encourage Party officials to communicate in alternative ways, and to convey complex scientific communist concepts to the public more effectively at a socially acceptable level. The visual and performing arts also create memorable moments and a celebratory atmosphere by appealing to emotions. When the emotions are targeted, individuals pay more attention to a particular event and commit to the cause, storing information in their long-term memory. According to Triandis' *Theory of Interpersonal Behaviour*, emotions significantly influence behaviour, which is moderated by moral beliefs, and cognitive limitations. Such an experience can create psychological ambiguity, in which the basic ideas and norms are broken down or weakened, and this disorientation allows individuals to reposition their points of view. The incorporation of art into communication of scientific communism can appeal to emotions and form a celebratory atmosphere."

"Excellent. Do it. This is not a request. And I will fund your research projects for five years. You wanted a Chair. It's yours."

It is amazing how a simple change of context highlights the true nature of what STEM people are up to in their appropriation of art. The original text you see came from a paper that was proposing to use art as a psychological tool to manipulate people's emotions and opinions. Someone once said that the exploitation of art (artists) is evil. Now you know why they said that.

People of the world - we have a problem. It is an old one. What are we going to do about these secular priests? Dewey's monster is here.

**151**

The West does not have a *Theory of Knowledge* anymore, just a science that thinks that it is the *Theory of Knowledge*.

Now it is time to see what the Encyclopaedia Britannica has to say about Leonardo. Just an examination of some of the things that they say, bearing in mind that here too in this apparently reputable publication, the Leonardo myth is alive and well, living among the many words, which include a fair amount of speculation and embellishment, which no serious scholar would accept at face value. But if you are a member of the art-science and art-technology mutual appreciation societies, serious scholarship is probably not on your mind. Is it?

We will also examine what they, the Encyclopaedia Britannica, do not say! So, here one more time, is the spokesperson for the Encyclopaedia Britannica:

"Leonardo was the illegitimate son of a Florentine notary and landlord. He grew up on his family's estate where he was treated as a legitimate son. He received as such a person, a basic education in reading, writing, and arithmetic. He did not seriously study Latin, the key language of traditional learning, until much later, when he acquired a working knowledge of it on his own. He also did not apply himself to higher mathematics - advanced geometry and arithmetic - until he was 30 years old, when he began to study it with diligent tenacity."

What they do not say, is whether Leonardo ever used Latin as a working language in his written works. What they do not say is whether

Leonardo was a good mathematician. They are not very specific about what mathematics he studied. What we do know however, which they do not say, is that science and engineering in Leonardo's day had yet to be mathematicised. This began with Kelper, followed by Galileo, and then, significantly by Newton, who was a mathematician. What they do not say is that much of the mathematics used in modern science and engineering had yet to be developed when Leonardo was alive.

People wanting to know more can of course look at the various Leonardo documents, those famous notes and sketches and see whether he wrote in Latin or the vernacular, and to see if they are any equations or calculations therein contained. It would be an interesting exercise!

Back now to what the Encyclopaedia Britannica has to say. Their spokesperson again:

"Leonardo received training in painting and sculpture as well as the technical-mechanical arts. In 1472 Leonardo was accepted into the painter's guild of Florence. Meanwhile he remained in his teacher's workshops for five more years, after which he worked independently in Florence until 1481. There are a great many pen and pencil drawings from this period, including, many technical sketches - for example, pumps, military weapons, mechanical apparatus - that offer evidence of Leonardo's interest in and knowledge of technical matters even at the outset of his career."

What they do not say is what constitutes the technical-mechanical arts, which presumably includes pumps, military weapons, mechanical apparatus, and such forth. Nor do they say why Leonardo sketched such things, nor do they say that he was inventing these things. They also do not say what a painter's guild was, what these guilds represented in that era and their history, or what being a member of such a guild meant in terms of social status. You could find out about the latter. It would be an interesting exercise!

So what are we saying?

**152**

The science of the Western world is based on a *Theory of Knowledge* entirely Western in character.

Then the Encyclopaedia Britannica tells about Leonardo's first Milan period, working in the service of the Duke, who we have already mentioned. Their spokesperson again:

"The count here is 17 years in total. He worked as an artist, sculptor, and designer of court festivals."

The spokesperson could have gone on to tell about the artistic work undertaken, or commissioned, telling you also about a statue in bronze, of the founder of the Sforza dynasty, about which we have actually already heard. They say, do the Encyclopaedia Britannica, that the statue is probably the main reason why Leonardo was invited to Milan! They say also that Leonardo was frequently consulted as a technical adviser in the fields of architecture, fortifications, and military matters, and he served as a hydraulic and mechanical engineer.

What they do not say, is what architecture, fortifications, and military matters he was consulted about, nor what consulted actually means. Neither do they say what work he did, if any, as a hydraulic and mechanical engineer. We would suggest that you find out for yourself, and that it would be an interesting exercise to do so, only you may find the task rather difficult if not impossible.

Here is the spokesperson yet one more time:

"In 1500, Leonardo is back in Florence, where he is received with acclaim and honour. In that very same year he was appointed an architectural expert on a committee investigating damages to the foundation and structure of church San Francesco al Monte. He seems to have been concentrating on mathematical studies rather than painting."

They do not say if this appointment to the committee was just an honorary one, or if he gave any advice, or what advice, nor if any other architects and engineers had been appointed for the investigation. Nor do they say what mathematics Leonardo was studying, if indeed he was studying mathematics, for the wording is not certain.

Encyclopaedia Britannica spokesperson:

"In 1502 Leonardo was working in the service of Cesare Borgia, surveying territory and sketching city plans and topographical maps, creating early examples of aspects of modern cartography. Then in 1503, he was in Florence, to make an expert survey of a project that attempted to divert the Arno River behind Pisa, to deprive the city of Pisa, then under siege by the Florentines, access to the sea. This led him to reconsider an idea first advanced in the thirteenth century, to build a canal that would connect Florence by water to the sea.

"He produced panoramic artistic landscape sketches in the period while in Florence that he also undertook dissections at a hospital, made systematic observations of flying birds, and he studied the physical properties of water."

Have you ever tried making systematic observations of flying birds? Without the aid of binoculars! Try it sometime. We did. Then come to your own conclusions about this claim.

What is not said is whether Leonardo worked alone on his surveying and map making work, or if indeed he wanted to do the work and the reasons why he did. What is also not said is why he spent his time on a canal project, the idea for which was an old one, if he had any interest in its implementation, whether in fact it was a sound plan, and if anyone was interested. Also, what they do not say, in connection with the studies, which they believe were scientific (but without taking into account the meaning of such a phrase in Leonardo's time), what was already known in this period. They do not say either whether what he studied resulted in anything that was new by way of discovery, or if indeed if was correct about the things he studied.

So what are we saying?

**153**

The Western *Theory of Knowledge*, projected, through the West's imperial power, on to the non-European world, has a name - cultural colonialism. You just cannot stop being Ancient Romans can you!

Then the Encyclopaedia Britannica, say things about Leonardo's second period in Milan. Here is their spokesperson:

"His patrons were Charles d'Amboise and King Louis XIII of France, and his duties were largely architectural. Tangible evidence exists for this claim in the form of plans for a palace-villa for Charles, and, it is *believed* that he made some sketches for an oratory for the church of Santa Maria alla Fontana. He also *looked into* an old project idea of linking Milan with Lake Como by water. He collaborated with a famous anatomist from Pavia - Marcantonio della Torre - and he mapped out a plan for an overall work that would include detailed reproductions of the human body and its organs, including a comparative anatomy and the whole field of physiology.

The spokesperson for Encyclopaedia Britannica also has some words about Leonardo's last years:

"He left Milan and went to Rome. Here, he kept a low profile and worked in his studio on mathematical studies and technical experiments. He surveyed ancient monuments. He magnificently executed a map of the

Pontine Marshes, which *suggests* that Leonardo was at least a consultant for a reclamation project that was initiated in 1514. He also made sketches for a new spacious residence for Medici in Florence."

What they do not say is why anyone would entrust an architectural project to Leonardo. Was he an architect? We know that Brunelleschi was an architect, as there is evidence for this. But what evidence exists that Leonardo was able to function as an architect and if he actually did so? There is more to architecting than making sketches! What they do not say either is what this collaboration with the anatomist involved. Perhaps we will encounter this again? Also, the Encyclopaedia Britannica do not say what mathematical studies and technical experiments Leonardo undertook. They do not say what monuments he surveyed, or what survey means in this case. Nor do they say for certain that he was involved in the reclamation project, or in what capacity he was working, when he made those sketches for a spacious residence.

So what are we saying?

**154**

If Europe had never been, and if the science that developed in another civilisation - China - had matured into a modern type of science, then would we have a different *Theory of Knowledge*? Do you not think that this would be the case? Or do you believe in determinism, in the one best way?

Then the Encyclopaedia Britannica has something to say about Leonardo's achievements.

"Between 1490 and 1495, he began a great programme of writing, and he developed an interest in two fields - the artistic and the scientific - which built towards a creative dualism that sparked his inventiveness in both fields."

A dual creative process! Or perhaps not!

The Encyclopaedia Britannica speaks too of his architectural achievements. They also say that according to Leonardo's own estimate, he dissected 30 corpses. They also say he invented machines.

What they do not say is why they refer to two distinct fields, artistic and scientific, when in Leonardo's time the meaning of science was quite different from what it means in the modern world. What they do not say is why, having admitted that his effectiveness was limited to the role of advisor, they still think of him as an architect even though there is no evidence for his personal participation as an architect. What they do not say is why they take this stance, when, they could have described him as being, for example, a draughtsman - making drawings for an architect.

What they do not say too, is whether Leonardo actually dissected human corpses himself, or was just present when others did this; nor do they say whether he also dissected animals. What they do not say is how they know that he invented machines, because making drawings and sketches of machines, and making notes, is not the same thing as inventing machines.

What they do say however, which is quite interesting, is that Leonardo was seeking to elevate the status of painting above that of the other arts, and also above the mechanical arts. He, they say, gave authority to the eye, in the context of *saper verdure* - which means *able to see* or *knowing how to see.* According to the Encyclopaedia Britannica, Leonardo believed that a painter, endowed with subtle powers of perception and the complete ability to pictorialise them, was the person best qualified to achieve true knowledge, as he could closely observe and then carefully reproduce the world around him. Hence they say that Leonardo conceived the staggering plan of observing all objects in the visible world, recognising their form and structure, and pictorially describing them exactly as they are.

This of course is description not explanation.

They, also, in drawing to an end, make an interesting statement, which is in summary: through his process of visual observation and drawing, he developed his own, *Theory of Knowledge* - one that was unique.

So what are we saying?

**155**

Of course, Western STEM people, being as they are, *people of the book*, and therefore caught-up in the notion of the sole truth, cannot imagine a *Theory of Knowledge* different to what they now have, if, that is, any Western STEM people, with their fragmented and reductive minds, know what a *Theory of Knowledge* is. Can you imagine a different *Theory of Knowledge*? How are you feeling now?

So what are we saying?

Now, while on this matter of *Theory of Knowledge,* it is a good time for Imanuel Kant to speak. And this is what he wants to say:

"We have on transcendental principles good ground to assume a subjective purposiveness in nature, in its particular laws, in reference to its comprehensibility by human Judgement and to the possibility of the connexion of particular experiences in a system. This may be expected as possible in many products of nature, which, as if they were devised quite specially for our Judgement, contain a specific form conformable thereto; which through their manifoldness and unity serve at once to strengthen and to sustain the mental powers (that come into play in the employment of this faculty); and to which therefore we give the name of beautiful forms.

"But that the things of nature serve one another as means to purposes, and that their possibility is only completely intelligible through this kind of causality for this we have absolutely no ground in the universal Idea of nature, as the complex of the objects of sense. In the above-mentioned case, the representation of things, because it is something in ourselves, can be quite well thought *a priori* as suitable and useful for the internally purposive determination of our cognitive faculties; but that purposes, which neither are our own nor belong to nature (for we do not regard nature as an intelligent being), could or should constitute a particular kind of causality, at least a quite special conformity to law, this we have absolutely no *a priori* reason for presuming. Yet more, experience itself cannot prove to us the actuality of this; there must then have preceded a rationalising subtlety which only sportively introduces the concept of purpose into the nature of things, but which does not derive it from Objects or from their empirical cognition. To this latter it is of more service to make nature comprehensible according to analogy with the subjective ground of the connexion of our representations, than to cognise it from objective grounds.

"Further, objective purposiveness, as a principle of the possibility of things of nature, is so far removed from necessary connexion with the concept of nature, that it is much oftener precisely that upon which one relies to prove the contingency of nature and of its form. When, for example, we adduce the structure of a bird, the hollowness of its bones, the disposition of its wings for motion and of its tail for steering, etc., we say that all this is contingent in the highest degree according to the mere *nexus effectivus* of nature, without calling in the aid of a particular kind of causality, namely that of purpose (*nexus finalis*). In other words, nature, considered as mere mechanism, could have produced its forms in a thousand other ways without stumbling upon the unity which is in accordance with such a principle. It is not in the concept of nature but quite apart from it that we can hope to find the least ground *a priori* for this.

"Nevertheless the teleological act of judgement is rightly brought to bear, at least problematically, upon the investigation of nature; but only in order to bring it under principles of observation and inquiry according to the analogy with the causality of purpose, without any pretence to explain it thereby. It belongs therefore to the reflective and not to the determinant judgement. The concept of combinations and forms of nature in accordance with purposes is then at least one principle more for bringing its phenomena under rules where the laws of simply mechanical causality do not suffice. For we bring in a teleological ground, where we attribute causality in respect of an Object to the concept of an Object, as if it were to be found in nature (not in ourselves); or rather when we represent to ourselves the possibility of the Object after the analogy of that causality which we experience in ourselves, and consequently think nature technically as through a special faculty. If we did not ascribe to it such a method of action, its causality would have to be represented as blind mechanism. If, on the contrary, we supply to nature causes acting designedly, and consequently place at its basis teleology, not merely as a regulative principle for the mere judging of phenomena, to which nature can be thought as subject in its particular laws, but as a constitutive principle of the derivation of its products from their causes; then would the concept of a natural purpose no longer belong to the reflective but to the determinant Judgement. Then, in fact, it would not belong specially to the Judgement (like the concept of beauty regarded as formal subjective purposiveness), but as a rational concept it would introduce into natural science a new causality, which we only borrow from ourselves and ascribe to other beings, without meaning to assume them to be of the same kind with ourselves."

**156**

Today we saw a raven. Not one in the Tower of London, which are in effect in captivity, but one in the wild. A raven in the County of Cheshire! That is a first for us. They are making a comeback after the efforts to exterminate them in the past. Recovery is possible.

So given all that has passed before your eyes concerning Leonardo, what conclusions can be reached in terms of that which actually was of interest? What was it that was of interest? Actually we did not say, because we did not know. Now we do – it is *Theory of Knowledge* that is important. Or perhaps this is not the case and we still are waiting to know. Let us therefore keep this open. No closing, no convergence.

So we now proceed, but not in the way you are probably expecting. We have invited some more people to speak about Leonardo!

First to speak is H.T. Pledge:

"Artists like Leonardo da Vinci (1452-1519) were, among the most enterprising workers in anatomy, optics, and mechanics. This lack of specialisation went with the absence of compact self-conscious scientific groups. And no one did for Leonardo what Halley did for Newton - made him write up his results in mechanics and saw them through the press. Leonardo's notebooks remained unpublished. In the interval between the two men the scientific world had already taken a long stride forward.

"From *Roger Bacon* (1214-94) until well into our period, discoveries were made and lost and made again. Leonardo's own break with Aristotle in mechanics had been anticipated by *Jean Buridan* (c. 1297-1358), and had to be re-made, this time finally, by Galileo. Like Roger Bacon, Leonardo urged the primacy of mathematics; but, also like Bacon, he did not in his own person end the isolation of that greatest legacy of Greece - from the vast fields of application which awaited it.

"Abstract clarity, the logic of *the enfant terrible*, is one of the virtues of the pre-scientific stage. In scholasticism there was often a strong clarity; and Leonardo da Vinci leapt at once beyond perpetual motion and the like to Newton's first and third laws and to Archimedes' clear conception of fluid pressure. Only, without mathematics, he failed to develop them.

"As regards the state of the individual sciences about 1450-1500, physics had not yet defined its fundamental units of thought. Its individual branches formed, therefore, a scattered collection. Sound was much occupied with a semi-mystical arithmetic of music. Magnetism, in spite of the stimulus of navigation, was in abeyance. Electricity was confined to a few mysterious observations on amber and the like. Heat was, in the main, part and parcel of alchemy.

"Like biology, optics and mechanics interested the great artists. To us though, their optics is little more than geometry - than perspective. But then both perspective and anatomical accuracy were part of the new phase of artistic naturalism which was setting in, with men like Botticelli (1444-1510), *Diirer* (1471-1528), and Leonardo da Vinci. Such phases are recurrent in the history of art and are commonly associated in discussions of the subject with that very consultation of nature which is one side of the scientific spirit. The geometrical side of perspective was studied by a series of artists, Brunelleschi (1379-1446), Leonardo, Alberti, and Diirer; and this study had important effects in mathematics and physics.

"Leonardo studied the eye, though he thought the liquid centre portion the main image former. But the old view that vision is a matter of tentacles coming out of, as well as light entering, the eye had still not lost its hold. For their movement-studies these artists called in the science of mechanics.

But they did little to advance it and Leonardo's work, being unpublished, produced little effect.

"*Nicholas of Cusa* (1401-64) had, like Leonardo, reached many modern views of motion in general, if indeed he had not reached beyond them. It is convenient to treat a few of these general points because abstract clarity about them has often, in history, been found easier before, than after, specific cases of them (such as the planets) have been investigated in full. Cusa asserted that as no two bodies are strictly alike, strict knowledge is impossible; he denied the strict truth of the then fundamental dogma of the circularity of the heavenly motions; he denied that the earth is the centre of the universe, or is exempt from the universal reign of motion. He had probably a clearer hold than Copernicus on that modern view of space which by itself almost makes the relativity of motion an inevitable concept."

Thanks that very brief overview of some aspects of your work. But to what extent did Leonardo break with Aristotle, and in what way, if at all?

So what are we saying?

**157**

*Theory of Knowledge*! We have mentioned it many times! But we have not finished with Leonardo. Now we will invite Joseph Needham to speak, once again:

"Contrasting Galileo, who must be considered the central figure in the mathematisation of natural science, with Leonardo da Vinci, we see that, despite the latter's deep insight into nature and brilliant experimentation, no further development followed because of his lack of mathematics.

"Leonardo was not the isolated genius that many have been led to suppose him; he was the most outstanding in a long line of men in the 15th and 16th centuries - artist-engineers and architects such as Brunelleschi; artist-metallurgists such as Cellini; gunners such as Tartaglia; surgeons such as Ambroise Paré; miners who found voice in Argicola; shipbuilders such as those of the Arsenal at Venice; gunpowder-millers and other chemical technologists whom Biringuccio represents; and instrument makers such as Robert Norman whose work greatly stimulated William Gilbert's work on the magnet. All these men busied themselves with the investigation of natural phenomena, and most of them produced experimental data ready, as it were, for the magic touch of mathematical formulation. Roughly speaking, they also had their Chinese counterparts. This is a very important point, for those who say, quite rightly, that China never produced a Galileo, a Vesalius or a Descartes, generally forget that

China did produce men of a type similar to Agricola, Gesner, and Tartaglia.

"People such as those named above can be referred to as *higher artisanates*. But they were also inhibited by their lack of hypothesis-making, as one can see in the relative theoretical backwardness of Leonardo. The higher artisans were essentially only experimental.

"Take Leonardo's achievements in matter in the gaseous state, where one can see that his ideas on air and fire, smoke and vapour, are so impregnated with medieval physics, that what he did and suggested, seems almost inexplicable. He was able to sketch a hygrometer, a helicopter, or a centrifugal pump, but he explained that the moisture of a wet rag has an intrinsic tendency to move to the fire, and that its less material parts accompany the ascent of that pure element towards the empyrean, for fire has a quasi-spiritual power of carrying light things up with it.

"There are other examples as well, such as, in relation to the idea of microcosmic-macrocosmic, which was very popular in the Middle Ages, Leonardo's attempt to calculate the size of the world lung.

"The importance of the point is that remarkable technical achievements may be effected without adequate scientific theory. This throws light on the Chinese situation and defines the point reached by the indigenous Chinese science and technology as Vincean, not Galilean.

"The situation in both Europe and China, before 1600, was that there were two groups which possessed portions of Galileo's method: the *higher artisanates* who experimented; and the *scholastic philosophers* who theorised."

So what are we saying?

**158**

*Theory of Knowledge* – it just keeps popping-up all the time. This must be what we are about for sure. This must be the thing that scriptovisualism is directed towards? Surely?

Now an Imam will read from the Holy Qu'ran, from Sûra 3, verse 7:

"He it is Who has sent down to thee the Book: in it are verses basic or fundamental clear in meaning; they are the foundation of the Book; others are not entirely clear. But those in whose hearts is perversity follow the part thereof that is not entirely clear. Seeking discord, and searching for its interpretation, but no one knows its true meaning except Allah. And those who are firmly grounded in knowledge say: 'We believe in it, the whole of it is from our lord:' And none will grasp the Message except men of understanding."

Thank you.

Evidently many men have not grasped its meaning. Richard Dawkins as well as the more obvious ones that sometimes are called, for example, by the BBC, *the so-called Islamic State.*

Now for a brief moment we return to Leonardo, and Tom Jones wants to say something:

"Many artists have worked, if only temporarily, in a cognitive mode. George Stubbs, the portrayer of the ideal English gentleman and his interests, is generally more to current taste than George Stubbs, the noted authority on animal anatomy. Our preferences should not obscure the fact that without his remarkable knowledge of animals, gained through detailed and arduous drawing of dissections, there would be no Stubbs' paintings as we know them. Such artists - and there have been many of them - evidently felt the need to research deeply into their subject matter. This is not to suggest that we should revise our ideas about artists and begin to see them as proto-scientists, but that a more balanced view of their achievements could be attained by taking their researches into account. In the case of Constable, it is known that he was more than interested in weather and has left numerous drawings to prove it. Though these constitute research, it would be as naïve to pretend that the resulting paintings are meteorological research, as it would be to regard them as *photographs* of Suffolk.

"The distinction between reference and research needs to be made. The important difference is in terms of time: our research today is someone else's reference tomorrow. In other words, research is firsthand enquiry while reference is second-hand. Here are some historical examples: when Delacroix studied Rubens' paintings in Antwerp (10 August 1850) to see how he dealt with shadows, he was referencing his own paintings in which he had encountered similar problems. In contrast, when Constable drew cloud formations and weather phenomena, he was researching in a way that forms a visual parallel to Gilbert White's work the *Natural History of Selborne.* The difference between the two processes is simple: Delacroix was selecting from someone else's knowledge, whereas Constable was absorbing original material from nature. It should be noted that this temporal distinction implies a progression in knowledge.

"While Leonardo da Vinci's drawings pioneered anatomical research, any work an artist does now can only be reference, the study of anatomy having progressed far beyond what can be observed by the unaided eye. Additionally, the medical skills now required are so specialized that they are unlikely to be possessed by any artist. Indeed, given current scientific understanding, it is difficult to conceive that much research into subject matter (in the sense in which it has been defined relative to Stubbs, Constable and Leonardo da Vinci) is possible nowadays. It is much more likely to be a matter of referencing the subject. While the major difference between reference and research is currently the same as in the past, it shows itself in a different way. To reference one's work is to select information necessary for its production from secondary sources, for example, books and paintings. In research, artists enquire for the same purpose, but here they are the sole authorities. It follows that the most likely area of research open to artists is not subject matter, but how that subject matter is perceived. The reason for this is that what used to be a subject of research is now, more often than not, the province of a science, for example, anatomy and biology."

This may not be entirely correct. Description is not explanation. We are back to the *Theory of Knowledge* once again. We do not have one that adequately encompasses what some artists do. The operative phrase here is: *some artists.*

*Ceteris paribus laws* – this too is something we need to address if we are to develop a new, post-European era, post-Enlightenment *Theory of Knowledge.*

**159**

Now we will paraphrase Karl Jaspers:

"Trade, which, as capitalism, in its enthusiasm for the salvation of mankind that it feels certain of securing, takes steps to mould the future by force of free markets, and trade as the idea of fulfilling the need for interchanges of materials and good, are alien to one another. The former consumes the man who surrenders himself to it in a faith that appears in the guise of science; it treats the unbeliever as a material available to manipulation by force. The latter casts no spell, lives in the present, requires the sobriety of reason, and the humaneness of unceasing communication. As long as trade demands are concretely visualised and thought out they remain within bounds. It is only when concrete reality is lost sight of and a fanatical paradise of man is presupposed as possible, that its demands become abstract and absolute. It is at this point that trade becomes an ideology. The demand for complete implementation in fact leads away from its fulfilment. Along a path of coercion it leads to servitude.

"Science which, as scientism, in its enthusiasm for the salvation of mankind that it feels certain of securing, takes steps to mould the future by force of total planning, and science as the idea of gradual realisation of improvements, are alien to one another. The former consumes the man who surrenders himself to it in a faith that appears in the guise of science; it treats the unbeliever as a material available to manipulation by force. The latter casts no spell, lives in the present, requires the sobriety of reason, and the humaneness of unceasing communication. As long as science demands are concretely visualised and thought out they remain within bounds. It is only when concrete reality is lost sight of and a fanatical paradise of man is presupposed as possible, that its demands become abstract and absolute. It is at this point that science becomes an ideology. The demand for complete implementation in fact leads away from its fulfilment. Along a path of coercion it leads to servitude."

What was it we were saying about the lack of a *Theory of Knowledge*? What was it we were saying when we read that extract from the Holy Qu'ran, the one from Sûra 3, verse 7?

**160**

So back to the matter of Leonardo - much more seems to be in the subconscious mind. It flows like a river - it is unstoppable!

Joseph Needham will speak:

"In Descartes the ideas of laws of Nature is as well developed as later in Boyle and Newton. Zilsel sees one essential component in the development of seventeenth century laws of Nature and in the empirical technologies of the sixteenth century. He points out that the higher craftsmen of that time, the artists and military engineers, of which Leonardo da Vinci was the supreme example, were accustomed not only to experimentation, but also to expressing their results in empirical rules and quantitative terms. He instances a small book written by Tartaglia in which quite exact quantitative rules were given for the elevation of guns in relation to ballistics. These quantitative rules of the artisans of early capitalism are, though they are never called so, the forerunners of modern physical laws. They rose to science in Galileo."

Rule of Thumb: A broadly accurate guide or principle, based on practice rather than theory. It was *Rule of Thumb* that so incensed Frederick Winslow Taylor. Thus did he develop the *Taylor equation,* which brings mathematics, and apparent exactitude, to the task of determining the speed and depth of cut on machine tools! Thus was the machinist's *Rule of Thumb* replaced by precise mathematics, as in science, only the equation is not exact for there are many other factors, not taken into account in the equation, that would, in *Rule of Thumb*, be factored in because there is also a direct relationship between the machinists via their hands and the machine tool, which provides the tactile feedback to know if the cutting parameters are correct.

We should speak with Zilsel. Perhaps we will. Later!

**161**

It is a truth not universally acknowledged that some artists have fallen in love with science, and some scientists have fallen in love with art. Thus do they stare at their beautiful reflection, and say quite strange things like:

"Niches for philosophy and religions are smaller with each day. Art is just the next fading trace of an age of great ignorance called humanism. The future is methodology - scientific method."

This of course has been said many times before. It was said in the nineteenth century. Evidently some people learn nothing from history.

Thus, in true Abrahamic style, there is no other way of knowing other than science, and an art that has become science is art no longer. There is no *Theory of Knowledge*, only knowledge produced by science.

We live in an age of great ignorance, for there is no one more ignorant that the scientist, and the artist, who express such views, and then jump around in celebratory glee at the achievement of this state of mind. It is not surprising that the Europeans have inflicted so much misery of the rest of the world, as well as upon themselves.

But neither scientist nor artist expected to see what takes shape in this book, although it is but a first step.

So back to the matter of Leonardo - much more seems to be in the subconscious mind. It flows like a river - it is unstoppable!

Joseph Needham will speak again, but not about Leonardo for the moment:

"To what extent was there any similarity between Chinese and European recognition of asterism and constellations? The answer is very little. The number of cases in which any parallelism of symbolic nomenclature can be made out is remarkably small, and the same groups of stars were not seen in the same patterns. Frequently a single European constellation appears on the Chinese planisphere as several different asterisms.

"Orion is one of the few examples where there some similarity, with the Chinese naming a group of stars as *Shen*, which has a parallel human figure symbolism to the European Orion. But there is no sword in the Chinese *Shen* constellation."

So people look and observe the same phenomena, but end up with different explanations - different stories. Or is that interpretations? Or perhaps description? Or perhaps prediction?

What do you mean prediction?

Scientific theory can predict. Art theory cannot predict. Different type of theory!

**162**

More about Leonardo, for it seems here too people see one phenomenon and arrive at different: explanations, or interpretations, or descriptions.

First though we have a Rabbi here who will read from *The Torah*:

"And the Lord God planted a garden eastward, in Eden; and there He put the man whom He had formed. And out of the ground made the Lord God to grow every tree that is pleasant to the sight, and good for food; the tree of life also in the midst of the garden, and the tree of the knowledge of good and evil."

Thank you for that. The reading was of verses 8 to 9 of chapter 2 of Genesis.

Now we will have another reading:

"And the Lord God commanded the man, saying: 'Of every tree of the garden thou mayest freely eat; but of the tree of the knowledge of good and evil, thou shalt not eat of it; for in the day that thou eatest thereof thou shalt surely die.'"

Thank you for that. The reading was of verses 16 and 17 of chapter 2 of Genesis.

So the matter of *Theory of Knowledge* crops up in the strangest of places, which brings us back to Edward O. Wilson, who is here once again to speak:

"Why no Descartes or Newton, under the Heavenly Mandate?"

Oh dear, that fragmented and reductive mind again. We will return to you, Edward O. Wilson in due course, we suppose. But now:

Joseph Needham will speak one more time, but again not directly about Leonardo:

"In one of the odes of Chia I it is said that:

> 'If water is swept along (*chi*) it becomes fiercely torrential (*han*). If an arrow is shot forth it travels far. Yet everything has a force reacting back (*wan wu hui po*) and vibrates in mutual opposition (*chen thang hsiang chuan*). Such is the Tao of nature.'

"This is reminiscent of Newton's Third Law. Now I will say some further words by way of comparison between Europe and China:

"So far as vehicles, projectiles, and engines of all kinds were concerned, Chinese mechanical practice was ahead of European, not retarded, down to the very time when the scholastics of the + fourteenth century were preparing the way for Galileo, and even later. In + 1400, though Europe had adopted during the preceding centuries many East Asian techniques, many were still to come – canal locks, iron-chain suspension-bridges, wagon-mills, sailing-carriages, rotary winnowing machines, square-pallet chain-pumps, and movable-type printing – to mention only a few."

So what does all this mean for Leonardo the inventor? What does it all mean for a *Theory of Knowledge*?

We saw today in the sky, a swift – the first one for many years!

**163**

So now we turn our attention to another matter, which also presses for our attention though Leonardo is still on our mind.

Boustrophedon: of written words; from right to left and from left to right in alternate lines. The Oxford English Dictionary gives the origin of the word in the early seventeenth century, being derived from Greek, literally 'as an ox turns in ploughing.'

And so now we turn to the matter of Asemic Handwriting, and here to speak about this are Tim Gaze and Michael Jacobson:

"We do not use adjectives such as *unreadable* and *illegible* when referring to the asemic, because the question of legibility and possible transference of meaning is precisely what is at stake with these writing traditions, which can be traced to many cultures. These writings are not completely *meaningless* or *illegible*, but challenge our common notions of reading, writing, and the meaningfulness of language."

We look at the constellation Orion, and project on to it, a meaning, even though it is but a number of stars, unrelated to one another, which just happen to look like a human figure - with or without a sword - in the mind of the observer.

So Orion: an explanation, or an interpretation, or a description, or the creation of meaning? And what does all this mean for a *Theory of Knowledge*?

Here to speak about C.P. Snow and his idea of *two cultures* is Paul Hartel:

"It is rather inaccurate to postulate, as Snow did, that in Western culture artists and scientists split into two absolutely segregated camps. Neither scientists nor artists represent monolithic orders. Scientists are divided into disciplines, methods and goals, just as artists are. Both groups function in the framework of the same historical context. The same historical background shapes the underling world views of artists and scientists. Furthermore, science and art are complementary metaphors of actuality."

Actually C.P. Snow was not talking about artists - he singled out only one group, which he called literary intellectuals.

Paul Hartel again:

"The problem of *two cultures* is itself a historical one. It is related to a process of specialisation and the division of labour as by-products of the industrial revolution."

Division of labour - nice to see that someone from the world of the arts is able to contemplate sociological knowledge - most art-science lovers do not!

Paul Hartel again:

"Those who tend to ignore the harmful side-effects of industrialisation and science-based technologies, such as: exploitation, alienation, crime, pollution, environmental deterioration, and the threat of nuclear war, often try to convince the public that science is a panacea."

Hartel is obviously talking about C.P. Snow! He is also talking about many of those art-science lovers as well. Scientism! Technocracy! Both very relevant words.

The European Commission's DG CONNECT's STARTS initiative is a case study in the failure of technocracy. Ultimately they will just engage in make-believe - art-science fakes! There are many. Perhaps we will encounter some as the days slip by, and new chapters are added? Who can tell?

Science has become the new religion, complete with temples, high priests, devotees, ideology and dogma, and it now needs art to legitimise itself, hence all this art-science nonsense; and the rewriting of art history that is necessary to make art fit the ideology.

**164**

So what would you feel if Karl Marx turned-up and started speaking? Would it be a matter relevant to a *Theory of Knowledge*?

Karl Marx is here and does want to speak by quoting from his work Das Kapital:

"A spider conducts operations that resemble those of a weaver, and a bee puts to shame many an architect in the construction of her cells. But what distinguishes the worst architect from the best of bees is this, that the architect raises his structure in imagination before he erects it in reality. At the end of every labour-process, we get a result that already existed in the imagination of the labourer at its commencement. He not only effects a change of form in the material on which he works, but he also realises a purpose of his own that gives the law to his modus operandi, and to which he must subordinate his will. And this subordination is no mere momentary act. Besides the exertion of the bodily organs, the process demands that, during the whole operation, the workman's will, be steadily in consonance with his purpose."

The matter of a *Theory of Knowledge* crops-up in the strangest of places. And that word once again - purpose!

Howard Rosenbrock will now speak:

"My own conclusion is that engineering is an art rather than a science and by saying this I imply a higher, not a lower status. Scientific knowledge and mathematical analysis enter into engineering in an indispensable way and their role will continually increase. But engineering also contains elements of experience and judgement and regards all social considerations and the most effective ways of using human labour. These partly embody knowledge which has not yet been reduced to exact and mathematical form. They also embody value judgements which are not amenable to the scientific method."

The matter of a *Theory of Knowledge* crops-up in many places.

Howard will speak one more time:

"In the field of computer-aided design I have cautioned against the automated design manual approach, where only minor design choices are left to the designer. This automated approach seems to me to represent a loss of nerve, a loss of belief in human abilities and further unthinking application of the doctrine of the Division of Labour. The designer is reduced to making a series of routine choices between fixed alternatives in which case his skill as a designer is not used and decays."

The matter of a *Theory of Knowledge* crops-up in many places.

The phrases anthropocentric, human-centred, and machine-centred, are interesting, and seem to be highly relevant to this matter of a *Theory of Knowledge*, and by corollary, to scriptovisualism.

**165**

William Shakespeare now reads from Romeo and Juliet:

> "Two households, both alike in dignity,
> In fair Verona, where we lay our scene,
> From ancient grudge break to new mutiny,
> Where civil blood makes civil hands unclean.
> From forth the fatal loins of these two foes
> A pair of star-cross'd lovers take their life;
> Whose misadventured piteous overthrows
> Do with their death bury their parents' strife.
> The fearful passage of their death-mark'd love,
> And the continuance of their parents' rage,
> Which, but their children's end, nought could remove,
> Is now the two hours' traffic of our stage;
> The which if you with patient ears attend,
> What here shall miss, our toil shall strive to mend."

And so to the matter of the two households: art and science, or art and technology, and those who now fall in love, to do … What exactly do they achieve, other than to become star-cross'd lovers?

When such love doth appear, that does not see what dwells beneath, to denial and delusion do they turn, and a life built on pretence comes to be, for the reformation of that which must be changed doth not happen. Thus does the world continue along trajectories that are but protections of the past into a future that must not be!

Will once again:

> "Let me not to the marriage of true minds
> Admit impediments. Love is not love

Which alters when it alteration finds,
Or bends with the remover to remove."

Here, we show to you that alteration, and let us see then how well this love affair endures when art's guiding hand does to this pen and brush deliver a different rendering.

**166**

To the matter of people, groups, organisations, societies, cultures and civilisations, projecting their values, beliefs, and ideologies on to others, do we now turn our collective attention. It is a very European thing to do, and much trouble has flowed from this practice and will in the future too, if these European people do not stop engaging in this evil practice.

First Karl Jaspers will speak:

"In earlier times, anyone who thought of universal history, because of the narrowness of his horizons, constructed a unity at the expense of restriction; amongst ourselves, for instance, he restricted himself to West, in China to the Central Empire. That which lay beyond had no part in it and was regarded as a life of barbarians, primitive peoples, which were certainly an object of ethnological interest, but not of history. Unity consisted in the presupposition of the tendency to cause all the still unknown peoples of the earth to participate, stage by stage, in one - namely one's own - culture, to bring them into one's own sphere of order.

"If faith presupposed one fundament and one goal, the idea sought to recognise these in real history. Assumptions of knowledge of unity, either given by divine revelation or intelligible to reason, were attempts at constructing the one history of mankind."

Mike Cooley will now speak:

"Our Western scientific methodology is based on the natural sciences. Within this, relationships are mathematically quantifiable. There has been a tendency to suggest that if you cannot quantify something it really doesn't exist. This is not without its political significance, for if the mass of ordinary people are incapable of providing *scientific reasons* for their judgement (which are based on actual experience of the real world), ruling elites can then silence their common sense with quantification. This has caused the French mathematician Jean-Louis Rigal to observe that quantification is the ultimate form of fascism. Rigal's concern about quantification is even more relevant when applied to the domain in which it evolved. Attempts to use this narrow, mathematically-based science in the much more complex and indeterminate social sciences and political activity give rise to very serious distortions, which are inevitable from the abstracted nature of the scientific method."

Remember the former chief scientific advisor to a former president of the European Commission. What was it she said? Oh yes: "people can have their own opinions, but they cannot have their own facts." And there is this thing called *Global Systems Science*, which is another form of so-called *Scientific Management* – both products of technocrats, of the fragmented and reductive mind. The ultimate form of fascism!

Now Shaun McNiff will speak:

"I would like to say something that relates to the above. The context is the village where I live, and a proposal to build a house on wetland that has been left undeveloped in the village, which has a colonial heritage, and peoples' attempts, via the conservation committee to prevent the building of the house. My neighbours argued from the perspective of history, aesthetics, and the ecological importance of protecting an area that naturally filters water run-off before it passes to a cove nearby. As the session progressed it became clear that the committee activity was conducted exclusively within the context of scientific and engineering data.

"The problem also occurs in the area of the justification of art as research. When something occurs outside the paradigm within which someone works and views reality, they seek justification according to that paradigm and its standards of research. This is an old rhetorical trick through which you demand that the perspective of the adversary be proven within the framework of reality that you construct and which denies the essential premises of the other's point of view. If the *outsider* complains that this imposition of an external measure of reality distorts the position being presented, the weight of the contemporary bias towards scientific truth is then used in yet another rhetorical move, *reductio ad absurdum*, which likens the complaint to a condition of absurdity, because who after all can question the absolute veracity of *scientific evidence*?"

We refer you once again to the former chief scientific advisor to a former president of the European Commission.

So are you beginning to understand? And how are you feeling?

The *Theory of Knowledge*, yet again!

Now to the saying of what we now will write: art as research stands on its own merit. Efforts to connect art to, science or technology, and the speaking of the usual benefits – communication, visualisation, creativity enhancements – are the acts of cultural colonialists and conquistadores.

You will notice that there is no enthusiasm for doing the same with the social sciences! Perhaps those that, do sing and dance in celebratory glee, have not the wish to be the subject of social sciences examination and critical thinking - too many holes, too many flaws, perhaps, to risk such a thing!

This book is not about art-science, nor art-technology. As for the cultural colonialists and conquistadores, and their collaborators, we decided to use them as the dull stone, against which we sharpen the sword. Have you not read our work, the one we called *STARTS - Science, Technology and the Arts: The Artistic Voices that DG CONNECT silenced*? More sharpening of the sword will no doubt in time appear!

DG CONNECT's STARTS Platform is a vicious attack on the arts by a failing organisation - the final acts of a dying system.

**167**

Creation: the action or process of bringing something into existence. So creation is a process!

Did you realise that in *The Torah,* in Genesis, chapter 1, what is described is a process where the universe and life emerges in stages. It does not just appear in a single moment! Now why might that be the case, given that people say of the Personal God of *The Torah,* that it is omnipotent, presumably therefore with the power to create everything in an instant?

The word you are looking for is hermeneutics. How many scientists and artists do you think possess skills in the hermeneutic way of knowing? How many of those among this lot - that self-proclaim their atheism and speak words such as *superstitions* - do you think, have ever read the *The Torah?* How many of them do you think, know *how to see*? *Saper Vedere!*

So we are back to Leonardo once again and the *Theory of Knowledge*.

There is more to seeing than using your eyes, more than that which is visual.

"In the blind man's world, the one eyed man is king."

That was John Wyndham reading a line from his novel, *The Day of the Triffids*.

So what then of those, who have two eyes, or three! Of course if you are one of the *Vainglorious Enlightened Ones* you will think that the third eye is just superstitious nonsense. That is because you a part of a culture that does not understand being human and never will. Even if you experienced the effects of the third eye, you would not know how to use it. Most likely you would seek to use it for your own gain, and would therefore lose it. The third eye: gateway to higher levels of consciousness. Did Leonardo da Vinci have the third eye – knowing how to see?

Our Rabbi is here once again to read from *The Torah*:

"And they heard the voice of the Lord God walking in the garden towards the cool of the day; and the man and his wife hid themselves from the presence of the Lord God among the trees of the garden. And the Lord God called unto the man, and said unto him: 'Where art thou?' And he said: 'I heard Thy voice in the garden, and I was afraid, because I was naked; and I hid myself.' And He said: 'Who told thee that thou was naked? Hast thou eaten of the tree, whereof I commanded thee that thou shouldest not eat?' And the man said: 'The woman whom Thou gavest to be with me, she gave me of the tree, and I did eat.' And the Lord God said unto the woman: 'What is this thou has done?' And the woman said: 'The serpent beguiled me, and I did eat.'"

Thank you for that. The reading was of verses 8 to 13 of chapter 3 of Genesis.

So once more the matter of *Theory of Knowledge* crops up, and in the strangest of places!

**168**

*Theory of Knowledge* is on our mind. So let us take a look at several ways of knowing. Shaun McNiff will describe some of them:

"Nine different research cultures may fit with the interests of art therapists. The first culture is *phenomenological studies* in which the researcher strives to understand the objects of inquiry and experiences as they are lived while minimising a priori assumptions and theories about these things.

"The second culture is *heuristic studies* which investigate phenomena with a focus on the inner experience and discovery of the researcher.

"Third comes the culture that is *hermeneutic studies* which emphasise the interpretive dialogue between the researcher and the objects of inquiry while accepting the inevitable influences of personal, cultural and historic biases.

"The fourth culture is *ethnographic studies* immersing the researcher in the particular cultural environment being studied.

"The fifth culture is *empirical/analytic studies* which quantify and measure the objects of inquiry according to the methods of scientific positivism.

"Sixth in this list of cultures is *action research* which uses the investigation as a way of encouraging change within a particular setting.

"Then there comes the seventh culture which is *comparative/historical research* examining phenomena over periods of time in keeping with the discipline of history.

"The eighth culture is *theoretical research focusing* on the theory as the object of inquiry and striving to create a new theory and/or a critique of existing paradigms.

"Finally, the ninth culture, which is *evaluation research* which generally uses the behavioural sciences' quantitative and qualitative methods to access the effectiveness of programmes."

So Edward O. Wilson, he who writes a book with the subtitle *Unity of Knowledge*: anything to say? No! Why we wonder?

*Theory of Knowledge* crops up everywhere, even in the observation of minds that claim to be addressing it! Now why is that?

**169**

Past: Belonging to a former time. Future: time regarded as still to come. History: the study of past events. Foresight: the ability to predict what will happen or be needed in the future.

Our Rabbi is here once again to read from *The Torah*:

"And the whole earth was of one language and one of speech. And it came to pass, as they journeyed east, that they found a plain in the land of Shinar; and they dwelt there. And they said to one another: 'Come, let us make brick, and burn them thoroughly.' And they had brick for stone, and slime had they for mortar. And they said: 'Come, let us build us a city and a tower, with its top in heaven and let us make us a name; lest we be scattered abroad upon the face of the whole earth. And the Lord came down to see the city and the tower, which the children of men builded. And the Lord said: 'Behold, they are one people, and they have one language; and this is what they begin to do; and now nothing will be withholden from them, which they propose to do. Come let us go down, and there confound their language, that they may not understand one another's speech.' So the Lord scattered them abroad from thence upon the face of all the earth; and they left off to build the city. Therefore was the name of it called Babel; because the Lord did there confound the language of all the earth; and from thence did the Lord scatter them abroad upon the face of all the earth."

Time is a singularity!

So once more the matter of *Theory of Knowledge* crops up, and in the strangest of places!

**170**

Divide: form a boundary between two people, or two things, or two ages.

More now about *Theory of Knowledge;* Joseph Weizenbaum wants to say some words about science:

"Surely, much of what we today regard as good and useful, as well as much of what we would call knowledge and wisdom, we owe to science. But science may also be seen as an addictive drug. Not only has our unbounded feeding on science caused us to become dependent on it, but, as happens with many other drugs taken in increasing dosages, science has been gradually converted into a slow-acting poison. Beginning perhaps with Francis Bacon's mis-reading of the genuine promise of science, man has been seduced into wishing and working for the establishment of an age of rationality, but with his vision of rationality tragically twisted so as to equate it with logicality. Thus have we very nearly come to the point where almost every genuine human dilemma is seen as a mere paradox, as a merely apparent contradiction that could be untangled by judicious applications of cold logic derived from a higher standpoint. Even murderous wars have come to be perceived as mere problems to be solved by hordes of professional problem-solvers."

**171**

We have not heard from John Dewey for a long time, so here he is, and this is what he wants to say:

"It is simply an impossibility that anyone today should experience the Parthenon as the devout Athenian contemporary citizen experienced it, any more than the religious statuary of the twelfth century can mean, aesthetically, even to a good Catholic today just what it meant to the worshipers of the old period. The *works* that fail to become *new* are not those which are universal but which are *dated.*"

So, highly relevant to the understanding of Leonardo and the notion of a *Theory of Knowledge*!

Jonathan Culler will now speak about the contrast between poetics and hermeneutics:

"Poetics starts with attested meanings or effects and asks how they are achieved. So what makes a passage in a novel seem ironic? What makes us sympathise with a particular character? Why is the ending of a poem ambiguous? Hermeneutics, on the other hand, starts with texts and asks what they mean, seeking to discover new and better interpretations.

"Poetics is derived from the model of linguistics. Hermeneutic models come from the field of law and religion, where people seek to interpret an authoritative legal and sacred text in order to decide how to act."

*Theory of Knowledge* yet again!

"Anything to say, Edward O. Wilson? No, of course not! Too busy reducing!"

Joseph Weizenbaum once again:

"The corruption of the word *problem* has brought in its train the mystique of *problem solving*, with catastrophic effects on the whole world. When every problem on the international scene is seen by the *best and brightest* problem solvers as being a mere technical problem, wars like the Vietnam War become truly inevitable. The recognition of genuinely but legitimate interests of coexisting societies – and such recognition is surely a precondition to conflict resolution or accommodation – is rendered impossible from the outset. Instead, the simplest criteria are used to detect differences, to search for means to reduce these differences, and finally to apply operators to *present objects* in order to transform them into *desired objects*. It is in fact, entirely reasonable, if *reason* means instrumental reason, to apply American military force, B52s, napalm, and all the rest, to communist dominated Vietnam as the operator to transform it into a *desirable object*, namely a country serving American interests.

"The mechanisation of reason and of language has consequences far beyond any envisioned by the problem solvers I have considered. Horkeimer long before computers became a fetish and gave concrete form to the eclipse of reason, gave us the needed perspective: 'Justice, equality, happiness, tolerance, all the concepts that were in preceding centuries supposed to be inherent in or sanctioned by reason, have lost their intellectual roots. They are still aims and ends, but there is no rational agency authorised to appraise and link them to objective reality. Endorsed by venerable historical documents, they may still enjoy a certain prestige, and some are contained in the supreme law of the greatest countries. Nevertheless, they lack any confirmation by reason in its modern sense. Who can say that any one of these ideals is more closely related to truth than its opposite? According to the philosophy of the average modern intellectual, there is only one authority, namely, science, conceived as the classification of facts and the calculation of probabilities. The statement that justice and freedom are better in themselves than injustice and oppression is scientifically unverifiable and useless. It has come to sound as meaningless in itself as would the statement that red is more beautiful than blue, or that an egg is better than milk.

"The reification of complex systems that have no authors, about which we know only that they were somehow given to us by science and that they speak with its authority, permits no question of truth or justice to be asked."

*Theory of Knowledge!*

Edward O. Wilson has of course nothing to say for his is the voice of instrumental reason, the voice of science, just engaging in a *scientific reduction to ignorance.*

**172**

Today is the summer solstice and scriptovisualism is pressing for attention, so, duly observing this surrealist way of writing, we now turn our attention to the matter of art.

Did you know that there is a great interest in the world of STEM in working with artists? We suppose you did, for it is a matter that has already flowed from our pen. Did you know that there is a difference between using art in research, and using artists? We think that you most likely did not know this.

We have some interest in using artists in research, but even more interest in using art, for which we may or may not use artists. We have no interest in art-science and art-technology movements, practices, for they are just more ideologies - religious cults. Our only interest there, is the observation of human behaviour, the collection of material for satire, and also learning what is not of any interest - most of it! Much of that which forms these religious cults is - the past! The time of art-science and art-technology has passed. Now is the moment for the creation of something new, something that is not art-science or art-technology. This is very much at the core of scriptovisualism.

Now Jonathan Culler will speak for a second time:

"I want to say something about the supposed death of theory and art and literature in a post-human world!

"Strangely, an interest in the post-human may have contributed to the revival of aesthetics, which has been pushed aside by literary and cultural theory of the late twentieth century. The reasons for the eclipse of aesthetics are not hard to understand. Traditional aesthetic concepts, such as artistic genius, the autonomy and universality of art, and its inherent spiritual value, were inextricably tied to a particular conception of the subject that major strains of theory, such as Marxism, psychoanalysis, structuralism, and post-structuralism, were engaged in combating. The triumph of theory and the widespread assumption that the concepts of aesthetics belonged to an outmoded elitist and universalising conception of art left open a space, a vacuum of sorts, that permitted, and even seemed to demand, a return of aesthetic issues in a new guise.

"What is sometimes called a *new formalism* or a *new aestheticism* denotes a renewed attention to literary artistic form in the context of the theoretical developments that seem to have rendered as reactionary the traditional aesthetics or approaches to literary form. Without aesthetics, Jacques Rancière has argued, there is no art: without specifically aesthetic values or perspectives, so-called art will merge with everything else – into a sea of consumer objects, we might say. A discrete understanding of art is needed to frame things as art and to keep art as distinctive.

"Similarly, literary works are not just language but products of specifically literary practices and systems of conventions which need to be understood: how does poetic metre work, for example? Even though literature is a social product and practice, entwined with ideology, it ultimately poses for critics and other thinkers the question of the specificity of the literary objects: are there distinctive features of literary works and the experience of literary objects, or is that an illusion? How should we conceive of literary or artistic invention? There is an interest in the singularity of the literary work, as a distinctive event, and in the ways in which works may disclose a world. What sort of role does literary form play in the effects literature achieves? For instance, what about pleasure? Pleasure had not previously been in the forefront of literary theory, although now it is, and the question of pleasure goes along with an interest in other sorts of feelings or *affects*."

**173**

Joseph Weizenbaum wants to speak again. First he wants to tell us about the imperialism of instrumental reason:

"Instrumental reason has made out of words a fetish surrounded by black magic. And only the magicians have the rights of the initiated. Only they can say what words mean. And they play with words and they deceive us. When Skinner contrasts science with common sense and claims the first to be much superior he means his *behavioural science* and he means the *common* in *common sense* pejoratively. He does not mean a common sense informed by a shared cultural perspective, or a common sense that, for no reason at all, balks at the idea that freedom and dignity are absurd and outmoded concepts.

"The technologist argues again and again that views and concerns that I have expressed about science and technology are anti-technological, anti-scientific, and finally anti-intellectual. He will try to construe all arguments against his megalomaniac visions as being arguments for the abandonment of reason, rationality, science, and technology, and in favour of pure intuition, feeling, drug-induced mindlessness, and so on. In fact, I am arguing for rationality. But I argue that rationality may not be separated from intuition and feeling. I argue for the rational use of science and technology, not for its mystification, let alone its abandonment. I urge the introduction of ethical thought into science planning. I combat the imperialism of instrumental reason, not reason."

Joseph now wants to tell us about acting against the imperialism of instrumental reason:

"The teacher of computer science must have the courage to resist the temptation to arrogance and to teach, again mainly by his own example, the validity and the legitimacy of softer knowledge. Why courage in this connection? For two reasons: The first and least important is that the more he succeeds in so teaching, the more he risks the censure of colleagues who, with less courage than his own, have succumbed to the simplistic world views inherent in granting imperial rights to science. The second is that, if he is to teach these things by his own example, he must have the courage to acknowledge, in Jerome Bruner's words, the products of his subjectivity."

**174**

All the issues of relevance that have been discussed, are largely views on what might happen if the United Kingdom leaves the European Union. This leaves us all in a position of choosing between what looks like the certainty of the *status quo*, and the uncertainty inherent in leaving. To *stay or to go* is probably the hardest question than any electorate in the United Kingdom has ever been asked to vote upon, made so by the fact that it is a single issue, unlike in a General Election when we vote on just about everything.

Yes, we are referring to the United Kingdom's June 2016 referendum on continuing membership of the European Union.

We want now to give people an insider's view, having been involved with Brussels for 30 years. But first a quote: 'In earlier times, anyone who thought of universal history, because of the narrowness of his horizons, constructed a unity at the expense of restriction; amongst ourselves, for instance, he restricted himself to West, in China to the Central Empire. That which lay beyond had no part in it and was regarded as a life of barbarians, primitive peoples, which were certainly an object of ethnological interest, but not of history. Unity consisted in the presupposition of the tendency to cause all the still unknown peoples of the earth to participate, stage by stage, in one - namely one's own - culture, to bring them into one's own sphere of order.'

These are the words of the German philosopher Karl Jaspers (in his book *The Origin and Goal of History*), writing in the aftermath of the National Socialists, the Second World War, and the Holocaust. These words very accurately describe how the European Commission thinks and operates; this organisation is very European, in the most negative sense of the meaning of such a phrase.

We first became involved with Brussels, namely the European Commission, in 1986. We did so because we knew that the Commission were not pursuing the right research agenda in relation to manufacturing industry. It is a long story, but now here in 2016, 30 years on, the same issues are still there, which is why we are now starting to break our connections with the European Commission. In the process of doing so we have come to call these people Prometheus and sometime also Narcissus, along with those that gather around the European Commission - the EC (Emperor's Court). Everyday they just reinvent themselves in the form they were the day before, but have so fallen in love with themselves that they are unable to see this.

They say that power corrupts, and that absolute power corrupts absolutely - the European Commission is the case study that demonstrates the first to be true, and may, if we do not do something about this monster, prove also that the second part is also correct. What it will take to prove the latter is the creation of and the alignment of the sort of social, political and economic conditions that lead people to look for simplistic ideologically based solutions - the type of conditions that the modern world is now busy bringing about. This simplistic ideologically based thinking is already taking shape in the European Commission and in the European Union - people from the world of STEM, along with their artists in residence, will play a large role in this future European nightmare.

People think that the European Commission is a bureaucracy. It most certainly is, but the problem is worse than this. The European Commission is also a technocratic organisation. It believes that it can intervene in matters complex, in business, in economy, in research, and so forth, and worse, that it actually knows what those interventions should be, hence they largely work top-down, rather than bottom-up, and most often fail, as is demonstrated by the fact that Europe has the lowest growth rate of any continent. The so called STARTS Platform is an example of this failure. So is the euro.

The European Commission has vast powers delegated to it by the various treaties and the money to pursue its technocratic fantasies. It has also become the target of the *priestly cast* of technocrats called scientists, technologists and engineers, who are now also working behind the scenes to gain power and influence.

This is in addition to powerful corporations who are increasingly, as the European economy continues to weaken, exerting their influence. The two - science and business - are closely allied with each other, as science is now primarily perceived as an activity to be pursued for economic reasons. Hence the appeal to the European Commission, which has been desperately trying, and failing, for 30 years and more to deal with the economic decline of Europe. The European Commission is unable to understand that it is part of the problem.

The scientific technocrats are advocating a form of *reason* that is not grounded in humanity, common sense, or instincts. Richard Dawkins is one of the *high priests*, but there are more such people, who hold in contempt anything that is not grounded in the scientific method. They are seeking what we have come to call *Scientific Government* and in the European Commission they find a willing partner in this madness. It is a marriage made in hell. It is also a relationship founded on the elitism of Ancient Greece - free people think, slaves work.

Some years ago, under the reign of a previous president, the man concerned decided to take unto himself a person to be his chief scientific advisor. The appointment was done in secret and there were no vetting processes. The appointee then proceeded to use her position to advocate that science was so important that scientists needed to have a seat in government. Many in the world of science support her views. All have conflicts of interest and they all, along with their institutions, stand to gain financially from this. The conflicts of interest of the chief scientific advisor were never independently established, nor were any disciplinary actions taken against her, for what in a democratic setting would have been seen as an abuse of a public appointment.

She was condemned by a number of MEPs. She herself, who was a mere employee of the European Commission, felt that it was within her power and her right to take these elected representatives to tasks for their views, and she used these people to further her argument which reduces to this statement: *in our system of government, politicians are told what to do and how to do it, and any improvement made to the instructions given to them is fatal to success.* Elsewhere one can also find people saying that policy decisions of a scientific or technical nature, for example, those relating to the use of GM crops, should be taken out of the hands of politicians! Juncker it seems was not prepared to inherit this troublesome *high priestess* and she was discarded and not replaced. Instead, an internal person, who is also a technocrat, was appointed as an innovation advisor. This though is just a foretaste of what could transpire in Brussels, as the European Union's economic circumstances deteriorate further, and *world events* trigger decisions that in hindsight, will be seen by the survivors, as acts of deluded people caught-up in dangerous fantasies.

We are dealing here with ideologues who have taken a valuable means of knowing the world - science - and turned it into a perverse ideology that consumes the mind, just as the Soviets did with the works of Karl Marx, right-wing free market extremists have done with the works of Adam Smith, some Christians have done with the Holy Bible, and some Muslims are now doing with the Holy Qur'an. This is a *one hell of a conflict* that you are in the process of creating.

Some may have read Hayek's book *The Road to Serfdom*. The European Commission has already built *The Road to Serfdom*, and it uses the money that it takes from taxpayers, via the Member State National Governments, to keep the serfs in line. It has created a dependency culture, whereby it decides what it will do, and then it seeks experts, all of whom will benefit from the Commission's proposal, to support their proposals. Then it often uses these experts to confirm that the money has been well spent. Dissenters are not tolerated and are sent into exile - that is to say they are dismissed as suffering from some deficit and never invited back. ICT-ART CONNECT - the STARTS Platform - is a prime example of this form of technocratic corruption, and there is a book that explores this. It is called *STARTS - Science, Technology and the Arts: The Artistic Voices that DG CONNECT Silenced*.

The European Parliament is aware of these problems but is largely powerless to do anything about it. People like Juncker share the technocratic mind-set so will do nothing to put an end to this madness.

What people do not understand is that the European Commission is, in effect, an unelected government, not a civil service. The peoples of the

European Union have not shared sovereignty, but have instead transferred sovereign power to a body over which there is no democratic control and which is not accountable to an electorate. When the European Commission fails, as it often does, the people of the European Union cannot evict the European Commission from office, nor is anyone punished for this failure - no one resigns in disgrace and no departments are exposed for wasting public money. The failures are covered-up so that you, the people, never learn about what goes on behind closed doors in Brussels. Cover-up is the normal practice in Brussels. In the United Kingdom, what they do would be called misleading Parliament, misleading ministers, and misleading the public.

As a government, the European Commission has diplomatic representations all around the globe. The president of the European Commission has the status of a head of state. The Treaty of Lisbon gave the European Commission power over foreign policy and the European Commission now has a foreign policy. It is one that is best described as cultural colonialism. So we have an unelected government preaching and lecturing to the world about European values, and seeking in an imperial way to project these values on to the non-European world and to enforce these values thorough trade and cooperation agreements. And the European Commission seems to be genuinely surprised about the backlash that has come in the form of the non-European world (what some Europeans think of as the barbarians in need of European civilisation) contesting this and taking their own steps to project their values and beliefs on the world. The seeds of conflict are being sown!

It can be said that in this very traditional European policy where there is no recognition of the genuinely different and legitimate interests of other societies. This is the fate that awaits Turkey and all from outside Europe who want to join the European Union - they will have to become European and this will probably create a violent response against Europe, which will just reconfirm the European view that those in the non-European world are barbarians. Hints of this have already been seen in the aftermath of the Paris and Brussels attacks. People in Brussels actually referred to the Brussels' terrorists as barbarians. Two days after the Brussels bombs, in The Hague, a man was sentence to 41 years in prison for war crimes and his part in the genocide of 8000 Muslim men in the Srebrenica enclave. No one called Radovan Karadzic a barbarian - he was one of us so he is a war criminal.

Let us be clear - when you start calling people barbarians, you classify and separate, and begin to dehumanise them. And in doing so you are preparing people for the eventual outcome, which you will *reason* your way towards, because the sort of *reason* that you have now come to value is devoid of any humanity. This is a path that only has one destination - it is a very European destination and you know what it is!

Efforts have been made to reform the European Commission, but the monster is very adept at finding ways around these reforms. The European Parliament and the Member State politicians seem to be powerless against the European Commission and its power grows with every new treaty that is signed. It already has a currency, which is in effect the German Deutsche Mark, which is why now the European Union is dominated by one nation - the Germans. It will one day have fiscal policy under its control. It will one day have a police force under its control. It already has the beginnings of such in the form of its internal security department which investigates leaks of confidential information from European Commission officials, and in the form of a fraud investigation body. It will one day have an army under its control as well. By that time, the Member State Governments will have been reduced to something similar to what in the United Kingdom we call Local Authorities, and will not have the power to resist the final step, which will be a European State, with an unelected government. Once a government with a police force and an army comes into being, that is not subject to democratic control and is accountable to no one, it has the means to do what it wants to do - you cannot resist such a government. Such governments have a name - they are called totalitarian.

Back in the early 1990s, there was idealism in Brussels. People talked about integration, working together, about a social Europe. No one has any idealist visions any more. Social Europe is dead. Everything now is about money, about the economy. The European Union now pursues a neo-liberal agenda, and has abandoned the people of Europe to the power of money, just as the United States government has done with its people. The so called Single Market is about the marketisation of the whole of Europe, and the giving of power to corporations, while removing it from ordinary people and from trade unions - just as in the United States. Anyone who speaks out against the European Union and what it is becoming is branded a right-wing populist, even when they are left-wing!

We get the feeling that the people of Europe are sleepwalking once more into yet another one of those very European nightmares.

It may be that the only hope left is that China and India will find a way to balance economy with humanity and nature and that they will be the ones providing the future model for civilisation. We have come to the conclusion that the days of the West are drawing to an end, and that the future lies in convincing the Chinese and the Indians to walk a different path to the European world by looking to their cultural heritage as a source of new inspiration, because what Europe offers is just the past masquerading as the future.

The European Commission is an affront to democracy. It is becoming a threat to democracy and the right of people to choose. The only way we can remove the European Commission (meaning Juncker and his College of Commissions and all the bureaucrats that work for them) from our lives is by voting to leave the European Union. This is the circumstance that people such as Schuman and Monnet have bequeathed to us. The irony is that by seeking to avoid the past, they may have laid the foundations for the past once again to become the future, which is Europe's history in a nutshell.

We cannot predict the future, but history does provide us with powerful lessons that we seem not to want to pay attention to. Back in 1933 if anyone had predicted what was to happen in Europe they would have been dismissed as a lunatic. Yet knowing how Europeans have behaved since the time of the Roman Empire, what happened between 1933 and 1945 can be seen as the past reinvented, but on a far more destructive scale.

"Probably it is true that the very magnitude of the outrages committed by the totalitarian governments, instead of increasing the fear that such a system might one day arise in this country, has rather strengthened the assurance that it cannot happen here. When we look at Nazi Germany the gulf which separates us seems so immense that nothing that happens there can possess relevance for any possible developments in this country. And the fact that the difference has steadily become greater seems to refute any suggestion that we may be moving in similar directions. But let us not forget that fifteen years ago the possibility of such a thing happening in Germany would have appeared just as fantastic, not only to nine-tenths of the Germans themselves, but also to the most hostile foreign observers however wise they may now pretend to have been."

The words are Hayek's, written in 1944, and taken from the book that we have already mentioned.

Already it seems we have forgotten this recent past and are now willing to trust our future to a *dictatorship of technocrats* in the belief that this will not happen again. It will, but in a different form. Now is the chance to take steps to prevent this.

The history of Europe is a history of European Unions.

**175**

Today is the day that the British people voted in the referendum on continued membership of the European Union.

Now, Pablo Picasso will speak:

"When we invented cubism we had no intention whatever of inventing cubism. We wanted simply to express what was in us. No one drew up a plan of campaign, and our friends, the poets, followed our efforts attentively, but they never dictated to us. Young painters today often draw up a programme to follow, and apply themselves like diligent students to performing their tasks."

Thank you for that, which must surely be a comment relevant to scriptovisualism.

Now Pablo will speak one more time:

"Everyone wants to understand art. Why not try to understand the song of a bird? Why does one love the night, flowers, every-thing around one, without trying to understand them? But in the case of a painting people have to understand. If only they would realise above all that an artist works of necessity, that he himself is only a trifling bit of the world, and that no more importance should be attached to him than to plenty of other things which please us in the world, though we can't explain them. People who try to explain pictures are usually barking up the wrong tree."

It is all about *Theory of Knowledge* in some way. In what way, we are, at the moment, not sure!

"I had not a dispute, but a disquisition, with Charles Dilke upon various subjects; several things dove-tailed in my mind, and at once it struck me what quality went to form a *Man of Achievement*, especially in literature, and which Shakespeare possessed so enormously - I mean *Negative Capability*, that is, when a man is capable of being in uncertainties, mysteries, doubts, without any irritable reaching after fact and reason."

That was the poet, John Keats, speaking.

**176**

Today is the day after the day that the British people voted in the referendum on continued membership of the European Union. The British people decided that the United Kingdom should leave the European Union. A triumph of democracy over vested interests and technocracy!

Now we are going to explore a word that we have already encountered. It is a word most interesting, and the word is - divide.

Michael Piore and Charles Sabel are here to speak, in one voice, about divides:

"Moments in time when the path of technological development itself is at issue we call industrial divides. At such moments, social conflicts of the most apparently unrelated kinds determine the direction of technological development for the following decades. Although industrialists, workers, politicians, and intellectuals may be only dimly aware that they face technological choices, the actions that they take, shape economic institutions for long into the future. Industrial divides are therefore the backdrop or frame for subsequent regulation crises.

"In our view the first industrial divide came in the nineteenth century. At that time, the emergence of mass-production technologies - initially in Great Britain and then in the United States - limited the growth of less rigid manufacturing technologies, which existed primarily in various regions of Western Europe. These less rigid manufacturing technologies were craft systems: in the most advanced ones, skilled workers used sophisticated general-purpose machinery to turn out a wide and constantly changing assortment of goods for large but constantly shifting markets. Moreover - in contrast to mass production - economic success in these industries depended as much on cooperation as on competition: unless the costs of permanent innovation were shared among firms, and between capitalists and workers, those who stood to lose from change defended their interests by blocking it. And the sharing of costs depended, in turn, on institutions that protected the vulnerable in the name of the community as a whole.

"Under somewhat different historical conditions, we argue, firms using a combination of craft skill and flexible equipment might have played a central role in modern economic life - instead of giving way, in almost all sectors of manufacturing, to corporations based on mass production. Had this line of mechanised craft production prevailed, we might today think of manufacturing firms as linked to particular communities, rather than as the independent organisations - barely involved with the neighbours - that, through mass production, seem omnipresent."

Thank you for your insights, which can be summarised as: *we had a choice*. So much for that strange Abrahamic belief known as *technological determinism*!

**177**

Today is the second day after the day that the British people voted in the referendum on continued membership of the European Union. The British people decided to leave the European Union. It is called democracy! You will not find democracy at work in the European Union.

Now, back to Leonardo, and the myth that he has become, and to do this we will examine another myth – the one surrounding Galileo.

Arthur Koestler will speak:

"The personality of Galileo, as it emerges from works of popular science, has even less relation to historic fact than Copernicus'. In his particular case, however, this is not caused by benevolent indifference towards the individual as distinct from his achievement, but by more partisan motives. In works with a theological bias, he appears as someone with a hidden agenda; in rationalist mythography, as the Maid of Orleans of Science, the St George who slew the dragon of the Inquisition. It is, therefore, hardly surprising that the fame of this outstanding genius rests mostly on discoveries he never made, and on feats he never performed. Contrary to statements in even recent outlines of science, Galileo did not invent the telescope; nor the microscope; nor the thermometer; nor the pendulum clock. He did not discover the law of inertia; nor the parallelogram of forces or motions; nor sun spots. He made no contribution to theoretical astronomy; he did not throw down weights from the leaning tower of Pisa, and did not prove the truth of the Copernican system. He was not tortured by the Inquisition, did not languish in its dungeons, did not say *eppur si muove*; and he was not the martyr of science."

A *Theory of Knowledge* must surely have something to say about mythology, whether it be of the ancient kind, or the projection of meaning on to stars and their groupings, or the strange beliefs of people, including scientists.

**178**

So to the matter of divides again!

First, Jeremy Rifkin will speak:

"It is too early to say for sure whether Europe is leading the world into a second Enlightenment. Certainly its multilateral agreements, its internal treaties and directives, and its bold cutting-edge initiatives suggest a radical re-evaluation in the way science and technology are approached and executed. The increasing reliance on the precautionary principle and systems thinking put Europe out in front of the United States and other countries in re-envisioning science and technology issues in a globally connected world. Still a world of caution is in order. The old power driven

Enlightenment science remains the dominant approach in research, development, and the market introduction of most new technologies, products, and services in Europe, America and elsewhere in the world. Whether the European Union government can effectively apply new-science thinking in its regulatory regime to old-science commercial applications in the marketplace remains to be seen. In the long run, a successful transition to a new scientific era will depend on whether industry itself can begin to internalise the precautionary principle and systems thinking into its R&D plans, creating new technologies, products, and services that are, from the start, ecologically sensitive and sustainable."

It is a good job that, back in 2004, when you wrote that Jeremy, you added the words of caution for we now know that the old Enlightenment science is alive and well and threatening to destroy both planet and humanity. Your mistaken optimism comes from the fact that you *do not know how to see.* You are a child of the Enlightenment, as is just about everyone else in Europe and the United States, and as a consequence, you are not able to see beyond its narrow horizons. You have spent too much time advising presidents of the European Commission and have become like them. You have seen the world from within the confines of Berlaymont; we on the other hand have seen the world from the tiny offices of faceless technocrats. You have seen shining towers from up high, in the *City of the Golden Stars;* we the grubby reality of the petty minds that run the lives of the peoples of Europe. You advise presidents, we advise no one. What is more, unlike you, we do not seek to do so. You and they are the darkness; we are the light that shines in that darkness.

Now Arthur Koestler will speak:

"The symptom that a particular branch of science or art is ripe for a change is a feeling of frustration and malaise, not necessarily caused by any acute crisis in that special branch - which might be doing quite well in its traditional terms of reference - but by a feeling that the whole tradition is somehow out of step, cut off from the mainstream, that the traditional criteria have become meaningless, divorced from living reality, isolated from the integral whole. This is the point where the specialist's hubris yields to philosophical soul-searching, to the painful reappraisal of his basic axioms and the meaning of terms which he had taken for granted; in a word, to the thaw of dogma. This is the situation which provides genius with the opportunity for his creative plunge under the broken surface."

**179**

Now it is time to reflect a little on the European Union and the dreaming of an American who has fallen in love with this very dangerous thing. Then we will be able to send Jeremy Rifken back to where he belongs, the United States, where few would be willing to tolerate the grossly undemocratic thing that is the European Union and its institutions, in particular the European Commission.

"The new generation of scholars was leery of overarching grand narratives and single-minded utopian visions that attempted to create a unified vision for human behaviour. By locking humanity into the *one right way* of thinking about the world, postmodernists contended, modern thought became dismissive of any other points of view and ultimately intolerant of opposing ideas of any kind. Those in power - be they capitalists or socialists, conservatives or liberals - continue to use these meta-narratives to keep people contained and controlled, argue postmodernists. Modern thought, according to the critics, has been used to justify colonial ventures around the world and keep people divided from one another and in conditions of subservience to the powers that be.

"It was the stifling nature of these all-encompassing grand visions and single-minded utopian ideas about how people were expected to behave and act in the world that the sixties generation rebelled against. The postmodernists provided the rationalisation for the revolt, arguing that there is no one single perspective, but, rather, as many perspectives of the world as there are individual stories to tell. Postmodern sociology emphasises pluralism and tolerance of the different points of view that make up the totality of human experience. For the postmodernists, there is no one ideal regime to which to aspire, but, rather, a potpourri of cultural experiments, each of value.

"The postmodernists engaged in an all-out assault on the ideological foundations of modernity, even denying the idea of history as a redemptive saga. At the end of the postmodern deconstruction process we arrive at modernity reduced to intellectual rubble and an anarchic world where everyone's story is equally compelling and valid and worthy of recognition.

"If the postmodernists razed the ideological walls of modernity and freed the prisoners, they left them with no particular place to go."

Thank you for that Jeremy, now off you go, and as you are making your way back home, we say that: what you have just described about modernity is a description of the European Union, which is just another manifestation of an overarching grand ideology, of a single minded utopian vision, of a one best way, of keeping people divided from one another and in conditions of subservience to the powers that be. Why would it be anything else? After all, we are here dealing with the *Vainglorious Enlightened Ones*. Time is a singularity!

It is all about *Theory of Knowledge*.

**180**

The contemplation of the eternal is the aim of philosophy, as the contemplation of the mysteries is the aim of religion. It is time to mention religion.

Arthur Koestler will now speak:

"For Pythagoras as for Kepler, the two kinds of contemplation are twins; for them philosophy and religion were motivated by the same longing: to catch glimpses of eternity through the window of time. The mystic and the savant jointly satisfied the dual urge of allying the self's cosmic anxiety and of transcending its limitations; its dual need for protection and liberation. They provided reassurance by explanation, by reducing threatening, incomprehensible events to principles familiar to experience: lightening and thunder to temperamental outbursts of man-like gods, eclipses to the greed of moon-eating pigs; they asserted that there was rhyme and reason, a hidden law and order behind the seemingly arbitrary and chaotic flux, even behind the death of a child and the eruption of a volcano. They jointly satisfied man's basic need and voiced his basic intuition, that the universe is meaningful, ordered, rational, and governed by some form of justice, even if its laws are not transparent.

"Apart from reassuring the conscious mind by investing the universe with meaning and value, religion acted in a more direct manner on the unconscious, pre-rational layers of the self, providing it with intuitive techniques to transcend its limitations in time and space by a mystical short-circuit, as it were. The same duality of approach - the rational and the intuitive - characterises the scientific quest. It is therefore a perverse mistake to identify the religious solely with intuition and emotion, science solely with the logical and the rational. Prophets and discoverers, painters and poets, all share this amphibial quality of living on the contoured drylands and in the boundless ocean. In the history of the race as of the individual, both branches of the cosmic quest originate in the same source. The priests were the first astronomers; the techniques of hunting, fishing, sowing, and reaping were imbued with religious magic and ritual. There was division of labour and diversity of method in the symbols and techniques, but unity of motive and purpose."

**181**

Rabi Jonathon Sacks has something to say about science and religion. He will be speaking later.

Arthur Koestler will speak again:

"The first open conflict between Church and Science was the Galileo scandal. Unless one believes in the dogma of historical inevitability - a form of fatalism in reverse gear - one must regard it as a scandal which could have been avoided; and it is not difficult to imagine the Catholic Church adopting, after a Tychonic transition, the Copernican cosmology some two hundred years earlier than she eventually did.

"The Galileo affair was an isolated, and in fact quite untypical, episode in the history of the relations between science and theology, almost as untypical as the Dayton monkey-trial was. But its dramatic circumstances, magnified out of all proportion, created a popular belief that science stood for freedom, the Church for oppression of thought. That is only true in a limited sense for a limited period of transition. Some historians, for instance, wish to make us believe that the decline of science in Italy was due to the *terror* caused by the trial of Galileo. But the next generation saw the rise of Toricelli, Cavalieri, Borelli, whose contributions to science were more substantial than those of any generation before or during Galileo's lifetime; the shift of the centre of scientific activity to England and France and the gradual decline of Italian science, as of Italian painting, was due to different historical causes.

"Never since the Thirty Years War has the Church oppressed freedom of thought and expression to the extent comparable to the terror based on the scientific ideologies of Nazi Germany or Soviet Russia.

"The contemporary divorce between faith and reason is not the result of a contest for power or for intellectual monopoly, but of a progressive estrangement without hostility or drama, and therefore all the more deadly."

The arrival of Richard Dawkins on the scene heralded a change. Now it is science, and more generally the world of STEM, which does now contest, both for power and for intellectual monopoly - supremacy is the word. Yet another form of colonialism, by scientific conquistadores, the existence of which Edward O. Wilson denied. We will be cross examining him more, later on in this odyssey - probably!

**182**

Now, we enter the world of art for a moment.

Words come drifting across cyberspace, from the domain of Facebook; it is an artist speaking:

"The main reason I embraced Cybersemiotics so whole heartedly is because the framework is grounded on the integration of human knowledge and consciousness through the four main aspects of life: natural, biological, phenomenological, semiotic. Cybersemiotics is not the final whole picture because we don't know what is coming in our future, but it offers a broader view that allows us to understand that there are many connecting parts with the potential of many more. In other words, all parts/views are potentially valid."

Then came a plea to share thoughts, thus we, in response, very insecurely, and within a frame of mind rooted in not knowing, offered this:

"Adopting Piore and Sable's term - divide - there came to our mind long ago, a view that we are now living through the age of a divide, not the industrial divide that interested them, but a divide that separates the world in the form that it is now, from something that has not yet a name or a clear definition. It is a divide between the age that people have been living through for the past several thousand years, which includes the age of empires of many different kinds, including in its latter stages, the European domination of the world, which is still being played out in a very dangerous game, and something that will, regardless of attempts to block it, replace this age.

"Within this context - which we plan at some stage to write about - the ice cap that is the dogma of Western STEM, is beginning to fracture and fall apart, and thus it becomes possible to see into the depths that lie beneath, and to glimpse the beginnings of a new synthesis.

"So, while the character that we have created that we call *Hawkins Dawkins Wilson Wolpert* engages in his dance of imperial celebratory glee, unseen to him, the world about softens and becomes, once more, malleable to the sculpture's shaping hands. And into those dark depths go those who dare, while the rest, like modern day scholastics, do to their crumbling world cling, often now it seems with an artist in residence by their side, for here too the crumbling dogma can be found.

"One cannot enter this dark place and survive to shine a light, if to old restrictions and definitions one does adhere. This place is not one where people self proclaiming to be atheists, positivists, or scientists, or whatever, can successfully operate, for all such things in this unknown world, are up for overturning and redefinition. And at this early stage, which has taken over 40 years to reach, we find in our lap, a jigsaw puzzle, which it seems has many thousands of pieces. So we look to Cybersemiotics and wonder what it might do to help in our strange lifelong quest to assemble this into something approaching meaningfulness.

"We also look to artistic and literary practices, which place our mind into a poetic and expressive mode, in the expectation that, to a cognitive mode it will lead. For it seems to us that somewhere deep down, something has been seen, and the challenge is to discover what it is. Thus does art, to the subconscious mind, and places that lie beyond, provide a gateway, out of which flows what we suspect most will think to be, a madness of lunatic proportion, being as they are, so caught-up in being what they are, whatever that may be.

"So it was to the pen we turned, for this is the way that we know and explore the world, and thus began to make work, scriptovisual in kind, trying to understand what this itself could be, and what it might be able to do, and if it could provide a stepping stone to something else. And in that process of exploration we came to see that, for us, it was a dual creative process that allows us to work with both descriptive and plastic expression, crucially, on the same theme. Through this came to mind the understanding that what we are doing is beginning the construction of a *Theory of Knowledge* for a post-European science, which is not based on the notion of nature and humans as machines; nor even, based on reaching back to an earlier time when there was an anthropocentric conception; but which looks forward, to the age that is slowly coming into being and the nature of what it is that we would like it to be."

The journey is only just beginning ...

**183**

This is the moment - the point where we are half way from the start, or half way to the end, or both.

Rachel Carson, who we did not forget, will soon speak. For the moment we will listen in on a conversation between Lotte Darsø and Margaret Wheatley. First to speak is Lotte, who is asking Margaret a question:

"When you look at management development and education, what do you see happening right now?"

"I see in the field of management development and education that we are trying to educate people to be technicians rather than leaders; that we are giving people a lot of metrics and formulas and really acting from this assumption that numbers express reality and that good managers are people, who can work with metrics and measurements. This is all over the world, I see this everywhere; any time I mention the *march of metrics*, people just nod their heads. We are training people to be good at believing that numbers convey reality, that with numbers you can understand your organisation, or your health, or your life.

"If you think about how everyone today knows how many grams of protein they should have, or what their cholesterol level should be - we have metricised, if I can use that word, how we look at reality. This is not only a tragedy, but it creates the kind of management that we are seeing, which is that people are too hurried, too rushed and too overwhelmed to even notice what is missing: too hurried or too overwhelmed to take time for relationships or for getting to know people; too hurried to appreciate the complexity of a problem; too overwhelmed to think.

"I have seen it going on for many years, but what makes it much more intense now is that people are doing so much more and have so many more tasks to do. The use of numbers is a quick way to understand something and when you are looking for speed and just moving on to the next task, well I think this is one of the reasons why measurement has now become the dominant mode of managing."

Very true, but you will also find many artists also embracing this view, and then singing and dancing about it in celebratory glee.

We have a really big problem when artists adopt this stupid mindless way of behaving. So much for the art-science, art-technology lovers!

These are the *Vainglorious Enlightened Ones* - destroying humanity, destroying the planet through numbers. The European Commission is the archetype of this nonsense. DG CONNECT and the STARTS Platform - technocrats and their artists in residence! We walked away from these people for the right reasons. Time is a singularity. Europe, the European Union, the European Commission, DG CONNECT, art-science, and art-technology - deluded people dancing on the edge of doom. Reduced to ignorance by STEM! Better rewrite the history of art to make it support the rule of science and reason. This is what is now happening - the rewriting of art history to match the ideology and dogma of STEM. This rewriting is the favourite pastime of those involved in this religious cult known as art-science, art-technology.

**184**

This is the moment - the point where we are half way from the start, or half way to the end, or both.

Rachel Carson, who we did not forget, will soon speak. For the moment it is to Arthur Koestler that we once again turn:

"For Galileo the book of nature is written in the mathematical language. Without its help it is impossible to understand a single word of it. But Galileo's chief mathematician is called Nature, not God, and his references to the latter sound like lip-service. Galileo takes the hypostatization of mathematics a decisive step further by reducing all nature to size, number, and slow or rapid motion, and by relegating into the limbo of the subjective or secondary qualities everything that cannot be reduced to these elements - including, by implication, ethical values, and the phenomena of the mind.

"The division of the world into primary and secondary qualities was completed by Descartes. He further reduced primary qualities to extension and motion, which form the realm of extension - *res extensa* - and he lumped together everything else in the *res cognitians,* the realm of the mind housed in a somewhat ungenerous manner in the tiny pituitary gland.

"Animals for Descartes are robot mechanisms, and so is the human body; and the universe (with the exception of a few million pea-sized pituitary glands) was now so completely mechanised that he could claim 'give me matter and motion and I will construct the world.'"

And now we know why China will in the end be shown to be better off for not having a Descartes!

Any comments Edward O. Wilson?

It seems that you have gone silent.

**185**

Now a surprise! Francis Bacon has arrived and is asking to speak. So, let it be so:

"They who have presumed to dogmatise on nature, as on some well investigated subject, either from self-conceit or arrogance, and in the professional style, have inflicted the greatest injury on philosophy and learning."

It sounds like you are talking about the European Commission's DG CONNECT and those STEM people from prestigious institutions that we met many chapters back. And that word dogma just keeps cropping-up. Over and over and over ...

Morris Weitz now speaks:

"Theory has been central in aesthetics and is still the preoccupation of the philosophy of art. Its main avowed concern remains the determination of the nature of art into a definition of it. It construes definition as the statement of the necessary and sufficient properties of what is being defined, where the statement purports to be a true or false claim about the essence of art, what characterises and distinguishes it from everything else. Each of the great theories of art - Formalism, Voluntarism, Emotionalism, Intellectualism, Intuitionism, Organicism - converges on the attempt to state the defining properties of art. Each claims that it is the true theory because it has formulated correctly a real definition of the nature of art.

"Is aesthetic theory, in the sense of a true definition or set of necessary and sufficient properties of art, possible? If nothing else does, the history of aesthetics itself should give one enormous pause here. For, in spite of the many theories, we seem no nearer our goal today than we were in Plato's time. Each age, each art-movement, each philosophy of art, tries over and over again to establish the stated idea only to be succeeded by a new or revised one, rooted, at least in part, in the repudiation of preceding ones. Even today, almost everyone interested in aesthetic matters is still deeply wedded to the hope that the correct theory of art is forthcoming.

"I want to plead for the rejection of this problem. I want to show that theory - in the requisite classical sense - is never forthcoming in aesthetics, and that we would do much better as philosophers to supplant the question; *what is the nature of art?*; by other questions, the answers to which will provide us with all the understandings of the arts there can be."

Interesting observations!

**186**

By way of a continuation of the previous chapter, Morris Weitz will now survey some aesthetic theories:

"First we deal with Formalist theory, which is mostly about painting. The essence of painting is plastic elements in relation. Its defining property is significant form, that is to say, certain combinations of lines, colours, shapes, volumes - everything on the canvass except the representational elements - which evoke a unique response to such combinations. Painting is definable as plastic organisation. The nature of art, what it really is, so the theory goes, is a unique combination of certain elements, the specifiable plastic ones, in their relations. Anything that is art is an instance of significant form: and anything which is not art has no such form.

"Second we deal with Emotionalism, and to the above the Emotionalist replies that the truly essential property of art has been left out. Tolstoy,

Ducasse, and other advocates of this theory, find that the requisite defining property is not significant form but rather the expression of emotion in some sensuous public medium. Without projection of emotion into some piece of stone or words or sounds, etc. there can be no art. Art is really such embodiment. It is this that uniquely characterises art, and any true, real definition of it, contained is some adequate theory of art, must so state it.

"Third we deal with Intuitionism where people would disclaim both emotion and form as defining properties. Here one would see art not being identified with some physical, public object but with a specific creative, cognitive and spiritual art. Art is really a first stage of knowledge in which certain human beings, artists, bring their images into a lyrical clarification or expression. As such it is awareness, non-conceptual in character, of the unique individuality of things; and since it exists below the level of conceptualisation or action, it is without scientific or moral content.

"Fourth there is Organicism, which states that all art is really a class of wholes consisting of distinguishable, albeit inseparable, elements in their causally efficacious relations which are presented in some sensuous medium. What is claimed is that anything which is a work of art is in its nature a unique complex of interrelated parts - in painting, for example, lines, colours, volumes, subjects, etc., all interacting upon one another on a paint surface of some sort.

"Finally comes Voluntarism which calls into question traditional aesthetics. Significant form, expression, intuition, and objectified pleasure are fallacious, either because, while true art, they are also true of much that is not art, and hence fail to differentiate art from other things; or else because they neglect some essential aspect of art. Art needs to be defined by a complex of characteristics, including: embodiment of wishes and desires imaginatively satisfied, language, which characterises the public medium of art, and harmony, which unifies the language with the layers of imaginative protections. Thus art is the provision of satisfaction through the imagination, social significance, and harmony."

And in Katja Kwastek's work one also encounters a later understanding of aesthetics which comes about by virtue of the digital, enabling interactive art.

"Interactive art places the action of the recipient at the heart of its aesthetics. In interactive art, the recipient becomes a performer."

Thanks Katja. In other words, the boundaries between the visual arts and the performing arts blur. This is most certainly relevant to scriptovisualism, as are the art theories that Morris Weitz briefly explained. But is it true that interactive digital art places the recipient at the heart of its aesthetics?

**187**

Arthur Koestler speaks:

"To science, the parting of the ways from religion seemed at the beginning to be an unmitigated boon. Freed from mystical ballast, science could sail ahead at breathtaking speed to its conquest of new lands beyond every dream. Within two centuries it transformed the mental outlook of *homo sapiens* and transformed the face of his planet. But the price paid was proportionate: it carried the species to the brink of physical self-destruction, and into an equally unprecedented spiritual impasse. Sailing without ballast, reality gradually dissolved between the physicist's hands; matter itself evaporated from the materialist's universe.

"This uncanny vanishing act began with Galilieo and Descartes. In his book *The Assayer*, Galileo banished the qualities which are the very essence of the sensual world - colour, sound, heat, odour and taste - from the realm of physics to that of subjective illusion. Descartes carried the process one step further by paring down the reality of the external world to particles whose only quality was extension in space and motion in space and time. At first this revolutionary approach to nature looked so promising that Descartes believed he would be able to complete the whole edifice of the new physics by himself. His less sanguine contemporaries thought that it might take as much as two generations to wrest the last secret from nature."

Unity: the state of being united or joined as a whole.

**188**

Morris Weitz will speak again:

"The very expansive, adventurous, character of art, its ever-present changes and novel creations, makes it logically impossible to ensure any set of defining properties.

"Any definition of art, for example, art is significant form, cannot be accepted as a true real definition of art; and most certainly it actually functions in their aesthetics as a redefinition of art in terms of the chosen condition of significant form. But what gives it its aesthetic importance is what lies behind the formula.

"Once we, as philosophers, understand that distinction between the formula and what lies behind it, it behoves us to deal generously with the traditional theories of art; because incorporated in every one of them is a debate over and argument for emphasising or centring upon some particular feature of art which has been neglected or perverted. If we take the aesthetic theories literally, they all fail; but if we construe them, in terms of their function and point, as serious and argued-for recommendations to concentrate on certain criteria of excellence in art, we shall see that aesthetic theory is far from worthless. Indeed, it becomes as central as anything in aesthetics, in our understanding of art, for it teaches us what to look for and how to look for it in art.

"What is central and must be articulated in all the theories are their debates over the reasons for excellence in art – debates over emotional depth, profound truths, natural beauty, exactitude, freshness of treatment, and so on, as criteria of evaluation – the whole of which converges on the perennial problem of what makes a work of art good. To understand the role of aesthetic theory is not to conceive it as a definition, logically doomed to failure, but to read it as summaries of seriously made recommendations to attend in certain ways to certain features of art."

*Unity of knowledge* does not imply consilience; it does not need consilience.

**189**

Today we woke in a different place, a different country – Germany. We walk to the Goethe University of Frankfurt. As we approach, we sense that there is something different about the buildings – they have a 1930s feel to them, yet most of Frankfurt, in and around the central area, was destroyed by bombing during the Second Word War. Later we learn that we are in the buildings that were once the headquarters of the infamous IG Farben – this was the chemical company that formed close links with Hitler and the Nazi Party and eventually the SS, becoming involved in the use of slave labour, among other war crimes. Here is a spokesperson for the Holocaust Education and Archive Research Team, referring to the trial of the executives of IG Farben at Nuremberg:

"The prosecutors were convinced that their material would provide ample evidence that even long before Hitler's accession to power, the managers of IG Farben had desired a dictatorship that would be *able to act without taking the whims of the masses into consideration* and that they had wanted to *dominate the entire European chemical industry, if possible also outside Europe*. Even before Hitler seized power, IG Farben had entered agreements with him for government support for the expansion of their synthetic gasoline facilities."

Now is a good time to speak with Rachael Carson, or to let her speak - same thing looked at in different ways - because she has just arrived and is waiting patiently in the wings, ready to step forward. Her she is:

"The whole process of spraying seems caught-up in an endless spiral. Since DDT was released for civilian use, a process of escalation has been going on in which more toxic materials must be found. This has happened because insects, in a triumphant vindication of Darwin's principle of the survival of the fittest, have evolved super races immune to the particular insecticide used, hence a deadlier one has always to be developed - and then a deadlier one.

"Future historians may well be amazed by our distorted sense of proportion. How could intelligent beings seek to control a few unwanted species by a method that contaminated the entire environment and brought the threat of disease and death even to their own kind? Yet this is precisely what we have done.

"Under primitive agricultural conditions the farmer had few insect problems. These arose with the intensification of agriculture - the devotion of immense acreages to a single crop. Such a system set the stage for explosive increases in specific insect populations. Single-crop farming does not take advantage of the principles by which nature works; it is agriculture as an engineer might conceive it to be. Nature has introduced great variety into the landscape, but man has displayed a passion for simplifying it. Thus he undoes the built-in checks and balances by which nature holds the species within bounds. One important natural check is a limit on the amount of suitable habitat for each species. Obviously then, an insect that lives on wheat can build up its population to much higher levels on a farm devoted to wheat than on one in which wheat is intermingled with other crops to which the insect is not adapted.

"It is not my contention that chemical insecticides must never be used. We however live in an era dominated by industry, in which the right to make a dollar at whatever cost is seldom challenged. When the public protests, confronted with some obvious evidence of damaging results of pesticide applications, it is fed little tranquillising pills of half-truth. We urgently need an end to these false assurances, to the sugar coating of unpalatable facts. It is the public that is being asked to assume the risks that the insect controllers calculate."

Sugar coating - artists in residence!

So we are talking about ethics for sure, but we are also talking about aesthetics! And what about the relationships between academics and those very big agrichemical companies? No doubt Rachel will be back later to tell us about this hidden aspect of the world of STEM.

And the conclusion of the Flagship evaluation panel is entirely negative - there is nothing special about the Flagships and they are not, and will probably never, deliver on the much sought for innovation. The European Commission - DG CONNECT in particular - does not understand innovation.

Thus will we write a report that is entirely negative! But as we leave the former headquarters of IG Farben the European Commission's DG CONNECT representative mentions with some concern the negative conclusions of the panel. We do not respond for we know that the negative report will be manipulated and will in the end hide the truth about the failure of the Flagships. This is what DG CONNECT will do for the STARTS Platform - hide its failure and present it as a success.

In the United Kingdom this would be known as misleading ministers, Parliament, and the public! Evidently the lessons arising from the case of IG Farben and the Nazis have not been learned.

**190**

Now we are going to hear from a scientist by the name Lord Rees. Yes he is British - they do love their titles!

"This authoritative and balanced report offers enlightening reading for all policy makers; its well judged recommendations should be heeded."

He is referring to a Royal Society report entitled *Reaping the Benefits: Science and the Sustainable Intensification of Global Agriculture,* published in October 2009. The report describes how the prudent application of recent and prospective biological advances can contribute to the *sustainable intensification* of agriculture. It argues that a multi-pronged approach is needed. Improvements in farming practices and crop management are essential, but modern genetics must be utilised too.

It is in fact a very poorly disguised and biased effort by a vested interest group, bio-tech scientists, to pursue their genetic modification research agenda.

Abraham Lincoln now speaks:

"You can fool all the people some of the time, and fool some of the people all the time, but you cannot fool all people all the time."

The Royal Society report uses the phrase *good science*. Is it good science to write a science-based report that advocates an inclusive approach, including crop management practices, but which fails to mention that in the European Union, a Framework Directive was on the verge of being issued (had actually been under development since 2006) which deals with the matter of pesticide use reduction, through the use of Integrated Pest Management? Is it good science to give the impression that Integrated Pest Management is something for the developing world, for that is the effect of discussing it within the context of developing countries, and not at all to mention that this is also what European farmers also need to adopt (and are now doing as a result of this Framework Directive)?

We could say more about this report, but will defer the matter to another book.

**191**

Rachel is back.

"I would first like to quote a Dutch scientist, C.J. Briejer: 'We need a more high-minded orientation and a deeper insight, which I miss in many researchers. Life is a miracle beyond our comprehension, and we should reverence it even where we have to struggle against it. The resort to weapons such as insecticides to control it is a proof of insufficient knowledge and of an incapacity so to guide the processes of nature that brute force becomes unnecessary. Humbleness is in order; there is no excuse for conceit here.'

"We stand now where two roads diverge. But unlike the roads in Robert Frost's familiar poem, they are not equally fair. The road we have long been travelling is deceptively easy, a smooth superhighway on which we progress with great speed, but at its end lies disaster. The other fork of the road - the one less travelled by - offers our last, our only chance to reach a destination that assures the preservation of our earth.

"Most of those best fitted to develop natural controls and assist in putting them into effect have been too busy labouring in the more exciting vineyards of chemical control. The major chemical companies are pouring money into universities to support research on insecticides. This creates attractive fellowships for graduate students and attractive staff positions. Biological-control studies, on the other hand, are never so endowed - for the simple reason that they do not promise anyone the fortunes that are to be made in the chemical industry. Biological controls are left to state and federal agencies, where the salaries paid are far less.

"This situation also explains the otherwise mystifying fact that certain outstanding entomologists are among the leading advocates of chemical control. Inquiry into the background of some of these men reveals that their entire research programme is supported by the chemical industry. Their professional prestige, sometimes their very jobs, depends on the perpetuation of chemical controls. Can we expect them to bite the hand that literally feeds them? But knowing their bias, how much credence can we give to their protests that insecticides are harmless?"

How much credence can we give to those who say that GM crops are harmless?

We are talking about ethics and aesthetics again.

Rachel could also be talking about artists who have fallen in love with science and technology – are they going to bite the STEM hand that feeds them? DG CONNECT for example? How much credence can we give to such artists?

**192**

So we need a more high-minded orientation and a deeper insight. Yes indeed we do.

Now speaks Marielle Bruijnis, Vincent Blok, Elsbeth Stassen, and Bart Gremmen, collectively we presume, as one voice:

"Worldwide, billions of day-old male chicks are killed annually. Although this instrumentalised use of animals is developed in response to the demand for affordable animal protein as food source, this large scale killing of young animals caused societal resistance in the Netherlands and Germany. Alternatives to the killing of day-old male chicks have been put forward to solve this problem, but each alternative leads to new dilemmas. At first sight, it does not seem logical to use genetic modification to solve such a problem as GM is rather controversial. However, due to the disadvantages of the other alternatives, the GM innovation might be the lesser of two evils. Therefore it is good to have a better understanding of the arguments related to GM in animal food production and whether these arguments change in the context of new innovative developments. So we study how it is possible that an application of GM in animal food production might be the lesser of two evils using a normative technology assessment on the ethical and social aspects, using existing literature and expert opinion as input."

We do indeed need a more high-minded orientation and a deeper insight as the question posed by the speakers highlights the nature of

European culture: a bag of bones and entrails dressed up with a little good meat. Ghandi is supposed to have said when asked what he thought of the idea of European civilisation, that 'I think it would be a good idea.' And when we do have a European civilisation, people will see the question posed about GM and one-day old chicks, for what it is – the product of a people that live life like a plague of locusts, striping the world of its assets and destroying most of what is of value, rather than living life like a lotus. This is not a personal attack, for we are sure all of you are very nice people, but it is time to stop the type of behaviour that leads to such stupid dilemmas. Nor is our statement a rejection of genetic modification, which many European minds will assume given their love of either/or thinking (another peculiar characteristic of the modern European mind). Man's intellect does recede!

So we are back with the matter of ethics and aesthetics again!

**193**

Rachel is back.

"The question is whether any civilisation can wage relentless war on life without destroying itself, and without losing the right to be called civilised."

We do indeed a more high-minded orientation, and we are back once more to the matter of ethics and aesthetics.

Rachel now continues:

"The system by which the Food and Drug Administration establishes maximum permissible limits of contamination, called *tolerances,* has obvious defects and creates an unjustified impression that safe limits have been established and are being adhered to. A laboratory animal though, living under controlled and highly artificial conditions, consuming a given amount of a specified chemical, is very different from a human being whose exposures to pesticides are not only multiple but for the most part unknown, unmeasurable and uncontrollable. Even if seven parts per million of DDT on the lettuce in a luncheon salad were *safe,* the meal includes other foods, each with allowable residues, and the pesticides on this food are only a part, and possible a small part, of total exposure. This piling up of chemicals from many different sources creates a total exposure that cannot be measured. It is meaningless, therefore, to talk about the *safety* of any specific amount of residue.

"Between the major groups of insecticides, and between them and other chemicals, there are interactions that have serious potentials. Whether released into soil and water or a person's blood, these unrelated chemicals do not remain segregated; there are mysterious and unseen changes by which one alters the power of another for harm.

"For the population as a whole, we must be concerned with the delayed effects of absorbing small amounts of the pesticides that invisibly contaminate our world. Responsible public health officials have pointed out that the biological effects of chemicals are cumulative over long periods of time, and that the hazard to the individual may depend on the sum of the exposures received throughout his lifetime.

"We are accustomed to look for the gross and immediate effect and to ignore all else. Unless this appears promptly and in such obvious form that cannot be ignored, we deny the existence of hazard."

**194**

Rachel will speak again (probably), but not now. Something else is on our mind, so someone else will now speak instead. It is Amitva Biswas. He works, or worked, we are not sure which, in the Computer Science Department at Texas A&M University.

"I would like to say that collaborations between art and technology, in academic settings, face several challenges due to differences in: the professional cultures and traditions; nature of professional training received; dominant personality types and cognitive styles of the practitioners; professional expectations; practice methods; difficulty in communication; institutional factors; paucity of appropriate project management methods and software tools."

We agree. We know this from personal experience. Who though is willing to admit this, especially when some funding body has money to throw around?

Please continue Amitva.

"Software engineering methods were matured for developing applications where requirements can be specified beforehand to a reasonable extent. It is difficult and often inappropriate to solicit system requirements from creative artists. This is because most often they do not plan or visualise their performance to the level of details as required for engineering design of the systems; they mature their visualisation as they work on it. So accurately specifying an unknown system is tricky. This causes severe problems for system development or software engineering activities. Requesting artists to speak out their requirements too often may distract them from their creative trajectories."

Yes indeed. Please carry on, for it is so refreshing to hear someone speak who is not doing so from within the love affairs that are art-science and art-technology.

"Art and technology work methods are so different that a suitable production process is necessary to insulate artists and technologists from one another and yet facilitate communication among them to protect both - artistic and technical creativity. Traditional engineering methods are not appropriate for art projects. Hence technologists have to creatively adapt their work methods to serve these projects."

Well, that is one perspective (which we do not agree with). There is of course that which people call Agile Software development methods, or more precisely, a spiral lifecycle model. Back to aesthetics once again! The whole point though, is not to isolate people from one another.

Now listen to us - Julia and Paul talking:

"We should define this very precisely, so that there is no ambiguity," said Paul.

"Why do you want to define and order everything? Let's leave things in an ambiguous state," responded Julia.

"Why do you two think and behave in the ways that you do?"

"Who was that speaking?" asked Julia and Paul speaking with one voice.

"It was me, the social scientist part of us," came the reply from the disembodied voice that has no name.

**195**

Now it is time for something completely different.

"Biologists have refined reductionism into a high art."

"It was on a dreary night of November, that I beheld the accomplishment of my toils. With an anxiety that almost amounted to agony, I collected the instruments of life around me, that I might infuse a spark of being into the lifeless thing that lay at my feet."

"They can achieve partial synthesis at the level of molecule and organelle."

"It was already one in the morning; the rain pattered dismally against the panes, and my candle was nearly burnt out, when, by the glimmer of the half-exhausted light, I saw the dull yellow eye of the creature open; it breathed hard, and a convulsive motion agitated its limbs."

"Even if complete cells are still beyond them, they know they can reconstruct some of the elements over time."

"How can I describe my emotions at this catastrophe, or how delineate the wretch whom with such infinite pains and care I had endeavoured to form?"

"They foresee no need for an overarching grand explanation as a perquisite for creating artificial life."

"His limbs were in proportion, and I had selected his features as beautiful. Beautiful! - Great God!"

"An organism is a machine."

"His yellow skin scarcely covered the work of muscles and arteries beneath; his hair was of a lustrous black, and flowing; his teeth of pearly whiteness; but these luxuriances only formed a more horrid contrast with his watery eyes, that seemed almost of the same colour as the dun-white sockets in which they were set, his shrivelled complexion and straight black lips."

"The laws of physics and chemistry are, most believe, enough to do the job."

"The different accidents of life are not so changeable as the feelings of human nature."

"The machine the biologists have opened up is a creation of riveting beauty."

"I had worked hard for nearly two years, for the sole purpose of infusing into an inanimate body. For this I had deprived myself of rest and health. I had desired it with an ardour that far exceeded moderation; but now that I had finished, the beauty of the dream vanished, and breathless horror and disgust filled my heart."

"Where the nucleic acids are the codes, the proteins are the substance of life, making up half of the animal's dry weight. They give form to the body, hold it together by collagen sinews, move it by muscle, catalyse all its animated chemical reactions, transform oxygen to all its parts, arm the immune system, and carry the signals by which the brain scans the environment."

"Now I need an artist in residence."

"I am passionate about science."

"Together we can engage in downstream public engagement."

"There is money to be made."

Who was speaking? Actually two people spoke! Who are they, these unnamed ones? Wait and see. Or was it three people speaking?

**196**

Now, one more time we paraphrase Karl Jaspers:

"European cooperation which, as the ideology that is the European Union, in its enthusiasm for the salvation of Europeans that it feels certain of securing, takes steps to mould the future by force of bureaucratic and undemocratic planning and technocracy, and European cooperation as the idea of gradual realisation of improvements, are alien to one another. The former consumes the man who surrenders himself to it in a faith that appears in the guise of a union of all Europe under one banner; it treats the unbeliever as a material available to manipulation by force. The latter casts no spell, lives in the present, requires the sobriety of reason, and the humaneness of unceasing communication. As long as European cooperation's demands are concretely visualised and thought out they remain within bounds. It is only when concrete reality is lost sight of and a fanatical paradise of Europeans is presupposed as possible, that its demands become abstract and absolute. It is at this point that European cooperation becomes an ideology. The demand for complete implementation in fact leads away from its fulfilment. Along a path of coercion it leads to servitude."

This we must say, from long engagement with this process of European cooperation, and being, as it were, on the inside while remaining on the outside, is what is happening in Europe. This is why, accepting some short term discomforts, the United Kingdom will in the end endure and prosper, and why the European Union will one day, in the not too distant future, become once again the past. Instead of one small nation being a problem, the problem will be escalated upwards to the level of a whole continent. Then once more the streets of Europe will be soaked in the blood of innocent people, and all because a few, an elite few, those who believe that free people think, slaves work, will have once again rendered ordinary people as mere objects.

"What is life? Life is the nation. The individual must die anyway. Beyond the life of the individual is the nation."

This we have seen already with the European Union and its treatment of the Greek people in the course of the European Council seeking to stave off the destruction of the euro currency, as a result of Greece's sovereign debt problems. Problems which the European Commission were aware of for many years but did nothing to resolve. Problems that are associated with the European Commission's technocratic failure known as the euro. Now you know what type of organisation the European Commission is, and the true nature of the European Union.

This is Jean-Claude Juncker speaking:

"What is life? Life is the European Union. The individual must die anyway. Beyond the life of the individual is the European Union."

He, by-the-way, operates under the delusion that he has been elected president of the European Commission. This is what happens to those who pursue megalomaniac schemes, as history well demonstrates.

And now you are asking: who was it that first spoke? We did not give you a name, we know. This was a deliberate compositional choice. You will have to discover for yourself who spoke. It is easy enough to do with Google search.

This is art. This is literary art. Actually this is scriptovisual art. We make no apologies for the comparison.

## 197

Jade Spell now speaks:

"It has been 200 years since Mary Shelley planted the seeds that became the novel Frankenstein, and her ominous warning sounds louder than ever: do not interfere with what you do not understand."

Rachel Carson would agree. Sorry for that interruption. Please continue.

"In 1999, world-renowned synthetic biologist Craig Venter set the ball in motion by exclaiming 'Shelly would have loved this!' when he announced plans to create the first synthetic biological genome. Later, in 2009, academic philosopher Henk van den Belt, in a paper published in *Nanoethics,* questioned whether synthetic biology could be accused of 'Playing God in Frankenstein's Footsteps' by attempting to manipulate life.

"Synthetic biology may seem the most obvious example, but as far back as the 1950s, another major experimental field has also been criticised for trying to play God – Artificial Intelligence (AI). In 1950, Isaac Asimov coined the infamous *Frankenstein complex* in his novel *I, Robot,* with one of the first predictions of a robotic Frankenstein's monster. Several other authors and mainstream movie directors have since followed suit.

"So why are we, still obsessed with Frankenstein? Because even with the strictest ethics and high standards of care, things can, and do, go wrong. With advances in science and technology announced every day, the potential for disaster seems closer than ever."

So we are back once again to the matter of a higher-mined approach, to ethics yes, but also to aesthetics.

In the course of working with the FET Flagship evaluation panel, we learned that the European Commission has a transparency web site, where minutes of the panel's meetings were being posted. We did wonder why all the important and relevant information was being removed from our minutes by the European Commission, before these - censored and sanitised - minutes were made public! Now why do this, if transparency is the aim?

**198**

We have mentioned aesthetics quite a lot recently in this strange work. Why not then let someone who has written about aesthetics, in particular eco-aesthetics, speak. Hence here is Malcolm Miles to say just a few words:

"My own view is that art cannot change the world - it is part of it, and the conditions of its production are always present in an artwork, even when it evokes other responses - but it might contribute, in the cases I examine in my book on eco-aesthetics, to facing the forces and trajectories which appear to bring the world to the edge of destruction. In some cases, this is achieved obliquely, through the use of the uncanny and the unexpected, or through interruptions of routine and predictability. There are other ways too, but the boundary between art and activism is not rigid but fluid and endlessly negotiable."

We understand your point, but do not fully agree for this is only valid if one limits attention to art objects. If one were to ask questions about the processes, beliefs, and values of those in the world of STEM, and to seek to use art to change these, to develop different processes, beliefs and values, different theories even, then what opens is the prospect for different world; a world that people want, not one determined by forces that are routed in the minds of elite groups who by virtue of what they do or have, hold power over the bulk of humanity. But it is certainly goodbye-time for grand overarching narratives.

Billy Klüver now speaks:

"Originally I believed that engineers should simply work for the artists. Once I gave a talk and made the point that an engineer should just be another tool for the artist. But Rob Rauschenberg very specifically said 'No! It has to be a collaboration!' I immediately understood what Bob was saying. The one-to-one collaboration between two people from different fields always holds the possibility of producing something new and different that neither of them could have done alone."

Thanks Billy for confirming that all this co-creation talk than one hears from artists, especially those involved in the love affairs that art-science and art-technology have become, is indeed very 1960s. So what comes next? So back once more to the reason for scriptovisualism, and the attraction of theory.

Sorry about this, but, paraphrasing Mark Twain: "the reports of my death as an author have been greatly exaggerated." The reports of the death of theory; these too have been greatly exaggerated.

Beth Venn of the Whitney Museum now speaks about the 1998 exhibition *Alternating Currents: American Art in the Age of Technology*:

"It is important to realise that not all art that concerns itself with issues of technology is *high-tech.* Indeed, many of the exhibition's works are drawn or painted, differing only stylistically from the portraits and landscapes of centuries past. The issues they expose can be foreboding or optimistic, offering as many types of opinion as there are types of artist. Despite all the technology displayed in the exhibit, however, the foremost by-product of the Information Age – the computer – is all but absent."

Thank you for confirming the validity and relevance of scriptovisualism to the world of STEM.

**199**

Rachel is back.

"An example of a carcinogen belonging to a group of new organic pesticides is a chemical widely used against mites and ticks. Its history provides abundant proof that, despite the supposed safeguards provided by legislation, the public can be exposed to a known carcinogen for several years before the slowly moving legal process can bring the situation under control. The story is interesting from another standpoint, proving that what the public is asked to accept as safe today may turn out tomorrow to be extremely dangerous. In effect, what happens when concerns are aroused, is that testing takes place, usually on animals, but in the meantime the chemical remains in use, thus in effect making humans act as guinea-pigs, testing the suspected carcinogen along with laboratory animals."

Thanks for that. Now a spokesperson from the Francis Crick Institute will read a paragraph from its strategy document:

"The *Crick* will conduct discovery science, oriented towards solving human health challenges. Our translation and innovation strategy will focus on maximising the value that can be generated from this science, measured in terms of improvements in the lives of people in the United Kingdom and internationally, and in new economic opportunities. We will explore opportunities for new and improved therapies, diagnosis and prevention strategies. In many cases, this will be accompanied by opportunities to form new and grow existing businesses; boost the supply of highly skilled and mobile people across the United Kingdom economy; and develop other activities that reinforce the United Kingdom's position at the forefront of global innovation in the life sciences."

Very good! So more wealth, which is about making money from trying to solve humanity's problems. Now what does your document have to say about ethical standards?

"Yes indeed. The strategy document says that the *Crick* will set and maintain the highest research standards."

But this is not what we asked about. Address ethics please.

"Oh right. Indeed, the strategy document says that at the heart of the *Crick's* philosophy is a commitment to the highest quality science. We believe that the necessary quality can only be delivered through a focus on the best and most imaginative scientists, which in turn requires a broad and flexible approach to our scientific programme. As we move closer to opening, the broad programmatic details contained in this document will be updated to reflect the strengths of scientists moving to the *Crick*, the latest research strategies of our founders, the developing engagement with our university founders, promising new areas arising from our horizon scanning activities and the mechanisms needed to facilitate effective translation of *Crick* science."

But what about ethics? We asked specifically about ethics. Please use the word in your answer.

"Ethics! Yes, of course! The strategy document states that we are permeable to outside talent and ideas and committed to transparent science for the public good that meets the highest ethical, animal welfare and bio-security standards."

So you are saying that an institute with a strategy to make money out of science, only uses the word ethics once in its strategy document?

"Yes."

How many times does the word *wealth* appear, in the sense of money?

"Six times!"

Well what we have to say about that is: *VW emissions scandal*!

So once more we confront the matter of a high-minded approach, which every time we mention the phrase is because we doubt its presence, especially as science is now about making money. What was it that Abe Lincoln said?

And we wonder about all those artists in residence taking money from all the *Cricks* in the world dedicated to making money from science. We look forward to seeing the art that comments on this attitude of mind and the potential for more *VW emissions scandals*. Or perhaps more IG Farbens!

**200**

So let us now mention Experiments in Art and Technology (E.A.T.) and in particular that we will call *pure E.A.T.* as defined by Billy Klüver. First an explanation about *Nine Evenings* that very famous event, held back in 1966. Nilo Lindgren will explain.

"Actually the experiment in art and technology that grew into the *Nine Evenings: Theatre and Engineering* at the Armoury in New York was originally scheduled as part of a festival of art and technology in Stockholm; but the project to go to Sweden grounded on the very issue that, according to some interpreters, imparted to the *Nine Evenings* their unique significance and their claim to success in the face of acerbic critical response.

"At issue was the relationship between the artists and engineers. Stockholm wanted to supply engineering support in Sweden; the artists were, in effect, to give in their orders and the engineers would carry them out. But the New York group opted for what constitutes not only a different mode of doing business but a fundamental difference in aesthetic outlook as well. The artists wanted the engineers to be in on the groundwork and planning, to engage in an ongoing dialogue with the artists, from the very beginning. To engineer Billy Klüver and artist Robert Rauschenberg, the two major driving forces of the *Nine Evenings*, the exploration of the process of collaboration constituted the fundamental experiment. It was even suggested at one point that the audience be allowed to come to the Armoury during the days to see the preparatory work in progress.

"Thus, the upshot of the dispute with Stockholm on the relationship between artists and engineers - which art critic Brian O'Doherty noted at the time as a 'prophetic friction!' – was to lead to a pioneering venture. The New York artists wanted, as O'Doherty wrote, 'to work with the technicians – follow their thinking, involve them in the artists' ways of thinking. This makes great demands on patience, mutual understanding, and above all, time. Ideally, it would acquaint the artists with their new materials, since art often arises out of the nature of materials themselves. This democratic idea is especially typical of Rauschenberg's equal respect for all materials, which in his case included people.'

"Rauschenberg himself speaks of the waste of possibilities in using technology only to carry out some pre-planned work, for the technology is then like a dead medium rather than an encounter. 'I do think there are, definitely two attitudes toward technology, and the one of using technology as a servant I find very dull and outdated. It seems to me closed-minded. However, in an open aesthetic investigation, where it almost doesn't matter whether one makes art out of it or not, in which an art object may not even be produced, the purpose is not to move in a single direction. Instead of taking the materials and just executing something preconceived, if from the beginning, if you have some curiosity about what would happen if you could do this, and then you share that curiosity with someone else and make him curious, then that's the beginning of the piece, and it grows and grows. The whole world then is the medium, if you can relate in your aesthetic to something other than the finished product.'"

There are other modes of operations as well, some less sophisticated, some even more sophisticated than *pure E.A.T.*, hence scriptovisualism. See, everything is connected, even in this subconscious driven, surrealist process of writing our odyssey. We must add however, that there are conflicts associated with *Nine Evenings* too, but these are usually dropped from the art-science and art-technology lovers' monologues – art science, art-technology fakes. This is what happens when cults, complete with ideologies, are formed – no room for reality!

If you know about the art of Robert Rauschenberg then you will know that *Nine Evenings* was not really a pioneering venture! Its uniqueness lay in its scale. It was entertainment – in case you were wondering or had been told otherwise.

## 201

We did not tell you something back in chapter 195. Who was speaking? Three people spoke, as we said. The first was Mary Shelly reading from her novel Frankenstein and the second was Edward O. Wilson reading from his book Consilience. And now you ask: which is which, for it is not always so obvious? Indeed!

But what about the third person?

Indeed, what about the third person!

Billy Klüver will now speak for a second time:

"Art technology collaboration has many appeals, which include: new types of people to work with; new horizons of ideas; a different way of looking at engineering; the satisfaction of carrying something through outside of the goal structures of a profit-making organization and the accompanying sense of special responsibility that one may miss in a usual engineering team effort; the opportunity to learn about art at first hand; the opportunity to learn how artists may work differently from engineers; and the contrary experience, finding out how much the techniques and processes of art and engineering may be alike."

One cannot look at the matter of these collaborations through the intellect alone, nor seek only rational and objective explanations and goals. Now you are starting to see something of the importance of scriptovisualism.

So a different way of looking, of seeing, is what you have said, Billy. How about looking at technology in a way that you, being one of the *people of the book*, have never seen it before?

Here is Joseph Needham:

"Taoist technology and *laisser-faire*, Mohist science and religion, Legalist uniformity and Hedonist self-cultivation, were all in a sense causalities of the paramount necessity for the centralised Confucian bureaucratic State."

Taoist technology – now what is that?

Engineers and technologists tend to have a naïve belief that technology is neutral, and that it is not the product of values, that is to say, that it is value free. The theory of the *Social Shaping of Technology* says otherwise.

Donald MacKenzie and Judy Wajcman now speak with one voice:

"We hypothesise that technology is far from neutral, but is the product of the prevailing values of those who develop these technologies. The belief that technology is neutral is known as technological determinism."

Taoist technology – now you have some further insight into why we pursue scriptovisualism.

**202**

Carlos Moedas, European Commissioner for Research, Science and Innovation in the (first) Juncker Commission now wants to speak.

"I strongly believe that the time has come for a European initiative on research integrity."

You are correct, for an ethical meltdown is underway in the world of STEM, but few people want to talk about it. But, Commissioner Moedas, you need to put your own house in order first. The European Commission has a research programme called Responsible Research and Innovation, which deals with ethics among other things, yet it attracts unethical proposals! And what of the extent of the unethical proposals submitted to other programmes?

Then there is the STARTS initiative - a tale of woe from the START. What do we mean by ethics? We are talking about the corruption that comes with unbridled and disingenuous efforts to acquire public funds to support research groups - and now also artists. We are also talking about the corrupt relationship that exists between the European Commission and the experts that it uses - we refer you to the former chief scientific advisor to a former president of the European Commission for fuller details of the practices she rightly criticised.

We also refer you to STARTS for a current case study. No chance at all of achieving anything approaching research integrity when the funding body engages in practices that are detrimental to research and detrimental to the trust that people place in the results of research. No trust is evidently now possible. For this we can blame the European Commission. A high-minded approach is needed. Or will we have to wait for another IG Farben scandal before action is taken?

Now a spokesperson from Wellcome Trust will read an extract from the assessment report of the Trust's SciArt Programme:

"A minority of artists and arts producers who considered themselves to be working at the more critical and controversial end of arts practice believed that the Trust occupied quite a conservative position in relation to mainstream science and that this made it more risk-averse as an arts funder than organisations that were more independent of the science sector, such as Arts Council England."

The above is related to the matter of the distortion of art practices that result from programmes like SciArt and STARTS, and the tendency towards presenting to scientists one particular, mind-set reinforcing type of art, which stands in sharp contrast to what Billy Klüver eventually came to see as the value of E.A.T.

Here is Günther Oettinger, European Commissioner in charge of the Digital Economy and Society (and responsible for DG CONNECT) in the (first) Juncker Commission:

"No restriction to artistic freedom should be applied as it might compromise the creative process and thus, the essence of innovation as the valuable out-of-the-box thinking of artists is rooted in their artistic independence."

It is interesting that, at such a late stage in the process, after 4 years of DG CONNECT's engagement with artists, you felt the need to make this statement. Perhaps it has something to do with the critique of STARTS that came forth from our minds, among others. Yet you still think that the value of working with artists lies in correcting an imaginary creativity deficit. You should have spoken to social scientists.

Our critique of STARTS comes in the form of a book - actually something more than a book - which makes DG CONNECT the subject of the work; it instrumentalises DG CONNECT, and exposes the nature of this organisation and the relationships that exist between the European Commission (EC) and those who gather around the Emperor's Court (EC) - it is a corrupt relationship as the European Parliament well knows. The book goes by the title *STARTS - Science, Technology and the Arts: The Artistic Voices that DG CONNECT Silenced*. We are sure that Oettinger greeted it with great enthusiasm given his stated commitment to ensuring artistic freedom and independence! We are also sure he will greet the revelations contained in this book with great enthusiasm!

**203**

Nilo Lindgren now speaks again:

"One of the most serious and most intriguing possibilities is that engineering and engineers will find, through working with artists, new goals. Engineering goals have never been fixed: in its deepening interaction with society, modern technology has imposed on it new goals - cold war systems, space systems, biomedical investigations - and it undoubtedly will continue to find new goals, as in transportation, and in ways of confronting the urban crises. The promise that art holds for engineers in this respect seems strong indeed."

Actually you mean people, and not technology, have imposed new goals. And one of those new goals could now be Taoist technology! Engineers may also learn to tolerate higher levels of uncertainty and vagueness, and hence change their behaviour and not be prime fodder for those with ideological agendas. And this notion of using artists for invention - well we are now in the realms of STARTS - art-science and art-technology fakes.

Robert Rauschenberg:

"In any really critical important development, it is important to give up ways of working in which you have been comfortable. To enrich one's environment and one's life, one must take risks. That is why the more arduous and difficult path of collaboration is preferable to the artists' mere exploitation of the given technology."

Fred Waldhauser:

"If the artist and engineer understand one another, then the artist can use the real technology rather then just using it as a found object."

And what is really interesting is what happens when the two meet in one mind, and the differences are understood. Then perhaps one can go beyond *pure E.A.T.* towards ...

Now we are getting closer to understanding what scriptovisualism could become.

The European Union is the last of the twentieth century's grand overarching narratives and single minded utopian visions. It is doomed to go the same way as all the other grand overarching narratives and single minded utopian visions. STEM too. And much of the art that it now appropriates, which manifests itself in STARTS - is totalitarian art. They too are doomed.

**204**

Byproduct: An incidental or secondary product made in the manufacture or synthesis of something else; an unintended but inevitable secondary result.

Now we return to C.P. Snow and his supposed *two cultures*. He published three main pieces of work dealing with this matter. The first one appeared in 1956. This was his New Statesman article entitled: *The Two Cultures*. The second was his Rede lecture of 1959 entitled: *The Two Cultures and the Scientific Revolution*. The third work appeared in a revised edition of the *The Two Cultures* book, published in 1964, with an addition called *The Two Cultures: A Second Look,* it being a follow-up taking into account the mostly negative responses to the *Rede Lecture*.

It would be interesting to count how many times Snow uses certain words in these documents, so we counted. We used pdf files enabled for optical character recognition and the Adobe software called Acrobat Professional, which allows this counting to be done. Here are the results:

Science: mentioned 66 times.

Scientist(s): mentioned 113 times.

Engineering: mentioned 9 times

Engineer(s) mentioned 19 times.

Industry: mentioned 14 times.

Industrial/industrialisation/industrialise: mentioned 55 times.

Technology: mentioned 6 times.

Mathematics: mentioned 8 times.

Mathematician(s): mentioned 3 times.

Art(s): mentioned 17 times.

Artist(s): mentioned 4 times.

Humanities: mentioned 0 times.

Intellectual(s): mention 63 times.

Culture(s): mentioned 130 times.

Literature: mentioned 20 times.

Literary: mentioned 50 times.

Writer(s): mentioned 24 times.

Here is an artist quoting C.P. Snow:

"Sixty years ago, C.P. Snow famously described the arts and the sciences as being *two cultures* separated by 'a gulf of mutual incomprehension, hostility and dislike'."

No he did not. Tell us what you said Charles:

"I said this: Literary intellectuals at one pole - at the other scientists, and as the most representative, the physical scientists. Between the two, a gulf of mutual incomprehension - sometimes (particularly among the young) hostility and dislike, but most of all lack of understanding."

What C.P. Snow is supposed to have said or written is just like a game of Chinese whispers!

If we were the police, investigating a crime, we would be very suspicious, given the above numbers and all this misquoting, and we would certainly investigate further, which we will, in due course, because we want to know what C.P. Snow was actually saying. We sense a misunderstanding! Another myth! Intuition is a wonderful thing!

When will we do this? We do not know, for we will have to wait for the subconscious mind to decide. This is the surrealist writer's way!

Later we may encounter someone who will inform us about British thought on the relations between the natural sciences and the humanities in the period 1870 to 1910.

**205**

We have not forgotten about Leonardo da Vinci. For a brief moment we will return to him for he has something to tell us. It is something to do with painters and writers. So please go ahead.

"You have set painting among the mechanical arts! Truly were painters as ready equipped as you are to praise their own works in writing, I doubt whether it would endure the reproach of so vile a name. If you call it mechanical because it is by manual work that the hands represent what the imagination creates, your writers are setting down with the pen by manual work what originates in the mind. If you call it mechanical because it is done for money, who fall into this error if indeed it can be called an error more than you yourselves? If you lecture for the Schools do you not go to whoever pays you the most? Do you do any work without some reward?"

Oh, what have we here? Surely all we now need is an Italian Renaissance C.P. Snow to appear. And lo and behold here we have one, for Senor La Neve is here:

"Literary intellectuals, writers, at one pole - at the other artisans, and as the most representative, painters. Between the two, a gulf of mutual incomprehension - sometimes (particularly among the young) hostility and dislike, but most of all lack of understanding."

Have we stumbled across something here? The police investigation will take this new evidence into account.

Anything you want to say about poetry and painting, Leonardo?

Yes it seems he does want to speak about this matter, so go ahead.

"When the poet ceases to represent in words what exists in nature, he then ceases to be the equal of the painter; for if the poet, leaving such representation, were to describe the polished and persuasive words of one whom he wishes to represent as speaking, he would be becoming an orator and be no more a poet or a painter. And if he were to describe the heavens he makes himself an astrologer, and a philosopher or theologian when speaking of the things of nature or of God. But if he returns to the representation of some definite thing he would become the equal of the painter if he could satisfy the eye with words as the painter does with brush and colour, [for with these he creates] a harmony to the eye, even as music does in an instant to the ear."

This probably has something to do with scriptovisualism.

We may return to Leonardo later.

**206**

Other modes of working with artists are possible, or put another way, other modes of working with STEM people are possible. We will list them: service mode; independent transference; pure E.A.T.; transcendence; unity of opposites.

"We do not understand."

That was some STEM people responding.

"We do not understand."

That was some artistic people responding.

"Yes that is right! You may know what we mean by *pure E.A.T.*, although it is also possible that you do not, or might think you do, but in fact mean something that is not meant by what we mean by *pure* E.A.T., but as for the rest - well it is very unlikely that you understand the rest, given that we use our own terminology here and in some case are referring to things that come from the mysterious East.

"Somewhere else, in a different book, we mentioned that we had been developing models. Well we did develop models. These models are specific to us. We are not asking for your approval for these models. We will explain, perhaps, but not now."

That was Julia and Paul speaking.

"Who are Julia and Paul?"

That was some notional reader entering into the script. Yes a script!

Now what is next?

It is Arthur Koestler again:

"Where on the three-centuries journey from natural philosophy to exact science did the estrangement from reality begin; at what point was the new version of Plato's curse uttered: 'Thou shalt think in circles'? If we knew the answer, we would of course also know the remedy; and once the answer is known, it will again appear as heartbreakingly obvious as the sun's central position in the solar system. 'We are indeed a blind race,' wrote a contemporary scientist, 'and the next generation, blind to its own blindness, will be amazed at ours'.

"Later I will address two examples of this blindness."

Rachel Carson again:

"To have risked so much in our efforts to mould nature to our satisfaction and yet to have failed in achieving our goal would indeed be the final irony. Yet this, it seems, is our situation. The truth, seldom mentioned but there for anyone to see, is that, nature is not so easily moulded and that the insects are finding ways to circumvent our chemical attacks on them.

"The impossible is now happening on two broad fronts. By a process of genetic selection, the insects are developing strains resistant to chemicals. The broader problem is that our chemical attack is weakening the defences inherent in the environment itself, defences designed to keep the various species in check. Each time we breach these defences a hoard of insects pours through.

"New problems also arise as insects once present only in insignificant numbers have increased to the status of serious pests. By their very nature chemical controls are self-defeating, for they have been devised and applied without taking into account the complex biological systems against which they have been hurled. The chemicals may have been pre-tested against a few individual species, but not against living communities.

"The balance of nature is not the same today as it was in Pleistocene times, but is still there: a complex, precise, and highly integrated system of relationships between living things which cannot safely be ignored any more than the law of gravity can be defied with impunity by a man perched on the edge of a cliff. The balance of nature is not a status quo; it is fluid, ever shifting, in a constant state of adjustment. Man, too, is part of this balance.

"Two critically important factors have been overlooked in designing the modern insect-control programmes. The first is that the really effective control of insects is that applied by nature, not by man. Populations are kept in check by something the ecologists call the resistance of the environment, and this has been so since life was created. The amount of food available, conditions of weather and climate, the presence of competing or predatory species, all are critically important. The second neglected fact is the truly explosive power of a species to reproduce once the resistance of the environment has been weakened. The fecundity of many forms of life is almost beyond our power to imagine."

Well Rachel, the thinking that leads to this situation is science - the fragmented and reductive mind. We have met *Hawkins Dawkins Wilson Wolpert* already.

**207**

Back now to the matter of *Theory of Knowledge* and once more Karl Popper is here to speak, linking Western science with the Abrahamic culture that gave birth to it.

"The historicist revolution, like most intellectual revolutions, seems to have had little effect on the basic theistic and authoritarian structures of European thought.

"The earlier, naturalistic, revolution against God replaced the name *God* by the name *Nature*. Almost everything else was left unchanged. Theology, the Science of God, was replaced by the Science of Nature; God's laws by the laws of Nature; God's will by the power of Nature (the natural forces); and later God's design and God's judgement by Natural Selection. Theological determinism was replaced by a naturalistic determinism; that is, God's omnipotence and omniscience were replaced by the omnipotence of Nature and the omniscience of Science.

"Hegel and Marx replaced the goddess Nature in its turn by the goddess of History. So we get laws of History; powers, forces, tendencies, designs, and plans, of History; and the omnipotence and omniscience of historical determinism. Sinners against God are replaced by *criminals who vainly resist the march of History*; and we learn that not God but History (the History of *Nations* and *Classes*) will be our judge.

"It is this deification of history which I am combating.

"But the sequence *God–Nature–History*, and the sequence of the corresponding secularised religions, does not end here. The historical discovery that all standards are after all only historical facts (in God, standards and facts are one) leads to the deification of Facts - of existing or actual facts of human life and behaviour, including, I am afraid, merely alleged facts - and thus to the secularised religions of Nations and of Classes, and of existentialism, positivism, and behaviourism. Since human behaviour includes verbal behaviour, we are led still further to the deification of the Facts of Language. Appeal to the logical and moral authority of these Facts (or alleged Facts) is, it would seem, the ultimate wisdom of philosophy in our time."

Interesting observations, especially if one is seeking to develop a *Theory of Knowledge* grounded in Taoism, to develop a Taoist science, and a Taoist technology. So now you know some more about scriptovisualism.

**208**

The notion of purpose has been mentioned in several places as we have proceeded to create our odyssey. Joseph Needham will now say something about purpose:

"In nature man is a machine. If he is not a machine, then he is nothing at all."

Yes Joseph. You wrote this as part of a short book called *Man a Machine,* published in 1927, which was a response to another short book, called *Man Not a Machine,* which was published a year earlier.

People should be clear though why you made this statement, which was because, at the time, you thought that teleological aspects of living beings was proved not to be quantitative, and therefore of no possible scientific significance. For as you said in your book, science is the study of the quantitative relationships between phenomena of the world we live in. Anything that can be measured, anything that can be weighed, anything that can be numbered, anything that can be expressed mathematically - all these can be laid up in the treasure house that is science.

You also quoted D. G Ritchie's review of James Ward's Naturalism and Agnosticism. Ritchie is here to tell us what he wrote:

"'Either the universe' says Ward, 'is mechanical or teleological, it is not likely to be a mixture of the two.' May not the universe be both at once, through and through mechanical when regarded in its material or spatial aspect, teleological when regarded in its spiritual aspect, when that aspect is not being treated abstractly for the purpose of a quasi-natural science of psychology, but as the *Meaning* of the whole process, a meaning such as we have in our consciousness of the ends and significances of some part of our activities of thought and deed? If epistemology shows that nothing can ever be known to us as having any actual existence except as an object for thought it then becomes a reasonable philosophical faith, though it goes beyond the limits of possible knowledge, to suppose that the ultimate reality of all things for the one mind which *IS* the universe in its inner aspect."

Back now to Joseph Needham: after quoting Richie, you then said something very interesting. What was that?

"I said that subjectivation of mechanism and teleology is thoroughly Kantian, and I confess it very greatly appeals to me."

You also admitted that the mechanisation of life – a mechanical theory – is a methodological fiction.

Now we ask the question, what if a science based on purpose, rather than causality, which was also quantitative, could be developed? Actually it already has, but few – perhaps a small handful of people in the world – know this! We know about it. Who else does? We are struggling to find the name. So back once more to *Theory of Knowledge*, and Taoist science and technology.

Actually, the universe can also be teleological when regarded in its material or spatial aspect. Causal or teleological – it is a personal choice. Two stories, alike in dignity, but with different implications for humanity.

**209**

So about this thing called purpose. Here is Angela Breitenbach:

"The biological sciences are special within the realm of the natural sciences. They employ concepts that have long been taboo in physics and chemistry. Biologists may speak, for example, of the functions of biological traits and of genetic programmes that control biological processes. They may ask what a particular trait of an organ is for, or what purpose it has for the functioning of the organism as a whole. Expressions such as these sound unmistakably teleological. We are familiar with these concepts from the description of our own actions. We speak of a person, for instance, as acting for a purpose, as designing an object to perform a certain function, or as creating a programme to carry out a particular task. In the realm of human activity, purposes, functions and programmes thus involve the intentions of an intelligent agent, intentions that the agent aims to realise by means of activities in the world.

"What, however, do the same concepts mean in the context of biology? Do they, as in the case of human action, require the existence of an intelligent agent? A positive answer would be incompatible with our modern conception of the task of science as providing explanations for natural phenomena without thereby resorting to supernatural design or purpose. Can we, then, explain teleological concepts in biology as referring solely to mind independent features of nature? What seems so special about teleological descriptions is that the purpose, or the end of a functional process or programme, although they may lie in the future, somehow determine what goes on in the present. And yet, would this conception of a final causation not contradict the assumption that causes must precede their effects? Would it not conflict with our conception of science as explaining all natural phenomena by reference to efficient causes? How, then, can we make sense of the phenomenon of teleology in the biological sciences?

"The problem of how to understand the use of teleological concepts in the life sciences is one of the most widely and controversially debated problems in contemporary philosophy of biology. It is surprisingly closely related to Kant's discussion of teleology and the purposiveness of nature. Commentators seem to disagree about the question whether teleological concepts can be explained in purely naturalistic terms or whether they entail analogical associations with intentional goal directedness. The Kantian conception of teleology, though, is essentially analogical. The crucial contribution of the Kantian account is to argue both that teleology plays an important heuristic role in the search for causal explanations of nature and, more fundamentally, that it is for us an inevitable analogical perspective on living beings. The Kantian approach to teleology thus introduces a focus that goes beyond any empirical investigation of nature. This is the focus on the very possibility of experiencing the living part of nature. By clarifying the relation between this Kantian account and empirical science I have tried to show that the Kantian perspective is not only compatible with the modern life sciences, but that it can advance the debate about teleology in biology precisely because it does not interpret teleology naturalistically."

Now we repeat; one can have teleological descriptions. It is just a question of the mathematics one uses. Scientists do not want to use the mathematics that leads to teleological descriptions. Control engineers use them all the time. This is a consequence of feedback.

**210**

Arthur Koestler will now explain his first example of blindness.

"The materialist philosophy in which the average modern scientist was reared has retained its dogmatic power over his mind, though matter itself has evaporated; and he reacts to phenomena which do not fit into it much in the same manner as his scholastic forebears reacted to the suggestion that new stars might appear in the immutable eighth sphere.

"Thus for many years, an impressive body of evidence has been assembled under strict laboratory conditions which suggests that the mind might perceive stimuli emanating from persons and objects without the intermediary of the sensory organs; and that in controlled experiments, these phenomena occur with a statistical frequency which invites scientific investigation. Yet academic science reacts to the phenomena of *extra-sensory perception* much as the Pigeon League reacted to the Medicean Stars; and it seems to me, for no better reason.

"If we have to accept that an electron can jump from one orbit into another without traversing the space between them, why are we bound to reject out of hand the possibility that a signal of a nature no more puzzling than Schroedinger's electron-waves should be emitted and received without sensory intervention? If modern cosmology has a single comprehensive lesson it is that the basic events in the physical world cannot be represented in three-dimensional space and time. Yet the modern version of scholasticism denies additional dimensions to the mind, or brain, which it readily accords to the particles of a piece of lead.

"I am not playing on the word *dimension* as a mechanical - after the manner of a *fourth dimension* of occult quacks. I am merely saying that since the space-time framework, the concepts of matter and causality as understood both by classical physics and commonsense experience, have been abandoned by modern physics, there seems to be no justification in refusing to investigate empirical phenomena because they do not fit into that already abandoned philosophy."

**211**

Gavin Parkinson speaks again about C.P. Snow:

"In spite of the remarks made for the sake of even-handedness about scientists who were uninterested in Dickens, Snow's main target is spelled out in the title of the second section of his four-part lecture: *Intellectuals as Natural Luddites.* By this he means to refer chiefly to literary figures opposed to advances in science and technology, who idealised the era preceding the Industrial Revolution."

C.P. Snow:

"It is bizarre how very little of twentieth century science has been assimilated into twentieth century art."

But Charles, you are wrong to say that, and here is why. Gavin Parkinson once again:

"A more complex situation arose in France in the 1920s, 1930s and 1940s. Against a background of intellectual eclecticism and mature interdisciplinary debate in these decades, there were a variety of responses from within and around Surrealism to the epistemological crisis set in train by the physics of Relativity and quanta. There were artists who deployed ideas from the new physics and its philosophy in ways that would have met with Snow's approval.

"This was not a stable situation but one modified by broader political, social, economic and cultural changes. Art history provides a socio-political account of Surrealism's engagement with modern physics, which encompasses Surrealism's approval, opposition, use and abuse of the language and discoveries of physics, which stands in contrast with the oversimplified model offered by Snow of a steadily increasing mutual incomprehension over sixty years, for which he gives no historical reasons."

"He gives no sociological reasons either!"

That was us, Julia and Paul speaking.

Back now to Gavin:

"In a famous footnote to the preface of *The Accursed Stare: An Essay on General Economy*, its author, the French intellectual and literary figure Georges Bataille underlines his continued interest in modern physics in a debt of gratitude, declaring that his book 'is also in large part the work of (the physicist) Georges Ambrosino', and expressing the hope that they will soon resume together their study 'of the movement of energy on the surface of the globe'. Theirs was a genuine exchange, contesting C.P. Snow's pessimistic vision of *two cultures* that 'can't talk to each other'. Indeed, through comparison with Eddington, Bataille presents a much more complex relation of rejection, interpretation, negotiation and agreement between arts and science than Snow ever conceived."

We could also mention here the name: Conrad Waddington. We could also mention the name Marcel Duchamp.

We think that Snow was not only being reductive, being simplistic, but also that he has been widely misquoted, widely misunderstood. He was evidently saying something important, but it went over the heads of people in the world of STEM, and even the world of the arts, but not those in the world of the social sciences, some of whom would recognise a familiar sociological observation that has been made before in this book.

Time to move on! We will return we suppose to this matter in due course.

**212**

Joseph Weizenbaum will now speak again:

"I now come to a most troublesome point concerning what computers can and cannot do. An effective procedure is a set of rules which tells us in precise and unambiguous language what to do from one moment to the next. I have argued that a language is precise and unambiguous only if its alphabet and its transformation rules can themselves be explicated in precise and unambiguous terms. And I have repeated Church's (and Turing's) thesis that, to every such explication of whatever language, there corresponds a Turin machine that can be imitated by a universal Turin machine. I have asserted further that virtually every modern computer is a universal Turin machine. Leaving to one side everything having to do with formally undecidable questions, interminable procedures, and defective procedures, the unavoidable question confronts us: Are all the decision-making processes that human's employ reducible to effective procedures and hence amenable to machine computation?

"Michael Polyanyi says that there are things we know that we cannot tell. So our question now is: What can we tell computers? We have taken telling to mean giving an effective procedure. And the question we are presently entertaining is: Can anything we may wish to do be described in terms of an effective procedure? To now assert that there are things we know but cannot tell is not to answer the question but to shift our attention from the concept of telling, where until now we have tried to anchor it, to that of *knowing*. This is a very proper and a crucially important shift, that the question of what we can get a computer to do is, in the final analysis, the question of what we can bring a computer to know."

So we are back one more time to the matter of a *Theory of Knowledge.*

**213**

Arthur Koestler will now explain his second example of blindness.

"A second example of the *hubris* of contemporary science is the rigorous banishment of the word *purpose* from its vocabulary. This is probably an aftermath of the reaction against the animism of Aristotelian physics, where stones accelerate their fall because of their impatience to get home, and against a teleological world view in which the purpose of the stars was to serve as chronometers for man's profit.

"From Galileo onwards, *final causes* (or *finality* for short) were relegated into the realm of superstition, and mechanical causality reigned supreme. In the mechanical universe of indivisible hard little atoms, causality worked by impact, as on a billiard table; events were caused by the mechanical push of the past, not by any *pull* from the future. That is the reason why gravity and other forms of *action at a distance* did not fit into the picture and were regarded with suspicion; why ethers and vortices had to be invented to replace the occult pull by a mechanical push.

"The mechanistic universe gradually disintegrated, but the mechanistic notion of causality survived until Heisenberg's indeterminacy principle proved its untenability. Today we know that on the sub-atomic level the fate of an electron or a whole atom is not determined by its past. But this discovery has not led to any basically new departure in the philosophy of nature, only to a state of bewildered embarrassment, a further retreat of physics into a language of even more abstract symbolism. Yet if causality has broken down and events are not rigidly governed by the pushes and pressures of the past, may they not be influenced in some manner by the *pull* of the future - which is a manner of saying that *purpose* may be a concrete physical factor in the evolution of the universe, both on the organic and unorganic levels.

"In the relativistic cosmos, gravitation is a result of the curvatures and creases in space which continually tend to straighten themselves out - which, as Edmund Whitaker remarked, 'is a statement so completely teleological that it would certainly have delighted the hearts of the schoolmen.' If time is treated in modern physics as a dimension almost on a par with the dimensions of space, why should we *a priori* exclude the possibility that we are pulled as well as pushed along its axis? The future has after all, as much or as little reality as the past, and there is nothing logically inconceivable in introducing, as a working hypothesis, an element of finality, supplementary to the element of causality, into our equations. It betrays a great lack of imagination to believe that the concept of purpose must necessarily be associated with some anthropomorphic deity.

"These are matters of speculation and possibly quite beside the point; but we have learnt from the past that impasses in evolution can only be overcome by some new departure in an unexpected direction. Whenever a branch of knowledge becomes isolated from the mainstream, its frozen surface had to crack up and thaw before it could be reunited with living reality."

And the world of science, of STEM more generally does need to be reunited with reality. But Arthur, the reason why it does not embrace purpose is for reasons of dogma.

Here is the philosopher Isaiah Berlin to explain the Enlightenment philosopher David Hume's observations on causality:

"Hume's chief service in his attack on the *Enlightenment* - and he did not appear to himself to be mounting such an attack - consisted in doubting the proposition that the causal relationship was something that we directly perceived, or indeed knew to exist at all.

"Hume also doubted the proposition that we are able to logically demonstrate that there is an external world at all. You cannot prove with mathematical certainty that anything exists. In the end one has to resort to beliefs and trust in the beliefs."

In other words one can observe the cause of something; one can observe the effect; but one cannot observe the causality! The causality is, we can say, unobservable. One can say that it is hidden - a kind of hidden variable. People are just assuming that it is there, trusting that it is there, or, put another way, they are projecting this property on to the universe. The appropriate phrase is: *constructing reality*. They could also project something else - purposefulness, but they choose not to. That is because they have not understood purposefulness, and choose to believe that it means one thing, when it could actually mean something else. This is about interpretation. So again we have another example of *I believe* science. Big mistake!

Is Hawkins keeping-up with philosophy we wonder?

**214**

Bardo is a strange word. It is used in Buddhism, and in the West often is interpreted as meaning an intermediate or inbetween state. Here is Sogyal Rinposche to explain:

"Bardo is a Tibetan word that simply means transition or gap between the completion of one situation and the onset of another.

"The whole of life and death can be presented together as a series of constantly changing transitional realities known as *bardos*. The word bardo is commonly used to denote the intermediate state between death and rebirth, but in reality *bardos* are occurring continuously throughout both life and death, and are junctures when the possibility of liberation, or enlightenment, is heightened.

"The bardos are particularly powerful opportunities for liberation because there are certain moments that are much more powerful than others and much more charged with potential, when whatever you do has a critical and far-reaching effect. I think of *bardo* as being like a moment when you step towards the edge of a precipice; such a moment, for example, is when a master introduces a disciple to the essential, original, and innermost nature of his or her mind. The greatest and most charged of these moments, however, is the moment of death."

So we have arrived back at the matter of scriptovisualism, and here now is the understanding of why this thing that people seem to want to call *the nexus of science, technology and the arts* is – not a nexus. We seek transitions, transcendence of boundaries, we seek unity of opposites. Why? The answer is, partly, in Sogyal Rinposche words.

**215**

So, now to answer that question: how many versions of quantum theory exist? Do you remember that question? We asked it a long time ago. Did you think we had forgotten about it? First Gavin Parkinson will speak again:

"Like everyone else excited by, interested in and anxious to understand the new physics, the Surrealists had to gain access to a very specialised field. Although theoretical physics since Newton has been practised largely through the medium of mathematics, the main problem with the transmission of the ideas arising out of physics in the twentieth century has been the increased level of mathematisation of the discipline, widening the gulf of specialisation between expert and layperson. Heisenberg and others stated flatly that events in space-time and at the sub-atomic level could not be conveyed in either ordinary language or imagery because their effects are not perceived by our senses."

Now Howard Rosenbrock will speak:

"There are three approaches to quantum mechanics: the first is the orthodox approach, which starts with Schrödinger's equation as one of the axiomatic bases upon which the theory rests."

This is theory number one which is the accepted theory; the one that most scientists abide by and do not dare to question. Sometimes it is called the Copenhagen orthodoxy. Shall we now use a special word? What word would that be? The word is dogma. Oh you doubt this do you?

Howard Rosenbrock again:

"Stochastic mechanics is well known to give the same results as the orthodox approach."

Now we would like invite the physicist Edward Nelson to speak about how to derive the Schrödinger equation from Newtonian mechanics:

"I have shown that the radical departure from classical physics produced by the introduction of quantum mechanics was unnecessary. An entirely classical derivation and interpretation of the Schrödinger equation using a line of thought that is a natural development of reasoning used in stochastic mechanics and in the theory of Brownian motion. It appears that the phenomena which first led to the abandonment of classical physics admit a simple classical interpretation which is only in a limited sense equivalent to quantum mechanics."

Now for another theory, number three, a different type of theory, one developed by the French physicist Maurice de Broglie, which is related to that which Edward Nelson just stated, as both have what are called hidden variables, which is why physicists do not like these theories. Yet they accept causality even though it cannot be observed! Strange people! And on the matter of hidden variables, there is the approach of the physicist David Bohm - Bohmian Mechanics.

Hidden variables - possible or impossible? Controversial for sure! It seems that local hidden variables are not possible, but non-local ones are.

Howard Rosenbrock again:

"Quantum mechanics can also be arrived at using a variational principle."

We will probably come back to this approach based on a variational principle later.

In an Abrahamic science there is no room for heretical views, for alternative beliefs. There is one truth, or perhaps we should say one interpretation of the universe. Recall what our friend Karl Popper had to say. And others too!

And yet this truth just keeps on changing, or alternative truths keep appearing. Who knows what we do not know, and what we have not been able to do, simply because of sole truth! Can we afford to go on like this?

So water is water is it? Well, H2O is H2O, but water is not always H2O.

It is not that observations are wrong. H2O can seem to be always H2O, but we know that H2O can have many meanings, many interpretations, many manifestations, as we experience water in our lives in many different ways. Even in science, for water is not always water: try D2O - heavy water, also know as deuterium oxide.

**216**

Physicists have been trying very hard for decades to unify Relativity and quantum mechanics. They have not been successful. Perhaps they are looking in the wrong place?

"The coincidence of opposites provides a method that resolves contradictions without violating the integrity of the contrary elements and without diminishing the reality of the force of their contradiction. It is not a question of seeing unity where there is no real contrariety, nor is it a question of forcing harmony by synthesizing resistant parties. Coincidence as a method issues from the coincidence as a fact of opposition that is resolved in and by infinity."

That was H. Lawrence Bond speaking about Nicholas of Cusa approach to the coincidence of opposites.

"Coincidence of opposites is also one of fundamental organising principles in Karl Jung's thought. Key concepts such as the self, the god image, the collective unconscious, wholeness and synchronicity are said to be instances of the coincidence of opposites.

"Coincidence here refers to simultaneity, not to chance or randomness. Two phenomena coincide when they occupy the same space, be it logical, imaginative or material space. In popular use this notion of coincidence is often missed."

That was David Henderson speaking. Now Mark Trevor Smith will speak:

"To call something a coincidence implies that it should not really happen and thus allows us to dismiss the coincidence as an unexpected, unusual anomaly that, according to probability, will not happen again. The history of the word in English, as chronicled in the Oxford English Dictionary, reveals that this dismissive meaning gradually creeps in, as the word is linked with *causal* and *undesigned*. The root meaning of the Latin word, which was used in the seventeenth century as a verb in English in its Latin form, *coincidere*, is simply to occur together."

Coincidence: a remarkable concurrence of events or circumstances without apparent causal connection; the fact of corresponding in nature or in time of occurrence; the presence of ionizing particles or other objects in two or more detectors simultaneously, or of two or more signals simultaneously in a circuit.

Here is a spokesperson for the Encyclopaedia Britannica to inform us about their, as it turns out, limited account of unity of opposites:

"The natural forces of fire and water, which evidently exclude each other, are brought together in a unity of opposites in the world views of early archaic civilizations. Both forces are purifying as well as protective and are viewed by many as being connected with the cosmic powers of the sun and moon. Where they are truly combined, often genetically, fire (as the sun) is usually male, and water (as the moon) female. Where the fire is included more into the chthonic (earthly) sphere, it may also receive a feminine character (for example, fire in the earth, preserved in the womb); where rain is viewed as the semen of heaven, which is usually personified as male, it takes on a male character."

We think we will return to these matters. Unity of opposites occurs in Taoism. It is most definitely relevant to scriptovisualism which deals with two opposites - descriptive expression and plastic expression. It is relevant too, to the aim of developing a post-European era, post-Enlightenment science and technology.

**217**

Howard Rosenbrock speaks again:

"Alternative, mathematical descriptions of nature, based on purpose rather than causality, but equivalent to a causal description are possible. Both the purposive and the causal descriptions can be seen as myths. Neither of these myths expresses the reality of the world, for this task is reserved to the equivalence class that contains both myths.

"The present situation is that by an act of faith the causal myth is almost universally believed to be the one *true* and *real* account of nature. It is not regarded as a human construct, intended to explain the world and give us power to control it. Instead, the myth is regarded as an accurate description of causal machinery, by which the world functions and by which all events in it are produced. This belief permeates our science and our technology, and our whole outlook upon the world and our behaviour in it. To demonstrate how deep this influence is, and how pervasive, the simplest course is to give a view of the world as it would appear if, by a different act of faith, we accepted the purposive myth with the same tenacious belief in its truth as we now give to the alternative."

This is exactly what Howard did! Now he continues:

"On the causal view, nature has no purpose. Human beings appear to have purpose, by direct introspective knowledge, and living things by extension may also have purpose. But this purpose exists only in a shadowy form. Upon analysis it dissolves into causal behaviour, and reflects no deeper underlying purpose. At most, it is the result of natural selection acting upon chemically determined and chance-mutated matter, and is the natural characteristic of whatever survives: it is the badge of the survivor.

"In such a world man is a stranger, but a liberated stranger. He has no duty, because duty implies an ultimate purpose. He has no responsibility to the future, because the future is as meaningless a concatenation of events as the present. Meanwhile his own purpose exists to himself; and the world, lacking any purpose, is subject to his will to the extent that he can impose it. The world exists to be owned and used and exploited, and in the pride of imposing his will he can forget the mental anguish of inhabiting such a world.

"This describes a sickness, deeply felt perhaps only by a few. But if we regard the way in which modern societies act, we shall find that it is in full accordance with the view described. We are exploiting the present without care for the future, and exploiting the future itself to the extent of our power.

"This is the view of the world corresponding to one myth. The other myth describes the world by means of the purpose which it follows. This is not something known *a priori,* or imposed theologically. It is the result of our attempt to explain the world, and the form which it takes discloses no hidden meaning. We have to accept it as a human construct, resembling in that respect the causal myth, and equivalent to it.

"But though it is equivalent, the purposive myth engenders a very different climate of belief about the world from the causal myth. Human purpose is not something alien, not just an appearance created by the evolutionary pressure in those organisms which succeed in surviving. It is part of an inherent purpose in the world, shared with animate and inanimate nature. We are at home in the world, but do not own it: we belong to it, but it does not belong to us.

"Consequently we are no longer free, as the causal myth suggests, to do as we wish with the world, up to the limit of our power. As part of an organic whole we have a responsibility for the whole.

"The two views of the world will govern our behaviour in it."

So we are back once more to the topic *Theory of Knowledge,* as it relates to purposive descriptions and Taoist science and technology - that which will come after the age of European domination and Abrahamic thinking comes to an end (very soon).

So we have two stories alike in dignity, but with very different implications for humanity and our relationship with the natural world, and our fellow human beings.

## 218

Michael Polanyi was mentioned a few chapters back, so he is now here, one more time, to speak, and to tell us why positivism is discredited:

"Tacit knowing, I have shown, can account for: a valid knowledge of a problem; the scientist's capacity to pursue it, guided by his sense of approaching its solution; and a valid anticipation of the yet indeterminate implications of the discovery arrived at in the end.

"Such indeterminate commitments are necessarily involved in any act of knowing based on indwelling.

"This kind of knowing solves the problem of the *Meno* by making it possible for us to know something so indeterminate as a problem or a hunch, but when the use of this faculty turns out to be an indispensable element of all knowing, we are forced to conclude that all knowledge is of the same kind as the knowledge of a problem.

"We must conclude that the paradigmatic case of scientific knowledge, in which all the faculties that are necessary for finding and holding scientific knowledge are fully developed, is the knowledge of an approaching discovery.

"To hold such knowledge is an act deeply committed to the conviction that there is something to be discovered. It is personal, in the sense of involving the personality of him who holds it, and also in the sense of being, as a rule, solitary; but there is no trace in it of self-indulgence. The discoverer is filled with a compelling sense of responsibility for the pursuit of a hidden truth, which demands his services for revealing it. His act of knowing exercises a personal judgement in relating evidence to an external reality, an aspect of which he is seeking to apprehend.

"The anticipation of discovery, like discovery itself, may turn out to be a delusion. But it is futile to seek for strictly impersonal criteria of its validity, as positivistic philosophies of science have been trying to do for the past eighty years or so. To accept the pursuit of science as a reasonable and successful enterprise is to share the kind of commitments on which scientists enter by undertaking this enterprise. You cannot formalise the act of commitment non-committally. To attempt this is to exercise the kind of lucidity which destroys its subject matter. Hence the failure of the positivistic movement in the philosophy of science. The difficulty is to a stable alternative to its ideal of objectivity. This is indeed the task for which the theory of tacit knowing should prepare us."

So *Theory of Knowledge* yet again.

**219**

Unity of opposites! Coincidence of opposites! Here now is a spokesperson from the Stanford Encyclopaedia of Philosophy:

"Heraclitus' flux doctrine is a special case of the unity of opposites, pointing to ways things are both the same and not the same over time. He depicts two key opposites that are interconnected, but not identical. Heraclitus sometimes explains how things have opposite qualities: 'Sea is the purest and most polluted water: for fish drinkable and healthy, for men undrinkable and harmful.'

"Barnes thinks Heraclitus gets his doctrine of the universal coinstantiation of contraries through fallaciously dropping qualifiers (such as: 'for fish,' 'for men'). But Heraclitus is perfectly aware of them, and we might rather say that he understands them tacitly even when he does not utter them. When he says, 'Collections: wholes and not wholes; brought together, pulled apart; sung in unison, sung in conflict; from all things one and from one all things', he does not contradict himself. There are perfectly good contexts in which everything he says is true. One can divide a collection into its parts or join the parts into a unified whole.

"Most tellingly, Heraclitus explains just how contraries are connected: 'As the same thing in us are living and dead, waking and sleeping, young and old. For these things having changed around are those, and those in turn having changed around are these.' Contrary qualities are found in us *as the same thing*. But they are the same by virtue of one thing changing around to another. We are asleep and we wake up; we are awake and we go to sleep. Thus sleep and waking are both found in us, but not at the same time or in the same respect. Indeed, if sleeping and waking were identical, there would be no change as required by the second sentence. Contraries are the same by virtue of constituting a system of connections: alive-dead, waking-sleeping, young-old. Subjects do not possess incompatible properties at the same time, but at different times.

"In general, what we see in Heraclitus is not a conflation of opposites into an identity, but a series of subtle analyses revealing the interconnectedness of contrary states in life and in the world. There is no need to impute to him a logical fallacy. Opposites are a reality, and their interconnections are real, but the correlative opposites are not identical to each other."

This one can say is a Western perception, an Ancient Greek perception of unity of opposites. The Taoist one we have already encountered: in the stillness there is motion and in the motion there is stillness.

More about the Taoist view of the unity of opposites will probably come later.

**220**

While the spokesperson for the Stanford Encyclopaedia of Philosophy is here we can learn something about semiotics, specifically Peirce's Sign Theory. The spokesperson now speaks:

"Peirce's Sign Theory, or Semiotic, is an account of signification, representation, reference and meaning. Although sign theories have a long history, Peirce's accounts are distinctive and innovative for their breadth and complexity, and for capturing the importance of interpretation to signification.

"For Peirce, developing a thoroughgoing theory of signs was a central philosophical and intellectual preoccupation. The importance of semiotic for Peirce was wide ranging. As he himself said, '[…] it has never been in my power to study anything —mathematics, ethics, metaphysics, gravitation, thermodynamics, optics, chemistry, comparative anatomy, astronomy, psychology, phonetics, economics, the history of science, whist, men and women, wine, metrology, except as a study of semiotic.' Peirce also treated sign theory as central to his work on logic, and as the medium for inquiry and the process of scientific discovery.

"In one of his many definitions of a sign, Peirce writes: 'I define a sign as anything which is so determined by something else, called its Object, and so determines an effect upon a person, which effect I call its interpretant, that the later is thereby mediately determined by the former.'

"What we see here is Peirce's basic claim that signs consist of three inter-related parts: a sign, an object, and an interpretant. For the sake of simplicity, we can think of the sign as the signifier, for example, a written word, an utterance, smoke as a sign for fire etc. The object, on the other hand, is best thought of as whatever is signified, for example, the object to which the written or uttered word attaches, or the fire signified by the smoke. The interpretant, the most innovative and distinctive feature of Peirce's account, is best thought of as the understanding that we have of the sign/object relation. The importance of the interpretant for Peirce is that signification is not a simple dyadic relationship between sign and object: a sign signifies only in being interpreted. This makes the interpretant central to the content of the sign, in that, the meaning of a sign is manifest in the interpretation that it generates in sign users.

"Things are, however, slightly more complex than this, but one needs to look at these three elements in more detail to see why."

So *Theory of Knowledge* yet again. Description, explanation, interpretation! Or should this be descriptions, explanations, interpretations?

**221**

Now it is time for György Kepes to speak about his book *The New Landscape in Art and Science*:

"The book was meant to be looked at more than read. It is a picture book, arranged to bring attention to a newly emerged aspect of nature, hitherto invisible but now revealed by science and technology. The *text* of the book is not its message. Primarily, a body of material is presented, rather than scientific information or aesthetic theory. The material is organised to help the reader to see, with the hope that he will grasp significant connections for himself.

"The method - for which the preparing the book served as a kind of laboratory experiment - fuses visual images and verbal communication in a common structure. The visual images - the pictures brought together in the book - are the content. The verbal statements - comments and documents - are illustrations. They do not constitute a connected systematic account. The quotations touch the subject from one angle, the comments from another, with the visual images forming the basis of the interrelated structure that alone tells a story.

"Thus the visual and verbal statements neither parallel one another in exact correspondence nor follow one another in a strict causal chain. They complement one another in an interwoven sequence. The structure builds, then, as the observer proceeds from one experience to another, and finally a new aspect of perspective is outlined. Scientific and artistic insight fuse, creating by the interplay a basic type of experience - neither science nor art, but a comprehension that has both the character of information and the qualities of poetic vision."

Interesting and relevant to scriptovisialism! And this idea of something that is neither science nor art is in the area for which scriptovisualism is being created - a transcending of boundaries, towards a unity of opposites.

Interestingly György Kepes started work on his book in 1947 and it was, in the main, completed at the end on 1952, but was not published until 1956, owing to contractual difficulties which delayed the publication and led to the book being finally published by a different publisher to one originally intended.

Also of note is that there are contributions in György Kepes' book from both the world of art and the world of STEM, thus one again we must call into question this *two cultures* notion put forward by C.P. Snow. Evidently he was saying something of importance, which is not, it seems, what most people think he was saying.

We will come back to György Kepes later. Perhaps!

**222**

Now we are going to listen to Stephen Wilson speaking:

"There is much confusion because artists approach science and technology in a variety of radically different ways. Even though they are sometimes interrelated it is useful to highlight the differences.

"First I will deal with continued modernist practice. Some artists seek to appropriate contemporary technologies to create new kinds of images, sounds, installations and performances - for example, digitally processed photography, computer music, or computer controlled sculptures. They see new technologies as tools that give profound new ways of doing what artists have historically done. Certainly new issues must be addressed, but the agenda is not radically different from historical practice. Artists focus on creating objects and events in specialized art settings such as museums and galleries, which they hope will enter the world of art discourse and the art marketplace. Even technological art such as interactive computer media, immersive virtual reality, and web art can be easily assimilated to the fundamentals of the model.

"Second I will address critical practice. Some artists believe that the centrality of science and technology requires a radically different response from the arts. In this they continue and update traditions of conceptual, performance and situationist non-object based art. In this view the arts main role should be to deconstruct cultural patterns of integrating science and technology to clarify underlying meanings ignored in the over-hyped flow of normal technological and commercial life. Artists acquire expertise in the technical worlds in order to understand them better and to use the technologies to subvert and analyze them. These artists often seek to place their art in everyday technological mediated settings rather than specialized art locales.

"Finally there is art as research. Some artists believe the most powerful response is to become researchers themselves. They attempt to enter into the heart of scientific inquiry and technological innovation to address research agendas ignored by the mainstream and to integrate commentary and play into the research enterprise. I believe this opens up enormous opportunities for the arts."

So here again we have some modes of operation.

**223**

Now we are going to further explore unity of opposites. Here to speak is Jerauld McGill:

"Although the unity of opposites and other dialectical principles have suffered various *refutations,* and certain interpretations and misapplications have quite properly been laid to rest, dialectic is very much alive today. The Platonic-Aristotelian tradition continues, and both the Hegelian and the Marxian dialectic have many followers. The unity of opposites, which Lenin described as the most important of the dialectical principles, states that a thing is determined by its internal oppositions.

"The principle was first put forward by the Milesian philosophers of the sixth century BC, and by their cotemporary, Heraclitus of Ephesus. It held its own through centuries of philosophical thought, though it took different forms which were seldom clearly distinguished. We have separated various forms of the unity of opposites principle, to show that they are of unequal importance and that their consequences are very different.

"The principle is not a complete method or philosophy. If one forgot all about the complementary principle that a thing is determined by its field or milieu, the result would be a one-sided distortion.

"From the time of Heraclitus it has been pretty well agreed that change involves a unity of opposites, of being and non-being. It is not true to say that the world is (being) or that it is not (non-being), said Heraclitus, but that it is becoming. 'You cannot step twice into the same river, for new waters are ever flowing in upon you.' The supposed contradiction involved led the Eleatics to deny the reality of change, and has been employed to disparage change, ever since. The Hegelian dialectic, on the other hand, accepted the contradiction as a real aspect of the world, which is continually overcome and continually renews itself in the process of change.

"Another famous historical example of the unity of opposites is the *One-Many* problem. The *One,* Plato has Parmenides say, must be many because it has parts, and the *Many* must be *One,* that is to say, *One-Many.* This paradox became the foundation of the system of Plotinus and was also essential to the Christian doctrine of the trinity. The same problem reappears in contemporary discussions of the foundations of mathematics.

"Thus Bertrand Russell, in his earlier treatment of classes in *The Principles of Mathematics*, finds it necessary to distinguish the class as one from the class as many. 'In the class as many,' he says, 'the component terms, though they have some kind of unity, have less than is required for a whole, they have in fact just so much unity as is required to make them many, and not enough to prevent them from remaining many.'

"Later on he concludes that it is more correct 'to infer an ultimate distinction between a class as many and a class as one, to hold that the many are only many, and are not also one.' But this raises the question what we are talking about when we refer to the class as many. It seems to be one in some sense. Although Russell ventures an answer, he nevertheless confesses that there are 'puzzles in this subject which I do not yet know how to solve.' This and other difficulties were avoided in *Principia Mathematica* by a solution which involved the denial of the existence of classes, and by the development of the theory of logical atomism. Russell's new position permitted statements about classes, even the null-class, though classes were interpreted as fictions. Apparently the *One-Many* problem was resolved only by translating statements about classes into statements about individuals similar to a given individual."

**224**

So, while we are focusing attention on the unity of opposites, this is a good time to let György Kepes explain the motivation behind his book *The New Landscape in Art and Science*. So György speaks again:

"Rapid expansion of knowledge and technical development has swept us into a world beyond our grasp; and the face of nature is alien once again. Like the forest and mountain of medieval times, our new environment harbours strange menacing beasts; invisible viruses, atoms, mesons, protons, cosmic rays, supersonic waves. We have been cast out of the smaller, friendlier world in which we moved with the confidence born of knowledge. We try to cope with the expanded scale of things without the standards which would enable us to evaluate them.

"To convert this new environment into a human landscape, we need more than a rational grasp of nature. We need to map the world's new configurations with our senses, dispose our own activities and movements in conformity with its rhythms and discover in it potentialities for a richer, more orderly and secure human life. The sensed, the emotional, are of vital importance in transforming its chaos into order. The new world has its own dimensions of light, colour, space, forms, textures, rhythms of sound and movement - a wealth of qualities and sensations to be apprehended and experienced. If we relate experience to experience, image to image, we can bring our environment into focus and become aware of the new order on the sensed and emotional levels rather than on the rational level alone. Reoriented, we shall then be in a position to cope with the new world of forms.

"Science appears to be the most confident and vigorous unifying force today. It has produced a new foundation for our material existence, but, as we all know, has not yet solved our human problems. So far, our recently acquired knowledge, with all its precision and power, has brought us as much ugliness, discomfort and danger as it has sanity and order. Science is only one component of the understanding that we need for a well-balanced attainment of human ends. In our chaotic and directionless world, it gives us two-edged weapons - powerful tools and ideas with which we may either create or destroy.

"Where our age falls short is in the harmonising of our outer and our inner wealth. We lack depth of feeling and the range of sensibility needed to retain the riches that science and technique have brought within our grasp. The images and symbols which can truly domesticate the newly revealed aspects of nature will be developed only if we use all our faculties to the full - assimilate with the scientist's brain, the poet's heart, the painter's eyes. It is an integrated vision that we need; but our awareness and understanding of the world and its realities are divided into the rational - knowledge frozen in words and quantities - and the emotional - the knowledge vested in sensory images and feelings. Artists and poets on the one hand, scientists and engineers on the other, appear to live in two different worlds. Their common language, their common symbols, do not exist."

We will comment upon this later. For the time being we note that those who fall in love tend to be blind to each other's failings, which is why we do not participate in the love affairs that art-science and art-technology have become. Our purpose is to change science, technology and engineering, and for that we need art, and other humanities subjects, but also the social sciences. We need a new *Theory of Knowledge* and for that we need scriptovisualism.

Now you ask: why do we need scripovisualism? Now you ask: what is scriptovisualism? Now we reply: we do not know! A state of not knowing! Look what comes forth from such a way of knowing! The cup that is full cannot be filled!

**225**

Now we are going to explore mathematics; in particular a subject called optimisation. Here is David Burley to explain:

"The problem of optimising assets has been an important consideration over many centuries. Before the advent of the calculus some problems could be solved, or had known solutions, but there were serious technical barriers to further progress. Once the new calculus had been established, however, most of the important results on necessary conditions for extrema were quickly established. Over the following two hundred and fifty years there was a slow but steady development of the subject; increasingly complex situations were considered, for example, problems with constraints and variational problems, and these were developed to high levels of sophistication. However, these classical results, carefully studied over the preceding centuries, foundered on one major difficulty, that these exact methods gave final numerical problems which were often quite intractable.

"With the advent of the electronic computer a new injection of life came into the subject. The computer provided just the means of resolving many of these numerical problems, but it was often found that these classical methods did not give the best computational methods. Consequently a set of algorithms for solving various classes of such problems developed.

"The main stimulus for much modern work came from the space industries. In such very expensive fields, even small improvements of one or two percent can give enormous cash savings. Considerations of this sort led to much effort and energy being put into the related fields of optimisation and control. In general, the large scale industries with expensive plant, chemical engineering, control engineering, steel, aircraft, have benefited most from this work.

"A special type of optimisation problem is one where the desired optimum is not a number or a set of numbers, but a function. This type of problem is called a *variational problem*."

We will now provide a few important words of explanation. First control engineering. This is the activity of designing and implementing feedback control systems, so it sits within a topic known as cybernetics. Second, the term *variational problem*. This is related to a part of mathematics known as the *Calculus of Variations* or *Variational Calculus*. Third, this type of calculus can be used in control engineering to design optimal control systems. An optimal control system has a control law that is aimed at achieving an optimisation, for example, following a trajectory that is the shortest distance between two points - a purpose, where the present is determined by the future.

The central problem of the *Calculus of Variations* is: which function(s) make the functional $J$, stationary with respect to variations of the function? Such functions are called *extremals*. This problem is solved by considering the first variation of the functional. As in ordinary calculus, a stationary point of $J$ can be a maximum, a minimum, or an inflexion point. Such classification conditions depend on the second variation of the functional.

The *Calculus of Variations* has many applications. In quantum mechanics variational methods can be used to find solutions to the Schrödinger equation, for example, computing the ground state energy of a helium atom. It can also be used in Relativity to compute the path of a planet in the neighbourhood of a star, or the path of a light ray as it passes close to a star, which deflects it owing to the relativistic effects. It can of course also be used in classical Newtonian mechanics to compute the trajectory of an object, thus the interest for optimal control.

*Calculus of Variations* can also be used to do something very Kantian, as we shall probably see in due course.

**226**

Moving on, we now explore the structure of organisations, and we have Arthur Koestler to help us do this:

"If we look at any form of social organisation with some degree of coherence and stability, from insect state to Pentagon, we shall find that it is hierarchically ordered. The same is true of the structure of living organisms and their ways of functioning - from instinctive behaviour to the sophisticated skills of piano playing and talking. And it is equally true of the process of becoming - phylogeny, ontogeny, the acquisition of knowledge. However, if a branching tree is to represent more than a superficial analogy, there must be certain principles or laws which apply to all levels of a given hierarchy, and to all the varied types of hierarchy just mentioned - in other words, which define the meaning of *hierarchic order*.

"The first universal characteristic of hierarchies is the relativity, and indeed ambiguity, of the terms part and whole when applied to any sub-assembly, sub-system etc. It is the very obviousness of this feature which makes us overlook its implications.

"A *part*, as we generally use the word, means something fragmentary and incomplete, which by itself would have no legitimate existence. On the other hand, a *whole* is considered as something complete in itself which needs no further explanation. But *wholes* and *parts* in this absolute sense just do not exist anywhere, either in the domain of the living organism or of social organisations. What we find are intermediate structures on a series of levels in an ascending order of complexity; sub-wholes which display, according to the way you look at them, some of the characteristics commonly attributed to wholes and some of the characteristics commonly attributed to parts.

"The members of a hierarchy, like the Roman god Janus, all have two faces looking in opposite directions: the face turned to the subordinate levels is that of a self-contained whole; the face turned upward towards the apex, that of a dependent part. One is the face of the master, the other the face of the servant. This *Janus* effect is a fundamental characteristic of sub-wholes in all types of hierarchy.

"But there is no satisfactory word in our vocabulary to refer to these *Janus-faced* entities: to talk of sub-wholes, or sub-assemblies, sub-structures, sub-skills, sub-systems, is awkward and tedious. It seems preferable to coin a new term to designate these nodes on the hierarchical tree which behave partly as wholes or wholly as parts, according to the way you look at them. The term I propose is *holon,* from the Greek *holos* = whole, with the suffix *on* which, as in prot*on* or neutr*on,* suggests a particle or a part.

"'A man', wrote Ben Jonson, 'coins a new word, not without some peril; for if it happens to be received, the praise is but moderate; if refused, scorn is assured.' Yet I think the *holon* is worth the risk, because it fills a genuine need. It also symbolises the missing link - or rather series of links - between the atomistic approach of the Behaviourists and the holistic approach of the Gestalt psychologist."

*Holon* - this is something we can work with: In the whole there are the parts; in the parts there is the whole. *Theory of Knowledge* yet one more time!

**227**

Returning to the matter of the *medium being the message,* Marshall McLuhan will speak for a second time:

"Electric light is pure information. It is a medium without a message, as it were, unless it is used to spell out some name. This fact, characteristic of all media means that the content of any medium is always another medium. The content of writing is speech, just as the written word is the content of print, and print is the content of the telegraph. If it is asked: what is the content of speech?, it is necessary to say that it is an actual process of thought, which is in itself nonverbal. An abstract painting represents direct manifestation of creative thought processes as they might appear in computer designs. What we are considering however, are the psychic and social consequences of the designs or patterns as they amplify or accelerate processes. For the message of any medium or technology is the change of scale or pace or pattern that it introduces into human affairs."

Some of what you said, that about words and painting, is evidently relevant to scriptovisualism. Marshall now continues:

"Whether light is being used for brain surgery or night baseball is a matter of indifference. It could be argued that these activities are in some way the content of the electric light. This fact merely underlies the point that the medium is the message because it is the medium that shapes and controls the scale and form of human association and action. The content or uses of such media are as diverse as they are ineffectual in shaping the form of human association. Indeed it is only too typical that the content of any medium blinds us to the character of the medium.

"The electric light escapes attention as a communication medium just because it has no content. And this makes it an invaluable instance of how people fail to study media at all. The message of electric light is like the message of electric power in industry, totally radical, pervasive, and decentralised. For electric light and power are separate from their uses, yet they eliminate time and space factors in human association exactly as do radio, telegraph, telephone, and TV, creating involvement in depth.

"Unawareness of the psychic and social effects of media can be illustrated from almost any conventional pronouncement. In accepting an honorary degree from the University of Notre Dame, General David Sarnoff made this statement: 'We are too prone to make technological instruments the scapegoats for the sins of those who wield them. The products of modern science are not in themselves good or bad; it is the way they are used that determines their value.'

"That is the voice of the current somnambulism. Suppose we were to say, 'Apple pie is in itself neither good nor bad; it is the way it is used that determines its value.' Or, 'The smallpox virus is in itself neither good nor bad; it is the way it is used that determines its value.' Again, 'Firearms are in themselves neither good nor bad; it is the way it is used that determines their value.' That is, if the bullets reach the right people firearms are good. If the TV tube fires the right ammunition at the right people it is good. I am not being perverse. There is simply nothing in the Sarnoff statement that will bear scrutiny, for it ignores the nature of the medium, of any and all media, in the true Narcissus style of one hypnotised by the amputation and extension of his own being in a new technical form."

Science, technology, engineering and mathematics are also media. Too much do the art-science and art-technology lovers focus on the content, and miss the notion that these things - science, technology, engineering and mathematics - are media. We can change the media, and in doing so we will also change the content. This is what the world now needs, but to do this we must break free from the past.

**228**

On we go, and moving now to another matter, we have here Friedrich von Hayek, who will now speak about his book *The Road to Serfdom*:

"Contemporary events differ from history in that we do not know the results they will produce. Looking back we can assess the significance of past occurrences and trace the consequences they have brought in their train. But while history runs its course, it is not history to us. It leads us into an unknown land and but rarely can we get a glimpse of what lies ahead.

"Although history never quite repeats itself, and just because no development is inevitable, we can in a measure learn from the past to avoid a repetition of the same process. One need not be a prophet to be aware of impending dangers. An accidental combination of experience and interest will often reveal events to one man under aspects which few yet see.

"Where the common beliefs of our generation will lead us is a problem not for one party but for every one of us, a problem of the most momentous significance. Is there a greater tragedy imaginable than that in our endeavour consciously to shape our future in accordance with high ideals, we should in fact unwittingly produce the very opposite of what we have been striving for?"

Indeed! Of course we have heard similar themes from others in the course of our journey. And Friedrich von Hayek's words are particularly relevant to the European Union: in seeking to avoid the past, one can say, they are unwittingly producing the very opposite of what they have been striving for. Yet few will see this and dismiss as ridiculous such a notion. Yet we speak, as Friedrich von Hayek did, from a position which is an accidental combination of experience and interest.

**229**

Here is an explanation of *Aesthetic Computing*, provided collectively by those who authored the *Aesthetic Computing Manifesto*:

"The application of computing to aesthetics and the formation of art and design has a long history. With the integration of hardware, software and cybernetics in the 1960s, computer art emerged as a new art form. We propose to look at the complementary area of *applying aesthetics to computing*.

"Computing and its mathematical foundations have their own pre-existing aesthetics; however, there is currently a relative lack of variety of these aesthetics, in contrast with art, which has a long history containing a multitude of historical genres and movements. We wish to strike a balance between cognitive and material aesthetics. Software as written in text or drawn with flow-charts may be considered elegant, but that is not to say that the software could not be rephrased or re-presented given the more advanced media technologies that are available to us today. Such re-presentation need not compromise the goal of abstraction, which is a necessary but not sufficient condition for mathematics and computing, as meaning, comprehension and motivation may be enhanced if the presentation is guided by a pluralism of aesthetic choices and multiple sensory modalities.

"Computer programs and mathematical structures have traditionally been presented in conventional text-based notation even though, recently, substantial progress has been made in areas such as software and information visualization to enable formal structures to be comprehended and experienced by larger and more diverse populations. And yet, even in these visualization approaches, there is a tendency toward the mass-media approach of standardized design, rather than a move toward a more cultural, personal and customized set of aesthetics.

"The benefits of these latter qualities are: an emphasis on creativity and innovative exploration of media for software and mathematical structures; leveraging personalization and customization of computing structures at the group and individual levels; and enlarging the set of people who can use and understand computing. The computing professional gains flexibility in aesthetics as well as associated psychological attributes such as improved mnemonics, comprehension and motivation. The artist gains the benefits associated with thinking of software, and its underlying mathematical structures, as subject material for making art. With these benefits in mind, we have created a new term, *Aesthetic Computing*, which we define as the application of art theory and practice to computing."

Not sure if this is relevant to scriptovisualism – we will have to think about this. But what we can say is that computing is another medium, which is the content of another medium, collectively know as STEM. We seek to reshape these media, so for us, *Aesthetic Computing* could just be the content, and what we should be looking at is the medium. It is as though there is a hierarchy of media. Back to the *Janus effect* and *Theory of Knowledge*!

But was computer art a new art form? Do not just accept text; always apply hermeneutics.

You will in any case only do what your beliefs will allow you to do – your Enlightenment beliefs!

**230**

Third research paradigms! What are these? Yes, plural not singular! Martyn Denscombe will explain one of these third research paradigms:

"The mixed methods approach has emerged as a third paradigm for social research. It has developed a platform of ideas and practices that are credible and distinctive and that mark the approach out as a viable alternative to quantitative and qualitative paradigms. As a research paradigm, the mixed methods approach incorporates a distinct set of ideas and practices that separate the approach from the other main research paradigms. Contrasting the mixed methods approach with research paradigms that have favoured the use of either quantitative or qualitative methodologies, the defining characteristics of the mixed methods approach involve three features. The first is the use of quantitative and qualitative methods within the same research project. The second is the use of a research design that clearly specifies the sequencing and priority that is given to the quantitative and qualitative elements of data collection and analysis. Finally the third feature is the use of an explicit account of the manner in which the quantitative and qualitative aspects of the research relate to each other, with heightened emphasis on the manner in which triangulation is used, and pragmatism as the philosophical underpinning for the research.

"Within the mixed methods approach, there are differing views on how quantitative and qualitative elements of research should be used within a specific project and whether they can be integrated, combined, or used in tandem. For some writers the philosophical premises of quantitative and qualitative methodologies are considered to be incompatible, which leads them to advocate their use *in parallel,* each playing to its respective strengths. Other writers have sought to find common ground between the two alternatives and have made efforts to highlight the similarities between quantitative and qualitative approaches as a basis for combining their use.

"Questions have also been raised about the viability of using a simple quantitative-qualitative dichotomy. The point has been made on many occasions that the distinction between the notions of quantitative and qualitative is not watertight and that any simple quantitative-qualitative distinction hardly does justice to the variety of epistemological and ontological assumptions that underpin the terms.

"On some occasions pragmatism provides a fusion of approaches. Seeking to challenge what are regarded as sterile and unproductive dualisms, some mixed methods researchers favour a search for common ground - some compatibility - between the old philosophies of research. On other occasions pragmatism provides a basis for using mixed methods approaches as a third alternative - another option open to social researchers if they decide that neither quantitative nor qualitative research alone will provide adequate findings for the particular piece of research. Contrasting with this, there are times when pragmatism is treated as a new orthodoxy built on the belief that not only is it allowable to mix methods from different paradigms of research but it is also desirable to do so because good social research will almost inevitably require the use of both quantitative and qualitative research to provide an adequate answer. And then there are occasions when pragmatism is treated in the commonsense way as meaning expedient. There is a commonsense use of the word pragmatic that implies a certain lack of principles underlying a course of action. There is the danger then that the mixed methods approach gets associated with this understanding of the word and thus comes to be regarded as an approach in which anything goes. It should be stressed that this is not the philosophical meaning of pragmatism and it is not a meaning that should be associated with the mixed methods approach."

See how matters can be more complex than that which might be expected. This is one of the reasons why we reject to notion of polymaths.

**231**

Back now to the matter of *holons*.

The term was invented to describe the behaviour and organization of social systems, and means something that is simultaneously a whole and a part of the whole. In this respect it is very similar to DNA, since within every living cell is the whole DNA of a creature or plant, thus the saying, *the whole is in the parts and the parts within the whole.* Organizationally holonic models can provide behaviours where organizational elements co-operate to contribute to a larger task, while at the same time acting autonomously to achieve their own goals.

This notion of holonics has been adapted to the technical domain to create architectures for computer integrated manufacturing technologies and manufacturing systems; these architectures being made up of autonomous but co-operating units called *holons*. Thus, in this context, *holon* just means a unit or element that has a dual property - autonomy and co-operation. These *holons* display goal seeking behaviour and will self-initiate co-operation with other *holons* to achieve goals. This self-initiation without higher hierarchical command opens the door to self-adaptive behaviour and reconfiguration. These *holonic* architectures are not based on a physical structure, but on functions that need to be undertaken or goals that need to be achieved, that is to say, it is a description independent of hardware or software platforms.

So how does this concept carry over from a technical architecture to an organization? The *holonic* architecture is essentially composed of *holons*, which sit within *holons*, which sit within *holons*, and so on. Each *holon* has a function and a goal, and each *holon* is itself made up of other *holons* with their own functions and goals that are required to fulfil the overall function and goal. Note however, that there is no expression about how these goals and functions will be achieved. It is important to understand this point, because a *holon* and an organizational unit like a manufacturing cell or a department are not in general the same thing. Also, the principle of self-similarity applies, so all *holonic* elements appear in all subsequent *holons*. This means that whatever initial *holonic* description is adopted, this reproduces itself within each *sub-holon*.

Confused? Evidently we have taken this beyond that which Arthur Koestler originally proposed. What use is this idea for a *Theory of Knowledge*? We do not yet know.

**232**

The medium is the message says Marshall McLuhan. As we observed previously, science, technology, and engineering are also media, and, we here add: not value free media.

McLuhan highlights in his book *Understanding Media* that the conventional response to all media is that it is how they are used that counts. But this he says is the numb stance of the technological idiot. The content of a medium he argues can be likened to the juicy piece of meat carried by the burglar to distract the watchdog of the mind.

It is a truth, not universally understood, that, because the content of a medium, is another medium, we become distracted by what is visible, and ignore what is most important and which poses the greatest threat - the silent narratives. This is so with the world of STEM. We are busy looking at what it produces, all that really disturbing stuff, like GM, nuclear power, and so forth, that we forget that it is the media that science, technology, engineering and mathematics are, that are the problem. If we wish for a peaceful world, a sustainable civilisation, then we should change the media. But to do so means engaging in heresy. See here, the use of language taken from the world of Abrahamic thought! STEM - secular religions!

The problem with art-science and art-technology relationships as love affairs is that they distract our attention away from the media. In fact, these love affairs become media, because they too are ideologies, hence the importance of those different ways, mentioned by Stephen Wilson, that artists can work with science and technology. The world of STEM is largely exposed to art that is *continued modernist practice* - STARTS is an example of this. This leaves the world of STEM unchanged and just reinforces their existing ideologies, their existing dogma. Cognitive biases are confirmed!

It is time to bring to the fore something different: *continued modernist practice*, will continue, but there is no need for STEM research agencies to fund it. They need to be funding other, more controversial work that calls into question the world of STEM and seeks to transform it into something fitting for the post-European, post-Enlightenment era. Another heresy! How are you feeling?

**233**

*August Strindberg* was a Swedish playwright, novelist, and short-story writer, who, according to the Encyclopaedia Britannica, combined psychology and Naturalism in a new kind of European drama that evolved into Expressionist drama. He also engaged at some point in his career in painting. So another writer who also painted! But did he paint on the same theme as one of his literary works?

Strindberg's paintings were unique for their time, and went beyond those of his contemporaries for their radical lack of adherence to visual reality. The many paintings that are acknowledged as his work, were mostly painted within the span of a few years, and are now seen by some as among the most original works of nineteenth century art. Today, his best-known pieces are stormy, expressionist seascapes. Though Strindberg was friends with Edvard Munch and Paul Gauguin, and was thus familiar with modern trends, the spontaneous and subjective expressiveness of his landscapes and seascapes can be ascribed also to the fact that he painted only in periods of personal crisis.

So did he engage in a dual creative process in the sense of scriptovisualism? Probably, at this distance in time, it would be impossible now to discover whether this was so. Does it matter?

**234**

Thus begins the beginning of that which we spoke in time recent past, for it seems now we are ready to begin, and thus do we begin ...

We stand in Roundhouse – a kind of Globe – which is, as informing name tells, a circular place, where space is bent into curved form, and lying within there sits for some moments in time, another round construction known by the name of (Ron Arad's) Curtain Call, creating an inner space and an outer one too, within the curved space that we have just entered into. And what appears is what you want to appear, made so by the projecting power of technology which is a means of displaying creation and seeing sounds, as you will soon, by these revealing words, find.

*Kima Wheel* is a wave from the imagination that comes forth specifically to fall upon the 360 degrees of canvas called Curtain Call, which unlike a normal curtain is not a barrier to freedom of movement of performers or audience, who move in time and space as this is not a strict classical Ancient Greek separation. The wave this time is a wheel of poetry, and comes by virtue of Anelema Group, which one might say, is a different way of looking at that which passes through our lives daily, often unnoticed, but without which there could be no life at all, which, if you care to see, is more than just science and technology, or art made with such.

Thus now we have set forth the scene in kind of Prologue form, befitting of the theatrical context which now before you will unfold.

It is about emotions it seems – feelings, all very subjective and quite a contrast to most art-science and art-technology, being as they often are, though few can see, but manifestations of the a certain strand of *Enlightenment* thinking, the stinking corpse of which lies rotting all about. Yet Enlightened minds sense only reek, and do not see that they are the corpse that decays, which is giving off the decomposing smell.

In the world – another Globe – here and there, come glimpses of a new and different age slowly taking form, being illuminated by light from eastern directions originating – some dare to call it the *Age of Tao*. That is us, referring to ourselves. We are inclined to be self-referencing at times.

But there must surely be others as well, who can see that Europe's time - the age of the West - is done, and to go with it too, must be those Abrahamic secular beliefs - that is Western capitalism and Western science and technology, if you are wondering. Oh yes - the European Union too! Western communism is already just about gone, and what goes next, is the rest, and by our hand will their demise be hastened, and thus will strange beliefs be despatched to the past where they do belong. 'Tis about the power of the pen, which becomes even more powerful when combined with other forms of expression as you will see anon!

Now is the beginning of the age of the universality of humankind. So no more cultural colonialism please - are your listening you people in Brussels, Washington DC and all those other cities of Ancient Rome and Ancient Greece, where you still foolishly pursue those dated dreams of making all the world European? How does the thought of a Taoist science and technology make you feel? In the East there is the West, and in the West there is the East. It is not a binary, or a duality, but a unity - a unity of opposites, which your fragmented minds cannot easily digest. That goes for art-science and art-technology too, which are ever so binary because of the dominance of Western minds that see only such separations. Everything is connected - can you not see?

Where to look is the empty space so to speak, waiting to be filled by those who dare to pursue a different dream. You will also find that what comes forth will be very Kantian too! That is another dimension to it. 'Tis a very beautiful heresy built on a different foundation to that which Western minds construct.

Back to the curved space and then comes the moment - it is the aesthetics of participation - so the audience becomes, for a time, the creating force - let us not here bother with the rhetoric and the postmodern ideology for this is already history, and we are very much alive. This is the moment to put the world back together again in different form, and so do we raise from the dead, theory too. To see, to know, to understand, to find something new, and in doing so to become different people - these are the goals that we pursue.

Continuing (we are back to the curved space again): hesitantly at first, we begin to speak those words that come from our own shaping hand and with eye we see, images formed, while being immersed in our own sound, and that of others too. This imaging of sound we discovered before appearing is made possible by virtue of technology, of neural networks, produced from research of the type that leads to PhDs. And the emotional moment is then transformed, and cognitive becomes, as we know that once again we are creating from words, visual images, for whatever reason do we in this strange scriptovisual process participate. It is something to do with morphogenesis rather than morphostasis, if to cybernetic concepts you are inclined.

And then we are lost in our own world of the mind, afterwards thinking: was that the intention or is this serendipitous action? Interactive art is a place where the unexpected sometimes happens. To be or not to be that is the question, whether it is better to be highly scripted, or very open, or somewhere in between? It is after all, just like a play – *Act 1, Scene 1, enters ...* (but not necessarily linear). It all depends upon what you want to achieve is the answer to the question.

Then we want a pen and desk and to start exploring, but this is not the venue for such things, but evidently it becomes possible to compose images with words, hence our intuitions are confirmed, for: in the beginning was the image, and from the image came forth the word, and to image doth the word return. That is what the image called Genesis in Reverse (on the homepage of our web site) is saying. But why is the question? As we said it is something to do with morphogenesis and the making of something transformational – most art-science and art-technology is not, but we keep looking for something that might be, or have the potential for such. It is a long, long quest. Did we find sight of journey's end in Kima? Only time will tell, but Kima in its many different manifestations, bringing together at different times, music, performance, immersion, dance, poetry, technology, sound, image and synaesthesia, does make for something very different to the normal explorations of an art space that is now at least six decades old. And being so old it has become a source of much dissatisfaction within us, as indeed has literature which is even older – thus another reason deep within for the quest that we have embarked upon.

Thus then finding that we are not anymore able to compose, we take on a different role, exploring through vicarious interaction that which lies around. Then it is done and we are gone to another place in London, where some conceptual art from the past lies waiting to be explored, and there we do find, more information about the image and the word combined – another serendipitous moment in time.

*Intriguing possibilities* is the ending phrase that comes forth from this waltzing encounter with word and image, bound together in ways it seems that still lie unseen.

## II.1

*Enter Marshall McLuhan and Paul.*

PAUL: Were DG CONNECT not listening? No they were not, for it is the funding of art that thou should do, not the development of unconventional and compelling products with under-funded afterthoughts, otherwise known as artists, in tow. And you get what you get, but you can get some interesting things if you know how to create the infrastructure to enable this. It seems we do and DG CONNECT do not, which is one of the reasons we engaged in critiquing DG CONNECT and what they did, and much more as well.

MARSHALL: I see many people do quote me – thus do they say, quite often, that the medium is the message. But it seems that is all they say. Do they not understand media? The point, most important, that I also made, is that our conventional response to all media, namely that it is how they are used that counts, is the numb stance of the technological idiot. Feel free to quote me.

PAUL: And be it true that this thing that we now call STEM, that too is media?

MARSHALL: It be true as any truth be true. Art-science and art-technology too, for these are media as well. Better look therefore and see beyond the numb stance.

PAUL: We do.

*Exeunt.*
*Enter Julia.*

JULIA: Forgive us if we seem to be, that which you do not wish us to be, but the world of STEM, and its funding of art, are problems writ large for those who wish to see, for they have their own agenda grounded in morphostasis, and silent narratives too, which we are able to hear, for we walked among them unseen, and in preparation, spent much time while growing up, learning how to see – and what we see are little children playing in the rye, blindly unaware that they dance on the edge of doom. Now they dance with artists too. Better catch them lest they fall. So back to morphogenesis again! Thus we engage in a beautiful heresy and bid welcome to a time that surely must be what humanity will come to see as the inevitable turning, for no age lasts for an eternity.

*Exit.*

**235**

György Kepes again:

"East and West, the trend of centuries has brought imbalance of intellect and feeling. The East has, in the main, moved in one direction, the West in another. The anatomy of the world preoccupies us in the West; our culture centres upon objective understanding of substance, forces and structures, to the impoverishment of our subjective appreciation of patterns and rhythms. The East has concerned itself less with the world's structure than with its physiognomy, and has institutionalised, as the West has not, highly developed appreciation of its harmonies, its tastes and flavours.

"Today, with all lands knit together by new tools of communication and world-embracing economic patterns, it has become possible more than ever to exchange attitudes as well as ideas. A more fundamental interchange of Western structural disciplines and Oriental emotional disciplines could contribute to a cultural equilibrium - on a higher level, this time, with a heightening of our ability to understand the world and read its significance.

"The Far Eastern discipline of sensibilities grew out of the feeling that men lived most fully by opening themselves to the universal rhythm of Nature, becoming one with trees and stones and animals. Nature was approached and entered through rapt contemplation of its forms, to the end of visualising the world in terms not of likeness but of what the Chinese called *rhythmical vitality* - the essence of things in their characteristic of life movement. In the West, the visualisation of our experiences has been looked at mainly as the fashioning of representations of nature, likeness of the things around us. This is the Aristotelian tradition; and in particular since the Renaissance, it has been assumed that fidelity to the optical appearance of objects is the only means of artistic description of reality. Nevertheless, at all times master works of art have transcended mere representation; and during the past seventy-five years Western art has shifted towards ideas and methods corresponding to this Far Eastern vision.

"What the artist Piet Mondrian called the liberated and universal rhythm distorted and hidden in the individual rhythm of limiting forms, the great pioneers of modern art have tried to make visible. Today artists more often depict the inner world of man than the likeness of particular objects, making visible and external their experience of connection with the world around them. But the general public, most scientists included, still assumes that art is naturalistic representation."

We need to speak to Piet Mondrian again, but in the meantime, Billy Klüver wants to speak:

"Kepes came to us very early. We went up to MIT to see him and he totally applauded what we were doing - he loved it! But you have to remember the enormous difference in the aesthetics that was going on with the Bauhaus movement and what was going on in New York. There would be no artist that would work like the Bauhaus, which is where people like Kepes came from. They wanted me to go up there and teach them but I never did. Kepes was a very nice guy."

**236**

Performance: An act of presenting a play, concert, or other form of entertainment; the action or process of performing a task or function.

Performative: In philosophy and linguistics, relating to or denoting an utterance by means of which the speaker performs a particular act (for example, I bet, I apologize, I promise). Often contrasted with constative.

Constative: Denoting a speech act or sentence that is a statement declaring something to be the case.

Friedrich von Hayek speaks again:

"As the hope of the new generation came to be centred on something completely new, interest in, and understanding of, the functioning of the existing society rapidly declined; and with the decline of the understanding of the way in which the free system worked our awareness of what depended on its existence also decreased.

"This change in outlook was fostered by the uncritical transfer to the problems of society, of the habits of thought engendered by the preoccupation with technological problems, the habits of thought of the natural scientist and the engineer, and how these at the same time tended to discredit the results of the past study of society which did not conform to their prejudices, and to impose ideals of organisation on a sphere to which they are not appropriate."

There is a word for that which Friedrich von Hayek is referring to. Two words actually. The first is scientism, the second is technocracy. The two go hand in hand, and lead to a place where humanity should not go. The European Union is heading there, led by the European Commission. This is why STARTS is important: it is a case study in the failure of technocracy. STARTS will be a failure, but will be presented as a success, just as the FET Flagships are so presented, yet, as revealed in the meeting held at the former headquarters of IG Farben in Frankfurt, the FET Flagships are a failure. The European Commission though will engage in a cover-up.

Edward O. Wilson yet again:

"The unification agenda does not sit well with a few professional philosophers. The subject I address they consider their own, to be expressed in their language, their framework of formal thought. They will draw this indictment: conflation, simplism, ontological reductionism, scientism, and other sins made official by the hissing suffix. To which I plead guilty, guilty, guilty."

You are indeed guilty of scientism.

C.P. Snow:

"We cannot avoid the realisation that applied science has made it possible to remove unnecessary suffering from a billion individual human lives - to remove suffering of a kind, which, in our own privileged society, we have largely forgotten, suffering so elementary that it is not genteel to mention it. For example, we know how to heal many of the sick: to prevent children dying in infancy and mothers in childbirth: to produce enough food to alleviate hunger: to throw up a minimum of shelter: to ensure that there aren't so many births that our other efforts are in vain. All this we know how to do. It does not require one additional scientific discovery, though new scientific discoveries must help us. It depends on the spread of the scientific revolution all over the world. There is no other way."

So you too, C.P. Snow! Possibly also the - what now seems to be obligatory - artist in residence as well!

Here is Roger Malina to speak:

"In 1967 a group of scientists, artists and scholars in Paris, led by Frank Malina, founded the Leonardo organization. In 1968 the first issue of their first project, the journal *Leonardo*, appeared under the aegis of Pergamon Press. This was the first international peer reviewed scholarly journal addressing the *two cultures* divide. Marked by the traumas of World War II, the founders were dedicated to world peace through international communication and collaboration. They shared an underlying optimism that advances in the techno-sciences would bring about new visions of the world that would ensure social justice and universal economic development. They advocated a special role for artists and scientists working together."

So them too!

And what about Experiments in Art and Technology? Probably those people also!

Technocracy, scientism, and cultural colonialism too! Billy Klüver:

"We all know how technology has become part of our lives. And now we can see absolutely no reason why it should not become more so. No sound has been heard from another culture to oppose Western technology. The faster the underdeveloped countries can have it, the faster they want it."

Oh what a lack of vision and imagination! Did you not learn anything from working with Robert Rauschenberg? Did you learn nothing from György Kepes?

**237**

Theosophy is a collection of mystical and occultist philosophies concerned with, or which seek direct knowledge of, the presumed mysteries of life and nature, particularly of the nature of divinity and the origin and purpose of the universe. Theosophy is considered part of Western esotericism, which believes that hidden knowledge or wisdom from the ancient past offers a path to enlightenment and salvation.

Now we will hear from David Carrier:

"Leonardo da Vinci (1452–1519) is a natural namesake for *Leonardo*, a journal devoted to exploring the relationships between art and science. Many visual artists have taken a casual interest in science. They have been interested in anatomy, colour theory and perspective, those practical techniques for the working artist. Leonardo, however, is the only major painter who devoted serious prolonged attention to studies dealing with science and technology. He was interested in anatomy, hydraulics and optics and designed many very original inventions. These scientific concerns influenced both his paintings and his abortive sculptural projects."

Only one problem with this statement, which is that Leonardo was not a scientist, as well you know.

David Carrier speaks again:

"In Leonardo's time, long before Galileo and Isaac Newton, the ideal of an experimental science was not yet established. He is not really the ancestor of such scientists, for his drawings of natural processes did not contribute to the development of physics. In astronomy, he was concerned with the appearances of heavenly bodies, not the observations of Copernicus and his successors, which created present-day scientific astronomy.

"Furthermore, although Leonardo sometimes spoke sceptically about the claims of Scripture, his interest in fossils and the Deluge did not actually prepare the way for modern geology and Darwin's theory of natural selection. As for his flying machines, they were not genuine precursors of the airplanes of the Wright Brothers. Because he lacked suitable technologies, these machines could not have gotten off the ground. His fascination with military technology now seems more ominous, but here too there is a great gap between his claims and his accomplishments. These inventions did not revolutionize warfare.

"Leonardo's goal as scientist, to understand the world visually, proved to be essentially unworkable. Leonardo lacked the knowledge of mathematics and the interest in experimentation needed for real science. As for his anatomy, it was boldly original but played little role in the scientific study of the human body, which created modern medicine. In the seventeenth century, a number of scholars, Tomasso Campanella and Athanasius Kircher for example, developed extremely ambitious but totally flawed attempts to understand the natural order of things. These men were not precursors of modern scientists but late-medieval natural mystics who thought that they could deduce the structure of the world by observation. Leonardo has much more in common with these failed scientists than he does with Copernicus, Galileo and Newton. He was almost exclusively concerned with the questions of the bodies' physical appearance rather than with making measured observations of their behaviour. Nor did Leonardo have any connection to those inventors who developed the steam engine and the apparatus of industrial culture."

Now apply hermeneutics to what Carrier has just said. Start by noting that he is using a modern word – scientist – to refer to people from the distant past. Then continue – if you can that is!

Leonardo is a modern day myth. We will hear, perhaps, some more about this myth later.

**238**

Today we stood, for a second time, in the European Parliament. This time we were in a place where anyone can go – no special passes required.

Pasted on the walls of the European Parliament visitor centre, a place known as the *Parlamentarium,* a space of both information and propaganda, there are words. Let us now hear these words spoken.

"National sovereignty is the root cause of the most crying evils of our time and of the steady march of humanity back to tragic disaster and barbarism. The only final remedy for this supreme and catastrophic evil of our time is a federal union of the peoples."

That was Lord Lothian speaking in 1939.

Now someone else speaks:

"What, in concrete and practical terms, does the independence of nations mean in the world of today, a world of the closest economic and political interdependence, which makes the destiny of all mankind indivisible."

That was Julius Braunthal speaking in 1943.

"We must build a kind of United States of Europe. In this way only will hundreds of millions of toilers be able to regain the simple joys and hopes which make life worth living."

That was Winston Churchill speaking in 1946.

You are all wrong. Understand that you have not understood.

At what point did the European dream become the European nightmare? What we see today in Europe, is exactly the same problem as in the past, but at a continental level. Let us see what Lord Lothian's words sound like when we rephrase them in terms of European Union sovereignty.

"European Union sovereignty is the root cause of the most crying evils of our time and of the steady march of humanity back to tragic disaster and barbarism. The only final remedy for this supreme and catastrophic evil of our time is ..."

Something not quite right here! Something is missing. Let us try again.

"European Union sovereignty is the root cause of the most crying evils of our time and of the steady march of humanity back to tragic disaster and barbarism. The only final remedy for this supreme and catastrophic evil of our time is for the peoples of Europe to abandon their love of flags."

A blue flag with a ring of golden stars, or a tricolour, or an eagle image on a coloured background! What is the difference? They are still flags. They are all symbols of nationalism and are equally dangerous!

**239**

Time to look once more at this thing called purpose - yes more about purpose! It just keeps cropping up! This time it is Friedrich von Hayek who will speak about this matter:

"If we survey the different fields in which we are constantly tempted to describe phenomena as *purposive* though they are not directed by a conscious mind, it becomes rapidly clear that the *end* or *purpose* they are said to serve is always the preservation of a *whole*, of a persistent structure of relationships, whose existence we have come to take for granted before we understood the nature of the mechanism which holds the parts together. The most familiar instances of such wholes are the biological organisms. Here the conception of the *function* of an organ as an essential condition for the persistence of the whole has proved to be of the greatest heuristic value. It is easily seen how paralysing an effect on research it would have had if the scientific prejudice had effectively banned the use of all teleological concepts in biology and, for example, prevented the discoverer of a new organ from immediately asking what *purpose* or function it serves."

We will have more to say about purpose, we are sure. Nicely heretical!

**240**

Now Roger Malina, that distinguished professor of Art and Technology will speak:

"We are traumatised by endemic conflicts and wars. The effects of human activities on the earth's climate and ecologies make it clear that the way our societies are organized is unsustainable. Some wonder whether the very premises of the *Enlightenment Project,* with its framework of human measure in the natural world, lay the seeds of its own failure. If the *Enlightenment Project* is fundamentally flawed, then we must re-examine techno-science's cultural foundations. We need the best scientific understanding of the world. We need appropriate tools that use our available resources in sustainable ways. This needed recasting of techno-science is a cultural project, and artists, scientists, engineers and scholars (among other professionals), working together, have important roles."

We are sorry Roger, but what you have just proposed will not work, for you, and your artists, are deeply embedded in the *Enlightenment Project* as you call it, and it is deeply embedded in you. The *Enlightenment* - a prison cell with transparent walls and bars! The *Enlightenment* - the unbreakable, invisible, and silent chains that you all wear! You want to wear them! You want to be in your prison cell! You have been condemned by Zeus. Cursed by Nemesis too! Only Hercules can break those chains and set you free.

Here is Jeremy Rifkin:

"The reason Americans think different from Europeans has to do with our history 200 years ago. When our founders came over to America 200 years ago, it was the last stage of the *Protestant Reformation* in Europe, and it was the early stages of the *European Enlightenment*. Our founders took those two currents here in Europe at the time and we froze them for 200 years. If you want to understand how the American idea of reality differs from yours, then understand that our dream is simply a European transplant. We took those ideas and they were more fertile on American soil than they ever were on German soil.

"Now, many of you Europeans say, 'We don't understand how you Americans can be so religious and so into material things at the same time. Isn't that a contradiction?' Not in our minds. Because we see a common thread between the *Reformation* and the *Enlightenment* that you don't see. The *Reformation* and the *Enlightenment* were very different ideas. The *Reformation* theology said, 'Suffer with Christ in this world, and your reward is eternal salvation in the next world.' The *Enlightenment* philosophers said, 'No, pursue happiness in this world and your reward will be progress on earth.' Completely different ideas. But in the American mind, the common thread between both of those, the *Reformation* and the *Enlightenment* - the individual was at the core of both of those ideas. Your Martin Luther said, 'You stand alone and naked with your God. No Vatican, no priesthood, no intermediaries, just you and the Bible, alone with God.' Adam Smith, the great Scottish economist of the *Enlightenment*, overthrew 2000 years of economic history: remember, *economy* comes from the Greek *eukos*, which means *family, community*. *Economy* was always part of *community*. He said, 'No, you stand naked and alone to pursue your self-interests in the marketplace, alone'. John Locke, the great political philosopher of the *Enlightenment*, said, 'You stand alone against the forces of nature, harness the commons, be productive - alone.' It fit America. We were alone, it was a frontier. And so for us, the idea of the individual - self-sufficient, independent, an island - it helped us tame the frontier.

"In America, our identity is still pretty fixed and frozen with the *Enlightenment* - God and country."

Of course, as we have already seen, Europe is still deeply rooted in the *Enlightenment*, as is its science and technology, as well as those art-science and art-technology practices as well. It is also deeply Abrahamic, as is its science and technology.

Oh boy, have you lot - the *Vainglorious Enlightened Ones* - got a big problem on your hands, for now is the *Age of the Tao*. Americans, Europeans, it matters not. You are all in serious trouble.

Enacted biography: a term that refers to a psychological process whereby a biographical pattern, once it becomes known, younger, would-be-artists begin self-consciously to pattern their lives according to the biographical model. It then becomes constitutive.

The term appeared as the final two words in a joint work by Ernst Kris and Otto Kurz called *Legend, Myth, and Magic in the Image of the Artist*. Ernst Kris, who further dealt with the matter in a later book, is here to say a few words:

"One gains sometimes the impression that the actual behaviour of famous artists has been used as a prototype, frozen into a formula which is being handed down as such to succeeding generations."

He is clearly referring to the art-science and art-technology lovers, and giving yet another reason why Roger and his friends will be staying in their cell and wearing their chains for evermore. Kris is also explaining why evidence of the value of these art-science and art-technology collaborations is hard to find - the participants are just engaging in enacted biography, and then fantasying about the results, hoping that the little boy who told the emperor that he was wearing no clothes will not turn up. But he, or she, is here. We could say more but will hold off - perhaps another book is taking shape? What we will say however is the European Commission, in particular, DG CONNECT, are very naked! START fantasying all you want - it will not save you! STARTS will produce yet more anecdotes. This is the failure of technocracy. Imagine what further damage they will do ...

Art stands in its own right as a domain for learning and knowledge production. It does not need to be shackled to science and technology and the associated ideologies. This is one way to recognise artists of interest - they are not shackled to science and technology, and the vested interests of STEM people. Nor are they part of the anecdote production crew. Nor do they live enacted biographies. Thus they are not so easy to see.

**241**

Now we will hear about another *third* research paradigm. Brad Haseman will explain:

"We stand at a pivotal moment in the development of research. Established qualitative and quantitative research methodologies frame what is legitimate and acceptable. However, these approved approaches fail to meet the needs of an increasing number of practice-led researchers, especially in the arts, media and design. Within the binary of quantitative and qualitative research, these practice-led researchers have struggled to formulate methodologies sympathetic to their fundamental beliefs about the nature and value of research. In response to this, *performative research* has emerged as a significant practice-led research which should be understood as a research strategy within an entirely new research paradigm. Taking its name from J.L. Austin's speech act theory, performative research stands as an alternative to the qualitative and quantitative paradigms by insisting on different approaches to designing, conducting and reporting research. Once understood and fully theorised, the performative research paradigm will have applications beyond the arts and across the creative and cultural industries generally."

More enacted biography perhaps? It seems we need to speak with J.L. Austin.

What is the relationship between Haseman's third paradigm and Denscombe's? See how things are more complicated than they might at first seem. Would a polymath see this complexity? Would someone with a liberal arts education see it?

Here is Linda Candy to explain the difference between practice-based research and practice-led research:

"Practice-based research is an original investigation undertaken in order to gain new knowledge partly by means of practice and the outcomes of that practice. In a doctoral thesis, claims of originality and contribution to knowledge may be demonstrated through creative outcomes in the form of designs, music, digital media, performances and exhibitions. While the significance and context of the claims are described in words, a full understanding can only be obtained with direct reference to the outcomes.

"Practice-led research is concerned with the nature of practice and leads to new knowledge that has operational significance for that practice. In a doctoral thesis, the results of practice-led research may be fully described in text form without the inclusion of a creative work. The primary focus of the research is to advance knowledge about practice, or to advance knowledge within practice. Such research includes practice as an integral part of its method and often falls within the general area of action research."

**242**

Quantum mechanics once again is on our minds, and in particular, the matter of hidden variables, Bell's inequality and the Orsay experiment.

The results of the above go against the hidden variable hypothesis and the attempt to model quantum mechanics with classical probability theory. Therefore a classical probabilistic model of the phenomena of quantum mechanics is not possible.

However, one comment must be made. The above is only correct if one can be sure that the measurement of the left particle in the direction $\alpha$ cannot influence the measurement of the right particle, a little later, in the direction $\beta$. This assumption is called locality in physics. In the Orsay experiment, this assumption was fulfilled. Indeed, they performed the measurements on the left and then on the right in an interval of time which is smaller than the flight time of a photon between the two pieces of apparatus. Thus, by the causality principle of Relativity Theory, the first measurement cannot have influenced the second one.

If one admits the causality principle to be valid in quantum mechanics, then the hidden variable hypothesis must be rejected, as well as any attempt to model quantum mechanics with classical probability.

If the causality principle in quantum mechanics is abandoned then there is still some room for modelling the Orsay experiment in classical probabilistic terms. This is, for example, the case of so-called Bohmian quantum mechanics. But a very large majority of physicists nowadays considers the causality principle to be valid in quantum mechanics.

This type of science is called *I believe* ... It is very Abrahamic. We have already encountered it, and we will, no doubt, encounter it again.

**243**

C.P. Snow and Leonardo – we just find ourselves coming back to them. Here is what another artist says, this time stating something that, the artist believes, C.P. Snow proposed:

"C.P. Snow, the essayist who coined the *two cultures* cliché, proposed a simple solution to the problem of divided cultures. He argued that we needed a *third culture,* which would close the *communications gap* between scientists and artists. Each side, Snow said, would benefit from an understanding of the other, as writers learned about the second law of thermodynamics and scientists read Shakespeare."

Let us now see what Snow actually said. So, first, this *third culture* - tell us, C.P. Snow, what you said:

"I have been in close intellectual contact with social historians most of my life: they have influenced me a good deal: their recent researches were the basis for a good many of my statements. But nevertheless I was slow to observe the development of what, in the terms of our formulae, is becoming something like a third culture. I might have been quicker if I had not been the prisoner of my English upbringing, conditioned to be suspicious of any but the established intellectual disciplines, unreservedly at home only with the *hard* subjects. For this I am sorry.

"It is probably too early to speak of a third culture already in existence. But I am now convinced that this is coming. When it comes, some of the difficulties of communication will at last be softened: for such a culture has, just to do its job, to be on speaking terms with the scientific one. Then, as I said, the focus of this argument will be shifted, in a direction which will be more profitable to us all.

"There are signs that this is happening. Some social historians, as well as being on speaking terms with scientists, have felt bound to turn their attention to the literary intellectuals, or more exactly, to some manifestations of the literary culture at its extreme. Concepts such as the *organic community* or the nature of pre-industrial society or the scientific revolution are being dealt with, under the illumination of the knowledge of the last ten years. These new examinations are of great importance for our intellectual and moral health."

In other words, Snow is pointing to other disciplines, not those from the world of the arts, but from social history, and sociology, who are better placed than he, a scientist, to explain the social phenomena to which he was referring. Thus Snow did not propose a *third culture,* which would close the *communications gap* between scientists and artists. The phrase *communications gap* does not appear at all in Snow's three works, to which we have already referred.

There is in fact a third culture - it is called the social sciences. The humanities are not social sciences, but, as in the case of social history, are sometimes related.

Here is Patricia Olynyk:

"C.P. Snow did us a great disservice when he declared that there were *two cultures*. This notion largely addressed the political dimension of the British educational system, but Snow's ideas have since been misconstrued to imply a significant gap between the overall world views of scientists on the one hand and artists and humanists on the other."

But of course there are vast differences, and it is only those that have fallen in love who do not want to see these differences. And György Kepes did see a fundamental gap, as did John Dewey. Interesting that among all the talk of Leonardo, Snow, E.A.T., and so forth, when the lovers do talk, as though whispering sweet nothings into one another's ear, few mention the names of Kepes and Dewey and what they said. And you will find too, that the matter of the differences between art and science, they just sit there as central matters of concern to the founder of that art movement that is called Surrealism – the man whose name is André Breton.

We need to hear from the British botanist and educator Eric Ashby and also the Austrian historian and philosopher of science Edgar Zilsel. The British biochemist and sinologist Joseph Needham should also speak as well.

**244**

Religious: relating to or believing in a religion. Spiritual: relating to or affecting the human spirit or soul as opposed to material or physical things.

"What happens to a world which loses all its spiritual dimension? I know that many Europeans say they're personally on a spiritual trip, but I worry when I go to a city like Amsterdam and I look in the eyes of young people in that city that's supposed to be so postmodern, and I don't detect a sense of awe. I sense life that's so secular, that it's lost a frame of reference altogether. There is a difference between institutional religiosity and spiritual awe. When you lose the latter, you're in trouble."

That was Jeremy Rifkin speaking. Europeans are in trouble.

It is a truth not widely known that Richard Dawkins, in his book *The God Delusion*, not once mentions the words spiritual or spirituality. We noticed this when we read that book. But to be sure, we bought an ebook version and checked. Not one reference to those words! This is what science and reason does to people – he is the icon that no one should aspire to be – a man who has lost his humanity to science and reason, yet so consumed is he with his ideology and dogma, he is blind to this. Strange beliefs in the world of STEM can be found; strange superstitions too.

Here is what science and reason have done to people: Stalin's view of the Socialist state and how to bring it into existence and maintain its existence; Hitler's view of a National Socialist state and how to bring Aryan beings into existence, and to maintain their dominance and existence. He needed Genetic Engineering for this - hence all those experiment on living humans. Think what he would have done with today's science! Think what the *Vainglorious Enlightened Ones* are going to do with it!

You can reason your way to hell on earth. You will. You are doing this now. It only stops when you become fully human by understanding that humans are spiritual beings and when humans start behaving as such. Would you like us to prove this? We will as soon as you can prove the belief that is the Church-Turing thesis and the belief (of scientism) that all real knowledge is scientific knowledge and that that there is no rational, objective form of inquiry that is not a branch of science. We expect to have to wait a long, long, long time before having to offer our proof.

**245**

Today is the beginning of metrological autumn. Only 122 chapters left to write. We are two-thirds through the odyssey that is 366. This number is a temporal division of a circle as found in Ancient China. The other division of a circle is 360, as found in Ancient Greece, and this is spatial division of a circle. So 366 is a special number, giving a special division of a circle based on time. Once again time appears. It seems everything is about time - it is all about time.

Kenwyn Smith will now speak about philosophical problems in thinking about organisational change:

"As organisational actors we often observe that things need to be different from what they are. There are many reasons for this, but whatever signs we are seeing they all lead us to believe that we should take action to produce change. This is understandable, the problem is what we think these signs are telling us and how our interpretation shape action. Do we interpret the signs through a technological, a medical mindset, viewing problems as diseases which some magical medicine or technological innovation can cure? Or are they symptoms of some deeper ill, some social cancer that must be exorcised?

"All the perspectives which deal with the matter of organisational change, it appears to me, pose epistemological roadblocks, some of which are easy enough to blast a way through, others to circumvent. But most of them stop me cold. One of these is whether organisations and organisational actors can and do learn how to change and, in particular, whether these changes alter the very core of things, such as the governing values and master programs that shape actions – or whether they merely change appearances; the domain of the superficial. This is a deeply complex question and is at that heart of the topic that cybernetics calls the morphostasis-morphogenesis issue.

"Morphogenesis is applied to changes similar to those that occur in natural evolution. Here change is of the form that penetrates so deeply into the genetic code that all future generations acquire and reflect those changes. In morphogenesis the change has occurred in the very essence, in the core, and nothing special needs to be done to keep the change changed.

"Morphostasis encompasses two types of changes. First there are those that enable things to look different while remaining fundamentally as they have always been. These changes are not very enduring and have the propensity to disappear unless change effort is continually applied. The second type of change occurs as a natural expression of the development sequence. These changes are embedded in the natural maturation process and the boundaries of this type of change are contained within the instructions coded into the system."

Here, looking beyond the art-science love affair and its claimed benefits, we make a brief observation about the more generic issue of artistic interventions in organisations, and express doubt about the claims being made for the benefits of this activity. Why? We refer you to what Kenwyn just said.

But our main point is this: now do we see the nature of the beautiful heresy – it is about changing the codes – the codes of life in the widest sense, those upon which we build our lives, decide upon how we relate to others, and the codes which shape how we relate to everything else that lives, or breaths or is part of that which we live upon and in.

"We were the first to assert that the more complicated the form of civilisation, the more restricted the freedom of the individual must become."

That was Benito Mussolini speaking. The current version of this type of thinking is called *Scientific Government.* A major part of this will be founded on Global Systems Science, which is being developed for you, by the European Commission. At some future time people reading this will ask (assuming that the book has not been burned by the European Union Nationalists!): how was it that we allowed them to do what they did? Some may also ask: how was it possible that Julia and Paul knew what they were going to do? You have learned nothing!

Here is a spokesperson for the European Commission:

"We were the first to assert that the more complicated the form of European Union civilisation, the more restricted the freedom of European Union citizens must become. Only we are more sophisticated than the fascists of old, for in our new fascism, we will allow you to vote; only it will not make the slightest difference, for policies will be determined by us using science and reason. And quite evidently many of you are gullible enough to fall for this deception. Evidently though, citizens of the United Kingdom are not. But many British people saw through the deceptions of previous fascists, so no surprises there!"

**246**

Naturalism: the philosophical belief that everything arises from natural properties and causes, and supernatural or spiritual explanations are excluded or discounted.

Emily Sargent, curator of *States of Mind: Tracing the Edge of Consciousness*, will now speak:

"What is consciousness? On the face of it, this seems like an easy question. We all know what it is like to be conscious. It allows for our individual experience of the world, making possible the magic of our everyday inner life as well as our interaction with an external reality. But if we try to pin down the exact nature of consciousness it becomes more problematic. Questions about consciousness have been asked, discussed and reframed for hundreds of years by philosophers, poets and artists. Now neuroscience has joined the debate. Significant progress has been made identifying the relationship between brain activity and cognitive functions such as memory or decision making. And yet science still struggles to describe how the subjective experience of consciousness arises out of the objective tissue of a human brain."

Yes of course, a very *Enlightened* view. Yet, so long as you only look at consciousness as a process, as a property of the brain, confined to the limits of the human body, then you will never come to understand consciousness, but you will think that you do. That is your weakness. Whose weakness? Why the *Vainglorious Enlightened Ones* of course.

"As social, financial and political upheavals and ideologies swirl around us, the timeless questions remain unanswered: What is more ubiquitous than consciousness? What is less understood than mind? As for conscious experience itself, while there is nothing we know more intimately than our inner sense of being, there is nothing we can experience with less comprehension than the conscious states of others."

That was Roy Ascott speaking.

**247**

What do we know about the Ancient Greeks? This is what we know about the Ancient Greeks: they were patriarchal; elitist; disparaging towards people who undertook manual work; intolerant of that which did not conform with the normal; contemptuous towards people who were not Hellenic (those people otherwise known by the Ancient Greeks as the *untermenschen* that were the barbarians); obsessed with their bodies and inclined towards achieving genetic purity through the practice of eugenics implemented through infanticide. They treated women as second class citizens only needed for breeding and pleasure. We did not mention slavery! Now we have.

What do we know about Europeans? We have just told you. What is that you say - no slaves anymore! Aye it is true, but only just - quite a recent development. But now they are creating new slaves because they know that they are not allowed anymore to have human slaves. They call these slaves machines. Some have specific names - robots. They like their robots to look human so they have anthropomorphic robots. It reminds them of the days when they used to be allowed to have humans for slaves. Unfortunately these robots are not smart, so they are developing something they call artificial intelligence, so that their new slaves will be even more human. One day they will have fully restored slavery, and then they will be happy again because they will once more have dominion over the new *untermenschen*. Then when the slaves are of no further use, they will send them to special facilities to be exterminated - sorry we mean recycled.

Time now to examine what the Ancient Greeks did in the Victorian era to the Profoundly Deaf. Here is Oliver Sacks to explain:

"Edward Gallaudet was an open-minded man who travelled extensively in Europe in the late 1860s, touring deaf schools in fourteen countries. He found that the majority used both sign language and speech, that the sign language schools did as well as the oral schools as far as articulating speech was concerned, but obtained superior results in general education. He felt that articulation skills, though highly desirable, could not be the basis or primary instruction - that this had to be achieved, and achieved early, by Sign.

"Gallaudet was balanced, but others were not. There had been a rash of *reformers* who clamoured for an overthrow of the *old fashioned* sign language schools and for the introduction of *progressive* oralist schools.

"One of the most important and powerful of these oralists was Alexander Graham Bell, who was at once heir to a family tradition of teaching elocution and correcting speech impediments - his father and grandfather were both eminent in this - tied into a strange mix of deafness denied - both his mother and his wife were deaf, but never acknowledged this - and, of course, a technological genius in his own right. When Bell threw all the weight of his immense authority and prestige into advocacy of oralism, the scales were, finally, overbalanced and tipped, and at the notorious International Congress of Education of the Deaf held at Milan in 1880, where deaf teachers were themselves excluded from the vote, oralism won the day and the use of Sign in schools was officially proscribed.

"Deaf were prohibited from using their own natural language, and thenceforth forced to learn, as best they might, the (for them) unnatural language of speech. And perhaps this was in keeping with the spirit of the age, its overweening sense of science as power, of commanding nature and never deferring to it."

Now you know your future when *Scientific Government*, Global Systems Science, and other parts of the scientism agenda - some times called science and reason - take hold, and Ancient Greece is once more restored.

Today you know the infanticide by a different name - abortion!

This day, standing in the Musee d'orsey in Paris, we did speak these words in silence:

"Through time's window we did gaze, and, like Blake before us, did see Past, Present and Future, existing all at once, because time is a singularity."

**248**

It is time to meet J.L. Austin, who will now speak:

"What I shall have to say here is neither difficult nor contentious; the only merit I should like to claim for it is that of being true, at least in parts. The phenomenon to be discussed is very widespread and obvious, and it cannot fail to have been already noticed, at least here and there, by others. Yet I have not found attention paid to it specifically.

"The type of utterance we are to consider here is not, of course, in general a type of nonsense; though misuse of it can, engender rather special varieties of *nonsense*. Rather, it is one of a different class - the masquerade. But it does not by any means necessarily masquerade as a statement of fact, descriptive or constative. Yet it does quite commonly do so, and that, oddly enough, when it assumes its most explicit form.

Grammarians have not, I believe, seen through this *disguise,* and philosophers only at best incidentally.

"We shall take, then, for our first examples some utterances which can fall into no hitherto recognized grammatical category save that of *statement,* which are not nonsense, and which contain none of those verbal danger-signals which philosophers have by now detected or think they have detected (curious words like *good* or *all,* suspect auxiliaries like *ought* or *can,* and dubious constructions like the hypothetical): all will have, as it happens, humdrum verbs in the first person singular present indicative active. Utterances can be found, satisfying these conditions, yet such that: A. they do not *describe* or *report* or constate anything at all, are not *true or false;* and B. the uttering of the sentence is, or is a part of, the doing of an action, which again would not normally be described as, or as *just,* saying something.

"This is far from being as paradoxical as it may sound or as I have meanly been trying to make it sound: indeed, the examples now to be given will be disappointing.

"The first example is: 'I do (sc. take this woman to be my lawful wedded wife)' – as uttered in the course of the marriage ceremony. The second example is: 'I name this ship Queen Elizabeth' – as uttered when smashing the bottle against the hull. The third example is: 'I give and bequeath my watch to my brother' – as occurring in a will. The final example is: 'I bet you sixpence it will rain tomorrow.'

"In these examples it seems clear that to utter the sentence (in, of course, the appropriate circumstances) is not to describe my doing of what I should be said in so uttering to be doing or to state that I am doing it: it is to do it. None of the utterances cited is either true or false: I assert this as obvious and do not argue it. It needs argument no more than that *damn* is not true or false: it may be that the utterance *serves to inform you* – but that is quite different. To name the ship is to say (in the appropriate circumstances) the words 'I name, etc.'. When I say, before the registrar or altar, etc., 'I do', I am not reporting on a marriage: I am indulging in it.

"What are we to call a sentence or an utterance of this type? I propose to call it a performative sentence or a performative utterance, or, for short, *a performative.* The term *performative* will be used in a variety of cognate ways and constructions, much as the term *imperative* is. The name is derived, of course, from *perform,* the usual verb with the noun *action*: it indicates that the issuing of the utterance is the performing of an action – it is not normally thought of as just saying something."

Katja Kwastek will now add a remark:

"Austin acknowledged later that it is not possible to make a clear distinction between constative and performative utterances, because all linguistic expressions may have the capacity to change reality. As a consequence, subsequent speech-act theories no longer focused on Austin's model of the performative and instead on the general behavioural dimension of speech. Nevertheless, interest in the relationship between performative and constitution of reality did not fade."

**249**

Time now to meet another Ancient Greek from the Victorian era! We have already encountered this engineer. His name is Frederick Winslow Taylor. He will now speak:

"Our whole system of management is quite accurately typified by the substitution of an elaborate engine to drive and control the shop in place of the old fashioned foot power. There is no question that our human managing machine, which is required for the maintenance and the effective use of both standard shop details, and standard methods throughout the establishment, and for giving each workman each day in advance a definite task which must be finished in a given time, calls for many more non-producers than are used with the old style management having its two or three foremen and a superintendent. The efficiency of our engine of management, however, compared with the old single foreman is like a shop engine as compared with foot power or the drafting room as compared with having the designing done by the pattern maker, blacksmith and machinist.

"We propose to take all the important decisions and planning which vitally affect the output of the shop out of the hands of the workman, and centralise them in a few men, each of whom is especially trained in the art of making those decisions and in seeing that they are carried out, each man having his own particular function in which he is supreme, and not interfering with the functions of other men. In all this let me say again that we are aiming at true simplicity, not complications."

Thank you for this insight into your thinking from 1907 and further back in time, which we can summarise as free people think, slaves work.

The above is also the basis for *Scientific Government*, which will come about because of scientism, and because in Europe this is the technocratic mind-set that prevails in the European Commission. So, let us now see how the former chief scientific advisor to a former president of the European Commissions would like European society to be run.

"We propose to take all the important decisions and planning which vitally affect society out of the hands of the politicians and ordinary people, and centralise them in a few people based in Brussels, each of whom is especially trained in the art of making those decisions and in seeing that they are carried out, each person having their own particular function in which they are supreme, and not interfering with the functions of other people. In all this let me say again that we are aiming at true simplicity, not complications."

Thus speaks the Ancient Greek mind, the fragmented and reductive mind.

When was the last time you heard Richard Dawkins speaking about compassion and empathy? When did you hear anyone at all for the world of STEM speaking such words? When did you hear anyone at all from the world of art who have embraced this world of STEM speaking such words? Silence!

"In science man is a machine. If he is not a machine, then he is nothing at all."

That was Joseph Needham speaking, and his view in 1927 was that such a position is unacceptable, and hence he declared himself to be very Kantian. Later he would encounter a culture that would offer a different view, one that could save the West, the Europeanised world, from itself, and from the self-evident consequences of believing nature and humans are just complicated biological machines.

In the causality there is purpose. In the purpose there is causality.

**250**

Interaction: reciprocal action or influence.

Now it is time to speak the West's Abrahamic creed:

I believe in God, creator of heaven and earth;
I believe in positivism which I find very convenient even though positivism is discredited;
I believe that the history of all hitherto existing societies in a history of class struggle;
I believe in causality even though causality cannot be observed or scientifically proved;
I believe in the axioms of quantum mechanics;
I believe that humans and nature or just very complicated machines;
I believe in private greed, public good;

I believe in the Church-Turing thesis which cannot be proved but means that I do not have to contemplate the heretical belief that the human brain is not a computer;
I believe all real knowledge is scientific knowledge and that there is no rational, objective form of inquiry that is not a branch of science;
I believe in relativistic causality which means that there are no global hidden variables, which means that I do not have to pay any attention to the heretical views of David Bohm, or Howard Rosenbrock.
I believe that through science we will establish a paradise on earth;
I believe that free people think, and that slaves work.
I believe that those who do not agree with the sole truth are suffering from a defect that needs to be corrected - how matters not.

You are indeed a strange people with equally strange beliefs! What would the world be like if there had never been a Europe, with its roots deeply embedded in Ancient Greece and an Abrahamic religion, that itself has more of Athens about it than Jerusalem?

**251**

Theory: A supposition or a system of ideas intended to explain something, especially one based on general principles independent of the thing to be explained.

Theory, often dreary, sometimes even eerie, gives to the eagle in its eyrie, eyes to see the world weary, and thus to make a query.

That was Julia and Paul being poetic.

According to the online Oxford English Dictionary the origin of the word theory is late sixteenth century (denoting a mental scheme of something to be done): via late Latin from Greek theōria, meaning *contemplation, speculation,* from theōros *spectator.*

"Theory is born when practice becomes conscious, for theory is nothing more than practice carried out in the imagination. The techniques of tool and weapon making, weaving, basketry, ceramics and construction gave men models for thinking: association, disassociation, unity, rhythm, harmony, structure, shape, size, volume, motion, growth. As sensory images mainly, as words only partially and very primitively, these *ideas* became comprehensible. They assumed enormous generative force in raising human life to new levels of technical and intellectual accomplishment. Before priest and philosophers could conceive metaphysical question of form and substance, permanence and change, or physical scientists laws of matter and motion underlying changes of state and composition, it was necessary that ancient man be moved to thought through his eyes and hands."

That was György Kepes speaking.

Praxis: practice, as distinguished from theory.

Here is Edward Andrew:

"The doctrine of the unity of theory and practice, usually associated with the teaching of Karl Marx, is also found in the thought of Friedrich Nietzsche. Neither Marx nor Nietzsche provided a systematic elaboration of this important doctrine.

"The basic task to which Marx and Nietzsche addressed themselves was the creation of new men – the overcoming of what had hitherto been conceived to be human nature. Both wished to use the experimental methods of natural science for the purpose of creating new men; a science of man should replace moral philosophy as the branch of practical reason concerned with the matters of most importance to man."

Evidently we are on dangerous ground, especially in the Abrahamic West. Let us go east instead where we find that all religions are true. So back to this matter of unity of opposites, of balance and harmony, of opposing force working together, with one being counteracted by the other, so that neither come to dominate. Evidently also there is a need for a spiritual dimension, otherwise we are back to that space where the Abrahamic West has been before in a religious context, and also in a secular one, and where it is heading once again, though few can see this.

We refer you to Richard Dawkins and his version of *Mein Kampf.*

**252**

Literature: Written works, especially those considered of superior or lasting artistic merit.

"Conceptual art was a critical art not a contemplative one – not necessarily for looking at, but for analysing or for reading. Theorisation became a driving principle for many artists. Philosophy, linguistics and a use of language and text were used as art itself. Conceptual art was grounded here as an art of communication – of distribution and network. What counts as a work of art? What is it that an artist does? What is the role of the viewer? Artists such as the collaborative group *ART & Language* used text to position art as theoretical and critical debate."

That was a spokesperson from Tate Britain reading from the *Conceptual Art in Britain 1964-1979* exhibition leaflet.

Now it is time for Piet Mondrian to speak:

"Although the evil of oppression is clearly manifested in life, it is always clearly discernable in relation to its basic causes, for it is veiled by all manner of particular complications in the appearance of things. This also applies to art. If art follows life or nature in terms of representation, oppression disguises itself in all kinds of particular forms and relationships, and therefore seems non-existent. But in art, time and movement are fixed; thus contemplation is easier and oppression can be studied in a purer way.

"Because of its camouflaged appearances in life and art, there is a danger of accepting oppression; oppression does not always cause revolt. Does there not exist in human nature this powerful opposition: the desire for oppression and the desire to be free of it, this opposition which causes that long and continual struggle that life and art show until human equilibrium is neared?

"Plastic art is an abstract, a free domain of life; the causes and consequences of its expression are purely for study. It does not tolerate oppression and can resist it, for art is not bound by material and physical conditions.

"Writers and thinkers have deeply explored the facts and the causes of the tragic realities created by oppression. Art has been used for immediate and personal purposes: it describes events, persons, battlefields; war camouflage and propaganda were made. But the function of plastic art is neither descriptive nor cinematic. It is not merely a means of enjoyment amidst an incomplete life, or a simple expression of that life even in its beautiful aspect. All this is incidental.

"Art is the aesthetic establishment of complete life - unity and equilibrium - free from all oppression. For this reason it can reveal the evil of oppression and show the way to combat it.

"Plastic art establishes the true image of reality, for its primary function is to *show*, not to describe. It is up to us to *see* what it represents. It cannot reveal more than life teaches, but it can evoke in us the conviction of existent truth. The culture of plastic art can enlighten mankind, for it not only reveals human culture, but being free, advances it."

**253**

Balance: a situation in which different elements are equal or in the correct proportions. Harmony: the quality of forming a pleasing and consistent whole. Equilibrium: a state in which opposing forces or influences are balanced. Static Equilibrium: lacking in movement, action, or change, especially in an undesirable or uninteresting way; physics - concerned with bodies at rest or forces in equilibrium. Dynamic Equilibrium: of a process or system characterized by constant change, activity, or progress; physics - relating to forces producing motion.

William Blake will now read from his prophetic work *Jerusalem: The Emanation of the Giant Albion.* Proceed:

"I see the Four-fold Man. The Humanity in deadly sleep,
And its fallen Emanation. The Spectre and its cruel Shadow.
I see the Past, Present and Future, existing all at once
Before me; O Divine Spirit sustain me on thy wings!
That I may awake Albion from his long and cold repose.
For Bacon and Newton sheath'd in dismal steel their terrors hang
Like iron scourges over Albion. Reasonings like vast Serpents
Infold around my limbs, bruising my minute articulations."

And again please:

"The Spectre is the Reasoning Power in Man; and when separated
From Imagination, and closing itself as in steel, in a Ratio
Of the Things of Memory, It thence frames Laws and Moralities
To destroy Imagination, the Divine Body, by Martyrdoms and Wars."

Here is an explanation of the Four-Fold Man from E.R.D. Maclagan and A.G.B. Russell:

"Man is at once the stage and the protagonist in the drama with which Blake is concerned, the Fourfold Man, called symbolically by the name of Albion, 'our ancestor, in whose sleep or chaos creation began,'; and his state depends on the union and agreement of the four elements that are met in him. Beside the Humanity, or central personality of the individual, stand the Spectre, the reasoning power, and the Emanation (a word sometimes abridged into Eon) the emotional and imaginative life, with the Shadow, which seems to be desire, restrained and become passive, 'till it is only the shadow of desire.' When these are united, and especially when the Spectre and the Emanation, contraries in whose interaction all other contraries are involved, are balanced and at peace, Man is in the state of salvation, which Boehme called temperature; when Spectre and Emanation have parted, Man is in a fallen state, and can only be redeemed by their reconciliation. This fall into division, and resurrection into unity, is the main subject of *Jerusalem* and indeed of most of the Prophetical books; for the partition of Reason and Imagination is the great tragedy, through which the Spectre becomes cold and the Emanation weak, the Shadow turns cruel, and the Humanity is overcome by deadly sleep. A sleep, too, full of dreams, in which Man wavers between evil and good, drawn alternately by the male Spectre and the female Emanation, and so called by Blake hermaphroditic: a sleep from which only Christ, the Divine Imagination, can save the fallen Man, by reuniting him with Jerusalem, his Emanation, and saving him from the dominion of his Spectre, the great selfhood, called Satan."

**254**

We have invited a priest to read from the Old Testament, the Book of Psalms, the 115$^{th}$ Psalm:

"But their idols are silver and gold,
made by human hands.
They have mouths, but cannot speak,
eyes, but cannot see.
They have ears, but cannot hear,
noses, but cannot smell.
They have hands, but cannot feel,
feet, but cannot walk,
nor can they utter a sound with their throats.
Those who make them will be like them,
and so will all who trust in them."

Marshall McLuhan will explain:

"The Psalmist insists that the beholding of idols, or the use of technology, conforms men to them. The poet William Blake developed the Psalmists ideas into an entire theory of communication and social change. In his long poem *Jerusalem: The Emanation of the Giant Albion,* he sees man as fragmented by his technologies. But he insists that these technologies are self-amputations of our own organs. When so amputated, each organ becomes a closed system of great new intensity that hurls man into martyrdoms and wars."

William Blake again:

"If Perceptive Organs vary: Objects of Perception seem to vary:
If the Perceptive Organs close: their Objects seem to close also."

This goes too for the science and technology of the *Vainglorious Enlightened Ones*. Their perceptive organs are closed as are the objects – thus we turn to the *Tao*.

The *Tao* – undefinable, inexpressible, unknowable. It lies beyond the confines of language. Yet each can come to know.

Does scriptovisualism recognise the limitations of language? Does it recognise the power of a dual act of creation in both the descriptive and the visual? Is this driven by an awareness that the development of a Taoist science and technology needs such a tool to bring it closer to realisation, and in doing so, to bring about a transformation, a morphogenesis?

**255**

*Digital Harmony: On the Complementarity of Music and Visual Art.* This is the title of a book by the artist John Whitney. He will now speak about this matter:

"The nature of art has become uncertain, and form, one sizeable block of composers, that grand Pythagorean certainty – *the harmony* – has come apart. Even so, the foundation of my work rests first upon laws of harmony, then in turn, upon proof that *the harmony* is matched, part for part, in a world of visual design. This is my hypothesis. My book, a record of the formulation of that hypothesis, shows how it has evolved as the basis for a life of work. If hypotheses are rare in art, proof within a lifetime is still rarer. My book documents how the application of graphic harmony, in that real sense of ratio, interference and resonance, produces the same effect that these physical facts of harmonic force have upon musical structures.

"There remains however a need for confirming demonstrations of that hypothesis in a larger body of work. Art, unlike science, is proved by art alone, not by mock-scientific experiments in the isolated case. Meanwhile, more exploration of the hypothesis continues by the only means available to the artists in this medium – by more work on film, tape or videodisc."

And could this also be for sciptovisualism? It will certainly require more work in the media of the descriptive and the plastic.

Now Piet Mondrian will speak:

"Abstract art emphasises the fact that in plastic art the expression of reality cannot be similar to that of palpable reality. The dynamic movement established by the opposition of forms and their colours constitutes the expression of universal reality. In single forms, dynamic movement reveals itself through the opposition of their composing elements: volumes, planes, determined by lines and colours. For this reason, the work appears as living. But in relation to the environment, simple forms show a static balance. They appear as entities separated from the whole. In order to establish unity, their proper unity has to be destroyed: their particular expression has to be annihilated. In plastic art, the static balance has to be transferred into a dynamic equilibrium which the universe reveals.

"It must be emphasised that it is important to discern two sorts of equilibrium: static balance; and a dynamic equilibrium. The first maintains the unity of particular forms, single, or in plurality. The second is the unification of forms or elements through continuous opposition. The first is limitation and the second is extension. Inevitably dynamic equilibrium destroys static balance. Opposition requires separation of forms, planes or lines. Confusion produced a false unity.

"In art as in reality, the plurality of varied and similar forms annihilates the existence of forms as entities. Similar forms do not show contrast but are in equivalent opposition. Therefore they annihilate themselves more completely in their plurality."

"There has never been a better time to be alive!"

That was someone speaking in the year 1910.

"There has never been a better time to be alive!"

That was someone speaking in the year 1930.

"There has never been a better time to be alive!"

That was someone speaking from the year 2016. Actually it was President Barack Obama.

How is it, that such people are always so wrong?

So you have between 9 and 14 years … 2030!

**256**

The notion that we are making progress! We should examine this, so Vanessa Cox will now speak:

"Too often, rather than broadening people's viewpoints, modern technology reflects people's own opinions, views, and biases back to them, an effect known as the *echo chamber*. The team wanted to use technology to decrease isolation, broaden people's views, and promote empathetic responses to opposing viewpoints. In effect, the team proposed to use technology to allow people to escape the *echo chamber*.

"The *echo chamber* is the idea that by reflecting people's ideas and opinions back to them, technology is reinforcing people's personal views rather than exposing them to a wide range of viewpoints. For example, Internet search results and advertisements are customized based on previous websites visited. Similarly, social networks surround people with friends who generally share their world view.

"By creating an environment in which people encounter different viewpoints less frequently, the *echo chamber* can lead to decreased empathy. The team thinks this is a significant social problem because decreased empathy can inhibit people's ability to collaborate and solve complex problems, whether locally, nationally, or globally."

If you care to look - do you care to look? - you will find that echo chambers have always been with us. Why therefore was technology designed and developed without taking this matter into account?

Artists too live in echo chambers. Why are you surprised? They are human. They are not all open minded. They are not all highly knowledgeable. There are quite a few, too many in fact, who have fallen in love with science and technology, and do not ask questions or seek new horizons. They are morphostatic. Just tune into Facebook to see what we mean.

Oh dear what a world the *Vainglorious Enlightened Ones* are making. Well you will insist on being so Abrahamic, so Ancient Greek. When you try to overpower nature ...

Let us now hear from Paul. He is referring to his 2009 paper entitled *On the Social Shaping of the Semantic Web,* which deals with the problems of Abrahamic technologies.

"The above paper addressed the Social Shaping of the Semantic Web in the context of moving beyond the workplace application domain, which has so dominated the development of Information and Communications Technologies (ICTs), as well as the *Social Shaping of Technology* perspective. The paper addresses the importance of paradigms and the values that shape technology, along with the dominant utility value of ICT, this latter matter being somewhat central in the development of these technologies. The new circumstances of ubiquity and of uses of ICT beyond mere utility, as a means of having fun for example, lead to a notion of the Semantic Web, not just as a tool for more effective Web searches, but also as a *means of having fun* and for exploration. Given this possibility of the Semantic Web serving two very different audiences and purposes, a key matter is how to do this without resorting to the obvious and rather simple conceptual formulation of the Semantic Web as *either A or B.* Existing *Social Shaping of Technology* perspectives seem not to consider this matter and the new thought presented, was that what needs to be central to the development of the Semantic Web, is *both A and B.* Key here is an intelligent relationship between the Semantic Web and those that use it. Central to achieving this is the notions of the value of people, control over technology, and non-utility as a dominant design principle - the idea of things that do not necessarily serve a specific purpose, although purposefulness is important."

Well we are back to purpose and Taoist science and technology again, which of course we cannot define! How does that make you feel?

**257**

Interaction: reciprocal action or influence. Relationship: the way in which two or more people or things are connected, or the state of being connected; the way in which two or more people or groups regard and behave towards each other; an emotional association between people.

This term interaction is what happens between machines, when one machine interacts with another machine. It is also increasingly applied to people - they now interact! It is also applied to people using machines, especially computers - hence the term human-computer interaction. But in reality there is no such thing as human-computer interaction, for all there is in the Western Abrahamic world, is machine-machine interaction, because the *Vainglorious Enlightened Ones* believe that human beings are machines. That goes for artists too working in the field known as interactive digital art. This is what we refer to as *I believe science and technology,* which is just a collection of strange beliefs, the most dangerous of which is that people who hold these beliefs are behaving in a rational and objective manner.

There is nothing more irrational than believing that you are being rational, and nothing more subjective than believing that you are being objective.

What happens when all these restrictions, biases and strange beliefs are just forgotten about, transcended, and a new and empty spaced is entered? The answer is that a beautiful heresy takes shape.

**258**

China!

The Chinese President Xi Jinping will now speak:

"If there is any party, any state, or any people in this world that should feel confident in itself, it is the Communist Party of China, the People's Republic of China, and the people of China. Whether *socialism with Chinese characteristics* is good or not can only be judged by the facts, by Chinese people, not by subjective assumptions made by those looking through tinted lenses.

"History has no end, and it cannot come to an end. The party and the Chinese people should have full confidence in seeking out a Chinese solution in humanity's search for a better social system."

Yes indeed, it is time for *Chinese Solutions*. Taoist science and technology! How does that make you feel, all you people that are part of the West's Abrahamic, *Enlightenment*-bound, research and innovation system?

Li Yanzu, from Tsinghua University will now speak:

"The notion of ecological design originates from human thinking on the relation between man and nature. Throughout the long history before the industrial age of modern machineries, the relationship between man and nature had always been a major issue in man's life and creation. Because of the limited capability to change nature, the relation between man and nature had been generally in harmony and under control, whereas there is a lot of experience that should be carefully studied and used by people nowadays. Take China as an example. Chinese classical design was developed during a history of thousands of years with the tradition and notions of worshiping and respecting, befriending and learning from nature. The experience and underlying philosophy is worthy of study."

**259**

Anthony Giddens will now speak:

"Weber concluded his study of China by drawing an explicit comparison between Confucianism and Puritanism. There are two primary, although interrelated, criteria in terms of which the degree of rationalisation of a religion may be determined: how far magic has been eliminated, and how far there has developed an internally consistent and universally applicable theodicy. With regard to the first, ascetic Protestantism has been more radical than any other religion; in terms of the second, however, Confucianism ranks with Puritanism as having attained a high level of formal rationality. But the content of Confucian rationalism and consequently its relationship to the imperfections or irrationalities of reality, was quite different from that of rational Puritanism. Where as the Puritan ethic introduced a deep tension between religious ideals and the earthly world, that of Confucianism centred upon the harmonious adaptation of the individual to an inevitability given order."

Max Weber will explain:

"For the Confucian ideal man, the gentleman, *grace and dignity* were expressed in fulfilling traditional obligations. Hence, the cardinal value and goal in self-perfection meant ceremonial and ritualistic propriety in all circumstances of life. The Confucian demanded no other kind of *redemption* save that from the barbaric lack of education. As the reward of virtue he expected only long life, health and wealth in this world and beyond death simply the maintenance of his good name. As for truly Hellenic man any sort of transcendental anchorage ethics, and tension between obligations to a supra-mundane God and world of the flesh, any pursuance of a goal in the beyond, or conception of radical evil, was lacking. The relentlessly and religious systematised utilitarianism peculiar to rational asceticism, that is to say acetic Protestantism, to live *in* the world and yet not be *of* it, has helped to produce superior rational aptitudes and therewith the spirit of the specialised man which, in its last analysis, was denied to Confucianism. The contrast can teach us that mere sobriety and thriftiness combined with *acquisitiveness* and valuation of wealth, were far from representing and far from releasing the capitalist spirit, in the sense that this is found in the specialised economic man of the modern economy."

Weber was in agreement with Marx that one of the basic precepts of modern capitalism is the use of a technology, which is constructed and organised on the basis of rational principles. Mechanisation is the clearest manifestation of this.

Rational – it is a word that just keeps cropping-up, time after time.

**260**

Transcendence: existence or experience beyond the normal or physical level. If you are one of the *Vainglorious Enlightened Ones* you will of course not believe in such things. More of your rationality and objectivity, founded on ... *I believe*. Positivism failed, did you not know? Or is this failure too inconvenient, and thus, like Wolpert, you dismiss as *rubbish* the cross-disciplinary efforts of philosophy to explain science? More of your scientism!

The artist John Whitney now speaks:

"At some time in the mid-sixties ad hoc committees within the art world were being formed to sponsor art and technology. I was immediately elated: my creative needs might be recognised and fulfilled.

"Then, just as quickly, I was discouraged by obtuse, confused, or empty attitudes that developed among people of influence. Excitement grew over projects that were formulated with bandwagon haste around this subject. As a fad that came and went, along with so many others of that decade, the art-technology boom left in its wake as much prejudice as enlightenment."

Thank you, John, for sharing your experiences and reflections. You will be disappointed to learn that little has changed, and that now we have names for these fads - STARTS, STEAM and SEAD. And the art-technology and art-science love affairs are still there, with people whispering sweet nothings in one another's ears. Here is someone to prove the point:

"Much has been said and written about the two-culture paradigm separating the world between artists and scientists. On one side of this debate are those who accept this cultural art-science divide. On the other side are those who reject it altogether to promote a better integration of art-science practices. In a paper, I presented a network analysis of 40 papers submitted to the SEAD Network for Science, Engineering, Arts and Design and tested the hypothesis that the papers submitted by artists and scientists are disconnected in the corresponding graph, as predicted by the art-science separation. Rejecting this hypothesis will provide support for the alternative art-science integration.

"In my meta-analysis of the 40 White Papers submitted to the SEAD Network, I relied on network analysis, employing a wide range of statistical and graph-theoretical approaches similar to those I use routinely for the analysis of gene similarity data. The analysis is based on the self proclaimed status of the authors: scientists, artists, or art-scientists.

"Under the art-science separation hypothesis, nodes associated with artists and scientists should be more densely connected within each category than among categories. Moreover, the art-scientists should be equally connected to nodes representing either one of the other two categories.

"Art-scientist was the only category with observed values more extreme than those expected by chance alone. In other words, whereas all network indices could not discriminate between texts authored by artists or by scientists, the papers authored by art-scientists were clearly more clustered and connected to each other, as well as the other two categories.

"Thus the White Papers have spoken: contrary to Snow's paradigm, artists and scientists are not distinguishable, but art-scientists are a different breed - individuals who thrive in transdisciplinary contexts. This is, of course, what the SEAD Network is all about, but this meta-analysis provides statistical support for promoting such collaborative endeavours."

That was François-Joseph Lapointe speaking. The conclusion is of course a nice one to reach if to the funding of these collaborative endeavours you aspire. We suggest therefore that you speak to the European Commission's DG CONNECT, who are not bound by notions of fiduciary responsibility or due diligence, and will fund anything based on the *I believe* mindset. We also suggest that you submit your paper to a social science journal. What might you learn by doing this?

Perhaps someone, a reviewer, might criticise your paper from the perspective of a different discipline to your own. You might perhaps, who can tell, learn something about academic rigor in the social sciences, rather that which you most obviously apply in your own scientific discipline. Perhaps you might see that you have not developed sufficient sociological understandings of what C.P. Snow was saying. Perhaps this might make you think about critical thinking - of the social sciences type? Perhaps you might see that wandering away from your disciplinary safe zone is actually quite challenging and time consuming? Perhaps you know all this already. Who can tell, for it is not evident from your paper!

The whispering of those sweet nothings - proving that John Witney's comments are as true today as they were back in the mid-sixties.

Who among you can take seriously people who cannot even read or understand that which C.P. Snow wrote and was saying?

Here is John Witney again:

"I determined that the relationship of sight and sound would be served best if it were possible to compose both components of an aural-visual work with some common aesthetic such as harmony would offer."

Yes we have changed the subject. Let us now shift that statement from John's world of the harmony of the complementarity of music and visual art. What about scriptovisualism and the complementarity of words and visual art?

We could say that the relationship of words with visual art, the descriptive with the plastic, would be served best if it were possible to compose both components of a scriptovisual work with some common aesthetic. And then what could we do with this? Apply it elsewhere, to that which is called interactive digital art. Or even, to the task of creating a post-European era, post-Enlightenment, science and technology. Then we are way beyond the art-science and art-technology love affairs. Far beyond the bandwagon, the fad, the emptiness, the nonsense of people continually referring to an irrelevance called C.P. Snow, to the *da Vinci myth*.

**261**

Ignorance: lack of knowledge or information; not knowing.

We will now return to Anthony Giddens, who wants to tell us about Max Weber's observations on Hinduism.

This is Anthony speaking:

"In India an important salvation religion emerged – Hinduism. The Hindu and Buddhist prophets were exemplary, but they did not see themselves as entrusted with a divine mission which must be actively disseminated. Hinduism differs in some important respects from each of the other world religions. Hinduism is an eclectic and tolerant religion: it is possible to be a devout Hindu and yet to accept highly important and most characteristic doctrines which every denominational Christian would consider exclusively his own. But there do exist certain beliefs which are shared by most Hindus, and which are *dogmas* in the sense that they constitute truths the denial of which is considered to be heretical. The most important of these are those of transmigration of souls and compensation (*karma*)."

"All religions are true."

That was Mahatma Gandhi speaking.

Probably you do not know what a prophet is, or have a mistaken understanding. William Blake was a prophet, but if you have a limited or non-existent understanding, then your view of Blake is also a distorted one. We too could be considered as prophets, so enjoy the fruits of your ignorance. Or you could do something about your ignorance.

Bureaucracy: a system of government in which most of the important decisions are taken by state officials rather than by elected representatives; a state or organization governed or managed as a bureaucracy; excessively complicated administrative procedure.

The European Union is a bureaucracy. It is too late to do anything about this. It is a monster that needs to be destroyed, before it destroys democracy in continental Europe. It will destroy democracy in continental Europe. Fortunately it will not destroy democracy in the United Kingdom. There is another way forward for Europe, but it will be, once again, the road not taken. Prophecy! Ignorance – not knowing the true nature of the European Union and the European Commission and voting in the June 2016 referendum, to remain a member of the European Union. Denial: The action of denying something; at some future time, stating that you voted to leave the European Union when actually you voted to remain in the European Union. Prophecy!

Assertion: a confident and forceful statement of fact or belief. Thesis: a statement or theory that is put forward as a premise to be maintained or proved.

We will now return to H. Lawrence Bond, who wants to tell us more about Nicholas of Cusa's doctrines on the coincidence of opposites.

This is Lawrence speaking:

"Coincidence of opposites accomplishes seven tasks in Cusa's thinking: it unites opposites; it transcends analogy and comparison; it overcomes the limits of discursive reasoning; it exceeds composition and synthesis; it surpasses both affirmative and negative language; it frees the mind from quantitative concepts and enables it to achieve a comparatively pure abstraction; and, whether operating from theology, philosophy, mathematics, or geometry, it renders infinite concepts understandable and describable without violating their incomprehensibility or illimitability."

Jürgen Habermas now speaks:

"In its current form, the European Union owes its existence to the efforts of political elites who were able to count on the passive consent of their more or less indifferent populations as long as the people could regard the Union as being also in their economic interests, all things considered.

"The Union legitimised itself in the eyes of its citizens primarily through the results it produced rather than by fulfilling the citizen's political will. This can be explained not only from the history of its origins but also from the legal constitution of this unique formation.

"The European central Bank, the Commission, and the European Court of Justice have intervened most profoundly over the decades in the everyday lives of European citizens, even though these institutions are almost completely beyond the reach of democratic controls. Moreover, the European Council, which has energetically taken the initiative during the euro crisis, is made up of heads of government whose role in the eyes of their citizens is to represent their respective national interests in distant Brussels. Finally, at least the European Parliament is supposed to establish a bridge between the political battles of opinions in the national arenas and the momentous decisions taken in Brussels. But there is hardly any traffic on this bridge."

The European Union is a proto-fascist proto-state. What will it take to transform it into a fascist state?

**263**

Spin: give a news story a particular emphasis or bias. This happens in the world of STEM as well as in politics. It also happens in the world of the love affairs know as art-science and art-technology.

We will now return to Anthony Giddens, who wants to tell us about Max Weber's observations on the formation of disciplinary specialisms.

This is Anthony speaking:

"The advance of bureaucratic organisation in the modern world is directly associated with the expansion of the division of labour in various spheres of social life. It is basic to Weber's sociology of modern capitalism that the phenomenon of specialisation of occupational function is by no means limited to the economic sphere. The separation of the labourer from control of his means of production which Marx singled out as the most distinctive feature of modern capitalism is not confined to industry, but extends throughout the polity, army, and other sectors of society in which large-scale organisations become prominent.

"The efficiency of bureaucratic organisation in the performance of routinised tasks is the main reason for its spread."

Here is Max Weber:

"The fully developed bureaucratic apparatus compares with other organisations exactly as does the machine with the non-mechanical modes of production. Precision, speed, unambiguity, knowledge of the files, continuity, discretion, unity, strict subordination, reduction of friction and of material and personal costs – these are raised to the optimum point in the strictly bureaucratic organisation."

Bureaucracy is as old as civilisation. It is a rationalised structure that operates according to systematised rules of conduct. That word *rational* once again.

**264**

Metaphysics: The branch of philosophy that deals with the first principles of things, including abstract concepts such as being, knowing, identity, time, and space.

Here is a spokesperson from the Oxford English Dictionary to explain the origin of the word metaphysics:

"The word is mid sixteenth century in origin, representing medieval Latin metaphysica (neuter plural), based on Greek *ta meta ta phusika* the things after the Physics, referring to the sequence of Aristotle's works: the title came to denote the branch of study treated in the books, later interpreted as meaning *the science of things transcending what is physical or natural.*"

Attempts by positivist to do without metaphysics failed. Hence to describe oneself as a positivist is to associate one's self not only with a failed school of thought, but is to deny reality - science is not independent and does not exist outside the context of values and beliefs, which can be very strange indeed, as Western science well demonstrates.

Now we shall hear from Sir Edmund Whitaker, FRS, who will address the matter of the attitude of physicists to metaphysics:

"The criticisms of professional philosophers, such as Kant, have at times received a great deal of attention from the exponents of natural theology, and have played an important part in reshaping the formulation of arguments. In contrast to this, contact between philosophers and men-of-science has been comparatively slight. The general attitude of physicists during the last two hundred years has been, that there is no need for them to know anything about metaphysics, whether it is a superstructure built on physics or not, since physics theories can be constructed, understood, and believed by anyone, whatever his philosophy.

"Newton, though profoundly interested in theology, seems to have held that the physicist can give his undivided attention to investigating the laws of which enable him to predict phenomena, and can leave the deeper problems entirely out of account: he can make it his purpose to describe rather than to explain. This was one of the implications of his celebrated declaration *hypotheses non fingo*, and it determined the attitude of his successors: that is to say, men-of-science since Newton have generally held that correct (even if some aspects limited) knowledge of physics could be combined with any views whatever on the fundamental questions of Being and Reality: that part of the world could be rightly understood without reference to the whole: that natural philosophy was independent of metaphysics.

"In a somewhat narrow sense this attitude can be justified: the fact cannot be disputed that great discoveries regarding the processes of the external world have been made by workers whose investigations in their field of research were not related in their own minds to any interest or belief outside it. But the effect of such a segregated way of thinking has been to make science a departmental affair, having no influence on life and thought except indirectly through its application; and in recent years this isolationist attitude has weakened; partly from a desire to see science playing a greater part in life and thought; partly with a view systematising and perfecting the rough home-made metaphysics which is implicit in modern scientific writings; partly from a recognition that every branch of science depends on presuppositions which cannot be established by the methods of science itself, and also that science deals with universal, which are proper objects of metaphysical study; and perhaps most of all, at any rate in recent years, from a hope that philosophy may help to clarify the mystery that wave-particles are the ultimate constituents of the universe."

There is no doubt, as *Hawkins Dawkins Wilson Wolpert* well demonstrates, that STEM, and its practitioners, have become a serious problem, and to ignore this, to deny it, leads along a path to dark places where Europeans have been before, with each visit becoming more destructive and more monstrous than the previous visit. Art and artist have, on each visit, accompanied and helped in the creation of the monsters. Some are doing it again, although they will deny it. STARTS …

**265**

Returning to the matter of C.P. Snow and those opinions he expressed that have been widely misunderstood, misinterpreted, misrepresented, and elevated to the status of a theory, even a paradigm, but which are just – opinions. A failed scientist making sociological observations! An early example illustrating the treacherous waters that are – *x-disciplinary* working. We will in due course encounter, possibly, the matter of different types of proofs. Or not!

Rather than listening to C.P. Snow we will instead focus on the words of Sir Eric Ashby. Ashby was quoted by Snow, but if we ask how many have read Snow's three works on the supposed *two cultures,* and understood them, we could also ask how many went further and read Ashby's book, which is also based upon a series of lectures. We wonder what the answer to such questions would be. Is the answer: everyone who has ever mentioned Snow's *two cultures*? Or perhaps only a small number of people mentioning Snow's *two cultures,* have actually read those three works? Even less have read Ashby! What do you think?

Ashby now speaks:

"In the eighteenth century a new way of thinking took root in Europe. It was an age of reason, when men found that they could explain many things in the universe with the aid of a few simple abstractions, and what they could not explain they were content to reject.

"Science became a respectable leisure-time occupation for country gentlemen and townsmen of substance. At meetings of philosophical clubs and societies, such as the Lunar Society of Birmingham (founded in 1766) and the Literary and Philosophical Society of Manchester (founded in

1781), lectures on electricity, hydrostatics, gravitation, mechanics, and optics, with experiments, were given to fashionable and enthusiastic audiences. All over the country there were peripatetic lectures financed by public subscription, and for members of the middle classes who wanted a more systematic training in science there were courses offered by some of the dissenting academies. The universities shared little of this enthusiasm.

"It is true that science was taught in the Scottish universities, but most of it was suited to the level of schoolboys. The Universities of Oxford and Cambridge, sleeping behind their ancient statutes took even less account of this enthusiasm. Even at the threshold of the nineteenth century, 113 years after the publication of Newton's *Principia,* institutions for higher education in Britain were still making practically no contribution to scientific thought.

"In the accomplishment of the scientific revolution British scientists played a notable and distinguished part. But British universities, except fortuitously and incidentally, played no part whatever. They had allowed the revolution to pass over their heads and still, a century later, they were providing no lead in scientific thought. They reflected yesterday: they did not illuminate tomorrow."

And what changed this circumstance was a Government intervention in response to developments taking place in continental Europe, especially Germany.

Back to Ashby:

"Fewer than three generations lie between the impoverished science teaching, most of it without laboratories or equipment, in the British universities of the 1850s and the research schools of J.J. Thomson in Cambridge, Rutherford in Manchester, Baylis in London, and Robinson in Oxford. During these three generations the British universities learned not to drag their feet at the back of the procession of scientific thought. Today the procession has become a race, with universities as the pacemakers. They have adapted themselves to the contemporary environment; and none have accomplished the adaptation more efficiently than Oxford and Cambridge."

Thank you for that: now we know why Charles Darwin, a gentleman scientist, self funded, worked outside the system. Many people were forced to work outside the system: Copernicus and Kepler are further examples. People are still forced to work outside the system. Why is this?

Back to Ashby to whom we ask this question: there is a *but,* is there not?

"Yes indeed, there is a *but*!"

**266**

Today is the day of the Autumn Equinox. Soon the circle of time will be complete.

Unity of opposites is on our minds once again! David Henderson speaks:

"There are three distinct phases of Cusa's use of coincidence of opposites. The first is a unity of convergence, that is, a falling together, a unity geometrically conceived, but without quantity. It is a unity of substance without mingling and without obliteration of either substance. Other types of union of opposites include instances where one opposite supersedes the other, where two opposites are superseded by creation of a third, and where the elements of the opposites are mingled. The coincidence of opposites is beyond the reach of discursive reasoning. The coincidence of opposites is a unity to which neither otherness nor plurality nor multiplicity is opposed.

"The second phase takes coincidence of opposites to its limits, beyond itself. It is the device by which finite knowing and saying can grasp the incomprehensible and speak the ineffable. It is in fact a principle and therefore discoverable, but no merely invested or contrived as we might comparisons, metaphors, or analogies in ordinary language, sets forth the way God works, the order of things in relation to God and to each other, and the manner by which humans may approach and abide in God.

"In the third phase Cusa embraces the negation of knowing and at the same time, the affirmation of sight."

Carl Jung:

"When Lao Tsu says: 'All are clear, I alone am clouded', he is expressing what I now feel in advance of old age. Lao Tsu is an example of a man with superior insight who has seen and experienced worth and worthlessness, and who at the end of his life desires to return to his own being, into the eternal unknowable meaning. The archetype of the old man who has seen enough is eternally true. At every level of intelligence this type appears, and its lineaments are always the same, whether it be an old peasant or a great philosopher like Lao Tsu. This is old age and a limitation. Yet there is much that fills me: plants, animals, clouds, day and night, and the eternal in man. The more uncertain I have felt about myself, the more there has grown up in me a feeling of kinship with all things. In fact it seems to me as if that alienation which so long separated me from the world has become transferred into my inner world, and has revealed to me an expected unfamiliarity with myself."

The searcher after meaning behind symbols, becoming a symbol?

H. Lawrence Bond again:

"Iconographic language, acknowledging the utter transcendence and mystery of God and communicating the paradoxical truth that God is known, and seen, as made known. The coincidence method in service of mystical theology, therefore performs both an evocation and a descriptive function."

**267**

Propaganda: information, especially of a biased or misleading nature, used to promote a political cause or point of view.

Now we must deal once more with the matter of division of labour, with the idea of specialisation which is so loved by the *Vainglorious Enlightened Ones* – it will be their downfall.

We are back to Anthony Giddens who will now speak about Émile Durkheim's sociology:

"We know that the growth of division of labour goes hand in hand with the disintegration of the segmental type of social structure. For this to occur, relationships must have formed where none previously existed, brining erstwhile separate groups into contact. The differing modes of life and belief of such societies, once they are brought into contact with each other, breaks down the isolated homogeneity of each group and stimulates economic and cultural exchange. Division of labour thus increases as there are more individuals sufficiently in contact to be able to act and react upon one another. Durkheim calls the frequency of such contact moral or dynamic density. The growth of diversified contacts between individuals obviously must derive from some sort of continuous physical relationships. In other words, the growth of dynamic density is mainly contingent upon an increase in the physical density of population. We can then formulate the general proposition that: The division of labour varies in direct ratio with the volume and density of societies, and if it progresses in a continuous manner in the course of social development, it is because societies become regularly denser and generally greater in volume."

Now we to turn to Durkheim's biological explanation for the division of labour. Anthony will explain:

"Durkheim employs a biological explanation in the mode in which he seeks to analyse conflict as a mechanism, within a quasi-Darwinian framework, which accelerates the progression of the division of labour. Darwin and other biologists have demonstrated, according to Durkheim, that the struggle for existence is most acute between organisms of the same type. The existence of such conflict tends to generate complementary specialisation, such that organisms can exist side by side without the one hampering the survival of the other. Differentiation of function allows diverse types of organism to survive. A similar principal, Durkheim concludes, can be applied to human society."

In Durkheim's own words:

"Men submit to the same law. In the same city, different occupations can co-exist without being obliged mutually to destroy one another, for they pursue different objects. The soldier seeks military glory, the priest moral authority, the statesman power, the businessman riches, the scholar scientific renown."

Today we would recognise the biological explanation as Social Darwinism which is no longer acceptable following the Holocaust. The point though is to show that sociologists in the nineteenth century developed social explanations for the social organisation of knowledge, and it would be a folly to ignore such explanations in the creation of a new *Theory of Knowledge*. It seems though that many people are engaging in such a folly, such is the nature of the modern fragmented and reductive mind.

And in the morning, when we did rise, riding across the sky, Orion did we spy, for as foretold, he has returned, and winter comes in his wake, turning full circle, time does return to where it started, and our odyssey does with this turning move to final words – scriptovisualism.

**268**

Emancipation: the fact or process of being set free from legal, social, or political restrictions; liberation.

Now we return to Sir Eric Ashby:

"By the 1870s Oxford and Cambridge were ready to accept experimental science as an ingredient in education; but they were not prepared to jettison their cherished ideals of a liberal education in favour of the ideal of the university as a research institution.

"The vague but none the less attractive idea of the scholar-gentleman is still alive. Liberal education is still the aim of the Oxford tutor, and it has become the aim of the civic university lecturer; for the centripetal attraction of Oxford and Cambridge and their ideas remain a tremendous force in the provincial universities of Britain. British universities have never quite surrendered the idea of *Renaissance Man* to the idea of the *Research Worker*.

"These two powerful social forces - the influence of the utilitarian and the cult of the practical man on the one hand, and the influence of the classical humanists and the cult of the scholar-gentleman on the other - profoundly affected the manner in which British universities adapted themselves to the scientific revolution. Oxford did not respond as Göttingen had responded; Sheffield did not regard science as Munich had regarded it. *Wissenschaft* won its way into British Universities, but it never quite came to terms with the tough pragmatism of lay governors in civic universities or with the tenacious ideals of classical dons in Oxford and Cambridge. To this day the new has not become altogether assimilated into the old; adaptation is not yet complete: and here lies the cause of some of our present problems in universities."

Let us now put this matter another way: one the one hand there are the practical men such as Leonardo da Vinci, and on the other the humanists, the literary class. Or this way: on the one hand there are the practical men such as the applied scientists, and on the other the humanists, the literary class.

Now you know what C.P. Snow was actually saying: he was raising an objection, just as Leonardo had done 500 years before, about that Ancient Greek maxim: free people think, slaves (who use their hands) work. Hence also, the philosophy of Frederick Winslow Taylor in that nonsense called *Scientific Management*.

But what exactly is the nature of the gulf that exists among disciplines? Note that we did not refer to any specific field of work, or domain of knowledge.

The attitude of elitism, of *free people think, slaves work*, is still alive and well, as the case of the former chief scientific advisor to a former president of the European Commission, and the technocracy that is the European Commission, well demonstrate.

Anything further to say Eric?

"Yes. The antithesis between science and humanism has almost vanished, but it has been replaced by another antithesis, equally mischievous. The future historian will record that the unprofitable debates of the 1860s on the humanities versus science were followed in the 1950s by equally unprofitable debates on specialisation versus a liberal education.

"There is an assumption that specialisation and a liberal education are antithetic. What is needed is to challenge the assumption itself. The Oxford Dictionary defines liberal education as education fit for a gentleman. That is still an acceptable definition; it is the idea of a gentleman which has changed. A century ago when Britain awoke to the need for technological education, a gentleman belonged to the leisured class. Modern gentlemen do not belong to the leisured class. Many of them work something like a seventy-hour week, and more and more of them are finding that their business requires expert knowledge. A case could be made therefore for including technology among the ingredients of a liberal education.

"What then is missing in a scientific or technological education? It is not a smattering of art or architecture which is missing, nor is it an acquaintance with history or literature. Indeed it is not a primary lack of subject-matter at all: the fault lies with what Whitehead called a celibacy of the intellect which is divorced from the concrete contemplation of the complete facts. It is a preoccupation with abstractions from reality, an escape from the whole of reality. Thirty years ago in *Science and the Modern World*, Whitehead warned us that this would become the great danger of professional education. Each profession, he said, makes progress in its own groove of abstractions, but there is no groove of abstractions which is adequate for the comprehension of human life."

A.N. Whitehead will now sum up his proposed adaptation which a university needs to make in order to assimilate science and technology:

"There is something between the gross specialised values of the mere practical man, and the thin specialised values of the mere scholar. Both types have missed something; and if you add together the two sets of values, you do not obtain the missing elements.

"When you understand all about the sun and all about the atmosphere and all about the rotation of the earth, you may still miss the radiance of the sunset. There is no substitute for the direct perception of the concrete achievement of a thing in its actuality. A factory, with its machinery, its community of operatives, its social service to the general population, its dependence upon organising and designing genius, its potentialities as a source of wealth to the holders of its stock, is an organism exhibiting a variety of vivid values. What we want to train is the habit of apprehending such an organism in its completeness."

Time to move on!

The future historian will record that the unprofitable debates of the 1860s on the humanities versus science were followed in the 1950s by equally unprofitable debates on specialisation versus a liberal education, followed in the early twenty-first century by an equally unprofitable debate about STEM to STEAM.

**269**

More now about the Principle of Causality; speaking on this matter is Sir Edmund Whittaker, FRS:

"What is a cause? The word belongs to common speech, and doubtless referred originally to the experience that we can by our own spontaneous activity bring about changes in the world; the changes are the *effect*, our activity in producing them, the *cause*. As Earl Russell says, 'The conception of *cause*, however loath we may be to admit the fact, is derived from the conception of *will*.' The notion remained vague and unscientific until Aristotle took the matter in hand and gave his celebrated analysis of causality. His word for cause was somewhat wider than the English word, and signifying everything that contributes to the production of an effect.

"He observed that for the constitution of a new object, four things are requisite, namely, something out of which it is made - this is *Material Cause*; something to give it its specific nature - this is the *Formal Cause*; somebody or something to induce the formal cause in the material - this is the *Efficient Cause*; and a teleological principle or motive, on account of which the efficient cause acts: this is the *Final Cause*. Thus, in the production of a statue, the material cause would be the original block of marble, the formal cause would be the figure chiselled out, the efficient cause would be the sculptor, and the final cause would be the idea or likeness embodied in it.

"The Aristotelian doctrine of causality consists in determination of the relations which these types of cause bear to one another, in the light of the physical knowledge of the time. It confirms the connection of the idea of cause with the idea of will; for in Aristotle's cosmos, the efficient cause of the regular movement of the celestial spheres was affirmed to be intelligence, operating in a way analogous to human volition.

"We must however be on our guard against supposing that the term final cause normally carries with it the notion of an agent whose intention was the constitution of the object; indeed, the idea of final cause in general approximates nearly to that of *function* or *purpose*; thus we may say that the final cause of the eye is sight, without necessarily implying the existence of a designer who formed the purpose of vision. This caution is needed in view of the argument from final causes which is found in Natural Theology, and which becomes confused and faulty unless the connotation of the words *final causes* is correctly understood."

Anthropophagy: the eating of human flesh by human beings. Could it be relevant to scriptovisulism? Wait and see!

Now it is time to hear from Alan Sokal and Jean Bricmont - they are physicists if you are wondering.

"The goal of our 1999 book, *Intellectual Impostures: Postmodern Philosophers' abuse of Science,* was to make a limited but original contribution to the critique of the admittedly nebulous *Zeitgeist* that we have called *postmodernism.* We make no claim to analyse postmodernist thought in general; rather, our aim was to draw attention to a relatively little-known aspect, namely the repeated abuse of concepts and terminology coming from mathematics and physics. We also analyse in the book certain confusions of thought that are frequent in postmodernist writings and that bear on either the content or the philosophy of the natural sciences.

"We use the word abuse to denote one or more of the following characteristics:

"First, holding forth at length on scientific theories about which one has, at best, an exceedingly hazy idea. The most common tactic is to use scientific, or pseudo-scientific, terminology without bothering much about what words actually mean.

"Second, importing concepts from the natural sciences into the humanities or social sciences without giving the slightest conceptual or empirical justification. If a biologist wanted to apply, in their research, elementary notions of mathematical topology, set theory or differential geometry, she would be asked to given some explanation. A vague analogy would not be taken seriously by her colleagues.

"Third, displaying a superficial erudition by shamelessly throwing around technical terms in a context where they are completely irrelevant.

"Fourth, manipulation of phrases and sentences that are, in fact, meaningless.

"Perhaps the abusers imagine that they can exploit the prestige of the natural sciences in order to give their own discourse a veneer of rigour. And they seem confident that no one will notice their misuse of scientific concepts. No one is going to cry out that the king is naked.

"Our goal is precisely to say that the king is naked. We are not attacking philosophy, the humanities or the social sciences *in general*; on the contrary, we feel that these fields are of the utmost importance and we want to warn those who work in them against some manifest cases of charlatanism. In particular, we want to deconstruct the reputation that certain texts have of been difficult because the ideas in them are so profound. In many cases we demonstrate that if the texts seem incomprehensible, it is for the excellent reason that they mean precisely nothing. What we are criticising is the pretension of some celebrated intellectuals to offer profound thoughts on complicated subjects that they understand, at best, at the level of popularisations."

But what exactly is the nature of the gulf that exists among disciplines? Note that we did not refer to any specific field of work, or domain of knowledge.

**271**

Comparable: able to be likened to another; similar; of equivalent quality; worthy of comparison. Complementary: combining in such a way as to enhance or emphasize the qualities of each other or another. Domination: the exercise of power or influence over someone or something, or the state of being so controlled.

We will hear from Sir Eric Ashby, FRS, once again:

"John Dewey pointed out that progress in science ought to manifest itself as an emancipation of the mind enabling it to pursue new ends and new ideals; but in fact progress due to science has been confined to more efficient means of satisfying old ends and old ideals. This criticism is still valid."

Thanks for that observation which was made in 1958. The criticism is still valid in 2016. The problem is in fact now much worse. This of course is a point where you, the reader, start tearing your hair out. Or perhaps there is no hair left, as you have been doing this all the way through the book! If you have not, then congratulations - there is still hope!

Back to Eric:

"Adaptations needed to bring British universities into equilibrium with the age of technology are changes of curriculum; they could be accomplished through the normal channels of university administration and legislation. But they would not be successful unless accompanied by subtle adaptations in academic thought: professors of technology need to be persuaded that the pattern of curriculum under which they themselves were trained is inadequate for their students."

Elsewhere we have called this the Prometheus syndrome; they have been condemned by Zeus. We also call them Narcissus, because they have been cursed by Nemesis!

Back to Eric:

"Professors of arts subjects need to be persuaded that the presence of technology in universities puts them under an obligation to reconsider the emphasis in their own humanistic studies. It is at this point that universities look to their faculties of arts for help; and it is at this point that they are often disappointed. For faculties of arts have themselves become so deeply influenced by science that they seem unable to offer help towards the assimilation of technologists. Instead of contributing to the university what the Victorians understood by a liberal education, some of them are doing with grammar and documents what scientists and technologists can already do with formulae and instruments. This is doubtless profitable for the progress of scholarship in the humanities, but one cannot escape the consequences that humanities cease to be humanising when they are treated that way. It is a sort of treatment which leads to a celibacy of the intellect as inimical to a liberal education in arts as it is in science."

And the involvement of artists in science and technology, that is so much discussed among those who live in the echo chamber that is art-science and art-technology, do the same - very little sign of humanisation will you find! Which is one of the reasons why we do not engage with this noisy lot - and they are noisy. Very noisy!

Hint - flaws in art-science and art-technology, and the art that it gives rise to! If indeed it is art. How to tell? Easy - do not try. Instead ask different questions. We did.

**272**

We have already heard from Winston Churchill. According to the propaganda of the European Parliament, he is supposed to have said that:

"We must build a kind of United States of Europe. In this way, only will hundreds of millions of toilers be able to regain the simple joys and hopes which make life worth living."

What the European Parliament did not however say was that when he said this he was not referring to the United Kingdom, but to France and Germany. If you understand the history of Europe over the past few centuries you will understand why. Perhaps Churchill was tired of the constant need of the British to intervene (dating back to the days of the so-called *Sun King*) to deal with tyrants and dictators?

Here is Jürgen Habermas:

"Churchill is clearly aware of the situation - the ruined landscape of Europe, the mass murders and the crimes against humanity - and speaks first in the conditional by referring to a Europe united in sharing its common inheritance and in doing so creating conditions that will result in there being no limit to prosperity and glory. He then evokes present and future dangers in the form of dark horizons for the approach of some new peril, tyranny and terror, and then offering a surprising answer to the rhetorical question of what could rescue the situation: 'we must build a kind of United States of Europe.' But then Churchill becomes practical."

Winston is here to finish the speech:

"I am now going to say something that will astonish you. The first step in the re-creation of the European family must be partnership between France and Germany. The structure of the United States of Europe will be such as to make the material strength of a single state less important. Small nations will count as much as large ones. Why should they not be a European group which could give a sense of enlarged patriotism and common citizenship to the distracted people of this turbulent and mighty continent?"

Jürgen Habermas now comments on this:

"It may come as a surprise to today's readers that Churchill mentioned the vantage point from which he was speaking only at the end of his address - with his impassioned words, he was appealing as a Briton to Frenchmen and Germans, and to their neighbours. He was addressing those governments and peoples of the continent who would be called the *core* of Europe only half a century later, from the perspective of a friendly and helpful observer. As was still quite obvious at the time, he saw Great Britain, along side America and Russia, as sponsors of the unification process he was recommending."

Does the fact that the European Parliament, along with the European Commission, both engage in propaganda disturb you? It should do!

Here is Jürgen Habermas one more time. This time he is talking about the European Commission Communication called *A Blueprint for a Deep and Genuine Economic and Monetary Union*:

"The proposal of the European Commission accords the expansion of steering capacities priority over a corresponding enlargement of the basis for legitimation also in the medium term. Thus the delayed democratisation is presented as a promise in the manner of a light at the end of the tunnel."

Does the fact that the European Parliament, along with the European Commission, both engage in propaganda disturb you? It should do!

Through means technocratic they will set about creating a united Europe, then at the end they will create democracy. But by that time the European Commission will have convinced itself that democracy is not needed. This mind-set will be reinforced by the social, political and economic conditions that will have taken shape, brought about by a European culture, and its imposition in the non-European world, that makes a virtue of living life like a plague of locusts.

The European Commission - are they creating a United States of Europe, or a Union of Subservient State Republics? You will recall that the USSR had a parliament, and would have claimed to be a state governed by the principle of the Rule of Law. The European Union also has a parliament and would claim to be governed according to the principle of the Rule of Law.

By the time the peoples of continental Europe discover what their fate is to be, it will be too late for them to do anything about it, other than to resort to force, to civil strife, to bloody civil war - it will be called the *War of European Union*. Thankfully the people of the United Kingdom will not be part of it, but we will once again have to face the ruined landscape of Europe, mass murders, and crimes against humanity, and once again we will ask: what is to be done about the problem that is Europe?

Of course it does not have to be this way, just as the Second World War did not have to be.

The problem is this: you will not listen! You never do! There is no point is us trying! Even if we were a highly respected, highly experienced, elder Statesman, you still would not listen. This is why we are what we are. This is why we know no one, have no influence, advise no one. We do not need to!

**273**

Now to matters of the *Natural Law* we turn, and in particular to the *Natural Law* in the Chinese tradition. Here to speak about this is Hu Shih who was Fellow of the Princeton University Library, Curator of the Gest Oriental Library at Princeton University, Ambassador of China to the United States in the period 1938-1942, and President of the National Beijing (Peking) University from 1946 to1949. He is also the author of *History of Chinese Philosophy* and the *History of the Living Literature of China*:

"I consider here this question: Did China in her long history develop any moral or juridical concept or concepts which may be compared to what has been known as *Natural Law* or *the Law of Nature* in the European, and particularly the Anglo-Saxon juristic and constitutional tradition?

"My answer is this: None of the concepts developed in China was able to achieve the objective of checking or limiting the absolute powers of the unlimited monarchy. No concept of the Natural Law alone can however ever achieve that objective, in China or in any other country on earth. But the Chinese experience is worth knowing because it confirms and verifies an historical thesis, namely, that the concept or concepts of Natural Law or Natural Right have always played the historical role of a fighting weapon in mankind's struggle against the injustice and the tyranny of unlimited human authority."

Now one may ask which concepts? There are many. Some of you will have heard of them.

Hu Shih again:

"All social and political thinking usually begins as a criticism of existing government, laws and institutions which have become unsatisfactory, harmful or oppressive. In passing adverse judgment on time-honoured institutions sanctioned by the political authority of the state and in proposing new and possibly radical theories or measures of reform, it has always been necessary for critics and reformers to appeal to some authority higher and more trustworthy than the highest political or ecclesiastical authority of the time.

"This is true at least of the long history of Chinese social and political thought, in which the student can discern a number of superficially different but essentially similar patterns in an endeavour to appeal to a higher law or a higher authority.

"In the Chinese tradition, this appeal to a higher authority has taken these main forms: Sometimes it takes the form of appealing to the authority of an imagined and quite freely idealized antiquity - the Golden Age of the ancient sage-rulers; Sometimes it takes the form of appealing to the Will of God (*t'ien-chih* or *t'ien-i*) as the highest norm or law; Sometimes the appeal is made to the Way (*Tao*) of Heaven or Nature, which is the Law of Nature; Sometimes, especially under the long, long centuries of the vast unified empire, the appeal is made to the authority of the Canon (*ching*, meaning the invariable, immutable way) of the Sacred Scriptures of Confucianism as the highest authority on all matters of moral and political justice; Sometimes the appeal is made to Reason or Law or Universal Reason or Natural Law (*li*, or *tao-li*, or *t'ien-li*) as it is intuitively evident in the moral conscience of men – what may be termed in the words of Coke *the common right and reason* of man.

"All these bear some essential resemblance to the historical appeals in the Western world to Natural Law and Natural Right. Even the first of these patterns, namely, the idealisation of remote antiquity as the Golden Age, is not so strange or so unreasonable when one recalls the numerous *utopias* designed by the social and political thinkers of the West, and especially when one recalls that the doctrine of Natural Rights was originally conceived as rights of men *in the state of nature* before they entered into the Social Compact or Social Contract. The Chinese thinkers, notably the Confucianists, who read their ideal social and political order into the remote reigns of sage-rulers, were merely inventing their *utopias* and populating them with supposedly historical personages, such as Yao, Shun and Yi, whom traditional chronology placed in the third millennium BC.

"When Confucius said: 'If there was any ruler who did nothing (wu-wei), yet governed well – was it not Shun? For what in effect did Shun do? Religiously self-disciplined, he sat reverently on the throne, and that was all'; he was eulogising the political ideal of non-interference or *laissez-faire* and making it more real by projecting it into the ancient reign of Shun, of whom we know as little as of the state of nature of Locke or of Rousseau. And when Mencius and the other political philosophers of ancient China vividly and sometimes dramatically described how the great sage-ruler Yao handed down the throne and the empire, not to his own son, but to Shun, the wisest man of the age and the choice of the people; and how Shun, in his turn, handed down the throne and the empire, again not to his own son, but to the great Yiu who had controlled the Great Flood and was the choice of the people – they were not deliberately fabricating history, but were merely using their utopian ideals to voice their own criticism of the evils of the hereditary monarchy and were covertly advocating a new and radical system of selection of the worthiest men to be rulers.

"The above is a brief explanation of what may seem to have been a peculiarly Chinese appeal to the authority of the utopian antiquity. I take up these matters in greater detail in my paper, where I describe four concepts in Chinese thought which in my humble opinion have played an historical role not unlike that of the Natural Law concepts of the Western World. They are: the concept of the way (T*ao*) of Heaven or Nature as taught by Lao Tsu; the concept of the Will of God (*t'ien-chih*) as taught by Mo Ti; the concept of the Sacred Canon (*ching*) as developed in medieval China; the concept of Reason or Law (*li*) or Universal Reason or Law (*t'ien-li* or *tao-li*) – Natural Law in the sense of *common right and reason* – as developed in relatively modern times."

Thus do we appeal to Natural Law as a fighting weapon against the injustice and the tyranny of unlimited human authority that is coming into being, at this early stage in its formation, making a critique of existing European Union government, laws and institutions which have become unsatisfactory, harmful or oppressive. Now is the time for action - not after the events that will unfold. Thus also do we begin to deal with the problem that STEM and its practitioners are becoming, as well as those art-science lovers! Everything is connected!

**274**

As we are still in the European Parliament, at least in spirit, and some MEPs are available to speak, it is a good time to hear from representatives of two of the groupings that were formed following the 2014 election.

First to speak is a member of the group called Europe of Freedom and Direct Democracy, otherwise called EFDD:

"The EFDD group is the only genuine opposition within the European Parliament. Its members were clearly elected on their rejection of the integrationist policies pursued by the European Commission in Brussels. A European Union run by technocrats, experts, officials, judges and bankers is the beginning of a terrifying anti-democratic era. The EFDD group is opposed to the pernicious centralisation of power in the hands of the unelected in Brussels."

The Soviet Union had a parliament - the Supreme Soviet. So you can rest easy and know that the spokesperson for the EFDD group can be ignored!

Now we will hear from a representative of the European United Left/Nordic Green Left alliance, otherwise called the GUE/NGL:

"The GUE/NGL does not support the European Union leaders' austerity drive which continues to work for the rich and powerful by maintaining an economic system that puts the needs of financial markets ahead of social concerns. The austerity measures being applied by undemocratic means in the European Union countries worst-hit by the crisis reflect the failure of the European Union's neo-liberal architecture. We urgently need to reverse this policy and deliver solidarity and job creation within an economic system that serves people not profit. Austerity is avoidable; it is a political choice that puts the interests of the privileged above the interests of the majority.

"Free trade between two partners does not always make both better off. Liberalisation of trade has caused worldwide social and ecological destruction. We need to control free trade and the free circulation of goods. We are opposed to all international trade agreements that are shaped by the interests of big business. We must change how we define trade relations.

"The European Union must play an active role in the fight to eradicate extreme poverty and hunger in the developing world. We fight for sustainable development, access to education, improving maternal health, and we support the fight against malaria and AIDS. With European Union trade policy putting developing countries in a position where they are evermore subservient to European Union trade demands, we call for a trade policy that is based on development aims and not the other way round."

Thanks for that insight.

European United Left/Nordic Green Left alliance presents a vision of a European Integration process based on solidarity, workers' rights, peace, gender equality, civil liberties, democracy and environmental responsibility.

This will of course not happen because the European Union is not based on the implementation of the will of the people.

Here we mention that Jürgen Habermas is a supporter of the European integration process, of the idea of trans-national democracy. His critique of the lack of democracy and the growth of technocracy in Brussels concurs with our own concerns, as we explained in our book *STARTS – Science Technology and the Arts: The Artistic Voices that DG CONNECT Silenced*. We quote here the visual poem that we wrote for that book:

*STARTS*

*What STARTS*
*The road to serfdom*
*Is a technocratic failing*
*CONNECTED with the past*
*Which in DG CONNECT*
*Can be found predominating*
*So not in the present*
*Nor the future*
*Is it ever found operating*
*Thus DISCONNECTED DG CONNECT*
*Now stands condemned*
*At the END*

*ENDS*

Be in no doubt, STARTS is operated in support of the neo-liberal European Union that is the cause of so much suffering in Europe. Once again artists are helping those in power to construct a system that is unjust and oppressive. It is also riddled with corruption. We mention here the Panama Papers with its links to the names of European Commissioners, past and present. Junker himself is deeply implicated in the creation of the tax-haven that is Luxembourg.

Credibility: the quality of being trusted and believed in; the quality of being convincing or believable. Credibility gap: an apparent difference between what is said or promised and what happens or is true.

Unlike Jürgen Habermas we know that the situation is now too far gone to change course. And thus the next European nightmare takes shape, that future danger in the form of dark horizons for the approach of some new peril, tyranny and terror that Winston Churchill spoke of.

You build the democratic institutions first not last! When will Europeans stop behaving like Europeans?

The history of Europe is a history of European Unions.

**275**

Now we deal with the love affair that is art-science, but mostly art-technology. And we hear first some words of truth, for not always truth do lovers speak.

"The discourse on art-science has been dominantly influenced by the natural sciences. Whither social science in this articulation?"

'Tis true. That was Sundar Sarukkai speaking.

"Art problematises science on grounds similar to the social sciences; all the more reason to explore the contours of the social science response to science, in the context of the art-science discourse, perhaps creating a much-needed art-social-science complex."

'Tis true. That was Sundar Sarukkai speaking for a second time.

'Tis true also that social science would critique this love affair called art-science, which is mostly art-technology.

Now William Shakespeare reads from one of his works:

> "Two households, both alike in dignity,
> In fair Verona, where we lay our scene,
> From ancient grudge break to new mutiny,
> Where civil blood makes civil hands unclean.

From forth the fatal loins of these two foes
A pair of star-cross'd lovers take their life;
Whose misadventured piteous overthrows
Do with their death bury their parents' strife.
The fearful passage of their death-mark'd love,
And the continuance of their parents' rage,
Which, but their children's end, nought could remove,
Is now the two hours' traffic of our stage;
The which if you with patient ears attend,
What here shall miss, our toil shall strive to mend."

And some more please, Will, if you would:

"I Would. Then do I say this:

"ROMEO:

"Draw, Benvolio; beat down their weapons. Gentlemen, for shame, forbear this outrage! Tybalt, Mercutio, the prince expressly hath Forbidden bandying in Verona streets: Hold, Tybalt! good Mercutio!

"MERCUTIO:

"I am hurt. A plague o' both your houses! I am sped. Is he gone, and hath nothing?

"BENVOLIO:

"What, art thou hurt?

"MERCUTIO:

"Ay, ay, a scratch, a scratch; marry, 'tis enough. Where is my page? Go, villain, fetch a surgeon.

"ROMEO:

"Courage, man; the hurt cannot be much.

"MERCUTIO:

"No, 'tis not so deep as a well, nor so wide as a church-door; but 'tis enough, 'twill serve: ask for me to-morrow, and you shall find me a grave man. I am peppered, I warrant, for this world. A plague o' both your houses! 'Zounds, a dog, a rat, a mouse, a cat, to scratch a man to death! a braggart, a rogue, a villain, that fights by the book of arithmetic! Why the devil came you between us? I was hurt under your arm.

"ROMEO:

"I thought all for the best.

"MERCUTIO:

"Help me into some house, Benvolio, Or I shall faint. A plague o' both your houses! They have made worms' meat of me: I have it, and soundly too: your houses!

"I am done reading. 'Tis enough."

Now swim to another shore, another time, and comes forth George Orwell reading the last paragraph from Animal Farm:

"Twelve voices were shouting in anger, and they were all alike. No question, now, what had happened to the faces of the pigs. The creatures outside looked from pig to man, and from man to pig, and from pig to man again; but already it was impossible to say which was which."

Ay! Well do we see the point! Do you?

**276**

What is a proof? The Oxford English Dictionary defines the word proof as follows: evidence or argument establishing a fact or the truth of a statement.

Sir Edmund Whittaker, FRS, will now speak about proofs:

"It is proper at this point to make some comments on proofs as a whole. A proof has been defined as a process by which A induces in B a sense of the justification for a conviction. The highest type of proof is the kind that is found in pure mathematics, such as proof of Lagrange's theorem that any number can be expressed as the sum of four squares.

"These proofs depend on no premise whatever, except purely logical propositions, which are universally accepted as the necessary presuppositions of reasoning; they constitute, therefore, inescapable demonstrations which have a coercive power over the minds of sane persons.

"The proof of assertions regarding the results of scientific experiments, such as the statement that water is composed of hydrogen and oxygen, are of an entirely different character from those of pure mathematics. The proof of the former class involves experiments that can be repeated, from which we can conclude that the result happens invariably.

"The process by which a schoolmaster induces in his pupils the belief that Julius Caesar invaded Britain in 55 BC are obviously of a different character again, the credibility of testimony now being prominent."

Self evidently none of the above types of proof can be used to prove the existence of God. Attempts to prove the existence or non-existence of God demonstrate a lack of understand about God and religion. Any sane person would understand this. Why therefore did not someone not explain this to Richard Dawkins, along with those people who described his book *The God Delusion*, as magisterial? It is nothing of the kind. It is intellectual nonsense. It should be re-titled *The Science Delusion*! Perhaps better: *My Struggle against Religion*. It needs a shorter more pithy title! *My Struggle* will do. The German edition would be *Mein Kampf*.

Different disciplines have different methods, different approaches, different types of understandings of what constitutes proof, of what constitutes evidence. And there are different silent narratives as well.

See, there is more to disciplines, than just subject knowledge! Has no one ever told you this?

With the fact that there are different types of proof, established, we must now move on, for time is pressing, and towards the end we are inexorably bound - and the end is 366. Let he who has understanding count the number of chapters, for it is the limit.

**277**

Returning to the matter of scriptovisualism, there is a word that is of interest, and the word is *remediation*.

One meaning of this word is the representation of one medium in another - transformation. So we are in the inter-media space, which is where we want to be - the spaces between spaces.

Back to that word anthropophagy. Here is Brazilian poet Oswald de Andrade reading from his *Manifesto Antropofago*:

"The struggle between what we might call the Uncreated and the Creation - illustrated by the permanent contradiction between Man and his Taboo. Everyday love and the capitalist way of life. Cannibalism. Absorption of the sacred enemy. To transform him into a totem. The human adventure. The earthly goal. Even so, only the pure elites managed to realize carnal cannibalism, which carries within itself the highest meaning of life and avoids all the ills identified by Freud - catechist ills. What result is not a sublimation of the sexual instinct? It is the thermometrical scale of the cannibal instinct. Carnal at first, this instinct becomes elective, and creates friendship. When it is affective, it creates love. When it is speculative, it creates science. It takes detours and moves around. At times it is degraded. Low cannibalism, agglomerated with the sins of catechism - envy, usury, calumny, murder. We are acting against this plague of a supposedly cultured and Christianized peoples. Cannibals."

Here is Lesley Barrie to comment:

"The *Manifesto Antropofago* has, especially in the last twenty years, been widely cited in Brazil as a paradigm for the creation of a modern and cosmopolitan, but still authentically national culture. In the earlier (1924) *Manifesto da Poesia Pau-Brasil* (*Manifesto of Brazilwood Poetry*), Oswald had announced an *export-quality* poetry that would not copy imported aesthetic models but find its material in Brazilian history, popular culture, and everyday life. Brazilwood poetry will provide, Oswald says here, '[...] the counter-weight of native originality to neutralize academic conformity'. Opposing the avant-garde notions of poetry as *invention* and *surprise* to the erudite, imitative art he associates with the colony and the Brazilian Empire (1822-1889), Oswald unites the search for national identity with the modernist aesthetic project. In this schema, Brazilian cultural production becomes both native and cosmopolitan. Brazil's *wild wilderness*, far from generating second-rate copies of Continental models, will give rise to an agile and candid modern poetry. Brazilwood poetry thus offers a solution for Brazil's perceived cultural inferiority, and at the same time injects new life into the international cultural arena. The playful, polemical theory of cultural *cannibalism* Oswald develops in the *Manifesto Antropofago* is a radicalization of these ideas. The *Manifesto Antropofago* challenges the dualities - civilization or barbarism, modern or primitive, and original or derivative - which had informed the construction of Brazilian culture since the days of the colony. In the *Manifesto Antropofago,* Oswald subversively appropriates the colonizer's inscription of America as a savage territory which, once civilized, would be a necessarily muddy copy of Europe. The use of the cannibal metaphor permits the Brazilian subject to forge his specular colonial identity into an autonomous and original (as opposed to dependent, derivative) national culture. Oswald's anthropophagist - himself a cannibalization, not of Rousseau's idealized savage but of Montaigne's avowed and active cannibal - neither apes nor rejects European culture, but *devours* it, adapting its strengths and incorporating them into the native self."

Roberto Simanowski speaks:

"The fusion of text and image goes back as far as the illuminated manuscripts of the Middle Ages and figurative poems of antiquity. It has been exploited in the twentieth century not only in concrete poetry, graphic design and advertising but also in the *Art and Language* movement.

"Images can be read as text and texts can be read for their imagery. Transformation of text into image is something different and digital technology makes it easy to consume text by obscuring it or rendering it less legible, or even illegible."

One might therefore say cannibalism!

**278**

Today it is Surrealism that is on our mind. So to the matter of Surrealism we do turn. First we hear from Gavin Parkinson:

"A reworking of philosophical dualism under the guidance of quantum mechanics penetrated Surrealist circles by the mid 1930s. Questioning the Cartesian-Kantian separation of subject and object, it offered an alternative duality to the Hegelian dialectic resolved through synthesis, as in the unification of opposites demanded by Breton in the *Second Manifesto of Surrealism*."

André Breton is here to speak:

"Surrealism, although a special part of its function is to examine with a critical eye the notions of reality and unreality, reason and irrationality, reflection and impulse, knowledge and *fatal* ignorance, usefulness and uselessness, is analogous at least in one respect with historical materialism in that it too tends to take as its point of departure the *colossal abortion* of the Hegelian system. It seems impossible to me to assign any limitations – economic limitations, for instance – to the exercise of a thought finally made tractable to negation, and to the negation of negation. How can one accept the fact that the dialectical method can only be validly applied to the solution of social problems? The entire aim of Surrealism is to supply it with practical possibilities in no way competitive in the most immediate realm of consciousness. I really fail to see – some narrow-minded revolutionaries notwithstanding – why we should refrain from supporting the Revolution, provided we view the problems of love, dreams, madness, art, and religion from the same angle they do. Now, I have no hesitation in saying that, prior to Surrealism, nothing systematic has been done in this direction, and at the point where we found it the dialectical method, in its Hegelian form, was inapplicable for us too. There was, for us too, the necessity to put an end to idealism properly speaking, the creation of the word *Surrealism* would testify to this, and, to quote Engels' classic example once again, the necessity not to limit ourselves to the childish: 'The rose is a rose. The rose is not a rose. And yet the rose is a rose.' "

Letters no longer, it has been said, serve the purpose for which the alphabet was developed: to be used to create words with meaning. Thus then what do they become? So we are back to the matter of scriptovisualism. When words with meaning disappear, what must take their place, is meaning by other means.

**279**

Let us now return to the matter of C.P. Snow, the scientist - actually a scientist who gave up science - who chose to make sociological remarks without first understanding, it seems, what sociologists have to say, leading to a lot of nonsense that people in the world of STEM and the arts just repeat, parrot like, for evidently they have either not read Snow's documents or have read them and not understood, they too being people who have not bothered to address the knowledge that sociologists bring to the world. So much for the often talked about, but little realised mutltidisciplinary thinking, or crossdisciplinary thinking, or interdisciplinary thinking, or transdisciplinary thinking - we have already explained what these actually mean. Or might mean!

And the example that proves the nonsense of Snow's opinions is Surrealism - the art movement that engaged with Relativity and quantum mechanics. It is a history that has already been written about - try reading about it. As far art-science - nothing new there too! Here is Gavin Parkinson:

"In the 1930s the city of Paris was preparing for the massive *Exposition International des arts et des techniques* of 1937. Designed to give notice of France's complete recovery from the trauma of the Great War, the exposition placed France at centre stage in matters of art, science and technology. The physicist Jean Baptiste Perrin presided over a major congress in the year of the Exposition, on biology, chemistry and physics. Attended by twelve hundred scientists from sixty-five countries, it must have seemed as though France had made up all the ground lost since the International Congress of Physics met in Paris in 1900. Keen to *emphasise close links between art and science,* the Exposition included a vast exhibition entitled *L'art et la science* held at Perrin's creation in the Palais la Decouverte."

Thus we now despatch Snow to the dustbin of history, noting that he serves as a warning of the difficulties of crossing disciplinary boundaries, for such requires a vast amount of intellectual effort that Snow was not willing, or, perhaps was unable to undertake. He had no interest in doing so, for he was engaging in technocracy, his being a fragmented and reductive mind.

On television, in the evening, a programme - *A World Without Down's Syndrome?* Here is the actor and writer and presenter of the programme, Sally Phillips, who has a child with Down's Syndrome.

"I was expecting tragedy, but I go comedy. Like most people with a Down's Syndrome child, I feel that the experience has benefited my family. What baffles me is why the medical profession leads mothers who screen during pregnancy to regard the possibility of Down's Syndrome as a catastrophe, a worst-case scenario. My lived experience bears no relation to that.

"The idea that my child's life is not worth living angers me. For the health service what matters is giving parents choice, but for me it comes down to a different kind of choice: What kind of society do we want to live in, and who do we think should be allowed to live in it?"

Now it is 1936 and we are in Nazi Germany. Now it is a time long ago, thousands of years ago, and we are in Ancient Greece. We have come back to the matter of genetic genocide.

"Would-be parents who learn their unborn child has Down's Syndrome have an ethical responsibility to abort it and try again."

That was Richard Dawkins speaking. So now it is immoral not to kill an unborn Down's Syndrome child! Science and reason! Who will be next on the extermination list?

In the United Kingdom, a termination of an unborn Down's Syndrome child can be legally done at any stage up to full term. At close to full term, the baby is killed while in the womb, then the birth is induced. It is legalised murder. If it were done the other way round - induction followed by killing - it would be murder.

The medical profession is expected to undertake this legalised killing. It is not a nice thing to witness. Would Dawkins be willing to get his *hands dirty* and do the deed? Out of sight, out of mind! Or perhaps he would get his hands dirty - another opportunity for proselytising, perhaps with an artist in residence.

This is what the *Vainglorious Enlightened Ones* call progress. We call it genetic genocide.

**280**

Time now for some spirituality! We said spirituality not religion - that will be mentioned, no doubt, at some future time.

What is of interest to us today, is spirituality in art. Here is Wassily Kandinsky speaking about the art of spiritual harmony:

"Every work of art is the child of its age, and, in many cases, the mother of our emotions. It follows that each period of culture produces an art of its own which can never be repeated. Efforts to revive the art principles of the past will at best produce an art that is still-born. It is impossible for us to live and feel, as did the ancient Greeks. In the same way those who strive to follow the Greek methods in sculpture achieve only a similarity of form, the work remaining soulless for all time. Such imitation is mere aping. Externally the monkey completely resembles a human being; he will sit holding a book in front of his nose, and turn over the pages with a thoughtful aspect, but his actions have for him no real meaning.

"There is, however, in art another kind of external similarity which is founded on a fundamental truth. When there is a similarity of inner tendency in the whole moral and spiritual atmosphere, a similarity of ideals, at first closely pursued but later lost to sight, a similarity in the inner feeling of any one period to that of another, the logical result will be a revival of the external forms which served to express those inner feelings in an earlier age. An example of this to-day is our sympathy, our spiritual relationship with the Primitives. Like ourselves, these artists sought to express in their work only internal truths, renouncing in consequence all consideration of external form.

"This all-important spark of inner life to-day is at present only a spark. Our minds, which are even now only just awakening after years of materialism, are infected with the despair of unbelief, of lack of purpose and ideal. The nightmare of materialism, which has turned the life of the universe into an evil, useless game, is not yet past; it holds the awakening soul still in its grip. Only a feeble light glimmers like a tiny star in a vast gulf of darkness. This feeble light is but a presentiment, and the soul, when it sees it, trembles in doubt whether the light is not a dream, and the gulf of darkness reality. This doubt, and the still harsh tyranny of the materialistic philosophy, divide our soul sharply from that of the Primitives. Our soul rings cracked when we seek to play upon it, as does a costly vase, long buried in the earth, which is found to have a flaw when it is dug up once more. For this reason, the Primitive phase, through which we are now passing, with its temporary similarity of form, can only be of short duration.

"These two possible resemblances between the art forms of to-day and those of the past will be at once recognized as diametrically opposed to one another. The first, being purely external, has no future. The second, being internal, contains the seed of the future within itself. After the period of materialist effort, which held the soul in check until it was shaken off as evil, the soul is emerging, purged by trials and sufferings. Shapeless emotions such as fear, joy, grief, etc., which belonged to this time of effort, will no longer greatly attract the artist. He will endeavour to awake subtler emotions, as yet unnamed. Living himself a complicated and comparatively subtle life, his work will give to those observers capable of feeling them lofty emotions beyond the reach of words.

"The observer of to-day, however, is seldom capable of feeling such emotions. He seeks in a work of art a mere imitation of nature which can serve some definite purpose (for example a portrait in the ordinary sense) or a presentment of nature according to a certain convention (*impressionist* painting), or some inner feeling expressed in terms of natural form (as we say – a picture with *stimmung*). All those varieties of picture, when they are really art, fulfil their purpose and feed the spirit. Though this applies to the first case, it applies more strongly to the third, where the spectator does feel a corresponding thrill in himself. Such harmony or even contrast of emotion cannot be superficial or worthless; indeed the *stimmung* of a picture can deepen and purify that of the spectator. Such works of art at least preserve the soul from coarseness; they *key it up*, so to speak, to a certain height, as a tuning-key the strings of a musical instrument. But purification, and extension in duration and size of this sympathy of soul, remain one sided, and the possibilities of the influence of art are not exerted to their utmost."

**281**

Experiments in Art and Technology – E.A.T. We are standing in Arts Catalyst Centre for Art, Science and Technology in London, surround by displays about – E.A.T.

Billy Klüver speaks:

"I was born in Monaco and grew up in Sweden. After graduating from the Royal Technical University in Stockholm, I came to the United States, received a PhD in Electrical Engineering at the University of California in Berkeley, and joined the Communications Research Department at Bell Laboratories in 1958 doing theoretical and experimental research on free electron beams and lasers. As a student in Sweden, I had been interested in film, and I soon began to meet artists and filmmakers in New York.

"I was concerned about the social implications of science and technology, and I came to believe that a meeting of art and technology and science could only take place on a practical, physical level."

György Kepes was also concerned about the social implications of science and technology, as are social scientists. Yet you brush-off Kepes' approach and ignore social scientists. Why?

Already, reading the exhibition displays, we are sensing some differences between what is on display in the Arts Catalyst Centre and what is stated in other documents. The Arts Catalyst exhibition *Experiments in Art and Technology,* was being staged 50 years on from the event known as *Nine Evenings: Theatre and Engineering*.

A spokesperson from Arts Catalyst speaks:

"The descriptive panels presented in our exhibition are a selection from *The Story of E.A.T.*, an image and text series that chronicles the first attempts at initiating crossdisciplinary collaboration between artists and engineers."

Interesting word used here - *crossdisciplinary*. And - *first*! We may come back to the matter. Another interesting word used - collaboration. We may come back to that matter too. For now we will just note that what is said, about first attempts at initiating crossdisciplinary collaboration between artists and engineers - you are wrong about this!

So back to the Arts Catalyst Centre and we are wondering who wrote these panels? We, as we are want to, start asking questions, and then came this answer:

"This is all spoken in the voice of an engineer."

We note the stress on the fact that it is an engineer speaking.

The answer we obtain is interesting, but is not an answer to the question we asked, which was: who wrote the material?

Then we are told it was Billy Klüver's wife, who we are then told is an artist. So in fact they are not Klüver's words at all, but those of his wife, or Klüver's words filtered by his wife. The panels are part of the E.A.T. archive. The phrase, *retrospective illusion of a reconstructed coherence,* comes to mind.

Elsewhere we have encountered Klüver's words, filtered by a journalist - they are not entirely the same as the words in the Arts Catalyst exhibition. We photograph every panel with the intention of adding them to our own archive and reading them later. We did.

In the evening, we are in the University of Arts London, for the Arts Catalyst's event *Nine Evenings: Theatre and Engineering Revisited.* Actually it is a new performance commission, called *Side Effects,* by Robert Whitman, a participating artist in the original *Nine Evenings* event and co-founder of

*Experiments in Art and Technology* which, it has been stated, produced *Nine Evenings*. Only they did not, as, according to the exhibition material, from the E.A.T archives, E.A.T. was founded after *Nine Evenings.* Another inaccuracy! Nevertheless *Nine Evenings* was a collaboration among artists and people who may have been engineers, or possibly scientists, or both. We are being pedantic and why not?

We enjoyed the performance. Later we see some similarity with Whitman's contribution to the original *Nine Evenings.* Yet was the new piece based on collaboration between artists and engineers for there is nothing evident in *Side Effects* that required an engineer (or a scientist)? Technology has moved-on! What we see could have been created by a teenager with a mobile phone, web-cams, and a personnel computer with appropriate open-source software. Now we are doing what we do best – transcending boundaries, entering deep within their territory, putting the world together in a different way, putting the artists under the microscope. It is all about human behaviour. So we are back to the social sciences.

We enjoyed the performance. It was art. Dare we say it? Yes we will – it was good art for the 1960s. Why? But is it good art here in 2016, 50 years on from 1966?

Later, we read a brochure handed out at the performance. This has the title *Transdisciplinary Collaboration, Exhibiting Archives, Re-staging Exhibitions,* from which we learn that one way of remembering exhibitions is called *Revisiting,* and this involves recreating the original exhibition in a contemporary way by using the key concepts of the original exhibition but exhibiting different works, often commissioned brand new works – hence *Side Effects* by Robert Whitman.

We also learn from this brochure, that within the text, the word transdisciplinary is not used! We are told that *Nine Evenings* was one of the first large-scale cross-disciplinary collaborations between artists, engineers, and scientists. Then we are told that *Nine Evenings* was one of the first large-scale interdisciplinary collaborations between artists, engineers, and scientists. Then we see responses to questions about cross-disciplinary working, responses posed with the word *multidisciplinary.* This is the problem with adjectives – they are so easy to drop into any text.

The original *Nine Evenings* was also subject to criticisms by one of the engineers involved (or was he a scientist?), but no mention of this in the exhibition. The truth about collaborations, whether among people from the world or art and STEM, or among the latter and the social sciences, is rarely ever revealed, and thus little is learned about this most important of matters.

Here is a spokesperson from Tate Modern speaking in advance of the Robert Rauschenberg retrospective:

"Rauschenberg had a lifelong engagement with collaboration."

So what would E.A.T. have been without Robert Rauschenberg? Billy Klüver provided a possible answer to this question in an earlier chapter.

E.A.T. - another example too of the need to understand how different disciplines operate, in particular, the historian, who should not accept at face value, what is said in the E.A.T. archives, or Art Catalyst's interpretation. Another example too, perhaps, of looking at the past with a mind firmly rooted in the present?

Stephen Wilson speaks:

"Over the years I have been a developer for new technology companies, consultant to NSF research projects, and artist-in-residence at research centres. My own art works have often focused on emerging technologies still in their formative stages. The involvements have been judged valuable by me and my collaborators; yet it is often difficult to trace concrete results."

So what utility E.A.T.?

**282**

Today we stood in Tate Britain - we visited the Turner Prize Exhibition. It was clear who would win - Helen Marten.

As we recently mentioned religion and it is on our mind, let us mention religion. Rabbi Jonathan Sacks will now speak:

"The West owes its development to two cultures, Ancient Greece and ancient Israel, Hellenism and Hebraism, the heritages respectively of Athens and Jerusalem.

"Jesus spoke Aramaic and Hebrew, but every book of the New Testament was written in Greek. The New Testament is, through and through, a Greek document, not a Hebrew or Aramaic one. So we have a unique situation of a religion whose sacred texts are written in what to its founder would have been a foreign and largely unintelligible language.

"In the first decades after Jesus' death the Church might well have gone in either of two directions. It might have become a Jewish sect, the Jerusalem church, under Jesus' brother James. In the event however, it was Paul who found a ready audience among the Hellenistic Gentiles of the

Mediterranean. It was the Greek, not the Hebrew-and-Aramaic-speaking population that proved to be the fertile soil in which Christianity took root.

"The Christianity that eventually emerged from the tradition of Paul, Augustine, Anselm and Aquinas had strong Judaic elements, but it also contained strands that were undeniably Greek and in striking contrast to the way that Jews read the Hebrew Bible.

"One of these contrasts is *dualism*. To a far greater extent than Judaism, Christianity after Paul develops a series of dualisms between *body and soul, the physical and the spiritual, earth and heaven, this life and the next,* with the emphasis on the second of each pair. There is nothing like this in Jewish literature. To be sure there is the evil inclination, but no suggestion that because of our embodied condition we are slaves to sin. The entire set of contrasts - soul as against body, the afterlife as against life - is massively Greek with much debt to Plato and traces of Gnosticism."

So dualism and the notion of separation!

"My method of interpretation and analysis suggests ways in which artistic inquiry may enhance social science research methods."

That was Victoria Brattström speaking.

**283**

Is man a machine? Old terminology, so rephrasing the question: are human beings machines governed by the rule of cause and effect - by causality? The *Vainglorious Enlightened Ones* - strange people - *believe* that this is the case. Note the word that we use - *believe*!

So here now is a famous biologist, Jacques Monod, to convince you, that you must believe this, otherwise you are not thinking in a rational and objective way, and are just being superstitious. So here are the rational and objective superstitions of modern science:

"The cornerstone of the scientific method is the postulate that nature is objective. In other words, the systematic denial that *true* knowledge can be reached by interpreting phenomena in terms of final causes - that is to say of *purpose*. An exact date may be given for the discovery of this canon. The formulation by Galileo and Descartes of the principle of inertia laid the groundwork not only for mechanics but for the epistemology of modern science, by abolishing Aristotelian physics and cosmology. Certainly, neither reason, nor logic, nor observation has been lacking among Descartes' predecessors. But science as we understand it today could not have been built upon those foundations alone. It still needed the strict censorship implicit in the postulate of objectivity - pure and impossible to demonstrate. It is obviously impossible to imagine an experiment which could prove the *nonexistence* anywhere in nature of a purpose, of a pursued end."

Thus have you just demonstrated why Leonardo da Vinci was not a scientist, putting aside the fact that the word itself did not exist in Leonardo's day. But this is a side issue, for the main purpose of our invitation to speak, was to repeat the point, that the choice between causality and purpose is - a choice. It is the expression of a belief, based on values. Neither causality nor purpose can be proved to exist, or not to exist.

Please continue, Jacques.

"But the postulate of objectivity is consubstantial with science; it has guided the whole of its prodigious development for three centuries. There is no way to be rid of it, even tentatively or in a limited area, without departing from the domain of science itself."

Which of course, is not true! It is again but a *belief* being expressed, based on values. An article of faith so to speak, that has become part of the dogma of science - the sole truth.

We interrupted you for a second time. Sorry about that. Please now continue.

"Objectivity nevertheless obliges us to recognise the teleonomic nature of living organisms, to admit that in their structure and performance they act projectively - realise and pursue a purpose. Here therefore, at least in appearance, lies a profound epistemological contradiction. In fact the central problem of biology lies with this very contradiction, which, if it is only apparent, must be resolved; or else proven to be utterly insoluble, if that should turn out to be the case."

So to dualities and separation we once again arrive. Here is the resolution: in the causality there is purpose; in the purpose there is causality.

But wait a moment, did you not say that Jacques Monod was going to convince us that we must believe that humans are machines governed by the rule of cause and effect - by causality?

Yes we did indeed - seems like he is struggling to do so! Such problems are often a sign that there is something wrong in the theory and that a new theory is needed. Such events are often referred to paradigm shifts, or scientific revolutions.

**284**

Turning to the matter of Galileo once again, here is Stillmann Drake to comment:

"The philosophy of Galileo as set forth in *The Assayer* is best characterised by saying that he presented an open system of scientific inquiry in opposition to the closed systems of the schools. His parable of the man in quest of the causes of sound illustrates this point. It teaches that scientific knowledge is infinite in scope, that the more one knows the more one learns how much remains to be discovered, and that a well-disciplined scepticism must always be maintained in such matters. But this scepticism is by no means classical scepticism.

"The traditional schools accepted agreement with authority and consistency with a fixed system of beliefs. Galileo adopted as his criterion a logical unity in the sense of mathematical description. In his hands this is an aesthetic rather than a metaphysical criterion; nature is to be regarded as a great book written in mathematical characters. Unlike the rationalist school which Descartes was soon to found, Galileo's philosophy avoided overt metaphysical assumptions of simple underlying principles.

"Unlike the empiricist school that was to take form in the hands of Hobbes and Locke, it gave no final authority to the evidence of the senses. Galileo ridiculed his opponent Grassi for believing that the sense of sight could not be deceived, and later he praised Copernicus for having risen above apparent sensory contradictions of his theory and for having placed his faith in that overwhelming combination of observational data and mathematical reasoning beside which those contradictions were of trivial significance.

"Thus Galileo was no empiricist in the philosophical sense, even though he anticipated the development of philosophical empiricism when he distinguished primary from secondary qualities in the *The Assayer* and suggested an atomistic theory of sensation.

"To Galileo, the universality of mathematical proportion was something discovered in the background when one approached nature by means of experiment; it existed only as a limit. Galileo's revival of atomism is a good illustration of his anti-metaphysical approach. Where Democritus did not escape endowing his atoms with the very properties they were designed to explain, such as sharpness and roughness, Galileo stripped them of all such qualities and required their motion alone to account for all sensation.

"The substitution of methodology for metaphysics is the key to the open system which Galileo offered as a rival to the closed system of the ancient philosophers. It is a key that has not lost its usefulness in opening the doors of nature's mysteries."

And yet science remains a closed system, for it comes with a dogma that has outlived its usefulness!

Did Copernicus, though, use mathematical reasoning, and if he did, was it correct? His system is actually not correct!

**285**

Big data: Coordinates: 53.2581° N, 2.1274° W; Day: Tuesday; Sunrise: 07.29; Sunset: 18.20; Minimum temperature: 10 °C; Maximum temperature 12°C: Wind: East North Easterly, ranging 8–12mph; Sky: sunny intervals, turning to light cloud, becoming thick cloud, turning back to light cloud then sunny intervals, followed by light cloud, then light rain showers in the evening; UV at level 2; pollution low.

So it seems we have resumed our big data collection exercise. This is not the case! Instead we have invited some people to speak about *big* ***bad*** *data*. The interest is in systems that, feel as though they are working. Here is Jim Adler, head of data at Toyota Research Institute in Palo Alto.

"Big data can shed light on areas with historic information deficits, and systems that seem to automatically highlight the best course of action can be seductive for executives and officials. In the vacuum of no decision any decision is attractive. Policymakers will say, 'there's a decision here let's take it', without really looking at what led to it. Was the data trustworthy, clean?

"A survey of executives by the Chartered Institute of Management Accountants found that although 37 per cent said big data had helped them make decisions, 32 per cent said it had actually made things worse and 80 per cent said a strategic decision was made based on flawed information at least once in the last three years. JP Morgan's $6bn London Whale trading loss – partly the result of Excel errors in a financial model – demonstrates how costly the flaws of misinterpreted data can be.

"Inefficient business decisions are a concern, but a bigger worry is similar problems bleeding into areas like civic governance, healthcare and employment: all areas that are increasingly relying on data-led techniques. A landmark paper in 2001 showed that legalising abortion reduced crime rates – a conclusion with major policy implications. But in 2005 two economists at the Federal Reserve Bank of Boston showed the correlation was due to a coding error in the model and a sampling mistake. This example pre-dates the era of what we now see as big data. These problems are not new, but they show the dangers of quantitative models of society."

So in other words the old adage applies: garbage in, garbage out. The new big version of this is: big garbage in, big garbage out.

The analysis of data to take decisions is not new, neither is the statement that data will help us to make better decisions. Perhaps sometimes it can, but it requires knowledge, experience and wisdom to know when. This is not something that those who adhere to the quantitative and data-driven ideologies are able to demonstrate they possess, as is evident from the mistakes that they make. They will build, if they are allowed to, an intellectual prison from which humanity will not be able to escape - because there will not be the big data that will support such a decision. It is a kind a bind, made by the blind.

**286**

Michael T. H. Sadler will now speak about Wassily Kandinsky's book *The Art of Spiritual Harmony*:

"It is no common thing to find an artist who, even if he be willing to try, is capable of expressing his aims and ideals with any clearness and moderation. Some people will say that any such capacity is a flaw in the perfect artist, who should find his expression in line and colour, and leave the multitude to grope its way unaided towards comprehension. This attitude is a relic of the days when *L'art pour L'art* was the latest battle-cry; when eccentricity of manner and irregularity of life were more important than any talent to the would-be artist; when every one except oneself was bourgeois.

"The last few years have in some measure removed this absurdity, by destroying the old convention that it was middle-class to be sane, and that between the artist and the outer-world yawned a gulf which few could cross. Modern artists are beginning to realize their social duties. They are the spiritual teachers of the world, and for their teaching to have weight, it must be comprehensible. Any attempt, therefore, to bring artist and public into sympathy, to enable the latter to understand the ideals of the former, should be thoroughly welcome; and such an attempt is the book of Kandinsky's.

"The author is one of the leaders of the new art movement in Munich. The group of which he is a member includes painters, poets, musicians, dramatists, critics, all working to the same end - the expression of the soul of nature and humanity, or, as Kandinsky terms it, the *innerer Klang*."

Relevant to scriptovisualism? Yes. What aspect is relevant? Spirituality - expression of the soul of nature and humanity, perhaps.

## 287

Now it is time for Galileo to speak:

"I do not believe that for exciting in us tastes, odours, and sounds there are required in external bodies anything but sizes, shapes, numbers, and slow or fast movements; and I think that if ears, tongues, and noses were taken away, shapes and numbers and motions would remain but not odours or tastes or sounds. These I believe, are nothing but names, apart from the living animal - just as tickling and titillation are nothing but names when armpits and the skin around the nose are absent."

Here is Arthur Koestler to comment:

"Though anticipated by the Greek atomists, it is for the first time in the modern age that this distinction is made in such concise terms, the first formulation of the mechanistic view of the universe. But on most contemporary readers of the *The Assayer* the significance of that passage was lost."

Thus we have what Arthur called the parting of the ways, not just the separation of science and religion, but the emergence of modern science which had no use for feelings, these being secondary qualities, related only to human experience, and not relevant or necessary to our understanding of the universe. And now we pay the price. Nature pays the price. Our humanity pays the price. We are in danger of becoming the *intelligent apes* that the likes of Richard Dawkins dream of, a race of, to use John Dewey's words, economic monsters, restlessly driving hard bargains with nature and with one another, bored with leisure or capable of putting it to ostentatious display and extravagant dissipation.

What can we do to recover that which is lost? We have the answer - it is to reinvent science and technology, and, if you have not already understood, that is exactly what we have been weaving our golden threads towards in the 285 chapters that you read to arrive at this one. This is our beautiful heresy! And to do this we turn to art; not to art-science or art-technology, those strange star-crossed lovers whispering sweet nothings into one another's ears, but to art and the construction of a *post-enlightenment, post-European era Theory of Knowledge*.

Convention: a way in which something is usually done.

There are conventions in the world of research and publishing concerning the citation and quoting of the works of others. It was a compositional choice not to follow these conventions.

"Why?"

This we suppose is the voice of the reader. Our answer: understand according to your capacity to do so. But here is a clue: read all the works listed in the bibliography if you wish, but your cup is full. And the cup itself - it is the vessel of your beliefs shaped to accommodate these beliefs, and these alone.

**288**

Maxim: A short, pithy statement expressing a general truth or rule of conduct.

Here is an example of a maxim, an Ancient Greek maxim: free people think, slaves work! It is also the maxim of all those caught-up in scientism, who seek to establish technocracies and scientific government, using big data, complexity theory, and that ultimate scientific nonsense called *Global Systems Science* brought to you courtesy of the reductive scientific minds found in the technocratic organisation that goes by the name the European Commission - the EC - the Emperor's Court. The story of the Emperor's new clothes is a story of collective delusion and denial, which the EC has in abundance.

Now back to Jacques Monad:

"To make dialectical contradiction the fundamental law of all movement, all evolution, is still an attempt to systematise a subjective interpretation of nature, showing it to have an ascending, constructive creative intent, a purpose; in short to make nature decipherable and morally meaningful. This interpretation is not only foreign to science but incompatible with it."

This is what you believe. Yes, *believe*. But do continue.

"All ideologies, without exception, take an initial teleonomic principle as the *primum movens* of evolution, whether of the biosphere alone or of the entire universe. In the eyes of modern scientific theory all these concepts are erroneous for reasons of method, since in one way or another they imply abandonment of the postulate of objectivity. At the source of these errors is the anthropocentric illusion."

This is what you believe. Yes, *believe*. But do continue.

"Living beings are chemical machines."

This is what you believe. Yes, *believe*. But do continue.

"In both macroscopic structure and their functions, living beings are closely comparable to machines. On the other hand they differ radically from them in the manner of their formation."

To compare living beings - human beings - with machines, is to reduce humanity to the status of a mere machine, and demonstrates *celibacy of the intellect*. But do continue.

"We say that events are accidental, due to chance. And since they constitute the *only* possible source of modifications in the genetic text, itself the *sole* repository of organism's hereditary structures, it necessarily follows that chance *alone* is at the source of every innovation, of all creation in the biosphere. Pure chance, absolutely free but blind, at the very root of the stupendous edifice of evolution: this central concept of modern biology is no longer one among other possible or even conceivable hypotheses. It is today the *sole* conceivable hypothesis, the only one compatible with observed and tested fact. And nothing warrants the supposition or the hope that conceptions about this should, or ever could, be revised.

"There is no scientific position, in any of the sciences, more destructive of anthropocentrism than this one, and no other more unacceptable to the intensely teleonomic creature that we are."

And what implications stem from this, Jacques?

"The choice of scientific practice has launched the evolution of culture on a one-way path; on to a track which nineteenth century scientism saw leading infallibly to a vast blossoming for mankind, whereas what we see before us today is an abyss of darkness."

The answer to this lies in changing the nature of science and its practice; in the causality there is purpose; in the purpose there is causality - otherwise known as *Chinese solutions*.

Here is Howard Rosenbrock:

"Purpose and anthropocentricism are rejected *a priory* by science. The facts of nature, as interpreted by the scientist, confirm that nature itself rejects anthropocentricism and purpose. This is the causal myth. But there is another myth that we could choose to follow - the purposive myth."

The West, being as it is, deeply embedded in an *Enlightenment* culture, which is itself deeply influenced by Christian *Abrahamic* thought as well as *Ancient Greek* beliefs, sees the world in terms of dualities - there is causality and there is purpose, so choose. The Chinese do not see the world in such a way, and therefore they will follow both, without any worries about contradictions – there is a unity in opposites.

**289**

Let us now to the matter of Experiments in Art and Technology - E.A.T. - turn our attention one more time. Here is Billy Klüver:

"I developed the idea that one-to-one collaborations between an artist and an engineer or scientist could satisfy the artists' desire to work with new technology and produce works of art that could not be realised any other way. And the collaboration could also work both ways. Artists' projects could stimulate the engineer in new ways of looking at technology and influence technical developments in the future."

Is this actually true, or, is this just post-rationalisation? Or even someone else's interpretation? Or a mixture? Or, some facts mixed in with other things? Any examples, Billy, to justify your last statement? No? The problem with making such statements is that there are people who will believe what you have just said! Do continue.

"The *raison d'etre* of Experiments in Art and Technology is the possibility of a work which is not the preconception of either the engineer or the artist but which is the result of the exploration of the human interaction between them."

Any proof that this was achieved, Billy? No?

Remediation: the action of remedying something, in particular of reversing or stopping environmental damage.

What does remediation mean in the context of understanding new media? Here are Jay David Bolter and Richard Grusin to answer this question:

"Media critics are captivated by a modernist myth of the new: they assume that digital technologies such as the World Wide Web, virtual reality, and computer graphics must divorce themselves from earlier media for a new set of aesthetic and cultural principles. Our theory challenges this assumption. We argue that new visual media achieve their cultural significance by paying homage to, being a rival to, and refashioning earlier media, such as perspective painting, photography, film, and television. This process of refashioning is called remediation, which is also evident in earlier media which refashioned one another: photography remediated painting, film remediated stage production and photography, and television remediated film, vaudeville, and radio."

So does this have a relevance to scriptovisualism? Probably because one can consider such a work as a remediation of two distinct media - descriptive expression and plastic expression.

But you object? Why? Because there is more to Bolter and Grunsin's thesis than what we said above. Here is Mark Deuze to explain:

"The ambivalent and contradictory ways in which new digital media function is a result of the double logic of remediation that is embedded in all media. On the one hand, media make themselves known to us by remixing and multiplying their properties; on the other hand media work very hard to make themselves invisible through a logic of immediacy, which erases or automates their operation."

Well it seems we have encountered contradictory opposites once again, so perhaps the relevance to scriptovisualism still holds, as this has, at its heart, the notion of unity of opposites.

**290**

We know what STEM means, and STEAM too. So what does STS stand for? STS - Science, Technology and Society.

It is a truth, not universally understood, that scientists and engineers often distance themselves from the social implications or unintended consequences of their work. It is an event most unusual for these professions to address the potential problems associated with their work in advance of widespread dissemination. It is an event even more unusual for them to - not do something because their work will lead to problems. There are always choices - they could choose to do something else instead. But if you are caught-up in Abrahamic thought - the one best way, the sole truth - then you will not think about choices.

Sheila Jasanoff, professor of Science and Technology Studies, will now speak:

"An enormous amount of human thought and energy in the two centuries since the *Enlightenment* has gone into creating and sustaining the idea that scientific discovery and technological innovation are in and of themselves value neutral. It then stands to reason that if you incorporate values into science and technology, you have corrupted the process."

But science and technology is not value neutral. This is just a belief held by those whose minds are firmly planted in the past - an *Enlightenment* past, a Christian *Abrahamic* past, an *Ancient Greek* past. Richard Dawkins proves the point.

As we are hovering around the matter of *intellectual celibacy* once again, here are some people making comments on the STEM world's obsession with Big Data.

"A lot of things I see treat big data about society as if it is a natural phenomenon. There is an overly mechanistic understanding which leads to overconfidence that the data only means one thing."

That was Matthew Zook speaking, a professor of geography.

"Those using big data, particularly when it is about people, must remember that it is often *found data,* skimmed from the internet, rather than the carefully sampled representative datasets of the social sciences. *Found data* is not unusable, but its limitations need to be taken into account when designing analyses. Big data is a little blunt so we need to be careful and think about the implications rather then just assuming it is representative. We are in an age of engineering where we chase quick results and solutions without really understanding the phenomenon."

That was Daniel McFarkand, a professor at Stanford, working in the area of data-based sociology.

"The deeper approach to the world of trying to understand, rather then just correlate, falls by the wayside, because in the end, all these big data directions push towards correlation, eclipsing explanation. But explanation is an important element of the human experience and one that has longer term pay-offs than the purely correlational, predictive approach."

That was Mark Andrejevik, an associate professor of media studies, speaking.

**291**

It is time yet again to turn to the matter of C.P. Snow and his *two cultures* nonsense. Here to speak is the geometer, Donald Coxeter:

"I first met the artist M.C. Escher in September 1954, when an exhibition of his work was sponsored by the International Congress of Mathematicians, meeting that year in Amsterdam. Throughout the previous seventeen years he had been making designs in which a drawing of some animal (such as a fish or a reptile or a bird) is repeated as on wallpaper, with two remarkable innovations: the basic unit (usually a single animal, or one half of a symmetrical animal or two different animals juxtaposed) is repeated not only by translations but also by other isometries (or congruent transformations): rotations, reflections or glide-reflections; and the replicas ingeniously fit together so that there are no interstices. In the language of mathematics (a subject in which Escher resolutely claimed to be 'absolutely innocent of training or knowledge'), the basic unit is a fundamental region for a symmetry group.

"In a letter of December 1958 he wrote: 'Did I ever thank you for sending me *A Symposium on Symmetry*? I was so pleased with this booklet and proud of the two reproductions of my plane patterns! 'Though the text of your article on *Crystal Symmetry and its Generalizations* is much too learned for a simple, self-made plane pattern-man some of the text-illustrations gave me quite a shock. Since a long time I am interested in patterns with *motives* getting smaller and smaller till they reach the limit of infinite smallness. The question is relatively simple if the limit is a point in the centre of a pattern. Also a line-limit is not new to me, but I was never able to make a pattern in which each *blot* is getting smaller gradually from a centre towards the outside circle-limit. I tried to find out how your figure was geometrically constructed, but I succeeded only in finding the centres and radii of the largest inner-circles. If you could give me a simple explanation how to construct the following circles, whose centres approach gradually from the outside till they reach the limit, I should be immensely pleased and very thankful to you! Are there other systems besides this one to reach a circle limit? Nevertheless I used your model for a large woodcut (of which I executed only a sector of 120 degrees in wood, which I printed three times). I am sending you a copy of it.

"This was his picture Circle Limit I."

So C.P. Snow anything you would now like to say in retrospect?

"Yes there is - sorry!"

Good, now off you go because we do not want to hear from you again.

C.P. Snow - a scientist! Actually a failed scientist!

**292**

On the matter of European democracy, or more correctly the lack of it; it is like being once again in Ancient Greece: free people (the political class) think, slaves (the ordinary people of Europe) work (if they have a job that is).

Ulrich Beck und Daniel Cohn-Bendit will now speak:

"The *European Year of Volunteering for Everyone* is a counter-model to the top-down Europe, the Europe of elites and technocrats that has prevailed up to now that considers itself responsible for forging the destiny of the citizenry of Europe - if need be, against its will. For this is the unspoken maxim of European politics that is threatening to destroy the entire European project."

A spokesperson for Alternattiva Demokratika will now make an address:

"As a member of the European Greens, Alternattiva Demokratika will focus on several issues which include democratisation of the European Institutions. Alternattiva Demokratika will work towards a more direct and participative democracy. Executive bodies such as the European Commission and Central Bank must be held accountable for their actions."

Indeed they should, but you will not succeed in this aim for you have no power to bring this about. The only way forward is for people to turn against the European Union and its institutions, for people have no other way to go, given that the European Union's leaders do not want to change: why would they when they have a means of pursuing policies that no one voted for and for which they will never be called to account for? It is just like the Soviet Union! A capitalist's version of – that is to say, embryonic National Socialism! We call it European Union Nationalism.

Here is Paul H. Ray and Sherry Ruth Anderson:

"We were involved in 1999 in the design and analysis of a study of the role of values and concerns for ecological sustainability in American life, sponsored by the Environmental Protection Agency and the President's Council on Sustainable Development. Once the big surveys and focus group data were in, the picture of an emerging sub-culture was clear. We called them the *Cultural Creatives.* The core group are people who combine concern for their inner spiritual lives, with a strong penchant for social activism, including a commitment to a sustainable future. All *Cultural Creatives* have green values in terms of concerns for ecology and the well-being of the planet, but the core group is far more intense and activist about them. In addition the core group has strong values of personal growth and spirituality that are much less important to *Green Cultural Creatives*, who are a second group that are more secular and extroverted, but which tend to follow the opinions of the core group."

**293**

Now a spokesperson from the European Commission will speak about the 2015 Call for Research proposals to address the topic INT-3-2015: Europe's contribution to a value-based global order and its contestants (HORIZON 2020 – WORK PROGRAMME 2014-2015 Europe in a changing world: inclusive, innovative and reflective societies PART 13).

"The notion of and access to global justice have been at the centre of many debates about the objectives and legitimacy of global governance and international law. For some, global justice implies solidarity with all humankind on the basis of inter-culturally shared values. Others think that such a broad conception of justice is untenable, and that high levels of justice may, if at all, only be attained at the level of the nation-state. While the debate is open on what a just global order can be and which values it could or should comprise, one can identify certain conditions that global political action would need to fulfil to move closer to the ideal type of such an order. With a European view of global justice and human rights, European policy-makers regularly portray the European Union as an ethical global player promoting values like democracy and human rights in its external relations.

"Moreover, the European Union's engagement in the fight for sustainability and against poverty and conflicts in the world may be an indicator of a determination to spreading its values so as to render the world more just. However, the Union's real impact on global justice, also at the institutional level, remains, not only underspecified, but is often also *contested*."

And that you are being contested is right and proper, because you are either hypocrites or extremely deluded. The Member States of the European Union are democracies, but the European Union is not. It is more like the USSR, and is heading more towards being so, only it is not communist, but National Socialist in orientation, but without the extreme racism (for now). And you respect human rights only when it suits you, as the migrant crisis well demonstrated. How many innocent people in Syria had their human rights violated when the French bombed Syria in revenge for the Paris terrorist attacks in November 2015? We note the Belgians did not bomb Syria in revenge for the Brussels terrorist attacks in March 2016. Most likely that is because they lack the military capability to do so. Evidently you are not familiar with the political aims of groups within the European Parliament, spokespeople for which we have already encountered.

"Right, well, as I was saying, the research should first look into the criteria for global justice and contrast various understandings of it, exploring European as well as non-European and non-Western perspectives. It should then examine whether and how various criteria for a just and value-based global order are met by the European Union and European Union Member and Associated countries' policies, activities and regulations in issue areas that may be of particular importance for the promotion of inter-culturally shared values and justice at a global scale, such as trade and development policy, the defence of human rights, the promotion of gender equality, religious freedom, conflict prevention and resolution, and the protection of livelihoods.

"It should also investigate the impact of governmental, non-governmental and transnational actors who contest the European Union's activities and present alternative visions of global justice and how to obtain it. Research should ultimately come to statements about if and how Europe does - or does not - contribute to a value-based global order. It should identify facilitating and inhibiting conditions for European action that effectively promotes its values. The participation of international partners in proposals submitted to this call is strongly encouraged.

"Expected impact is as follows: the research on the European contribution to a just global order, especially through institutions and law, will advance the understanding of the Union as a value-based actor in global affairs, while uncovering the conflicts that exercising this function creates with other major global players. It is thus expected that this research will shed empirical light on the potential for and limits to the European Union's future place in a multi-polar global order. By identifying facilitating and restraining factors for effective European Union external activities aimed at the export of values, research will provide the bases for future-oriented and evidence-based reflections on the contents and conduct of these external activities."

Are you aware that there is another name for export of values? It is called *Cultural Colonialism*! You may have (for the moment?) abandoned the use of hard power in your campaign to make the world European, but you have not abandoned this aim. Instead you use soft power.

Perhaps it is time for a counter-initiative, a *Cultural Colonial* effort from the Chinese to make Europe Chinese. See how you would like it? Only the Chinese do not do such things. That is the big difference between their more advanced civilisation and your more primitive Ancient Greek civilisation. They do not consider you to be barbarians in need of civilisation.

**294**

And now the time has come, to once again tackle scientism. To refresh your memory we refer you to Richard Dawkins, who, for reasons better known to himself, made the extraordinary claim, that he had proved the improbability of God's existence. Part of his struggle! So now you are refreshed about scientism, and its delusory attractions, we will speak with W.J.M. Mackenzie about biological ideas in politics.

"I once thought that if one is to play with the idea of natural science of politics, it must be end-on to biology, or perhaps even be part of biology. Man is an animal; as Aristotle put it, he is a political animal. The natural history of man has made enormous and exciting steps; one of its main themes is that man has conquered the world because he excelled all other animals in social adaptivity.

"I can now see that I was confusing three different ways of applying any natural science, and biology in particular, in any field of social science. The first is to treat biology as analogous to the field of society. Probably every language uses biological metaphors in talk about society. These may be made explicit poetically in myths and fables; they may be abstracted to form symbolic models, sometimes even quantifiable, as in the relationship between the models of economies and those of ecology. The biological metaphor is irresistible and adds life to political rhetoric at many levels - but watch for it."

Someone should tell Richard Dawkins.

"The second thing to say to oneself - 'that is a very good and profitable science, let us copy its methods' - and this can happen unconsciously, or at least naïvely, as is easily seen when someone trained to think as a biologist or as an engineer, or in some other powerful discipline, begins to think methodically about human society.

"I have no claim at all to explain the methods of biology but perhaps it is worth saying that one learns fairly quickly to think of biological sciences, or the life sciences, in the plural, that is to say, there is a single focus on life, a puzzle which we seek to understand, but there are many techniques and methods. There is a range of scale, from that of ecology of planet earth to that of molecular chemistry; there is a range of method - library study, field study, laboratory work; there are various different kinds of outputs, such as taxonomy, or evolutionary theory, or mathematical modelling. There is not much in this that we can use or imitate directly, but it may help us to cultivate humility and patience."

Someone should tell Richard Dawkins.

"The third thing is to treat man as, in some sense, the subject of biology, the natural history of man as end-on to the study of other forms of life. I use the words political adaptivity as a catch-phrase or ticket for the appeal to biological science in that sense, and, in the present discussion, I am reverting to an old paradox that man is, and is not, part of nature. Man's cell biology and so forth must be regarded as parts of nature; yet there are characteristics of man for which no analogies or antecedents have yet been found in nature."

Someone should tell Richard Dawkins. But you want to say more by way of an ending:

"The story of Social Darwinism, the path from scientism to racism, is surely so familiar that the broken ground is fenced beyond risk of accident."

Would that this be true! So we arrive back to scientism yet again. Richard Dawkins - what a strange ideology - dogma - that he holds to, and encourages others to as well! Science and reason - who will save us from such madness? Some artists are also caught-up in the Dawkins nonsense, so do not look to these people for salvation.

**295**

With C.P. Snow despatched to oblivion, it only remains to invite the literary critic and professor of English Literature, Stefan Collini, to comment:

"Snow looked back to the inter-war period, and especially to the Cambridge of the 1930s, as a Golden Age of original scientific research, and he evidently imbibed a certain cultural conception of science that was especially powerful in those years, particularly among progressive scientists and radical spokesmen for science. He saw science as the great hope in a world which the traditional elites had mismanaged and led into economic depression and to the brink of a second devastating war. And in more parochial terms, the young Snow developed an antipathy to literary intellectuals, especially to what he identified as their snobbish and nostalgic social attitudes, which was never to leave him.

"He had an apparent hankering for the rule of a *scientific elite*. He was also a technocrat."

Today we find ourselves faced with a situation where it is the scientists that have become the elites, and it is they who are participating in a colossal mismanagement of the very fabric of that which we are, and the planet on which we depend. Not to mention scientism and the lure of technocracy which, the European Commission's DG CONNECT, embraces wholeheartedly. The STARTS initiative illustrates the point - a case study in the failure of technocracy.

**296**

Chapter 296! Time moves on, as do the chapters, and we are approaching the end, if there is such a thing. No, not the end, just a limit imposed by the structure of the year 2016 - 366 days and thus 366 chapters.

It is time now for the industrial psychologist, N. Jordan, to speak about something that in the world of Human Factors engineering is called allocation of functions:

"Allocation of functions is often not very relevant. The term allocation of tasks to people and machines is often meaningless because this assumes that people and machines are comparable. They are not comparable, rather they are complementary. The problem is not therefore one of deciding what each should do, but of designing a system so that the work is done by people and machines.

"This requires new formats for systems analysis and design. My suggestion is that designing from a tool perspective will be part of this new approach."

It appears to be almost a universal fact, that there is no new vision of any other way of tackling the human-computer relationship other than human-computer comparability - man is a machine.

In these days, when people are more aware of the dehumanising nature of *Enlightenment* technologies, it is common for researchers to state that people and computers are complementary, and often the term human-centred is used. But this statement is typically followed by an illustrative example of the form: computers are fast and consistent but lack flexible reasoning, while human information processing is slower but can handle events which are ill-defined. In other words they still adopt a machine-centric view!

The problem with such examples of the form given above is that they are based on the principle of human machine comparability not the principle of complementarity, because they make comparisons between people and computers. In other words comparability still prevails, even when complementarity is preached. This is because this conception of humans as machines is deeply embedded in Western culture and its science, technology, engineering and mathematics - such is the nature of *Enlightenment* technologies.

There is one way out of this dilemma - leave people to determine their own relationship with the computer and thus we should design them to be *skill-oriented technologies*.

Here is Sir Joseph Needham, FRS, to speak about *Chinese solutions*:

"In the volume of *Science and Civilisation in China* that deals with agriculture, Francesca Bray makes a distinction between mechanical technologies, like that of European Agriculture, and *skill-oriented* technologies, such as rice cultivation, and she concludes that the transition to capitalist relations of production whereby landowners evict their tenants in order to run large consolidated farms using cheap wage-labour, did not occur in China."

Agricultural practices such as permaculture and organic farming are *skill-oriented* approaches. They involve the intensification of human knowledge and skills and the de-intensification of other inputs such as capital and energy.

Let us now dig deeper and understand what the term *sustainable intensification* means. Here is Jules Pretty to explain:

"Intensification, one should understand, refers to making better use of resources in agriculture, meaning such things as nature, land, knowledge, cooperation, physical items such as plant and equipment etc, technology, and capital (which enable access to, or use of, other types of resources). The critical question is: what type of intensification?"

We might also ask: which resources do we de-intensify? Traditionally, the trend in the industrialised world has been towards increased technological intensity, often driven by increased scientific intensity, both of which often require higher capital intensity. But now the issue is whether technological intensity should be reduced (and with it, possible certain types of science).

Here be heretical thought! So to our beautiful heresy once again!

**297**

On the matter of European democracy, or more correctly the lack of it, we revisit the European Parliament. Recall that, pasted on the walls of the European Parliament visitor centre, a place known as the *Parlamentarium*, a space of both information and propaganda, there are words. Let us now hear these words spoken once again.

"National sovereignty is the root cause of the most crying evils of our time and of the steady march of humanity back to tragic disaster and barbarism. The only final remedy for this supreme and catastrophic evil of our time is a federal union of the peoples."

That was Lord Lothian speaking in 1939.

We shall now do what those who engage in propaganda sincerely hope no one will ever do. Normally they are correct in this assumption, for people are just inclined to accept what they are told. We saw this in abundance in during United Kingdom's Contunuing Membership of the European Union Referendum campaign. It is most likely that 99.99 percent of those who voted in the referendum have little idea what the European Union is, what its various institutions do, who holds the power, how the legislative process works, how many different legislative procedures exist, that amendments made by the European Parliament to Directives, can be rejected, and so forth. People even speak about the free movement of people, without realising that there is no free movement of people, only free movement of goods, services, capital and people, for the sole reason of creating a Single Market - economics! So we have a situation where just about everyone voted for the wrong reasons. This applies to those who voted to remain and those who voted to leave. Who was right? Figure it out for yourself.

Now to the matter of European Parliament propaganda and the doing of that which they hope no one will do - we researched where this quote came from, went to the British Library Reading Rooms in London, and read the publication. Oh dear!

What we learned from this exercise, is that the quote is taken from a pamphlet published by the Federal Union. The context of the pamphlet is the failure of the League of Nations, and that which, at the time the pamphlet was published - 1939 - what was already an inevitability. And the analysis is superficial and not about the creation of a federal Europe, but a federal world government which is what the Federal Union was seeking, and still does. The European Union has not abolished sovereignty, just made the problem worse.

What the European Parliament will also not tell you, is that Lord Lothian was a leading advocate of appeasement - the discredited policy that the British Government pursued as Hitler's military power increased and was used against ... Well, we all know the story! Lothian even met Hitler and announced to the world that Hitler wanted peace; a case of flawed judgement as well as flawed analysis. Another fragmented and reductive mind!

Here we also note that if you want answers to all those issues that we mentioned, about how the European Union works, do not bother visiting the *Parlamentarium.* In fact you will discover, if you care to try, that it is very hard to find out how the European Union works, for the vast majority of the information provided comes from the European Union itself, and is biased. The whole system of the European Union seems to have been designed to ensure that no one will ever understand it!

Once again we find ourselves in Ancient Greece: free people (the political class) think, slaves (the ordinary people of Europe) work (if they have a job that is).

A closing remark on this issue from Jürgen Habermas:

"The European Parliament has been pushed to the sidelines by the heads of government. The European Council has monopolised all power."

This is the outcome of the banking, financial and sovereign debt crisis. Thus does the European Council pursue the neoliberal agenda and the imposition of austerity measures on the crisis countries through means undemocratic. And what we find is people in the United Kingdom who oppose the Tories, because of their neoliberal and austerity policies, voting to remaining in the European Union which is pursuing neoliberal and austerity policies, but which cannot be changed by the electorate through the election of a new government – just like the situation in the former Soviet Union! So who exactly were the stupid ones?

One day in the not so distant future, when freed from the influence of European Union propaganda, people in the United Kingdom will come to see the European Union in a different light, as indeed those who supported Stalin (and Hitler) in the 1930s eventually did. But what now can be done to free the people of Europe from this travesty called the European Union? Perhaps the reformers will win, but the prospects look grim. We wish them well, for what could have taken shape in Europe was a dream and a vision worth pursuing. Now it is gone. It has become a distant memory. Perhaps one day, when Prometheus has been set free?

**298**

The *Enlightenment*; time to break the chains of the *Enlightenment*! It is still with us. This is 2016, not 1916, or 1816. Times change, the world changes, circumstances change. We need to change. How does that make you feel? This is the beautiful heresy.

The philosopher Isaiah Berlin will now say something about the *Enlightenment* and the roots of *Romanticism*.

"The *Enlightenment* can be boiled down to three propositions, which are the three legs upon which the whole of Western tradition rested. The propositions are roughly these:

"First, all genuine questions can be answered and any question that cannot be answered is not a question. Second all these answers are knowable, that they can be discovered by means which can be learnt and taught to other persons. Third, all answers must be compatible with one another, because if they are not compatible with one another, then chaos will result."

Sole truth! This is indeed very Abrahamic! Yet these are the general presuppositions of the rationalist Western tradition, whether Christian or pagan, whether theist or atheist. Richard Dawkins is the living proof of this.

"The particular twist that the *Enlightenment* gave to this tradition was to say that the answers were not to be obtained in traditional ways - not from revelation, not from tradition, not from dogma, not from individual self-inspection of the privileged, etc.

"There is only one way of discovering these answers, and that is by the correct use of reason, deductively as in the mathematical sciences, inductively as in the natural sciences. That is the only way in which answers in general, true answers to serious questions, may be obtained. There is no reason why such answers, which after all have produced triumphant results in the worlds of physics and chemistry, should not equally apply to the much more troubled fields of politics, ethics and aesthetics."

But the Chinese do not share this perspective. They have their own natural law, which leads to different conclusions, as indeed does Hindu culture.

"All religions are true."

That was Mahatma Gandhi speaking.

"Be the change that you wish to see in the world."

That too was Mahatma Gandhi speaking.

And the change we wish to see in the world is that people, all people, West and East, not only understand that all religions are true, but practice that belief. All religions, whether they be religious or secular! Back to the once more to our beautiful heresy. We need change our science, technology and engineering.

Back to the matter of the *Enlightenment* and one of the early attacks came from David Hume. Isaiah Berlin again:

"Hume doubted two propositions. In the first place he doubted whether the causal relationship was something which we directly perceived, or knew existed at all."

This is indeed true. So we are back to the matter of observability - causality cannot be observed. It is assumed to exist - a case of *I believe* science. We are also back to the artists Magritte, and his work *The Treachery of Images*. The image is most definitely not the thing. The equations are also not the thing, for we can replace causal equations with purposive ones and still predict the same results.

"Hume also doubted if we can deduce logically whether there is an external world, for there is no way of demonstrating that tables exist. Nothing can be proved with mathematical certainty to exist. Therefore the existence of the world is a matter of belief, something that must be taken on trust. The universe is not known by intellect alone, but also by faith. Belief is not the same as deductive certainty. Indeed, deduction does not apply to matters of fact at all."

Hume effectively broke the ideal of rationalism. It is all about probabilities, but not certainties.

"As for the person who struck the most violent blow against the *Enlightenment*, that person was Johann Georg Hamann."

And what he said is very profound. Here is Berlin again:

"Hamann drew the conclusion, that there was a flow of life, and that to cut this flow into segments killed it. The sciences were very well for their own purposes. If you wished to discover about how to grow plants; if you wished to know about some kind of general principle, about the general properties of bodies in general, whether physical or chemical; if you wished to know what climates would assist what kind of growth to develop in them, and so forth; then no doubt, the sciences were very well. But this is not what men ultimately sought.

"If you asked yourself what were men after, what did men really want, you would see that what they wanted was not what Voltaire supposed they wanted. Voltaire thought that they wanted happiness, contentment, peace, but that this was not true. What men wanted was to create, what men wanted was to make, and if this making led to clashes, if it led to wars, if it led to struggles, then this was part of the human lot.

"A man who had been put in a Voltairean garden, pared and pruned, who had been brought up by some wise *philosophe* in knowledge of physics and chemistry and mathematics, and in knowledge of all the sciences which the Encyclopeadists had recommended - such a man would be a form of death in life.

"The sciences if they were applied to human society, would lead to a kind of fearful bureaucratisation of thought. He was against scientists, bureaucrats, persons who made things tidy, smooth Lutheran clergymen, deists, everybody who wanted to put things in boxes, everybody who wished to assimilate one thing to another, who wished to prove, for example, that creation was really the same as the obtaining of certain data which nature provides and their rearrangement in certain pleasing patterns - whereas for Hamann, of course, creation was a most ineffable, indescribable, un-analysable personal act, by which a human being laid his stamp on nature, allowed his will to soar, spoke his word, uttered that which was within him and which would not brook any kind of obstacle. Therefore the whole of the *Enlightenment* doctrine appeared to him to kill that which was living in human beings, appeared to offer a pale substitute for the creative energies of man, and for the whole rich world of the senses, without which it is impossible for human beings to live, to eat, to drink, to be merry, to meet other people, to indulge in a thousand and one acts without which people wither and die. It seemed to him that the *Enlightenment* laid no stress on that, that the human being as painted by the *Enlightenment* thinkers was, if not economic man, at any rate some kind of artificial toy, some kind of lifeless model, which had no relation to the kind of human being whom Hamann met and wished to associate with every day of his life."

The political economist and theorist of the state, Adam Heinrich Muller has the last word:

"Science can reproduce only a lifeless political State; death cannot represent life, nor can stagnation represent movement. Science, utilitarianism, the use of machinery do not convey the State, which is not a mere factory, farm, insurance company or mercantile society; it is the intimate binding together of the entire physical and spiritual needs of a nation, of its entire physical and spiritual riches, of its entire internal and external life, into a great energetic, infinitely active and living whole."

**299**

We have not forgotten about Leonardo. Having dealt with C.P. Snow we turn our attention once more to what we call the *Leonardo myth* - that he was a scientist and a humanist.

To speak about the sociological roots of science and Leonardo, we have invited Edgar Zilsel to make an appearance:

"In the period from 1300 to 1600 three strata of intellectual activity must be distinguished: university scholars, humanists, and artisans. Both university scholars and humanists were rationally trained. Their methods, however, were determined by their professional conditions and differed substantially from the methods of science. Both professors and humanistic literati distinguished liberal from mechanical arts and despised manual labour, experimentation, and dissection. Craftsmen were the pioneers of causal thinking in this period. Certain groups of superior manual labourers (artist-engineers, surgeons, the makers of nautical and musical instruments, surveyors, navigators, gunners) experimented, dissected, and used quantitative methods. The measuring instruments of the navigators, surveyors, and gunners were the forerunners of the later physical instruments. The craftsmen, however, lacked methodical intellectual training. Thus the two components of the scientific method were separated by a social barrier: logical training was reserved for upper-class scholars; experimentation, causal interest, and quantitative method were left to more or less plebeian artisans. Science was born when, with the progress of technology, the experimental method eventually overcame the social prejudice against manual labour and was adopted by rationally trained scholars. This was accomplished about 1600 (Gilbert, Galileo). At the same time the scholastic method of disputation and the humanistic ideal of individual glory were superseded by the ideals of control of nature and advancement of learning through scientific co-operation. In a somewhat different way, sociologically, modern astronomy developed. The whole process was imbedded in the advance of early capitalistic society, which weakened collective-mindedness, magical thinking, and belief in authority and which furthered worldly, causal, rational, and quantitative thinking."

Thank you for clarifying that matter; now to the Leonardo myth. Please continue.

"The university scholars and the humanistic literati of the Renaissance were exceedingly proud of their social rank. Both disdained uneducated people. They avoided the vernacular and wrote and spoke Latin only. Further, they were attached to the upper classes, sharing the social prejudices of the nobility and the rich merchants and bankers and despising manual labour. Both, therefore, adopted the ancient distinction between liberal and mechanical arts: only professions which do not require manual work were considered by them, their patrons, and their public to be worthy of well-bred men.

"The social antithesis of mechanical and liberal arts, of hands and tongue, influenced all intellectual and professional activity in the Renaissance. The university-trained medical doctors contented themselves more or less with commenting on the medical writings of antiquity; the surgeons who did manual work such as operating and dissecting belonged with the barbers and had a social position similar to that of midwives. Literati were much more highly esteemed than were artists. In the fourteenth century the latter were not separated from whitewashers and stone-dressers and, like all craftsmen, were organized in guilds. They gradually became detached from handicraft, until a separation was effected in Italy about the end of the sixteenth century. In the period of Leonardo da Vinci (about 1500) this had not yet been accomplished. This fact appears rather distinctly in the writings of contemporary artists who over and over again discussed the question whether painting and sculpture belong with liberal or mechanical arts. In these discussions the painters usually stressed their relations to learning (painting needs perspective and geometry) in order to gain social esteem. Technological inventors and geographical discoverers, being craftsmen and seamen, were hardly mentioned by the humanistic literati. The great majority of the humanists did not report on them at all. If they mentioned them, they did so in an exceedingly careless and inaccurate way. From the present point of view the culture of the Renaissance owes its most important achievements to the artists, the inventors, and the discoverers. Yet these men entirely recede into the background in the literature of the period.

"Beneath both the university scholars and the humanistic literati the artisans, the mariners, shipbuilders, carpenters, foundrymen, and miners worked in silence on the advance of technology and modern society. They were uneducated, probably often illiterate, and, perhaps for that reason, today we do not even know their names. Among them were a few groups which needed more knowledge for their work than their colleagues did and, therefore, got a better education. Among these superior craftsmen the artists are most important. There were no sharp divisions between painters, sculptors, goldsmiths, and architects; but very often the same artist worked in several fields, since, on the whole, division of labour had developed only slightly in the Renaissance.

"Following from this, a remarkable professional group arose during the fifteenth century. The men we have in mind may be called artist-engineers, for not only did they paint pictures, cast statues, and build cathedrals, but they also constructed lifting engines, canals and sluices, guns and fortresses. They invented new pigments, detected the geometrical laws of perspective, and constructed new measuring tools for engineering and gunnery. The names of these people are familiar: Brunelleschi, Ghiberti. Leone Battista Alberti, Leonardo da Vinci, Vanoccio Biringucci, Benvenuto Cellini (1500-1571), and Albrecht Dürer.

"Many of the artist-engineers wrote - in the vernacular and for their colleagues - diaries and papers on their achievements. For the most part these papers circulated as manuscripts only. The artist-engineers got their education as apprentices in the workshops of their masters. Only Alberti had a humanistic education."

Thanks for these insights Edgar.

So Leonardo was one of these superior artisans, and not a humanist as someone from the world of STEM claimed in an early chapter. We wonder how they faired with their new patron Joseph Stalin!

Neither was da Vinci a scientist in the sense that we know the term. He was the carrier of practical methods, along with many other people like himself. These practical methods, when combined with the theoretical, the open-minded approach, the critical approach pioneered by Galileo, led to that which we now know as science. Leonardo da Vinci was no Galileo.

But now you ask: what about his inventions?

Indeed what about them. We need to find Joseph Needham, for he has some valuable information that may help to clarify this.

## 300

Only 66 chapters left after this one; our destination is in sight! But where have we been. We do not know, because we did not look back.

Scriptovisualism - we will today say some more about this. And we have invited to speak someone who has much to say about scriptovisual art. She is an artist. Her name is Elizabeth Chaplin and her theme is sociology and visual representation.

"Most feminist art is figurative, or scriptovisual, because works of this type have proved to be very effective in the short term as a means of confronting viewers with feminist ideas. Moreover, a semiotic perspective indicates that many effective communications have both visual and verbal components, that these components interlace, and the manner in which they do so is crucially important to the meaning of a communication, and therefore to its effect. This suggests that there is considerable scope for a scriptovisual critical practice.

"It can also be stated that for an increasing number of visual artists, including Victor Burgin, critique is social and cannot be separated from the practice of art itself, which suggests the notion of a socially grounded critical visual art practice underpinned by a theory of communication informed by principles of semiotics. It is a critical practice which understands the positive political implications of communicating via visual art while acknowledging that a social critique must nevertheless involve the verbal dimension.

"Victor Burgin urges a critical art practice which forms part of a wider critical discourse, or more accurately, a discursive formulation. This practice acknowledges that both visual and verbal media are implicated; and its work may include both pictorial and textual elements. The pictorial element addresses topics shared by other contributors to the discursive formation, rather than consisting of modernist abstract painterly marks which only refer back to themselves and the modernist rationale. The verbal element may include text, and is, therefore, not simply to be understood in the sense that a visual work is invaded by language during the course of interpreting it. And since the verbal permeates the art work anyway, there is no good reason to ban its formal presence on or alongside the image, so long as it remains in a tempered relationship with the image, such that the relationship enhances the total critique.

"Scriptovisual art transgresses the hallowed division between visual and verbal in the institutionalised system of representational practices. A critical visual art practice that consciously deploys both visual and verbal channels to produce critique, builds on a sounder base than one which merely offers verbal text."

And it is just possible that by engaging in scriptovisualism we might come to understand unity of opposites and how to fundamentally transform Western STEM, and to bring together West and East with a new way of knowing that is not grounded in the fragmented and reductive mind, in intellectual celibacy, in sole truth, in scientism, in ... Perhaps the list is too long to continue?

**301**

We will now hear from Mao Tse-Tung:

"There are two different attitudes towards learning. One is the dogmatic attitude of transplanting everything, whether or not it is suited to our conditions. This is no good. The other attitude is to use our heads and learn those things which suit our conditions, that is, to absorb whatever is useful to us. This is the attitude we should adopt."

Thus does this sentiment apply to finding a unity between West and East, for there is much that is good in both. Humanity's needs must come before the ideology of the West, which is still caught-up in its imagined superiority. It is time to abandon those dreams of making the world European, which is why the European Union has to be destroyed, by which we mean - destroy the European Commission. Slay that monster. It is already destroying itself. Let us therefore help it on it way! Then we can build a different Europe - one that is not European! A Europe of people and communities, not banks, technocrats, elitist politicians, markets, and all those things that are destroying humanity and the planet.

If we in the West do not approach our friends, our fellow human beings in the East with humility and a willingness to accept that we have made some terrible mistakes, and that we have nothing to learn, only to teach, then our future as a species must surely be in doubt. Now is the moment of universality – are you listening all you cultural colonialist in Brussels, or perhaps we should call this place Ancient Rome!

"All reactionaries are paper tigers. In appearance, the reactionaries are terrifying, but in reality they are not so powerful. From a long-term point of view, it is not the reactionaries but the people who are really powerful."

That was Mao Tse-Tung again.

It is indeed a truth that people are powerful. They are powerful enough to change the course of human history if they choose to do so. We are one people, one species, and we must act together to steer ourselves away from the path that we now follow which leads only to one place – the edge of doom. Thus do we embrace scriptovisualism!

**302**

Now to matters of feminism and science! First to speak is Evelyn Fox Keller:

"I refer to the ideology that asserts an opposition between (male) objectivity and (female) subjectivity and denies the possibility of mediation between the two. A first step in extending the feminist critique to the foundations of scientific thought is to re-conceptualise objectivity as a dialectical process so as to allow for the possibility of distinguishing the objective effort from the objectivist illusion. Piaget said that objectivity consists in so fully realizing the countless intrusions of the self in everyday thought and the countless illusions which result – illusions of sense, language, point of view, value, etc. – that the preliminary step to every judgement is the effort to exclude the intrusive self. Realism, on the contrary, consists in ignoring the existence of self and thence regarding one's own perspective as immediately objective and absolute. Realism is thus anthropocentric illusion, finality – in short, all those illusions which teem in the history of science. So long as thought has not become conscious of self, it is a prey to perpetual confusions between objective and subjective, between the real and the ostensible.

"In short, rather than abandon the quintessentially human effort to understand the world in rational terms, we need to refine that effort. To do this, we need to add to the familiar methods of rational and empirical inquiry the additional process of critical self-reflection. Following Piaget's injunction, we need to *become conscious of self*. In this way, we can become conscious of the features of the scientific project that belie its claim to universality."

Now Dona Haraway speaks:

"Social constructionists make clear that official ideologies about objectivity and scientific method are particularly bad guides to how scientific knowledge is actually made. Just as for the rest of us, what scientists believe or say they do and what they really do have a very loose fit."

This applies to artists too, including those who collaborate with STEM people. We are back once again to E.A.T.!

"The only people who end up actually believing and, God forbid, acting on the ideological doctrines of disembodied scientific objectivity – enshrined in elementary textbooks and techno-science booster literature – are non-scientists, including a few very trusting philosophers."

And we could also add some artists and members of the public. So we are back to Richard Dawkins again.

**303**

Dialogic: relating to or in the form of dialogue. Discursive: relating to discourse or modes of discourse.

Now we come back to the matter of the rational and the objective, of independence of experts, and such forth. To do this we will return to the matter of the United Kingdom's Continuing Membership of the European Union Referendum. In the aftermath of this we heard much about how the *Leave Camp* had misled the British public, but very little about how the *Remain Camp* had misled the British public.

We turn our attention to a BBC Newsnight special (live) programme that was broadcast on 23 May 2016. There was much discussion about countries like Switzerland and the matter of trade deals. The presenter, Evan Davis, asked one of the experts, a professor from Oxford University, to give an assessment of the ease with which Britain can sign trade deals. What follows is a transcript of the expert's response:

"The essence of trade deals is how much market have you got to offer the other side? So if Switzerland wants to have a trade deal with China, it says 'here's our market, quite small, your market's enormous, we need a trade deal with you.' And China says, as they have done, 'fine, you can give us complete entry into your economy, and we will open up for you trade in ours in 15 years' time. Now that's a deal, it's a deal you should do. So my point is that for Britain, if everybody in the world believed in a free market as a matter of religious faith, Britain might have a chance, but that is not how trade negotiations work. Countries say, 'how big's your market?' and that's the deal that we'll do with you. And that's why Britain as part of a European market of 500 million is part of some of some very successful trade deals with more than 60 other countries.

"Now if Britain wants to try to negotiate those alone there's two big problems: one it will always be the rule taker. The big economies will always just dictate the terms. And second, it takes a lot of time, it takes on average 28 months to come up with a trade deal. And the reason that's important is all these flourishing sectors that Dan Hannan talked about: biotech, education, they need investment. Now when investors decide they want to come to Britain, they want to know two things: they want certainty about what rules are going to be, and they want a big market share. So if Britain says 'well, we're not sure what rules are going to have, it's going to take us a few years to negotiate and we're not going to have access to a big market,' Britain's got a problem."

So Evan, any wiser about the ease with which Britain can sign trade deals? No, then we will enlighten you a little. International agreements always take a long time to prepare - this is just a fact of life when governments and international agencies are involved. But one of the reasons why trade deals take a long time to negotiate, is also because they are very specific and what is included is carefully defined, and there are many matters technical in nature that need to be addressed. There are other reasons as well, because they are often more than trade deals.

Here is an artist, on Facebook, in October 2016, who posted a *YouTube* clip, of this expert's speech:

"MUST-WATCH: Oxford University professor Ngaire Woods explains, in fewer than two minutes, trade deals, Brexit and why we must remain part of the European Single Market. Please SHARE this clear, independent expert analysis."

Independent is it?

This is someone from the Confederation of British Industry speaking about the China-Switzerland trade deal:

"In the China-Switzerland trade deal, as the much bigger partner, China has set the terms of trade. It is allowed more time to remove tariffs on Swiss goods - up to 15 years in some cases - than the Swiss, who have to let in Chinese goods tariff-free almost immediately. It is also limited in scope; it does not cover cars or financial services."

So already we have gone from *complete entry into your economy* to *limited entry to your economy,* and not necessarily immediately, and *from we will open up for you trade in ours in 15 years' time* to, *15 years in some cases.*

Now we wonder why the professor got this so wrong? Does she think the Swiss government is stupid and that they would negotiate a deal that was not to Switzerland's advantage?

Rather than using this very imprecise phrase *opening-up markets*, we will instead now refer to the real issue which is that of the elimination of customs duties for we will do what we always do, and look at the text of the trade agreement, where we find that there is a schedule, that places some Swiss goods and services in a category where China will eliminate customs duties within one year from the outset of the agreement, and another category where China will eliminate customs duties within 5 years from the outset of the agreement, and another category where China will eliminate customs duties within 10 years from the outset of the agreement, and another category where China will eliminate customs duties within 12 years from the outset of the agreement, and another category where China will eliminate customs duties within 15 years from the outset of the agreement, and ...

Actually it is complicated, but it is clear that both the CBI and the professor were wrong. We wonder why? We will leave you the reader to ponder on the value of other aspects of the professor's words.

No doubt we will be hearing from the Swiss and some analysts in due course. In the meantime, we remind people of those *warnings about experts* that we heard during the referendum campaign and to close the chapter we invite Paul to read from one of his short stories (*Encounter with a Wise Man*):

"What of the third force, the intellectual one? This is just as disappointing as the other two, but in a much more subtle way. First in science, then later in an activity that came to be known as engineering, and in more recent times technology, a very mechanistic and reductionist view of people would take hold, which ultimately would see humans regarded as little more than machines, albeit very complex ones. Worse still, the importance of scientific, engineering and technological facts would begin to dominate, rendering as secondary, or unimportant, human desires, aspirations and feelings. In short, the wishes of ordinary people would no longer be seen as important, or would be treated with contempt by experts, and the world would begin to descend into an existence where only facts produced by experts would be taken into account, and these experts, knowing best, would trample over the aspirations of ordinary people. Perhaps even worse, though, would be the selectivity of facts presented, with only information that supported the various vested interests being taken into account, and the rest, no matter how legitimate, being branded by the label of lack of right-minded thinking. The world would become dominated by those who advise in their own interests."

**304**

We return to the matter of the natural sciences and the humanities. Now we shall explore something that is more learned than that which one often encounters, for now we will hear an academic speaking about the matter of British thought on the relations between the natural sciences and the humanities. The person speaking is Roger Smith. There is much of course that can be said, but something specific is of interest, which is the use and meaning of the word *science.*

"The growth of authority of natural science during the Victorian period accompanied the spread, in English, of the now customary usage, equating science with natural science or learning claiming to follow the pattern of natural science methods and forms of understanding. Earlier writers referred to natural and moral philosophy, the latter indicating the sphere of learning about human actions. Usage was neither settled nor consistent. But is was more common towards the end of the nineteenth century, to discuss the relations between the sciences and moral philosophy rather than the relations among the sciences, in the way German-language literature discussed relations among the different branches of *Wissenschaft.* Yet it was still common to refer to humanities subjects such as history or political economy as science. However the term *man of science* firmly meant natural scientist.

"Some remarks by the Oxford historian E.A. Freeman well illustrate the changing meaning of the word *science* and the problems its gradual identification with *natural science* caused some scholars with a classical education. 'I use the word science at all simply to assert our right to use it if we please; for our own use I far better like such plain English words as *knowledge* and *learning.*' And: 'When I was young, science in the University of Oxford meant chiefly the knowledge of man's moral faculties, the lore which we learned from Aristotle. It has now taken on itself other meanings, and some special merit and dignity is often claimed for those branches on the strength of this unfair monopoly.'"

Thank you for providing some further insights into the use of and meaning of the word *science.*

And we see that Roger has something further he wants to say.

"Indeed I have. In 1899 the Royal Society of London, participating in a meeting to found an international association of academics, was embarrassed to expose the fact that there was no British body which could speak for learning beyond the natural sciences, which it represented. There was then a debate among British scholars about whether the Royal Society should create a section to include *literature, antiquities and philosophy*, or whether there should be an entirely new academy. The outcome was the new British Academy for the Promotion of Historical, Philosophical and Philological Studies, cementing the institutional separation of the natural sciences and other areas of scholarship. Lord Reay perpetuated the older usage of the word *science*, committing the academy to the study of the moral sciences.

"In 1870 there were hopes for a unified intellectual culture; by 1900, there were many specialists in different disciplines who got on with research without giving the issue much thought."

**305**

There is the matter of how those in the world of STEM think. There is the matter of all those initiatives of artists in industry and in research establishments, and what they actual led to. There is this thing called E.A.T. There is the matter of those who use art for image making - their own image, as in the case of the European Commission's DG CONNECT and its so called STARTS Platform, which we explored in our book: *STARTS - Science, Technology, and the Arts: The Artistic Voices DG CONNECT Silenced*.

Here is Edward Shanken to comment:

"In 1966, when Klüver and Rauschenberg organized *Nine Evenings: Theatre and Engineering*, the landmark event that led to the launch of E.A.T., there was greater disciplinary autonomy and insularity than exists today. John Cage, who participated in the event, claimed that the engineer was separate from artists and other people because of 'his very highly specialized knowledge.' Similarly, Klüver observed that, as a result of their training, engineers are 'locked into a very restricted way of looking at the world,' which prevents them from 'using their brains to change the environment, to make a more human environment, as they should.'"

Just how much has changed? The answer is given in the previous chapters.

Edward Shanken again:

"Over and above Klüver's specific goal of making materials, technology and engineering available to any contemporary artist, E.A.T. created an institutional infrastructure to facilitate and enable communication and collaboration between artists and engineers when practitioners in these disciplines had little or no access to one another either socially or professionally. Ultimately, Rauschenberg believed, the success of E.A.T. could be measured by the extent to which it had become a redundant organization – in other words, that artist-engineer collaborations would have become so commonplace that E.A.T. no longer was needed to facilitate them.

"To what extent has E.A.T. achieved its goals?"

That is a very good question Edward!

"What cultural changes and institutional formations have emerged over the last forty years that facilitate or promote interdisciplinary collaborations between artists, scientists and engineers? What lingering or new structural problems hinder such collaborations?"

Yes indeed, Edward. Who has unbiased answers to such questions?

"There has been many collaborations since E.A.T. was formed. There are some well known examples of big corporations inviting artists into their establishments."

Yes indeed Edward, and for reasons not entirely clear, and most of them have come to an end without any clearly documented record of what was done or achieved.

"Despite their largesse, it would be naïve to imagine that industry partners invite artists into their labs or provide funding for the sole purpose of research. In artworks such as *On Social Grease* and *Mobilization*, Hans Haacke poignantly revealed how supporting the arts can whitewash a company's tainted image, transforming it into a corporate good citizen. By accepting corporate sponsorship, Haacke later argued, artists and cultural institutions become complicit in supporting the interests of capitalism and globalization. Nonetheless, it is difficult to imagine a *pure* place of operations where artists can produce and exhibit work that is autonomous from economics. For some artists, whose practices demand access to and participation in the development of emerging technologies, there may be little option but to rely on corporate, public or institutional partners. At the same time, individual artists have succeeded in producing technologically complex work, and alternative spaces have presented it without relying on such support."

So once more we are back to our book *STARTS – Science, Technology, and the Arts: The Artistic Voices DG CONNECT Silenced*, where some of the above issues are also raised in *connection* with DG CONNECT and its appropriation of the arts.

Today is the day that the FET Flagship evaluation panel abandoned any pretence at being independent and agreed to write a report *helpful* to the European Commission.

**306**

The Swiss Government have been in touch, and want to respond to the Oxford University professor. So here, on behalf of the Swiss Government, is one of its representatives:

"A comprehensive bilateral Free Trade Agreement (FTA) between Switzerland and the People's Republic of China was signed by Federal Councillor Johann Schneider-Ammann and the Chinese Minister of Commerce on 6 July 2013. A bilateral agreement on labour and employment issues was signed simultaneously by Federal Councillor Johann Schneider-Ammann and the Chinese Minister of Human Resources and Social Security. Following initial exploratory contacts in November 2007, joint workshops in 2009 and a joint feasibility study in 2010, the negotiations were officially launched in January 2011. The agreement was then negotiated in nine rounds of negotiations and various inter-sessional meetings from April 2011 to May 2013. After the notification of the completion of the internal procedures of approval (on the Swiss side approval by the two chambers of Parliament) in April 2014, the FTA entered into force on 1 July 2014 and the agreement on labour and employment issues on 9 June 2014.

"The FTA will improve mutual market access for goods and services, enhance legal security for the protection of intellectual property and for bilateral economic relations in general. Together with the related agreement on labour, it contributes to sustainable development and promotes cooperation between Switzerland and China.

"For the vast majority of bilateral trade, the FTA will dismantle tariffs fully or partially, sometimes subject to transition periods. In the area of technical barriers to trade and sanitary and phytosanitary measures, sector-specific cooperation agreements aim at reducing non-tariff barriers to trade. For trade in services, more precisely defined rules compared with the agreement on trade in services GATS of the WTO will apply, for example, for approval processes, as well as improved market access commitments for various services.

"Regarding intellectual property, the level of protection in selected areas will be improved compared with the multilateral standards of the WTO, including in the area of enforcement. The FTA provides for coherent implementation, orientated towards basic principles of international relations and the objective of sustainable development. To this end, the preamble stipulates basic principles of the UN and of public international law, while other provisions of the FTA and the parallel agreement on labour and employment issues concluded simultaneously with the FTA contain rules on environmental and labour issues respectively. Finally, the FTA provides for the deepening of bilateral cooperation in various other fields."

So tell us something about the FTA for Switzerland:

"China is the world's second largest economy after the United States and one of Switzerland's most important foreign trade partners. China is the largest buyer of Swiss industrial products in Asia and the third largest worldwide (after the European Union and the United States). In 2013, Switzerland exported goods to the value of CHF 8.7 billion to China (4.1% of all Swiss exports), with Swiss imports from China totalling CHF 11.4 billion (6.1% of total imports). The main products exported by Switzerland to China include machines and instruments, watches and chemical and pharmaceutical products. Swiss imports from China include machinery, textiles and clothing, and watch-making and chemical products. Also trade in services is of importance. Many Swiss service providers operate in China (including banks, insurance companies, logistics companies, quality inspection companies, management consultants). Conversely, Chinese service providers show increasing interest in Switzerland as a business location.

"The agreements cover the following subjects: trade in goods (industrial and agricultural products), rules of origin, customs procedures and trade facilitation, trade remedies, technical regulations, sanitary and phytosanitary measures, trade in services, protection for intellectual property, competition, investment promotion, transparency in government procurement, trade-related environmental and labour issues, economic and technical cooperation, institutional provisions (Joint Committee, consultation process, dispute settlement).

"With such comprehensive coverage, the FTA establishes preferential trade relations for goods and services between Switzerland and China, facilitates two-way trade, strengthens the protection of intellectual property and generally improves legal security for economic exchange. Together with the Agreement on Labour and Employment, it will promote bilateral cooperation between Switzerland and China and contribute to sustainable development. From a Swiss perspective, the FTA enables a competitive advantage compared with countries which have do not have a FTA with China. At the same time, the FTA prevents discrimination against Swiss economic operators compared with China's existing and future free trade partners. The Agreement also establishes an institutional framework for cooperation between authorities for the monitoring and further development of the agreement and for resolving specific issues.

"Since the economic relations of Switzerland with the large and fast-growing Chinese market are very important for the diversification of Swiss export destinations as well as for the procurement of imports, the FTA between Switzerland and China will significantly contribute to strengthening the position of Switzerland as a business location."

And as regards the professor's comment about access to Chinese markets in fifteen years, what do you say to that?

"At the entry into force of the FTA, the remaining Swiss tariffs on Chinese industrial products will be abolished. This means that under the FTA, in addition to the industrial products that - on the basis of the autonomous Generalised System of Preferences for developing countries - can already now be imported, including from China, tariff-free into Switzerland, Chinese textiles and shoes will be granted tariff-free access to the Swiss market. Conversely the vast majority of Switzerland's industrial exports to China will enjoy full or partial tariff dismantling, either from the entry into force of the Agreement or with dismantling periods of 5, 10 (in a few cases 12 or 15) years. Transitional periods and partial dismantling relate to products for which China has claimed to have a specific need for adjustment given the sometimes substantially higher level of tariffs (for example, selected products in the watch-making, machinery and chemical sectors), exceptions from tariff dismantling relate to some particularly sensitive tariff lines (notably in the latter two sectors)."

Would you like to say what you think about the professor, given that you have just demonstrated that she was not speaking accurately about the facts of the agreement?

"I am afraid that diplomatic protocols prevent me from saying, that which I obviously need to say!"

Understood! Never mind, for we have with us instead, Mark Twain to make a comment:

"There are lies, damn lies and that which a certain professor from Oxford University said on a BBC 2 Newsnight programme in May 2016."

**307**

A cold night, with a clear sky, and on waking, there is Orion stalking across the sky, and on the car roofs, the first frost of the autumn season. And in the daylight we see, for it is a beautiful sunny day, the golden colours of the season, but still much deep green remains. The climate has changed, for only now at the beginning of November is autumn underway, yet travel back in time, to the 1970s, and this sight would have been seen much earlier in the year. In that decade, the seasons began to turn in early September. But not in these times, for September is most often sunny and warm and people speak of Indian Summers, which are not anymore a rare thing.

Returning to the matter of consilience, Edward O. Wilson is here once again to speak about this:

"Consilience is the key to unification. I use this word because it has only one meaning, while words such as coherence have several possible meanings."

Yes you have already said that. The word was used, as you have said, by William Whewell in his 1840 book *The Philosophy of the Inductive Sciences*. You say that it means literally a jumping together of knowledge by the linking of facts and fact-based theory across *disciplines* to create a common groundwork of explanation. You then quote Whewell. But is that what William Whitwell actually meant - across disciplines?

We have invited William Whewell to speak. He will tell us about consilience:

"The *Consilience of Inductions* takes place when an Induction, obtained from one class of facts, coincides with an Induction, obtained from another different class. This Consilience is a test of the truth of the Theory in which it occurs."

This is what Edward O. Wilson quotes in his book. So anything further to say William?

"Of course."

Then let us hear more about consilience - that which Edward O. Wilson, being engaged in reduction, reduced out of existence!

"The statement about consilience is actually an aphorism of the inductive sciences - one of 56 that I listed in the first volume of my work, the title of which you already know. I also discussed consilience in volume II of the work."

Let us reiterate this point: Edward O. Wilson neglected to mention that consilience is an aphorism of the inductive sciences! A fact that he just reduced out of existence! Consilience is not about the jumping together of knowledge by the linking of facts and fact-based theory across *disciplines* to create a common groundwork of explanation.

Tell us William what you said in volume II.

"Men cannot help believing that the laws laid down by discoverers must be in a great measure identical with the real laws of nature, when the discoverers thus determine effects beforehand in the same manner in which nature herself determines them when the occasion occurs. Those who can do this, must, to a considerable extent, have detected nature's secret; must have fixed upon the conditions to which she attends, and must have seized the rules by which she applies them. Such a coincidence of untried facts with speculative assertions cannot be the work of chance, but implies some large portion of truth in the principles on which the reasoning is founded. To trace order and law in that which has been observed, may be considered as interpreting what nature has written down for us, and will commonly prove that we understand her alphabet. But to predict what has not been observed, is to attempt ourselves to use the legislative phrases of nature; and when she responds plainly and precisely to that which we thus utter, we cannot but suppose that we have in a great measure made ourselves masters of the meaning and structure of her language. The prediction of results, even of the same kind as those which have been observed, in new cases, is a proof of real success in our inductive processes.

"We have here spoken of the prediction of facts of the same kind as those from which our rule was collected. But the evidence in favour of our induction is of a much higher and more forcible character when it enables us to explain and determine cases of a kind different from those which were contemplated in the formation of our hypothesis. The instances in which this has occurred, indeed, impress us with a conviction that the truth of our hypothesis is certain. No accident could give rise to such an extraordinary coincidence. No false supposition could, after being adjusted to one class of phenomena, exactly represent a different class, when the agreement was unforeseen and uncontemplated. That rules springing from remote and unconnected quarters should thus leap to the same point, can only arise from that being the point where truth resides.

"Accordingly the cases in which inductions from classes of facts altogether different have thus jumped together, belong only to the best established theories which the history of science contains. And as I shall have occasion to refer to this peculiar feature in their evidence, I will take the liberty of describing it by a particular phrase; and will term it the Consilience of Inductions.

"It is exemplified principally in some of the greatest discoveries. Thus it was found by Newton that the doctrine of the Attraction of the Sun varying according to the Inverse Square of this distance, which explained

Kepler's Third Law of the proportionality of the cubes of the distances to the squares of the periodic times of the planets, explained also his First and Second Laws of the elliptical motion of each planet; although no connexion of these laws had been visible before. Again, it appeared that the force of Universal Gravitation, which had been inferred from the Perturbations of the moon and planets by the sun and by each other, also accounted for the fact, apparently altogether dissimilar and remote, of the Precession of the equinoxes. Here was a most striking and surprising coincidence, which gave to the theory a stamp of truth beyond the power of ingenuity to counterfeit. In like manner in Optics; the hypothesis of alternate Fits of easy Transmission and Reflection would explain the colours of thin plates, and indeed was devised and adjusted for that very purpose; but it could give no account of the phenomena of the fringes of shadows. But the doctrine of Interferences, constructed at first with reference to phenomena of the nature of the Fringes, explained also the Colours of thin plates better than the supposition of the fits invented for that very purpose. And we have in Physical Optics another example of the same kind, which is quite as striking as the explanation of precession by inferences from the facts of perturbation. The doctrine of Undulations propagated in a Spheroidal Form was contrived at first by Huyghens, with a view to explain the laws of Double Refraction in calc-spar; and was pursued with the same view by Fresnel. But in the course of the investigation it appeared, in a most unexpected and wonderful manner, that this same doctrine of spheroidal undulations, when it was so modified as to account for the directions of the two refracted rays, accounted also for the positions of their Planes of Polarization a phenomenon which, taken by itself, it had perplexed previous mathematicians, even to represent.

"The Theory of Universal Gravitation, and of the Undulatory Theory of Light, are, indeed, full of examples of this Consilience of Inductions. With regard to the latter, it has been justly asserted by Herschel, that the history of the undulatory theory was a succession of felicitie. And it is precisely the unexpected coincidences of results drawn from distant parts of the subject which are properly thus described. Thus the Laws of the Modification of polarization to which Fresnel was led by his general views, accounted for the Rule respecting the Angle at which light is polarized, discovered by Sir D. Brewster. The conceptions of the theory pointed out peculiar Modifications of the phenomena when Newton's rings were produced by polarized light, which modifications were ascertained to take place in fact, by Arago and Airyi. When the beautiful phenomena of Dipolarised light were discovered by Arago and Biot, Young was able to declare that they were reducible to the general laws of Interference which he had already established. And what was no less striking a confirmation of the truth of the theory, measures of the same element deduced from various classes of facts were found to coincide. Thus the length of a luminiferous undulation, calculated by Young from the measurement of fringes of shadows, was found to agree very nearly with the previous calculation from the colours of thin plates.

"No example can be pointed out, in the whole history of science, so far as I am aware, in which this *Consilience of Inductions* has given testimony in favour of an hypothesis afterwards discovered to be false. If we take one class of facts only, knowing the law which they follow, we may construct a hypothesis, or perhaps several, which may represent them: and as new circumstances are discovered, we may often adjust the hypothesis so as to correspond to these also. But when the hypothesis, of itself and without adjustment for the purpose gives us the rule and reason of a class of facts not contemplated in its construction, we have a criterion of its reality, which has never yet been produced in favour of falsehood."

Thank you, William, for that very precise explanation of the Consilience of Inductions. So now we have dealt with the Consilience of Inductions. And it turns out that consilience has nothing to do with jumping together of knowledge by the linking of facts and fact-based theory across *disciplines* to create a common groundwork of explanation, but the jumping together of knowledge by the linking of facts and fact-based theory from a single discipline, being physics in the cases examined, to create a common groundwork of explanation.

Now Edward, why did you say, jumping together of knowledge by the linking of facts and fact-based theory across *disciplines* to create a common groundwork of explanation, when you should have known that Whewell was referring to *single disciplines*? Were you being too reductive? Were you too caught-up in scientism and hubris?

Do not go away for we have not finished with you yet!

The reader will of course be aware at this stage in the book, that in 1840, there were not the divisions across disciplines that had taken shape by the end of the nineteenth century. Nor were whatever divisions existed, seen to be as problematic as they are today. So consilience could never have been proposed as a way of dealing with this matter, which was not an issue at that time. Was Wilson proposing therefore that it could be used as such? Read his book and judge for yourself. Can it be used for such? It does not seem likely given that theories from the social sciences do not, for example, try to explain physical phenomena. But there can be other ways of seeking a unity of knowledge. This is what we are pursuing. It seems that the establishment is not – once again!

Now we will encounter a different view of the unity of knowledge. This view goes by the name cybersemiotics, and here to speak about this is Søren Brier:

"Cybersemiotics constructs a non-reductionist framework in order to integrate third person knowledge from the exact sciences and the life sciences with first person knowledge described as the qualities of feeling in humanities and second person intersubjective knowledge of the partly linguistic communicative interactions, on which the social and cultural aspects of reality are based.

"The modern view of the universe as made through evolution in irreversible time, forces us to view man as a product of evolution and therefore an observer from inside the universe. This changes the way we conceptualize the problem and the role of consciousness in nature and culture. The theory of evolution forces us to conceive the natural and social sciences as well as the humanities together in one theoretical framework of unrestricted or absolute naturalism, where consciousness as well as culture is part of nature. But the theories of the phenomenological life world and the hermeneutics of the meaning of communication seem to defy classical scientific explanations. The humanities therefore send another insight the opposite way down the evolutionary ladder, with questions like: What is the role of consciousness, signs and meaning in the development of our knowledge about evolution? Phenomenology and hermeneutics show the sciences that their prerequisites are embodied living conscious beings imbued with meaningful language and with a culture.

"One can see the world view that emerges from the work of the sciences as a reconstruction back into time of our present ecological and evolutionary self understanding as semiotic intersubjective conscious cultural and historical creatures, but unable to handle the aspects of meaning and conscious awareness and therefore leaving it out of the story. Cybersemiotics proposes to solve the dualistic paradox by starting in the middle with semiotic cognition and communication as a basic sort of reality in which all our knowledge is created and then suggests that knowledge develops into four aspects of human reality: Our surrounding nature described by the physical and chemical natural sciences, our corporality described by the life sciences such as biology and medicine, our inner world of subjective experience described by phenomenologically based investigations and our social world described by the social sciences. I call this alternative model to the positivistic hierarchy the cybersemiotic star."

Here we move beyond scientism and hubris; the beginning of something that is not tarnished by the past, which constantly impinges upon what the European world thinks and does. Yet Brier leaves science unchanged – it needs to be changed!

**309**

Now we have a spokesperson with us from the DEMOS think tank who will speak about the idea of upstream public engagement in science:

"For more than a decade, the language and methods of Public Understanding of Science (PUS) oozed across the face of United Kindom science policy. But instead of lubricating understanding, scientists gradually discovered that PUS was clogging the cracks and pores which might have allowed genuine dialogue to breathe. Implicit within PUS was a set of questionable assumptions about science, the public and the nature of understanding. It relied on a *deficit model* of the public as ignorant and science as unchanging and universally comprehensible. Partly as a result of PUS's failings, relations between science and society festered throughout the 1990s, and an occasional rash of blisters erupted (the BSE crisis, GM crops, mobile phones, MMR). It wasn't until 2000 that PUS was washed away, when an influential House of Lords report detected 'a new mood for dialogue'. Out went PUS, which even the government's chief scientific adviser now acknowledged was a rather backward looking vision. In came the new language of *science and society*, and a fresh impetus towards dialogue and engagement.

"The House of Lords report detected a new humility on the part of science in the face of public attitudes, and a new assertiveness on the part of the public. And this resulted in a perceptible change. Consultation papers, focus groups, stakeholder dialogues and citizens' juries were grafted on to the ailing body of British science, in the hope that they would give it a new lease of life. Every so often, a few drops of PUS still dribble out from a Lewis Wolpert or a Lord Taverne, but these voices are now a dwindling force. The science community has embraced dialogue and engagement, if not always with enthusiasm, then at least out of recognition that BSE, GM, and other controversies have made it a non-negotiable clause of their *licence to operate*.

"Yet despite this progress, the link from public engagement back to the choices, priorities and everyday practices of science remains fuzzy and unclear. Processes of engagement tend to be restricted to particular questions, posed at particular stages in the cycle of research, development and exploitation. Possible risks are endlessly debated, while deeper questions about the values, visions and vested interests that motivate scientific endeavour often remain unasked or unanswered. And as the GM case demonstrates, when these larger issues force themselves on to the table, the public may discover that it is too late to alter the developmental trajectories of a technology. Political, economic and organisational commitments may already be in place, narrowing the space for meaningful debate.

"But a new term has entered the lexicon of public engagement. Scientists and science policy-makers increasingly recognise the limitations of existing approaches, and there has been a surge of interest in moving engagement *upstream*.

"The task of upstream engagement is to remove some of the structures that divide the back-stage from the front-stage. It seeks to make visible the invisible, to expose to public scrutiny the values, visions and assumptions that usually lie hidden. In the theatre of science and technology, the time has come to dismantle the proscenium arch and begin performing in the round."

Indeed, but will science listen to what people want? Or will they just go ahead and do it anyway disregarding the views expressed by the public? Is upstream public engagement just a way of manipulating public opinion? Is it a way for the STEM community to determine the agenda, to frame discussions about risks within narrow terms that they find acceptable?

The questions are pertinent ones given the apparent willingness of organisations like the Royal Society and the Royal Academy of Engineering to produce biased reports that just represent the views of vested interests with conflicts of interest. We refer you to the Royal Society report *Reaping the Benefits,* to which we have already referred and the joint Royal Society/Royal Academy of Engineering report on *Fraking,* which opens by stating that the risks can be managed, but then proceeded to list many uncertainties, while ignoring the key matter of unintended consequences.

The spokesperson for DEMOS:

"Researchers at Lancaster University compared public attitudes towards GM and information technology. They found that controversies arise in areas where, although it is sensed that there is no knowledge, this is conveyed misleadingly in terms of reductionist scientific uncertainty. In such situations, the research team identified a *deep cultural dislocation* between the way that policy-makers and the public frame relevant questions. While the former tend to ask simply 'What are the risks?', the latter ask in addition, 'What might be the unanticipated effects? Who will be in charge of, and will take responsibility for, the responses to such surprises? And can we trust them?' Many public engagement processes, however well intentioned, get caught in this trap. Questions of risk – the known uncertainties – can easily dominate proceedings and squeeze out broader discussion of unknown or unanticipated consequences."

And it is these unknown or unanticipated consequences that are the theme of Paul's novel *Moments in Time*. The statement that *risks can be managed* is a culture, not a proper response to the legitimate concerns of the public.

And who among the artists in residence engage in upstream public engagement?

**310**

Time to learn something more about semiotics and here is Elizabeth Chaplin to explain from the perspective of sociology and visual representation:

"Charles Peirce's semiotic theory draws visual art and sociology, as forms of communication, even closer together. I will now briefly explain his typology of signs. In this, the *iconic* sign proposes a fitness of resemblance to the object it signifies, as a portrait resembles the sitter. The *indexical* sign has a concrete, actual relationship to the signifier – usually of a sequential, causal kind – in the sense that smoke is the index of fire. The *symbolic* sign signifies by virtue of a contract or rule – the equivalent of Saussure's arbitrary linguistic sign. In this triad, the iconic, indexical and the symbolic signs are not mutually exclusive. Rather they are three modes of a relationship between signifier and object or signifier and signified which co-exist in the form of a hierarchy in which one of them will inevitably have dominance over the other two. And each of these signs bears a different relation to the interpretant as well as to the object."

And what relevance does this have for scriptovisualism?

"Analysis of figurative paintings clearly shows how images signal iconically and symbolically. While an image of a table propose a fitness of resemblance to the table it signifies (iconic signal), this is not the only message that the image gives off. It may symbolise upper-class affluence and dinner parties or it may symbolise poverty and toil. Colour can also signify symbolically.

"But there is an area of common ground between pictorial and linguistic signs, because just as visual images signal symbolically, linguistic signs also exhibit iconic characteristics.

"Both forms of discourse, that is the visual and the sociological, contribute to society's cultural patterns, visual art works cannot be purely visual, and sociological texts cannot be purely verbal even when they contain no images. And while in visual texts iconic communication may predominate, whereas in sociological texts symbolic communication is stronger, both texts and images exhibit iconic and symbolic characteristics in different hierarchical combinations."

**311**

What path have we meandered along, to bring us to this point, this step along the journey that is 366 chapters long?

Now we will consult with an expert – yes one of those people again! This time it is an expert, a researcher, on Sino-European relations and economic diplomacy. He is based at the Norwegian Institute of International Affairs. His name is Marc Lanteigne. He now speaks:

"The successful completion of the Sino-Swiss free trade agreement marks a significant economic and diplomatic achievement for Switzerland. In sharp contrast to the European Union, Switzerland, in the tradition of its *Sonderfall* (exceptional case) politics and foreign policy, has sought to maximise the benefits of its agreement with China since the time it came into force. The agreement, in addition to removing bilateral tariffs and trade barriers on goods and services, also provides for intellectual property protection, which was a key demand from Switzerland since the start of the negotiations and a main reason why free trade talks between Switzerland and India experienced delays. The deal was also accompanied by parallel agreements on employment and labour. Other provisions of the agreement include investment promotion, technical cooperation and environmental protection, and overall the deal has provided a platform for future bilateral economic cooperation, notably in the area of banking.

"In July 2014, a three-year currency swap agreement was signed, worth approximately US$24 billion, between the People's Bank of China and the Swiss National Bank (SNB), also allowing for the SNB to purchase up to about US$2.4 billion in Chinese bonds. The swap agreement has provided China with improved access to the Swiss franc which has been viewed as a safe haven currency in comparison with the euro, especially since January 2015 when the decision was made by the SNB to allow the franc to appreciate against the euro after an exchange rate cap implemented in 2011 was abruptly removed.

"Also in January, after many months of anticipation, the decision was finalised to designate Zürich as a trading hub for China's currency the renminbi, joining a small number of cities, including London, Luxembourg, Toronto and Singapore, that hold this status. Shortly after that announcement was made, Swiss Finance Minister Eveline Widmer-Schlumpf announced that the China Construction Bank would open a Swiss branch during 2015 in the wake of the hub agreement, and the bank formally applied for a license in April 2015. Finally, in March, Switzerland joined a long queue of European economies seeking founding member status in Beijing's Asia Infrastructure Investment Bank, opening yet another door to economic cooperation.

"The trade volume between China and Switzerland has been growing rapidly over the past decade, with Swiss exports to China rising from CHF 2.5 billion to CHF 8.8 billion (US$2.7 billion to US$9.4 billion), between 2003-13, and Chinese imports in Switzerland jumping from CHF 2.5 billion to CHF 11.4 billion (US$2.7 billion to US$12.2 billion) during the same time period. It is too early to measure the degree to which the start of the Sino-Swiss free trade agreement will boost overall bilateral trade, but the trading relationship between China and New Zealand, which signed its own free trade agreement with Beijing in 2008, might be a useful indicator.

"In 2010, Sino-New Zealand trade stood as NZ$10 billion (US$7.6 billion), and that year the leaders of both states set a target to bring that figure to NZ$20 billion (US$15.1 billion) by 2015. That target was actually reached ahead of schedule, in June 2014. With the Sino-Swiss free trade agreement coming into effect, a similar rise in trade volume is expected in the coming years, to the benefit of many sectors. Even Swiss luxury watch sales in China, which softened during 2012-2013 as the incoming Xi Jinping administration announced sweeping government austerity measures, showed signs of recovery by 2014. According to a June 2014 report by Credit Suisse, China is expected to overtake Germany as Switzerland's largest trading partner by 2035, based on current figures.

"At the end of 2014, President Xi Jinping announced that China would accelerate its free trade strategy and seek out new free trade agreement partners in the coming years. The rise of China as an economic power and its growing influence in Europe, coupled with the initial successes in Switzerland in completing its free trade agreement and building new financial links with Beijing, should weigh considerably on the question of whether the European Union may more seriously consider adding China to its own list of free trade partners in the near future."

So a successful trade agreement for the Swiss and another exposé of experts who are willing to ... We leave you to decide what the Oxford University professor was doing.

Who is next on the list of European countries that China will sign trade deals with? Any guesses? There are not many to choose from as the Member States of the European Union are prevented by the Customs Union from negotiating their own trade deals.

**312**

The matter of the Nazis, does today come to mind, thinking as we are of the European Commission and Friedrich von Hayek, and his essay *The Road to Serfdom,* and what he has to say about scientists and engineers, especially in that chapter with the title *The Totalitarians in Our Midst.*

Here is Friedrich to speak:

"Possibly we have not given enough attention to one feature of the intellectual development in Germany during the last hundred years which is now in an almost identical form making its appearance in this country: the scientists agitating for a *scientific* organisation of society. Frequently conspicuous in the scientific specialist, the expert, is the intolerance of reason and an impatience with the ways of the ordinary man, and contempt for anything which is not consciously organised by superior minds according to a scientific blueprint. Nothing better illustrates the effects on a nation of a general and thorough shift of the greater part of its educational system from the *humanities,* to the *realities,* than Germany between 1840 and 1940.

"The way in which, in the end, with few exceptions, her scholars and scientists put themselves readily at the service of the new rulers is one of the most depressing and shameful spectacles in the whole history of the rise of National Socialism."

Here is R.A. Brady to comment:

"The scientist, *per se,* is, perhaps the most easily used and coordinated of all the especially trained people in modern society. The Nazis, to be true, fired a good many University professors, and dismissed a good many scientists from research laboratories. But the professors were primarily among the social sciences where there was more common awareness of and a more persistent criticism of the Nazi programmes, and not among the natural sciences where thinking is supposed to be more rigorous.

"Those dismissed in this latter field were primarily Jewish or exceptions to the generalisations made above, because of the equally uncritical acceptance of beliefs running contrary to Nazi views. Consequently the Nazis were able to coordinate scholars and scientists with relative ease, and hence to throw behind their elaborate propaganda the seeming weight of the bulk of German learned opinion and support."

It is therefore not very surprising that the FET Flagship evaluation panel just followed the lead that they had been given, and started rewriting their findings so that they agreed with what the European Commission wanted.

**313**

Now we are going to hear from someone connected with the European Commission, from DG CONNECT, who is going to say something about the STARTS Platform - that is, Science, Technology and the Arts!

First however Friedrich von Hayek will quote Julien Benda:

"There is a superstition of science held to be competent in all domains, including that of morality; a superstition which, I repeat, is an acquisition of the nineteenth century. It remains to discover whether those who brandish this doctrine believe in it or whether they simply want to give the prestige of a scientific appearance to passions of their hearts, which they perfectly know are nothing but passions. It is to be noted that, the dogma that history is obedient to scientific laws, is preached especially by partisans of arbitrary authority. This is quite natural, since it eliminates the two realities they most hate, that is to say, human liberty and the historical action of the individual."

This could have been a reference to Richard Dawkins, had it not been written in the early 1940s. He was in fact speaking about Richard Dawkins long before Dawkins made himself notorious. He could have been speaking about people from the European Commission. Here is one of them, a scientist - yes another one - informing you about the use of art in research and innovation, because he thinks he is competent to speak about art in this context:

"As you know we had art-science programmes for a long time and they were mainly about dissemination and communication. I think this was a viable goal and it has been a viable goal for thirty years. To have art as an ingredient in the knowledge creation and programme development process is a new idea. It is very different to say artists are there for dissemination, for explaining research to the general public or to say that the artists are an integral part of the development process. I don't think this is the same level of funding and certainly STARTS is of the second type."

Oh dear someone thinks that artists working in knowledge creation and programme development process is a new idea! Now where do people get these silly ideas? This is the problem with technocracy and technocrats. STARTS is a case study in the failure of technocracy for it includes the elements that render STARTS a worst practice case of study in the area of the arts and its participation in research and knowledge creation - which is not new. STARTS also demonstrates that DG CONNECT, along with many STEM people, are interested in totalitarian art!

Here is a spokesperson from the United States National Academy of Sciences to read from a report, called *Beyond Productivity*, detailing a study undertaken in 2002:

"The European Commission's first round of hybrid research networks was focused on intelligent information interfaces (I3) and linked art centres, corporations, and national laboratories; continued work is being planned to support high-quality digital content and cultural-heritage preservation efforts.

"The I3 networks attempt to join art, product design, and academic research and development with corporate product development and market research. In addition, research networks for digital arts began to appear in the 1990s, consistent with a turn away from the stovepipe separation of R&D and commercialization. The concept is captured in the European Commission's Fifth Framework Programme, which emphasizes cross-disciplinary work and features as a theme the *User-Friendly Information Society*."

So when it comes to art-science, art-technology, people from the European Commission, and those who gather around the pig-trough talking utter nonsense - it looks like they are taking illegal substances! How else can one explain their stupidity? Idiocy? Ignorance? Corruption? Or, perhaps, all three?

**314**

Today we wake up to the news that Donald Trump has been elected 45th President of the United States of America. Many are outraged about it, yet none seem to know what is happening. We did though tell you that something rather important is happening today in our world. The word fascism is being used by many, but do they really know what fascism actually is? Have they taken the trouble to look, to discover, for example, what actually, for example, National Socialism - the Nazis ideology - is actually based upon? They would be surprised if they looked. We looked, and found something very European. It is still, in part, with us today - in the European Commission!

"Why no Descartes or Newton, under the Heavenly Mandate?"

It is Edward O. Wilson again! Seems like a very typical comment from someone rooted in a European culture, with limited understandings of other cultures, and all too ready to dismiss other cultures and their histories as an irrelevance, and an inclination towards - superiority. The Narcissus Syndrome! Here he goes:

"The reasons were historical and religious. The Chinese had a distaste for abstract codified law, stemming from their unhappy experience with the Legalists, rigid quantifiers of law who ruled during the transition from feudalism to bureaucracy in the Ch'in dynasty (221-206 BC). Legalism was based on the belief that people are fundamentally antisocial and must be bent to laws that place the security of the state above their personal desires. Of probably even greater importance, Chinese scholars abandoned the idea of a supreme being with personal and creative properties. No notional Author of Nature existed in their universe; consequently the objects they meticulously described did not follow universal principles, but instead operated within particular rules followed by those entities in the cosmic order. In the absence of a compelling need for the notion of general laws - thoughts in the mind of God so to speak - little or no search was made for them."

That is one hell of a reductive statement that you have just made there, but you are a reducer, a simplifier, and what you just said is simple!

"Western science took the lead largely because it cultivated reductionism and physical law to expand the understanding of space and time beyond that obtainable by the senses."

And reductionism led to the fragmented and reductive mind, of which your comments are well illustrative.

"Reductionism, given its unbroken string of successes during the next three centuries, may seem today the obvious best way to have constructed knowledge of the physical world, but it was not so easy to grasp at the dawn of science. Chinese scholars never achieved it. They possessed the same intellectual ability as Western scientists, as evidenced by the fact that, even though they were far more isolated, they acquired scientific information as rapidly as did the Arabs, who had all the Greek knowledge as a launch pad. Between the first and thirteenth centuries they led Europe by a wide margin.

"But according to Joseph Needham, their focus stayed on holistic properties and on the harmonious, hierarchical relationships of entities, from stars down to mountains and flower and sand. In this world view the entities of nature are inseparable and perpetually changing, not discrete and constant as perceived by the *Enlightenment* thinkers. As a result the Chinese never hit upon the entry point of abstraction and break-apart analytic research attained by European science in the seventeenth century."

Now let us examine the reduction - the simplifications - that you just made. We will now tell you, dear reader, where the above came from.

Edward O. Wilson consulted (as you will discover from the notes section of Wilson's book), *The Shorter Science and Civilisation of China: An Abridgement of Joseph Needham's Original Text, Volume I,* prepared by Colin A. Ronan. Wilson did not consult Joseph Needham's work directly. We will explain:

The above mentioned abridgement is a reduction into a 306 page book, of two volumes of Needham's monumental project which still continues to this day, to research and document the Chinese contribution to the development *of Science, Scientific Thought, and Technology,* in various historical periods, ancient and medieval.

Needham points out in the Preface to Volume I, that for later periods, after the coming of the Jesuits to Beijing in the early seventeenth century, Chinese science gradually fused into the universality of modern science. This latter point of course Edward O. Wilson does not mention! That is reductionism for you!

Volume I of Needham's work is an introductory orientation (248 pages not counting end and title page material); Volume II is a history of scientific thought (583 pages not counting end and title page material). What Edward O. Wilson is reproducing in his book is some of the text from page 306 of Ronan's abridgement, which is largely a reproduction of pages 582-3 of Volume II of Needham's full work. Wilson is also reproducing some text from Ronan's introduction (page 2), which is in itself a reproduction of text from the Preface of Volume I of Needham's work.

Now Ronan will tell you what Edward O. Wilson left out from page 306 of the abridgement:

"Yes, for following on from the three reasons why conception of the laws of nature did not develop in China, I wrote (as indeed did Needham in Volume II of *Science and Civilisation in China*) that the Chinese world-view depended on a totally different line of thought. The harmonious cooperation of all beings arose because they were all parts in a hierarchy that formed a cosmic pattern. They obeyed the dictates of their own natures, not the orders of some superior authority. Modern science and the philosophy of organism have come back to this wisdom, fortified by a new understanding of cosmic and biological evolutions."

So, in other words, Chinese science was way ahead of Western (European) science, foreseeing the philosophy of organism, long before it emerged in the West in the twentieth century.

Colin, do you want to say something about this to Edward O. Wilson?

"Yes! We must be on our guard against the temptation of thinking that the whole of modern civilisation began with figures like Galileo and Vesalius in the sixteenth and seventeenth centuries, and the conclusion that Wisdom was born with us. There was a Chinese contribution to Man's understanding of Nature and his control over it, and it was a great one. No single people or group of peoples has had a monopoly in contributing to the development of science. All achievements should be recognised, and celebrated, if we are to move on our way to a universal brotherhood of Man."

Joseph Needham will now make an important point:

"Had conditions been more favourable to science, the opposing intellectual forces that I identified in Volume II might have been overcome. All we can say of a Chinese science of Nature which would have then developed, is that it would have been profoundly organic and non-mechanical."

It would have been non-Abrahamic as well, and one day science will change because of China, and to that day we look forward. It is not far away, so it seems that we are back to scriptovisualism again.

**315**

On the matters of European democracy, capitalists, STEM people and the Nazis. In 2008 a very important book was published. It goes by the title *Hell's Cartel: IG Farben and the Making of Hitler's War Machine*. The author's name is Diarmuid Jeffreys.

Here is an editor from Bloomberg News speaking about this book, which is a tale of how a company, whose Nobel-winning scientists discovered vital medicines, became a Nazi collaborator:

"In February 1941, a top manager of German chemical group IG Farben thought he had found the perfect location for a new factory. The site was out of range of Allied bombers, easily reached by rail, and had plenty of water and coal. Most important, there was a ready supply of cheap labour. The site was near a town in occupied Poland called Oswiecim, better known to history as Auschwitz.

"IG Farben went on to build a plant at Auschwitz. It was in effect a privately run death camp, where inmate workers' life expectancy was only a few months. That was hardly the only activity that put IG Farben at the centre of the Nazi nightmare. An IG Farben subsidiary produced Zyklon B, the pesticide repurposed by the SS for the gas chambers. And the conglomerate was the main supplier of chemical products essential to the Nazi war machine.

"Probably never in history has a major corporation been twisted to such evil ends. Yet what is most disturbing is that, for most of its past, the conglomerate and its predecessor companies embodied many of the qualities we admire in business today. Its managers were inventive and diligent, and its scientists discovered vital medicines, with quite a few earning Nobel prizes."

This is what the European Commission wants from its FET Flagships - Nobel prizes and innovation! But the people of Europe may get more than this - again!

"IG Farben traced its origins to the efforts of men such as Carl Bosch of BASF Group, who led the effort to mass-produce synthetic nitrates - a key ingredient in both explosives and fertilisers. Bosch was a man of conscience but also deeply patriotic. During World War I he became a national hero because of his efforts to develop the process used to produce the synthetic nitrates needed to produce explosives, given that natural supplies were no longer available because of the British navel blockade. Most notoriously, BASF chemist Fritz Haber came up with the idea of using chlorine gas as a weapon.

"That was only the first of the moral compromises that foreshadowed greater horrors. As the war drained manpower from the chemical factories, Bayer Chief Executive Friedrich Carl Duisberg lobbied for a novel solution: importing forced labour from occupied Belgium. Germany sent the workers home after vehement international protest, but the precedent was set.

"IG Farben came into being in the mid 1920s. Bosch convinced Bayer, BASF, Hoechst, and many smaller companies to form a cartel powerful enough to withstand turbulent times. Although Bosch, who became chairman, criticized Adolf Hitler publicly as late as 1939, he was unable or unwilling to prevent more opportunistic members of IG Farben management from joining the Nazi Party, providing financial backing to the regime, and taking an active role in secret war preparations.

"IG managers later insisted they knew nothing of Nazi crimes or said they had been coerced. But the Nazi military worked closely with the company: Had IG Farben been even marginally less compliant, Hitler would have struggled to get his war machine moving. About 35% of a German foot soldier's equipment was made of IG Farben materials. Even the keys to Enigma coding machines, the Nazi communications devices were made of IG's plastic.

"The final chapters of Jeffreys' book, becomes a courtroom drama as Jeffreys chronicles United States Army prosecutor Telford Taylor's efforts to convict IG Farben managers of war crimes. Shamefully, rightwing members of the United States Congress tried to impede Taylor's work, perhaps attempting to protect constituents in the chemical industry who had cooperated with IG Farben before the war.

"It is not a pretty history. But it is gripping and full of warnings about the potential of corporations to mutate into criminal enterprises."

It is a history that is now being repeated in the European Commission with its corrupt relationship with its experts - STARTS and also the FET Flagships.

**316**

Today is Armistice Day - the eleventh date, of the eleven month! On this day in 1918, at the eleventh hour, the Great War - the bloodiest conflict in human history - came to an end. And then began the slow walk to the next war, which in turn became the bloodiest conflict in human history. We are always in a state where the bloodiest conflict in human history has yet to happen. When will we arrive at the next one?

The European Union Flag - 12 gold stars, set on a blue background. Some people have been waving it excessively recently, which is not something we British do - wave flags. Some have been waving Union Jack flags too. Not many - because waving flags is distasteful to the British, except if you live in Northern Ireland and are pro the Union that is England, Scotland, Wales and Northern Ireland. The Scots have been waiving their flag much recently as well - the St Andrew's cross.

The European Union Flag - it is still a flag. Swap one form of nationalism, for another, but its still nationalism. It is still a flag. Now Wilfred Owen reads the last lines of his poem, *The Next War*:

"We laughed, knowing that better men would come
and greater wars; when each proud fighter brags
He wars on Death – for life; not men for flags."

Now Nicholas of Cusa speaks:

"We should acknowledge our ignorance of the most profound things. Discovering that we do not know who God is and what life is about is the learning of ignorance, ignorance about the very meaning and value of our lives. This is the starting point for a more grounded, open-ended kind of knowledge that never closes in on fixed opinions."

**317**

On the matter of European democracy, or more correctly the lack of it, we revisit the statement made by the European Commission official from DG CONNECT, for he wants to say some more about all the new things that they want to do with the STARTS Platform. Recall that he is a scientist! Here he goes:

"I repeat that to have art as an ingredient in knowledge creation is new."

Yes, so you said. You are rewriting history. This is what the Nazis did.

"There is a great willingness on our side to really invest into the arts as a means of production, of making technology more human-centred, of using art as a means of social innovation. Technology is a means for society to progress and I think a lot of what art can contribute is in helping technology to become useful for citizens. It is about making technology useful for humans."

So there you have it, technology is not useful to humans! Artists are going to make it so. What a strange belief! What a strange ideology. The art that DG CONNECT is interested in is totalitarian art!

As we mentioned the Nazis and their rewriting of history – yes they did have very strange ideologies and beliefs too; very sinister and evil ones – we will now hear again from Robert A Brady:

"Because of the naïve and credulous popular acceptance of the expert, the scientist, the university, the professor – an acceptance more pronounced among the rank and file in Germany than in any other capitalist country in the world – the propaganda value of co-ordinated science to the Nazis can scarcely be overestimated. By co-ordinating it to Nazi ends they could capitalise on the childlike faith with which science was accepted by both the educated and the ignorant. By calling any body of doctrine, however irrational and self-contradictory, science they have been able to cast a glow of impartial sanctity around that which they seek to promote. In this sense and to the same basic purposes they have exploited to the full the well-known advertising techniques of buying scientific testimonials, using esoteric scientific terms, and generally playing upon the *halo* and awe-inspiring mysticism which the popular mind has surrounded the white-frocked knowers and keepers of the secrets of the universe. And by controlling all media for the expression of scientific thought, they have been able to deflect or neutralise scientific investigation from examination of the thesis, the assumptions, the interests, and the very foundation of the Nazi programme itself."

The propaganda value of coordinated science to the European Commission can scarcely be overestimated. Now art too - coordinated - has become propaganda value.

So we are back once more to the matter of Global Systems Science, the product of the same mind that produced STARTS - the mind of a scientist. Fragmented and reductive! Totalitarian art!

And in the European Commission one finds exactly the sort of minds that one found in the Nazi Party, using propaganda to mislead - to mislead ministers, to mislead Parliament, to mislead the public. And that is not just misleading those in Brussels, but in every Member State. It is more than misleading - it is manipulation.

**318**

On the matter of European democracy, or more correctly the lack of it, we now hear some reflections on the European Commissions *i3* programme, that is to say, the Intelligent Information Interfaces programme, or as we like to call it, the STARTS I Platform:

"The *i3* programme supported interdisciplinary cooperation in art, design and technology. It was the first time that such a rich and complex, multicultural, multidisciplinary community cooperated. It was about listening to and developing technology for ordinary people. It was about involving people, end users, in the business creation and in the product and system evolution. For the first time researchers and companies in Europe were called to investigate and develop new human centred interfaces for interacting with information, aimed at the broad population. And one of the reasons is that the world of stuff is not enough; differentiation is needed, product distinctiveness, and this is achieved by the declaration of values, by combining products and services to suit individual needs, by providing the right customer experience, and through converged and connected services.

"The dialogue among researchers, artists, and designers was intended to stimulate intercultural cooperation to reflect on and improve life in a digital networked society. The *i3* programme promoted research focusing on interaction with and through the information technology by communities of users. The research contributed to a radical innovation of the software culture opening it to the media art and design culture and proposing a fundamental change in the software development process; from *interaction patterns to functions,* instead of the traditional *from functions to interfaces.* The most important perquisite for making the European research system useful is the successful cooperation of artists, computer scientists, designers, media theorists and cultural researchers. And *i3* demonstrated that innovation cannot go ahead without generating scenarios of human experiences to make sense of the technological development."

So, European Commission DG CONNECT official, you have much history to rewrite! And now you have an insight into the failure of technocracy, of its manifestation in the form of Global Systems Science. Time surely, to act against the European Commission? To stop it before it does what all past European Unions have done! But you will not listen, and thus do we walk together to the edge of doom. Are you listening fair lady with the beautiful soul?

**319**

We will now visit Oscar Wilde in his prison cell, for he has something important to reveal:

"People whose desire is solely for self–realisation never know where they are going. They can't know. In one sense of the word it is of course necessary, as the Greek oracle said, to know oneself: that is the first achievement of knowledge. But to recognise that the soul of a man is unknowable, is the ultimate achievement of wisdom. The final mystery is oneself. When one has weighed the sun in the balance, and measured the steps of the moon, and mapped out the seven heavens star by star, there still remains oneself. Who can calculate the orbit of his own soul?"

How true. Sometimes it takes the events that Oscar had to endure, the injustice that he experienced, to make one understand. What is science? What is Richard Dawkins? Here surely be a person who does think that he can calculate the orbit of his own ... not his soul! He is an intelligent ape and behaves like one, so he denies the existence of soul – that is what he has chosen. This is what he believes! More beliefs! We on the other hand do not want to be an intelligent ape – we want to be human, for we know what intelligent apes do. Nazis were intelligent apes.

There are quite a lot of intelligent apes in Brussels. Thus the scene is set. The world does not turn. Time is a singularity.

**320**

The artist John Witney wants to speak:

"Apart from the obvious traditions that music and painting are both wholly self-sufficient, why not a musical experience to be shared with, and equally involving the eye? Why not visual colour patterns which are so constructed as to weave with aural patterns in a fruitful complementarity of architecture? The interrelationship might be as elaborate and the consort as true as violin and piano which discourse in the typical partnership of all duo or multi-voiced compositions. This would be a partnership that is valid only if the combinations produce interest greater than the separate contribution of either the aural or the visual member."

And what goes for visual colour patterns and aural patterns, goes too for plastic expression and descriptive expression, for the image and the text, for the visual with its simultaneity and the written word with its linear sequential unfolding. This one might say is what we do when we pursue this thing that we have called scriptovisualism.

Now Robert A. Brady speaks again about the Nazis and how they operated:

"All *coordinated* and *planned* action pursues objectives. Armies are trained for wars. Workmen are organised to produce goods. Business is pursued for the making of profit. Police are established to keep the peace; law courts to settle disputes. The more careful the planning, the more definite the objectives must be. The more activities *coordinated* in the plan, the more each and every detail of organisation must be understood in terms of its usefulness in promoting the objectives for which activities are coordinated and planned.

"The Nazis make a fetish of coordinating. Their literature is replete with references to the *totalitarian state, uniformity of spirit*, the *Leadership Principle*. These terms have a central meaning which might be stated as follows: the Leader and his lieutenants will coordinate the totality of all persons and activities in Germany, in order that all, with an inner uniformity of spirit, will work towards the goals set by the leader for them.

Nothing, in short, is to escape the drag-net. All economic activity, all political, social, cultural life is to be *coordinated,* canalised, and directed.

"But whither? To what end? For what purposes? These questions must be raised with every plan, every programme, every institution, every activity coordinated. Unless one keeps an eye constantly focused on the purpose for which all things are coordinated, the product of his observations will be endless confusion."

This sounds much like the European Commission! He is talking about the so-called College of Commissioners!

"Determined to stamp out all opposition, it proposes for all *Gleichschaltung* (coordination)."

Here is Joseph Goebbels to explain:

"This coordination only outlines what will be the normal condition of Germany. This Germany is to have only one goal, one party, one conviction, and the state organisation is to be identical with the nation itself."

Sounds like the European Commission. He is talking about the so-called College of Commissioners!

Here is Fritz Saukel, Nazi Governor of Thuringa:

"In the name of this new Germany, I commend you now to show intolerance against all else. Henceforth there must be one political faith in Thuringa – National Socialism. Discussion of matters affecting our existence and that of the nation must cease altogether. Anyone who dares to question the rightness of the National Socialist world outlook will be branded a traitor."

And in Brussels anyone who dares to question the rightness of the European Commission, of the European Parliament, of the ideology and dogma of European Union Nationalism, is likewise branded a traitor, a right-wing populist – even those on the left.

Fritz Saukel was tried by the Allied Military Tribunal at Nurenberg and was found guilt of war crimes and crimes against humanity. He was executed on 16 October 1946. He died unrepentant and proclaiming his innocence.

**321**

And what has become of the other part of this odyssey, the plastic expression, the painting being created as part of this composition unknown? We were creating a circle, a time division of a circle, with 366 segments. Did we manage to make it? Yes we did, using Adobe Illustrator. But we had to develop a manual technique to overcome the assumptions, built into the software, which constitute limitations imposed on those that use the software, by those who design software. And the time division of the circle now lies within that which we created on a canvass. This is part of the story of creation that has been unfolding before you eyes, part of the strange work that now begins to draw to a conclusion, for 366 is in sight.

Joseph Weizenbaum:

"The computer metaphor has become another lamppost under which light, and only under whose light, men seek answers to burning questions."

Now John Witney speaks:

"Whether computers will, as Marvin Minsky believes - yes believes - ever write really good music or draw highly meaningful pictures; flatly I will express aloud my disbelief at the implications of these words. Computers will do no such thing, of themselves - not ever! Indeed art is a matter of judgement not calculation."

That they will produce some music or some pictures is not however in doubt, but John is stating good music, highly meaningful pictures, and such, done, unaided. He is right because all computers do is calculate, whatever way you look at it, they, in the end, only calculate - all binary operations reduce to calculations. And we do more than that, but the people under the lamppost have lost sight of that. Here is one such lost soul:

"The most recent foolishness along the lines that the brain is not a computer was penned by psychologist Robert Epstein. Knowing virtually nothing about Epstein, I am willing to wager that (a) Epstein has never taken a course in the theory of computation (b) could not pass the simplest undergraduate exam in that subject (c) does not know what the Church-Turing thesis is and (d) could not explain why the thesis is relevant to the question of whether the brain is a computer or not."

Hold on there a moment, did you mention the Church-Turin thesis? You did! Oh dear! Here is a representative of the Stanford Encyclopaedia of Philosophy, to tell all you people standing under the lamppost, that which you with your *Enlightenment* science - the brain is a computer (dogma) - have completely ignored. This is what happens though when dogma takes hold:

"A myth seems to have arisen concerning Turing's paper of 1936, namely that he there gave a treatment of the limits of mechanism and established a fundamental result to the effect that the universal Turing machine can simulate the behaviour of any machine. The myth has passed into the philosophy of mind, generally to pernicious effect. Some people believe that Turing showed that his very simple machine can specify the steps required for the solution of any problem that can be solved by instructions, explicitly stated rules, or procedures. Elsewhere people maintain that Turing has proven - and this is probably his greatest contribution - that his Universal Turing machine can compute any function that any computer, with any architecture, can compute; also that every task for which there is a clear recipe composed of simple steps can be performed by a very simple computer, a universal Turing machine, the universal recipe-follower. It is also asserted that Turing's results entail something remarkable, namely that a standard digital computer, given only the right program, a large enough memory and sufficient time, can compute any rule-governed input-output function. That is, it can display any systematic pattern of responses to the environment whatsoever.

"These various statements are typical of writings on the foundations of the computational theory of mind. It seems on the surface unlikely that the people who make such statements mean to restrict the general notions of *explicitly stated rule, instruction, clear recipe composed of simple steps, computer with any architecture, rule-governed function* and *systematic pattern* so as to apply only to things that can be obeyed, simulated, calculated, or produced by a machine that implements *effective* methods in Turing's original sense. But unless these notions are restricted in this way from the start, we should reject such claims.

"Turing did not show that his machines can solve any problem that can be solved by *instructions, explicitly stated rules, or procedures,* nor did he prove that the universal Turing machine *can compute any function that any computer, with any architecture, can compute.* He proved that his universal machine can compute any function that any Turing machine can compute; and he put forward, and advanced philosophical arguments in support of, the thesis here called Turing's thesis. But a thesis concerning the extent of effective methods - which is to say, concerning the extent of procedures of a certain sort that a human being unaided by machinery is capable of carrying out - carries no implication concerning the extent of the procedures that machines are capable of carrying out, even machines acting in accordance with *explicitly stated rules.* For among a machine's repertoire of atomic operations there may be those that no human being unaided by machinery can perform.

"The further proposition, very different from Turing's own thesis, that a Turing machine can compute whatever can be computed by any machine working on finite data in accordance with a finite program of instructions, is sometimes also referred to as (a version of) the Church-Turing thesis or Church's thesis."

The so called Church-Turin thesis is, as its name clearly states, a thesis not a mathematical theorem. It is widely agreed that is cannot be proved. People choose to believe Church-Turin thesis because they see that it is valid in the circumstances that they are able to observe, and then just assume that it is general and applies in all cases. This is what Karl Popper has to say about such inductive sciences:

"According to a widely accepted view, the empirical sciences can be characterised by the fact that they use inductive methods. According to this view, the logic of scientific discovery would be identical with inductive logic. It is usual to call inference inductive if it passes from singular statements, which are sometimes called particular statements, such as accounts of the results of observations or experiments, to universal statements, such as hypotheses or theories.

"Now it is far from obvious, from a logical point of view, that we are justified in inferring universal statements from singular ones, no matter how numerous; for any conclusions drawn in this way may always turn out to be false: no matter how many instances of white swans we may observe, this does not justify conclusions that all swans are white."

And what goes for white swans goes too for the Church-Turing thesis, for Relativistic Causality, even for causality itself. For evolution too! I believe …

In the causality there is purpose and in the purpose there is causality.

**322**

To the matter of stars and constellations we now do turn, and the constellation that we in the West refer to as Orion, which is Hellenic in its origin, for in Ancient China they had a different constellation and its name is *Shen* which does not include all the Hellenic stars.

The story of Ancient Chinese astronomy is a long one, which is reflected in Joseph Needham's work, *Science and Civilisation in China*, from which we learn that there developed a system of dividing the heavens along the equatorial circle into four main palaces: the Blue Dragon in the east; the Vermillion Bird in the south; the White Tiger in the west; and the

Black Tortoise in the north. And we find Shen, as one of 28 *hsiu*, located in the White Tiger, as *hsiu* 21.

The reference star for Shen is ζ Orionis (Zeta Orionis). The star is also known by the ancient name of Alnitak, which is located in what we in the West refer to as Orion's belt, it (Zeta Orionis) being the brightest of the three stars in the belt.

The constellation Orion and the constellation Shen, are just different ways of looking at the world, in this case a star pattern in the night sky. Causality and purposefulness are just different ways of looking at the world, they given different types of equations, of mathematisation, but they predict the same results. They are equivalent. Yet when one moves into the world of quantum mechanics something surprising might happen. Here is Howard Rosenbrock to explain:

"The rejection of purpose in favour of policy based on causal relations can only be regarded as dogma: a decision made before any explanation of the phenomena is sought. And dogmatic rejection of purpose can impoverish science: explanations of phenomena may be available from purposive descriptions which cannot be obtained from the causal description."

Which is what Howard, by developing purposive explanations for Relativity and quantum mechanics demonstrated; it is not very likely at all that you will find anything like this explored by those artists who claim to be working at the forefront of art and science collaborative practice – that is because they often are not working at the forefront, many of them being largely engaged in Public Engagement in Science (PUS). Back to the matter of totalitarian art! Others do other things.

**323**

Today we are back to C.P. Snow. We promised not to mention him again, but this is a composition unknown so we cannot know what we will write about, and he is back in our minds because we found someone from the world of the arts mentioning him, and we always like to follow-up on such things. So here is Snow, to speak on the subject of specialisation:

"On the matter of those literary persons I mentioned. They dismiss scientists as ignorant specialists. Yet their own ignorance and their own specialisation is just as startling. A good many times I have been present at gatherings of people who, by the standards of the traditional culture, are thought highly educated and who have with considerable gusto been expressing their incredulity at the illiteracy of scientists. Once or twice I have been provoked and have asked the company how many of them could describe the Second Law of Thermodynamics. The response was cold: it was also negative. Yet I was asking something which is about the scientific equivalent of: Have you read a work of Shakespeare's?

"I now believe that if I had asked an even simpler question - such as: What do you mean by mass, or acceleration, which is the scientific equivalent of saying, Can you read? - not more than one in ten of the highly educated would have felt that I was speaking the same language. So the great edifice of modern physics goes up, and the majority of the cleverest people in the Western world have about as much insight into it as their Neolithic ancestors would have had.

"The cultural divide is not just an English phenomenon: it exists all over the Western world. But it probably seems at its sharpest in England, for two reasons. One is our fanatical belief in educational specialisation, which is much more deeply ingrained in us than in any country in the world, west or east. The other is our tendency to let our social forms crystallise. This tendency appears to get stronger, not weaker, the more we iron out economic inequalities: and this is especially true in education. It means that once anything like a cultural divide gets established, all the social forces operate to make it not less rigid, but more so.

"Nearly everyone will agree that our school education is too specialised. But nearly everyone feels that it is outside the will of man to alter it. Other countries are as dissatisfied with their education as we are, but are not so resigned. Talk to schoolmasters, and they say that our intense specialisation, like nothing else on earth, is dictated by the Oxford and Cambridge scholarship examinations."

Now we could go on, but all we find here is C.P. Snow droning on about specialisation, and not providing any analysis of why such specialisation exists, and what it represents in sociological terms. He does not do this because he is a specialist himself, full of science hubris, and holding those who are not practical scientists in low esteem.

"C.P. Snow cites specialisation as the cause for the development of the (so called) *two cultures*."

That was someone from the world of the arts speaking. But Snow's analysis is flawed. The most important lesson that Snow teaches is that if you are going to transcend disciplinary boundaries, you need to put much effort into building the underpinning foundations of knowledge and methods, otherwise you end up making naïve remarks and observations, which is a lesson reinforced today by the European Commission's DG CONNECT and their STARTS platform.

It would be appropriate at this point, to hear from an artist who began working with new technology in the early 1940s and who was one of the first to work in an industry research lab. Here is John Witney to speak:

"The fallacy of mechanically translated previously composed music into some visual equivalent is established repeatedly in critical writing on the subject. A more fruitful, less mechanistic approach is possible today, for truly creative possibilities arise when the image structure dictates or *inspires* sound structure and vice versa, or when they are simultaneous conceptions. This obviously is best realised when both parts have common creative origins."

And this statement can also be repeated for scriptovisualism:

It is a fallacy that one can mechanically translate previously composed words into some visual equivalent. A more fruitful, less mechanistic approach is possible, for truly creative possibilities arise when the image structure dictates or *inspires* word structure and vice versa, or when they are simultaneous conceptions. This obviously is best realised when both parts have common creative origins.

**324**

Ancient China and the hexagram: a hexagram is a figure composed of six stacked horizontal lines where each line is either Yang (an unbroken, or solid line), or Yin (broken, an open line with a gap in the centre).

*I Ching – The Book of Change*. Thomas Cleary will now explain:

"The *Book of Change* is the most ancient and profound of the Chinese classics. It is the ancestor of Chinese philosophy, it is the source for the pragmatic mysticism of the *Tao Te Ching*, the rational humanism of Confucius, and the analytic strategy of Sun Tzu's Art of War.

"The *Book of Change* was originally composed over three thousand years ago by a king and his son in the tutelage of a Taoist sage. Six centuries later the book was analysed by Confucius, the great scholar and educator. The result of this work was the classic in the format known today, a compendium of advice on the causes behind the rise and fall of kingdoms and careers."

Beyond knowing that is indeed an ancient text, we have no way of knowing if, what has just been said, is true, but we can if we wish, explore further to see if it is accuarate. We can also look with our own eyes and see that within the *I Ching* there are 64 hexigrams. We see too that there is wisdom written in the *I Ching* and we begin to understand the balance.

Here is Joseph Needham to explain what the *I Ching* has done for the West:

"Among Leibniz's Chinese studies and discoveries was a mathematical interpretation of diagrams in the *Book of Changes*, the significance of which is still somewhat disputed.

"Without going into the history of this, which is complex, as it involves the role of other people, and would involve describing the twelfth century Chinese representation of the *I Ching* in the form of what is called the representational table, what one can say in summary is that Leibniz with the help of others came to see that the hexagrams in the *Book of Changes* correspond to binary numbers. The solid Yang lines correspond to the digit 1, while the broken Yin lines correspond to the digit 0. What Leibniz recognised was the binary notation he developed employed for the series of numerals 63 to 0 in the hexagrams of the *Book of Changes*, which in his day were universally believed to go back to at least -second millennium. It seems that Leibniz assumed that the Ancient Chinese had had an understanding of binary arithmetic, but had long forgotten this.

"There has been much speculation about this, but it is unlikely that any meaning can be attached to the Chinese creation of the hexagrams, that there was in fact no mathematical significance attached to the arrangements, and that no quantitative calculation process was implied, and it is most likely that the Chinese did not realise that they were executing simple binary arithmetical operations. There is no evidence for this, nor any evidence of the use of it for binary calculation. It is though possible that the hexagrams are related to a very ancient from of multiplication, and are a degeneration of this ancient form of multiplication.

"However the main point to note is the influence of upon Leibniz, who besides developing binary arithmetic, was also the founder of modern mathematical logic and a pioneer in the construction of calculating machines. It may be that Chinese influence was responsible, at least in part, for his conception of an algebra or mathematical language. And these developments are today, the foundations upon which computers operate."

And do you have anything to add about the *Book of Changes*?

"Modern scholarship points to Confucius not being a writer of any part of the *I Ching*. A branch of Confucianism eventually developed, under the influence of Taoists and the Yin-Yang experts, based on somewhat superstitious semi-political number mysticism and manipulations of the Five Element theories and the *I Ching* hexagrams. The *I Ching* itself represents a proto-scientific use of an elaborate symbolic structure. Its dating is highly disputed. There are also many views about its origin. It would seem to be a work that evolved over a long period of time, with its origins probably lying in peasant omen texts, and accumulating a mass of more sophisticated material used in the practices of divination, becoming in time more abstract until the system as a whole assumed the character of a repository of concepts, to which all concrete phenomena in Nature could be referred."

So not exactly the same as Thomas Cleary would have us believe! Yes it is that word once again!

**325**

Now to the matter of the rule of science and reason which is coming to Europe by virtue of its most ardent advocates – the European Commission. Here now is one of its leading intellectual thinkers – or not – to tell you what to expect. Professor C.H. Waddington, FRS, speaks:

"One cannot dismiss the Nazis, facists and communists all together for being, in their different ways, totalitarian; fairly soon we shall all be so, in some way or other. The totalitarians of today have taken, with the wrong foot foremost, a step which we shall have to take tomorrow. Totalitarianism is inevitable. I believe however that the meaning of totalitarianism is coloured by systems where totalitarian political systems have been set up – Nazis and fascist systems and also Russian communism. The objectionable features of these states are not inevitable accompaniments of totalitarian economic systems. Along with science and reason goes ethics, and scientists are very ethical."

There you have the words of the 1930s and 40s version of Richard Dawkins. Basically bonkers! Very dangerously bonkers! Who would be mad enough to put into practice the fantasies of C.H. Waddington and Richard Dawkins?

What is the difference between the fantasies in Hitler's *Mein Kampf*, and those of C.H. Waddington and Richard Dawkins? The answer is that we have seen the results of the former, and can expect nothing different from the latter.

Waddington speaks again:

"The rational economic system, at whose birth pangs we are already assisting, can only be fully utilised it if is infused by a culture whose method of approach is also rational, intelligent and empirical."

C.P. Snow would have agreed with you – did agree with you!

"Prim science has so far neglected to confess to the world that he has begotten such an offspring on the harlot humanities; but the infant culture is beginning to peep already – in its bastard vigour lies the only hope for an heir worthy of the civilisations of the past."

That was Waddington again.

And he that we have already heard from, Friedrich von Hayek, has already well demonstrated that your thoughts too, like those found in *Mein Kampf*, are nought but dangerous fantasies. And those who advocate scientific government, we note, do not form a political party and campaign for what they would do, but work behind the scenes to infiltrate the structures of power, so that they can by stealth bring about their fantasies, and in the European Commission, they have a willing ally.

"But I insist that we should examine in close detail both the Russian system and the Nazi system so we can learn from them and implement our totalitarian system in a different way."

That was Waddington again. Off you go. You are bonkers. And take Richard Dawkins and the former chief scientific advisor to a former president of the European Commission with you. You are all bonkers and extremely dangerous people.

And to speak about IG Farben, Diarmuid Jeffreys:

"The IG Farben story contains a clear warning about the risks inherent in any close relationship between business and state and what can go wrong when political objectives and the pursuit of profit become dangerously entwined."

And add to this, people from the world of STEM seeking power and influence, and bringing with them strange beliefs and tendencies towards totalitarianism and right-wing extremism, then all one needs are the right conditions to create an explosive cocktail that will lead once again to yet another European nightmare.

We have in Brussels, in the form of the technocratic European Commission and those to whom it has bound itself, a tendency to create this very dangerous entwinement, and coming soon are the social, political, economic and environmental conditions that will trigger once again the decent into hell. It does not have to happen! Yet it will if we do not act now to prevent it.

Have you understood yet? Or are you too caught-up in dangerous fantasies to understand?

Now Robert A. Brady will speak about the assumptions underlying the Nazis system which define their point of view and determine their actions:

"I will list 15 points:

"One - productive property and natural resources are to be privately owned; freedom of contract is guaranteed except for aliens.

"Two - individual initiative, the business entrepreneur, conduct of business for profit and ownership control are basic.

"Three - business men are to be free, if responsible, to fix by agreements prices, production totals and quotas, marketing areas, and the conditions and terms of purchase and sale.

"Four - stock and commodity exchanges, commission houses, brokers, and speculative transactions are inevitable and necessary for the conduct of organic business.

"Five - heavy industries, particularly those catering to the military and foreign trade, are encouraged; large scale units unless uneconomical are to be kept intact; cooperatives are to be broken up.

"Six - the social class structure of society is sanctified, strengthened, made semi-hereditary, and hardened into caste lines; the Middle Class are the Myrmidons of the *Ėlite* and, as such, the backbone of the state.

"Seven - employers have practically complete control over workmen in regard wages, hours, and working conditions. They must take care of their workmen.

"Eight - collective bargaining is completely abolished; strikes are illegal; trade unions are forbidden; requests for wage increases are *lése majesté*.

"Nine - control is completely from the top; there is and can be no such thing as control or discussion of policies from below; the leaders decide all things as they see fit; each holds appointed office for indefinite periods at the will of his superiors.

"Ten - National Socialist Party and the German State are one and inseparable, as spirit and body. Legislative, executive, and judicial authorities are fused together. The central government controls all local government and all activities in all their fields.

"Eleven - civil and military are fused together; as in the military there can be no freedom of speech, of assembly, of writing, of action, of *thoughts*.

"Twelve - Germany must be made self-sufficient at all costs.

"Thirteen - non-Germans cannot be citizens; as a corollary, all Germans residing outside Germany either belong or owe allegiance to the Third Reich.

"Fourteen - communism is the major enemy. There can be no such thing as equality or rights, opportunities, or income for the classes, races, or sexes. The broad masses are fools and must be duped and led to meet the purposes of the *élite*.

"Fifteen - all sciences and culture must be coordinated and made to serve the purposes of the *leader, total, corporate master* state. Propaganda is the method. Propaganda knows neither right nor wrong, neither truth nor falsehood, but only what it wants."

And just how many of the above are also the assumptions underlying the European Union and it institutions, especially the European Commission? Perhaps one day we may tell you. But for now you should just keep listening to their propaganda, that which we once helped them to produce - we came to call it misleading ministers, misleading Parliaments, misleading the public.

The European Commission makes a fetish for coordinating. What sort of totalitarian system will the European Commission bring into being in Europe?

**327**

Will they understand? Probably not!

It is certainly the case that stating that another form of science, one determined by different values, coming from a different culture to the Abrahamic European culture that gave birth to what we call science, opens up an avenue of criticism. It is possible that some may accuse us of being racist, for it is certainly true that the National Socialists had some perverse views about science. Yet, what we have said already should be enough to clarify that we are, in pursuing the development of a different science and technology, trying to produce something that deals with the problem that science has become - its hubris, the scientism, the hint of cultural superiority found in Western science and technology.

Science is indeed universal, but that does not mean that it is not possible to approach it with a different mindset. To deny this is ... Do we need to spell it out?

Here is Joseph Needham to explain:

"There is a fundamental point I wish to make. The contribution of China to the world may be particularly valuable because Chinese culture has never had this post-Renaissance *scientism*. Since modern science did not develop in China originally, but only in Europe, there has never been any necessity for a Chinese William Blake to oppose the single-vision of Isaac Newton's mechanistic conception of the universe.

"Chinese science has never really been tempted to regard the natural sciences as the sole vehicle of human understanding. This is the case because for twenty centuries an ethical system not based on supernatural sanctions had been completely dominant in Chinese society; and it was history, not theology and not physics which had been the *regina scientiarum*. This dominance of morality is perfectly represented in China by the slogan, *put politics in command*, for this essentially means human moral values, and that in practice means the health and well-being of your brother and sister at the bench, in the field, on the shop-floor, and next to you in the office or at the council table.

"In China you will not be tempted to regard him or her as nothing more than a behaviourist automaton, or a walking flask of amino-acids and enzymes, because by the wider wisdom of Chinese organic humanism, you know that he or she is a lot of other things as well, for the full reality of which mechanistic or reductionist scientific explanations, however successful for limited ends, can never suffice."

Are you listening Edward O. Wilson?

Joseph Needham:

"It may be that scientism, the idea that scientific truth alone gives understanding of the world, is nothing but a Euro-American disease, and that the great contribution of China may be to save us from the body of this death by restoring humanistic values based on all the forms of human experience."

Are you listening Edward O. Wilson?

This by the way, Edward O. Wilson, is the conclusion that Joseph Needham reached as a result of his life long work to research and document *Science and Civilisation in China*. Work that you could not be bothered to explore, because of your reductionism, your hubris, your scientism, your …

Goodbye Edward O. Wilson.

We watched a video posted on the Earth Sky Lab's Facebook page. The video is about a man who thinks we should not own hammers, but instead, every time we want to use a hammer, we request one from, we suppose a tool hire company, who will then deliver it to us, not using a delivery van, but by using a drone!

Bonkers! It is an example of technocentric thinking! All one does is to create a demand for drones. It is still consumption!

This is what Earth Sky Lab said about this daft idea:

"Yes, the future is shared ownership, but not of things delivered by magic anonymous zero energy drones to your isolated castle, but in community. The sharing economy will work much more effectively in a high-trust village. Shared tools, workshops, cars, gardens, etc. All this makes much more sense if you are sharing with friends and people you care for."

We agree. The idea of not owning a product, but just renting it when you need it, is not new. Sharing is also not new. How to make it work is what we need to think about, and how to develop it into a concept that can be something to be applied across the globe in many different situations and contexts. Somewhere there is always an economic transaction, even with shared ownership. It could be a money transaction or it could be an exchange transaction.

The idea of sharing a product in a service deliver context is something that we defined back in the 1990s as *Managed Consumption*, as part of a visioning exercise for manufacturing industries. This is what we wrote in a report in 1997:

"*Managed Consumption* is really a systems approach applied to each individual customer's consumption of goods and services. It is a way of resolving a paradox - how to achieve economic growth without the consumption of resources that has been associated with mass production. *Managed Consumption* does not render recycling redundant. Materials will still need to be returned to their original forms or reused in manufacturing. Neither does *Managed Consumption* replace remanufacturing. Cars will still be manufactured, and will need to be remanufactured to reduce waste. What *Managed Consumption* does is take environmental responsiveness to a higher level of sophistication by sharing resources between consumers, only delivering a resource to a customer when they need it."

Here in 2016 we can see some examples. An early stage of the development of a business model for *Managed Consumption* partly exists in places like London where you can hire a car for a few hours, rather than owning one - which is clearly a problem in big cities. There are many products - lawnmowers for example - which are used for a few minutes each week, and not at all in the winter months, but just about everyone with a lawn, owns a mower. It does not make any sense from a sustainability perspective, but is very sensible from a mass consumption one, which is the dominant paradigm - this is what government policies are designed to foster. With mass production goes mass consumption. It is the economies of scale thing. In the 1980s a new term appeared - *economies of scope* - which few people know about, as it has largely been forgotten. It was in any case largely considered in the context of manufacturing, rather than more generally.

The intellectual foundations for a different world already exist, but there is a great emotional attachment to the intellectual foundations of the past. We call this the Prometheus Syndrome. We need to break the invisible, silent and unbreakable chains that bind people to the rock of the past. Probably we do not need drones to achieve this! There is a simpler, non-technological answer - an organisational one. Science and technology are part of the problem, and just focusing on science and technology as the solution, when they are part of the problem, is just reinventing ourselves today, in exactly the same form we were yesterday - the Prometheus Syndrome.

**329**

Now Stephen Hawkins will say something silly:

"I believe that life on earth is at an ever-increasing risk of being wiped out by a disaster, such as a sudden nuclear war, a genetically engineered virus, or other dangers. Humans won't last another 1,000 years' on earth. Hence we should develop human space flight further so that we can leave the planet."

Bonkers! If such a calamity should befall humanity it will be because of people such as yourself, who, although seemingly intelligent, are unable to use that intelligence, and just make stupid remarks. We are going nowhere for there is nowhere to go. This planet, this place we call home, is all we have, and we need to start looking after it, and to start taking the actions necessary to ensure that STEM people never again are allowed to work on those very things that you believe will wipe us out. There is no place far enough away for humans to run to, for humans are the problem, as you well demonstrate. We need to change.

Now here is someone speaking.

"My research of past Nobel laureates shows that they engage in arts and crafts avocations significantly more than past Fellows of the Royal Society or past members of the National Academy of Science, who in turn engaged significantly more in arts and crafts than members of Sigma Xi or the United States public."

That was Robert Root-Bernstein.

We see that you list C.H. Waddington as a past Fellow of the Royal Society. He, you say, had an avocation in the visual arts. We have already encountered this nutty professor - he is the one who advocates the implementation of a totalitarian system! Evidently art does not prevent scientists saying stupid things and engaging in *scientism*.

You say also that he had an arts avocation, but we looked and found that he did not practice in the arts - he was just knowledgeable about the arts. We consulted his obituaries and found that he wrote extensively about many topics, including ethics, science and society, and yes, art.

Anyone can have an interest in art. Being a practicing artist and being interested in art is not the same thing. We could also say something similar about C.P. Snow. We know he wrote novels, but he too was into the *saying of ridiculous things* and engaging in technocracy.

There is of course another reason why Nobel Laureates might engage with the arts. If you have an interest in science, work in science, you would want to just get away from it, and pursue something different by way of a hobby - avocation means hobby or minor occupation. You may have proved that such people have a higher tendency to engage with the arts, but that is all you have proved. You have not, and cannot, prove a causal link between success in science and having an arts and craft avocation. Personally we would give more credence to research undertaken by social scientists, rather than a professed art-science lover. Oh that whispering of sweet nothings!

Albert Einstein - we know he played the violin, but elsewhere his success in rethinking physics and the development of special and general relativity has been attributed to his ability to think in visual terms, not playing the violin.

And there is another point, one which is made by Arthur Koestler in his study known by the title *The Act of Creation*. Sometimes it is necessary to just put the problems one is working on aside, and do something else. Somewhere deep in the subconscious the work goes on though. One day something surfaces. It could be that these Nobel Laureates that you have studied, through (for those who choose to take a critical look) not very reliable means, just understood the value of letting the subconscious do its thing and what better way of distracting oneself than with a violin?

Of course there is also a well established theory of creativity, grounded in social psychology, which people should take note of, rather than paying attention to art-science lovers.

**330**

Time is running out. The end is now coming into view - chapter 366.

About this thing called technocracy, which if you did not know, is re-emerging. Those who advocate this nonsense, along with those who pursue the rule of science and reason, seek to usurp democracy, through any means possible, mostly covert, and without the people knowing what is happening - by the time they discover it will be too late. If this sounds familiar, this is because it is - 1933.

Now a spokesperson for *The Economist* will report a few facts on technocracy. This is based on an article that appeared following the adoption of the technocratic approach in Italy and Greece in November 2011.

"Technocracy was once a communist idea: with the proletariat in power, administration could be left to experts. But the appliance of science to politics was popular under capitalism too. A fully fledged Technocratic movement flourished in America in the inter-war period: it believed in an economy based on measuring energy inputs rather than prices, and in what would now be called crowd-sourced solutions to political problems. *The Economist* first used the term technocracy in March 1933, when a book reviewer bemoaned the lurid prominence of the term. He derided its proponents as *half-scientist, half-charlatan,* and decried their *indefensible* conceptual basis, and ascribed their popularity to *extraordinary* American credulity."

Yes credulity - most STEM people who advocate technocracy, and the rule of *science and reason,* and its tools such as *Global Systems Science,* suffer from this problem of extraordinary credulity. Yes indeed they are gullible and naïve, and as we already seen from the Nazis. STEM people are also inclined towards hierarchy and order. It seems, as Waddington has already declared, that STEM people also want a totalitarian system, which is why we seek to change STEM and the people who work in this field.

Actually some of the above from *The Economist* is only partly true. There were technocrats before the 1930s and we have already met one - the arch-technocrat called Frederick Winslow Taylor. But the term itself came after him. We suppose we will have to find someone to speak about the origins of the term. In the meantime, we return to the spokesperson from *The Economist*:

"Jan Fischer, the Czech Republic's chief statistician, became an acclaimed prime minister in 2009 when the government collapsed in the midst of the country's six-month Presidency of the European Union. His main message, he says, was to tell everyone to 'protest against this kind of government'; it was a lamentable departure from normal democratic principles, justified only by the most serious circumstances.

"History suggests that technocrats do best when blitzing the mess made by incompetent and squabbling politicians. But the problem for the new leaders of Greece and Italy is that the source of their woes, the euro zone's design flaws, stems from mistakes made in Brussels - from people who are themselves unelected technocrats."

Technocracy does not work - it does not resolve political issues like the sharing of pain and gain, it does not guarantee competency, it does not prevent corruption or favouritism. Most of all, it fails because experts are less likely than non-experts, to recognise that many problems are too complex to admit solutions, and that often objectives cannot be fully determined. So we are back to STARTS - a case study in the failure of technocracy. A case study that demonstrates all the flaws - corruption, ignorance, incompetence, deceit ... One day we may write a book about it.

If you are engaged in using operations research to work out the best size for an allied convoy crossing the North Atlantic to minimise losses, or the best depth at which air-dropped depth charges should explode to maximise the destruction of enemy submarines, as C.D. Waddington once did, then you may find that your existing cognitive biases are reinforced, and you may emerge from the conflict being fully convinced that the rule of *science and reason* must be the way for governments to operate, and that a totalitarian approach is the best. But it would also be a dystopia, because you will have to silence, imprison and kill those who do not agree with your sole truth, as the Nazis and communists did. The rule of science and reason will do exactly the same, and that rule is taking shape in Brussels today. Missing are the conditions that will give this type of mind-set precedence over wisdom. These conditions are now taking shape in the modern world, and especially in Europe. But we know you will not listen and take heed of the warning signs. So we are back to STARTS, and also the FET Flagship evaluation panel as well.

331

Now we are going to explore some of those art-science lovers' whispering of sweet nothings in one another's ears. Robert Root-Bernstein now speaks:

"In my 2008 paper, entitled *Art Fosters Scientific Success: Avocations of Nobel, National Academy, Royal Society, and Sigma Xi Members,* published in Journal of Psychology of Science and Technology, I included a table of scientists who publicly exhibited and/or sold visual art or sculpture, published works of fiction, or publicly performed musical pieces. In the list of scientists who publicly exhibited and/or sold visual art and sculpture I included C.H. Waddington who was a Fellow of the Royal Society."

Well it happens that C.H. Waddington, who became a Fellow of the Royal society in 1947, is one of those people who we have summoned from the dead to speak, and we have already heard some of the silly things he had to say, so we decided to check this claim of yours. We did this because we have grave concerns about your method. You looked at Obituary Notices, and Biographical Memoirs published by the Royal Society, and other such material. We note that you are from the world of STEM. People from the world of the social sciences would understand that the material you used is largely subjective, unreliable, even biased, and cannot be used to undertake research of the type you have undertaken, which seems to be directed at achieving a desired result rather than exploring the psychological, sociological and historical contexts and conditions that may have contributed to their achievement of Fellowship of the Royal Society, or a Nobel Prize.

We checked to see if indeed, Waddington had done as you said. Here is Alan Robertson, who wrote the 49 page Biographical Memoir of Conrad Waddington:

"In the memoir I included some reminiscences by a close friend of Waddington from his undergraduate days at Cambridge University, Dr Barton Worthington. Worthington mentions that Waddington had had strong artistic leanings."

But did you, Alan Robertson, say that Waddington had publicly exhibited and/or sold visual art or sculpture?

"No. I mentioned that Waddington had once edited a poetry magazine while an undergraduate, he saying that he was more interested in poetry than science while an undergraduate. In this magazine I said that he had once included a poem by himself. This poem I include at the end of the memoir. I said however in the memoir that his main interest was the visual arts, and I included mention of his book on modern painting which is entitled *Behind Appearance: Study of the Relations Between Painting and Natural Sciences in This Century*."

Here is Waddington:

"The thoughts I have had on these subjects have arisen in the course of a lifelong interest in paintings, beginning with my friendship, from about 1930 onwards, with the English *avant-garde* painters of that time: John and Myfanwy Piper were my closest friends, and through them, and the architect Justin Blanco White, whom I married around that date, I came to know Henry Moore, Ben Nicholson, Barbara Hepworth, Ivon Hitchens, Sandy Calder, Moholy-Nagy, Gropius, and a few more - there were not so many of that kind at that time."

Interesting observation! Shows just how much nonsense is bound up in C.P. Snow's *two cultures*! What was Snow really up to?

It would seem that Waddington was not one of those intellectual celibates that Whitehead referred to! At least not always!

It would also seem that Waddington had a wide range of interests, not just artistic, but also philosophical and ethical. But he did not, as far as we can tell with the evidence that is available, that is to say his books and the published obituaries, engage in, beyond writing a single poem, artistic practices, nor did he exhibit any work.

Oh yes, we forgot to mention that he was interested in dancing - Morris dancing!

**332**

Continuing with the Robert Root-Bernstein theme! We now ask: why are we so critical? Why is it important to be critical?

First to say is that we do not think that looking at people from the world of STEM is not an interesting thing to do, and not worthwhile. Clearly that is not the case. There are two issues to address. The first is method. We have already questioned this. We note that Root-Bernstein in his paper also brings in statistics in the analysis of his results. So not only is he using unreliable (albeit interesting) evidence, but adding a touch of science and the numerical, by referring to statistical significance. It shows the problem of crossing disciplinary boundaries, and not understanding other disciplines' methods and what they regard as evidence and the conclusions that can be drawn from such evidence. The application of this kind of scientific mind-set to what is being studied, points to the problems that we have already encountered - that if something is not reduced to numbers is not scientific and therefore has no validity.

The second point, related to the first actually, is the attitude of mind, now often found in the world of STEM, that the arts are to be appropriated to the problem of addressing what those in the world of STEM have decided is a *creativity deficit*, but for which they advance no evidence of the form that they typically regard as evidence - numerical in nature derived from scientific studies.

Now we invite some teachers to comment, based on the observation that art is a way of knowing - art being the visual arts, drama and literature in this case.

"We are becoming aware that children can demonstrate classroom learning and understanding through art in ways that integrate intuition and cognition."

"Research shows that drama does enhance learning. Students learn by seeing, and by hearing, and by doing. Drama can compliment other modes of instruction and it transforms school from a place where we tell students to think to a place where we help them to experience thinking. The majority of thinking for drama entails analysis, synthesis, and evaluation."

"I identified four roles for literature in the classroom. Literature is a way to learn language, to explore content area, know the world, and to critique the world."

That was Karolynne Gee, Paul G. Heller, and Kathy G. Short, speaking in that order.

The important point to note is that the arts, as above defined, are not to be appropriated to the teaching of STEM, but are a way of teaching all subjects. Importantly also they are a way of dealing with the problems of STEM peoples' behaviour, their attitudes, and their tendency towards conforming and accepting, and their liking for authority, hierarchy, certainty, and sole truths. Arts therefore have a place also in higher education, especially towards deprogramming STEM mind-sets, which seem already to be well formed by the time they come to university.

Speaking as one who has experimented with learning techniques in a university setting, we can expect much resistance from staff and students. It seems that many are happy to be intellectual celibates, and even many who are not, still behave according to STEM ideology and dogma - C.H. Waddington serves to illustrate the point.

"Art is a way of knowing, a way of making contact with your feelings and intuition, of sensing your inner being and forging a path to the river of soul that runs below everyday life, and of becoming more alive in the process."

That was Pat B. Allen speaking.

**333**

Time to hear from a Nobel Laureate! The person in question is the psychologist Daniel Kahnenman. He was awarded the Nobel Prize for Economics in 2002. Yes, that is correct, a psychologist awarded a Nobel Prize for Economics, not for research in economics, but for research in psychology with fundamental implications for economics. Over to Daniel:

"I am going to speak about Fast Thinking and Slow Thinking. Psychologists have been interested for several decades in two modes of thinking and have offered many labels for them. I refer to them as System 1 and System 2. System 1 operates automatically and quickly, with little or no effort and no sense of voluntary control. System 2 allocates attention to effortful mental activities that demand it, including complex computations. The operations of System 2 are often associated with the subjective experience of agency, choice, and concentration.

"When we think of ourselves, we identify with System 2, the conscious, reasoning self that has beliefs, makes choices, and decides what to think about and what to do. Although System 2 believes itself to be where the action is, the automatic System 1 is where it actually is.

"System 1 effortlessly originates impressions and feelings that are the main sources of the explicit beliefs and deliberate choices of System 2. The automatic operations of System 1 generate surprisingly complex patterns of ideas, but only the slower System 2 can construct thoughts in an orderly series of steps.

"I now want to tell you about *cognitive illusions*. They are similar to visual illusions, but more stubborn. There are several, but one example is that which is grounded in skill, and the illusion that stems from being able to exercise high-level skills at a professional level, but which involve people making judgements that they *feel* confident to make, but which in fact they are not able to make, but are not aware that they are unable to make – they are ignorant of their ignorance. It is a problem that is reinforced by professional cultures.

"Overcoming cognitive illusions is very difficult. Because System 1 operates automatically and cannot be turned off at will, errors of intuitive thought are often difficult to prevent. Biases cannot be always avoided, because System 2 may have no clue to the error. Even when cues to likely errors are available, errors can be prevented only by the enhanced monitoring and effortful activity of System 2. As a way to live your life however, continuous vigilance is not necessarily good, and it is certainly impractical. Constantly questioning ones own thinking would be impossibly tedious, and System 2 is much too slow and inefficient to serve as a substitute for System 1 in making routine decisions. The best we can do is to compromise: learn to recognise situations in which mistakes are likely and try harder to avoid significant mistakes when the stakes are high.

"It is easier to recognise other people's mistakes than our own."

STARTS is the result of System 1 thinking and cognitive illusions.

The above is fundamental! It is another reason why technocracy does not work, and why the rule of science and reason will not work. It is also fundamental in another sense - the survival of humanity. Our ability to steer a course towards a sustainable future rests on recognising that, when the time comes, people are going to make the wrong decisions, because they will have the belief that they are doing the right thing - cognitive illusion! We need to have in place the safeguards to prevent these people from acting on what they believe is right! We have a name for this - we call it behavioural policymaking, behavioural decision-making. Its overriding assumption is that people will do the wrong thing and we need to demonstrate to the right people, that they must not listen to the experts and what they are saying. The experts in question will come from the world of STEM. They will be technocrats. They will work in places like the European Commission, and they will surround themselves with people who share their cognitive illusions, and together they will form a powerful and dangerous professional culture. The consequence of their decisions, will far exceed any horror and catastrophe yet seen in human history.

This is why the European Commission DG CONNECT's STARTS Platform is so important for it is not only a *worst practice* case study of the exploitation of art, as well as being a case study in the failure of technocracy, but also an example illustrating what is said in the paragraph above. We will use this case study to help others see, so that they can take steps to avoid the catastrophe that is waiting for humanity. We will walk with humanity to the edge of doom.

STARTS will be a failure, but DG CONNECT and its experts will conspire to present it as a success. So once more we return to the matter of IG Farben. DG CONNECT need to find a Giorgio Vasari to write a fictitious account of STARTS to mislead ministers, Parliaments and the public.

We are looking forward to the meeting in Brussels on December 9th when another conspiracy, another exercise in misleading ministers, Parliaments and the public, will draw to a close.

Have you understood yet?

**334**

We live in a world dominated by change - social, political, economic and environmental. It is a world of continuing change. It is a world of significant change. And we also live in a world that is increasingly uncertain and unpredictable. Change, uncertainty, and unpredictability! In such circumstances technocracy cannot work. In such a world, the rule of science and reason will be a disaster. It will only lead to the past - that place where humans did things that no one wants to see again. Yet they will if we listen to the siren voices of the technocrats, to the voices of the *Hawkins Dawkins Wilson Wolperts* of the world, of the art-science lovers of the world!

In the early 1990s we became involved is something that went by the name of Agility - the Agile Enterprise, Agile Manufacturing. We wrote the world's first book on Agility. It was called *Agile Manufacturing: Forging New Frontiers*. This is what Paul wrote in the Preface to that book:

"Agile Manufacturing is a relatively new term, and we have yet to fully define the concepts, yet it seems from the many articles that are now appearing in our journals and trade magazines, that our manufacturing enterprises are already practising Agile Manufacturing. But are they? It seems to me that Agile Manufacturing is being interpreted as lean production, or flexible manufacturing, or computer integrated manufacturing (CIM). This is wrong. My primary aim in writing this book was to contribute to the process of defining Agile Manufacturing, and to dispel any beliefs that Agile Manufacturing is just another way of describing lean production, flexible manufacturing or CIM."

Few people listened to this and the result was something of a circus and the creation of what we called, in another case - art-science and art-technology - a discredited zone of practice. Evidently we were not alone in thinking such, for ultimately the United States government prematurely terminated its agility research programme.

We however, continued to work on the matter and over an extended period came to see something important which would one day be useful to humanity - to save it from the mind-set, the beliefs, the values that we have been exploring in this book. And many years later in a chapter for an edited book, we wrote this:

"It is possible, in the context of significant changes, uncertainties and unpredictability, that companies may begin to fail when the assumptions built into their businesses, become invalid. These assumptions are typically based on the notion of infrequent changes and manageable uncertainties. To understand when these circumstances are occurring, it is necessary to distinguish between two types of change scenarios: structural and non-structural.

"Structural changes are those that involve modifications to the structure of the business environment, that is to say, in terms of such things as technologies, product concepts, competitors and competitive strategies, regulation, supplier choice and capabilities, techniques for product delivery and distribution, etc. Non-structural changes are those associated with more operational issues, such as varying order quantities, one-off requests, delayed supplier deliveries, material shortages, machine breakdowns, etc. However, sometimes non-structural changes can indicate structural changes are taking place in the business environment."

We then went on to explain what agility is and why it is important in a regime of continuing change, uncertainty, and unpredictability, when these are related to structural rather than non-structural change.

We would now like to formally welcome you to your world of significant structural changes - economic, social, political and environmental - and the massive uncertainty and unpredictability that goes with this. You do not understand what is happening and your institutions, beliefs and values will all fail you. You do not want to contemplate the scale of the disaster for humanity that this implies. But you do want to avoid it. You can. But you are not listening, nor do you believe what we are saying. As a result, what might have been an evolutionary advantage for early humans - System 1 and System 2 - will most likely be your undoing as a species!

It is not everyday that you encounter someone like us! You can choose to listen to us, or to *Hawkins Dawkins Wilson Wolpert* and technocrats. Which is it to be?

**335**

And it came to pass that technocracy, scientism, and the rule of science and reason - scientific government - did spread like a plague across the earth, and men - they are a big problem - thinking themselves superior to other creatures and in control, did walk to the edge of doom. And so it was that we walked with them for we cannot abandon humanity to its fate.

Now we are going to ask Joseph Needham if it would have been possible for ideas and inventions originating in China to have found their way into the notes and sheets of paper, sometimes referred to as the notebooks of Leonardo da Vinci!

"In Volume I of *Science and Civilisation in China,* I noted that there were detailed technical transmissions of textile technology from China to Italy. I also noted that a trade in slaves which provided domestic servants to Italian homes during the +fourteenth and +fifteenth centuries took place, and that further investigation was needed to understand if this had resulted in the transmission of ideas and techniques.

"In Volume II, Part 2, I discuss flywheels, and a problem addressed around +1490 in Europe of ways overcoming dead centre, doubtless because of the increasing use of cranks and crankshafts. Here I stated that it was by no means fanciful to connect this development with the Central Asian domestic slaves who were so numerous in +fourteenth and +fifteenth century Italy, and who may have been responsible for many transmissions of technological interest. What had changed between 1954 and 1965, the publication years of the above mentioned two volumes, was the work of Lynn White, who contributed more evidence to strengthen the case for this argument."

Thank you Joseph for that important insight! So, one conclusion is that it could be possible that Leonardo da Vinci recorded some of these transmissions in his notes. We have broken another taboo!

Now to a matter of disentanglement!

"As regards the practices of cultural agencies, there is a tendency to entangle evaluation, or even research, with advocacy, with the result that there is a hunt for proof of value, rather than an honest enquiry into what has taken place and the results produced."

That was François Matarasso speaking. He could also have been talking about the art-science, art-technology lovers. We refer you to Root-Bernstein and his activities - research or advocacy? Biased research for sure!

**336**

Sorry to say that we must now return to Robert Root-Bernstein and his paper that we have already mentioned. We have here with us, Conrad H. Waddington. Recall we have already discussed Waddington in connection with Root-Bernstein's paper. There is however, more to say on the matter.

Root-Bernstein says of Waddington, that he was an artist-dancer-poet. We know where the poet part comes from - he once wrote a poem. So did many children while at school! But what about the dancer bit? Here is Alan Robertson, once more to explain:

"In my Biographical Memoir of Conrad Waddington, I included, as I have already said, some reminiscences from a close friend of Waddington from his undergraduate days at Cambridge University, Dr Barton Worthington.

"Worthington says that Waddington devoted a good deal of attention to Morris dancing: he became an exponent of its techniques, squire of the Cambridge Morris Men, a group which tended to attract brains, rather than brawn, and he led dancing tours through the Cotswolds and other traditional centres. Indeed he collected one or two Morris dances and folk tunes which would otherwise have disappeared, and to see him dance one of these, *The Nutting Girl*, on a village green, accompanied by his wife on the accordion, could be a moving experience."

So he was a Morris dancer. If you are wondering what such a thing is, here is a quick explanation: Morris dancing is a form of English folk dancing. The dancers wear a uniform and have bells attached to their shins and carry sticks (which they often hit against the sticks carried by other dancers) and wave handkerchiefs about. You also may be interested to know that at the time Waddington was involved in Morris dancing, there was a revival of this folk dancing as part of an effort to preserve it for future generations - cultural heritage.

We do not think that this Morris dancing helped Waddington become a Fellow of the Royal Society. We read the Biographical Memoir. It seems that he was a precociously talented person, able to switch disciplines quite easily - he studied geology at first and worked for a PhD in this subject, but became interested in genetics and thus switched. It seems that in Cambridge in the 1930s one could do such things!

Now, having dealt with this matter, there is something that caught our attention in Root-Bernstein's paper. He says that Waddington advocated David Bohm's philosophy that creative process is a transdisciplinary link between the sciences and arts. Did he? Or are we mixing up evaluation (research) and advocacy? We look forward one day of finding the place in Waddington's books where he says that he advocates David Bohm's philosophy that creative process is a transdisciplinary link between the sciences and arts, or even just that creative process is a transdisciplinary link between the sciences and arts. One day ...

Root-Bernstein then provides a quote from Waddington, which is taken from Waddington's book *Biology and the History of the Future*. We obtained a copy of this book, and can confirm that the quote is *inaccurate*, in that it is only part of a dialogue with the artist John Cage, which is why we have invited Waddington and John Cage here now to speak.

So the scene is this: it is a small symposium, organised in 1969 by Waddington, on behalf of the International Union of the Biological

Sciences, with further financial help from UNESCO. There is a discussion taking place involving Waddington, John Cage, Margaret Mead, Ruth Sager, John Platt, Gunther Stent, and Carl-Goeran Heden. The subject is *The End of Faustian Man and the Limits to Progress*, in the context of the theme, *Towards a Post-Industrial World*.

At the start, Stent says that Faustian Man strives for power over his environment; but the achieving of power is satiating and will (in 30 years?) turn off the drive. The accelerating rate of progress means that we must approach certain limits (for example air transport) in a few decades. So the twenty-first century will be hippie, Polynesian, non-striving (The Beatles and the Beatniks are the first symptom). Limits of progress have been reached now, in arts and sciences, for a similar reason in both cases. Then John Cage speaks. Please tell us John what you said.

"I said, the progress may be the idea of *dominating* nature. But in the arts, it may be *listening* to nature. In the forties, I conceived the idea of a piece *with no sounds in it*, but I thought it would be incomprehensible in the European context. Five years later, I was inspired to do it by seeing the paintings of Robert Rauschenberg - one of which was a canvass with no paint on it. Charles Ives wrote a romantic essay about sitting in a rocking chair, watching the sunset, 'listening to your own symphony'."

What happened next John?

"Mead said, 'as in a Quaker meeting,' to which Waddington responded, 'I remember in a Quaker meeting in my childhood listening to the bees'."

What did you say next, John?

"I then said, I once went to a Quaker meeting - with silence - and found myself thinking of what I should *say* - that is, how to dominate the meeting (Faustian!) - and then I realised that was not the point - not to dominate, but to listen. And to listen to silence. By silence I mean the multiplicity of activity that constantly surrounds us. We call it *silence* because it is free of our activity. It does not correspond to ideas of order or expressive feeling - they lead to order and expression, but when they do, it *deafens* us to the sounds themselves."

Waddington then spoke in response to Cage. What did you say Conrad?

"I said that your silence, Cage, is an instruction to listen. An art object is always an instruction, to do or to experience, not a piece of information; and living things are organised instructions, not organised information."

We now leave you dear reader to ponder why Robert Root-Bernstein and his co-authors, wrote what they wrote, and linked the above assertion about an art object is always an instruction, and so on, to their statement that for Waddington, understanding how art was made was a way to understand his own field of embryology, because, an art object is always an instruction, to do or to experience, not a piece of information; and living things are organised instructions, not organised information. You have now the context is which this was said – *The End of Faustian Man and the Limits to Progress,* in the context of the theme, *Towards a Post-Industrial World.* What has this to do with art fosters scientific success? Advocacy?

We invented a new phrase: *art-science fakes.* We might say some more about this later. We may not say some more about this later. We may write a whole book on *art-science fakes.* We may not. There will certainly be plenty of material available, especially from STARTS. We may in such a book mention the highly fictitious account of Leonardo da Vinci, to be found in Giorgio Vasari's *Lives of the Most Eminent Painters, Sculptors, and Architects,* possibly also including in the book, mention of the nonsense of Michael Gelb's book *How to Think Like Leonardo da Vinci,* which perhaps would be better called *The Leonardo Gospel According to Michael Gelb.* The *Seven Da Vincian Principles* indeed! Absolute rubbish! For those able to see – those who want to see – it is evident that this self-styled *Renaissance* man, this *New Leonardo,* produced a book that is a violation of the thing that he claims Leonardo did: question accepted wisdoms. Evidently Gelb just believed what Vasari wrote about Leonardo, along with all the other mythical stories surrounding this artist, without questioning anything. As normal – no critical thinking evident! But to those who look – who know how to see – there is a different Leonardo to be discovered; a more useful one than the mythical character so beloved by the community of art-science lover's. What Leonardo? How is this different Leonardo useful?

We are not saying! Not yet at least.

Michael Gelb – enacted biography!

**337**

So what is failure? Did you know that failure can be defined? This is what the Oxford English Dictionary provides by way of a definition:

Failure: lack of success; the neglect or omission of expected or required action; the action or state of not functioning.

We have here with us two people who have studied failures – failures in systems. This is V. Bignall and J. Fortune defining what they mean by system failures:

"Failure occurs when there is a disappointment or a shortfall between what is and what could have been."

We have been studying such failures for a long time. The subject became of interest because of our long involvement in design. We learned that people in the world of STEM do not like to talk about their failures, of which there are many. This, we think, is very unprofessional, because only by being open about the shortcomings of STEM and the people who work in the field, can we hope to create confidence and improve. Instead, people prefer to hide facts from the public and employ artists in residence in image making exercises that are partly intended to fool the public by getting them involved in the madness that is taking shape in the world of STEM. It is a kind of madness that the National Socialists were able to bring about in the Germany of the 1930s. Soon you too will be able to stand in the dock with those STEM people, at the next Nuremberg, and deny all responsibility for what comes from such madness. Is this not so, all you executives of IG Farben, many of them STEM people, charged with war crimes and crimes against humanity?

Back in the 1980s we looked at failures in the area of investments in advanced manufacturing technologies and Information Technology systems. It turned out that there were quite a lot of these, manifesting themselves in the form of: budget overruns, equipment not working as expected, very long timescales to bring equipment into full production use, project overruns, underperformance in terms of achieving the expected benefits, etc.

We discovered many reasons for these failures:

- Technology oriented rather than business and market focused investment plans;
- Inappropriate or non-existent business or manufacturing strategies, or both;
- Failure to link business and manufacturing strategies;
- Inadequate assessment of the costs and benefits of investments in new technologies;
- Failure to understand customer requirements and to design and build products that customers will buy;
- Neglect of organisation and people issues.

Interestingly all these system failures involved people who we would call experts and would want to be regarded as such. They made mistakes! They were not competent to do what was required! They applied practices that were no longer relevant. But they were experts, so no one stopped them! Even more interesting was that much of this came to light because government agencies were investing heavily in manufacturing innovation at the time, but were not seeing the expected benefits, which set them asking questions and supporting research into why this was happening.

This will happen too, with STARTS, only DG CONNECT will organise a cover-up, and conspire with its experts to pretend that STARTS was a success. Never mind, for we will use it as a worst practice case study to help the Chinese.

As we have already said, technocracy does not work - it does not stop people from making mistakes, or being corrupt, or being incompetent, or from using outdated practices, or resolve the problem of *people who do not know what they do not know*. The age of the expert is indeed dead and gone, but someone needs to tell the experts this! We need also to tell this to people who think that experts know best! Surely you have understood, having reached this chapter that this is one of the messages of the book? No?

"There is a striking discrepancy between what participants in artistic interventions value from the experience, and the types of impacts expected by policymakers, lacking experience in the area, expect and wish evaluation studies to document."

That was Ariane Berthoin Antal and Anke Strauβ speaking. Let us now rephrase the above:

There is a striking discrepancy between what participants in STARTS artistic interventions in DG CONNECT projects value from the experience, and the types of impacts expected by DG CONNECT, lacking experience in the area, expect and wish evaluation studies to document. Thus will they engage in a cover-up. Thus will we then have more material to show the world the true nature of the European Commission and why action must be taken by the Member States against the European Commission.

**338**

Complete this sentence: Leonardo da Vinci was an accomplished ... Here you may insert whatever you want. Many people have! Here is a list to choose from: painter, sculptor, draftsman, poet, musician, athlete, writer, philosopher, chemist, biologist, botanist, geologist, palaeontologist, anatomist, engineer, inventor, designer, mathematician, chef, student of the classics, architect, humanist. Feel free to add more ... This conception of Leonardo da Vinci is truly ridiculous! You agree, yes? Or perhaps you too are an art-science lover, a devotee of this secular religion? Oh dear we seem to have ridiculed your divine figure-head. Heresy!

Now once more we return to the matter of technocracy and the European Union. Here is Claudio Radaelli:

"In 1999 I said that the political system of the European Union, notably the European Commission, is in the firing line. At stake is the allegation of being a political system ruled by technocrats who ignore the basic thrust of democracy. While democracy is based on legitimate consensus, free elections and participation, technocracy recognises expertise as the sole basis of authority and power.

"Complaints about the *mandarins of Europe* are not new, but in the past Brussels had been able to make public policy in relative isolation from ordinary citizens. The policy process was managed smoothly, and the citizen did not care much. Europeans differed in the degree of Euro-enthusiasm according to their nationality, but more fundamentally they were not obsessed with what was going on in Brussels and how democratic it was. The salience of *Europe* was indeed low.

"Things changed at the beginning of the 1990s. It all started with the ratification of the Treaty of Maastricht. Its political message was clear – a single currency for Europe. Citizens had to decide whether they were prepared to swallow it or not. In 1991 the Danes said no, and one year later, in France, the single currency proposal was accepted by the electorate by a narrow margin. The press commented at the time that the Danish Referendum exposed the gulf between the Europe planned by the technocrats and heads of states, and the Europe of the People. The European Commission president at the time, Jacques Delors, an arch-federalist, conceded this point in an article that appeared in the Financial Times in September 1992.

"The reason often given by the French who opposed the Treaty of Maastricht was opposition to technocrats in Brussels.

"The flaws of technocracy have been well analysed by the social sciences. Technocracy nevertheless is alive and technocrats still argue that there are a vast number of policy areas that should be insulated from the mess of democratic policy making. This claim is fundamentally flawed from a normative point of view. Politics has to do with values, and no algorithm will ever provide an answer to the puzzle of confronting values. In addition technical rationality is just one type of rationality, and democracy needs normative rationality as well. Not only do technocratic claims present normative drawbacks, they are also empirically weak.

"Technocracy does not work and its literature is full of exaggerations and bias. On the other hand one cannot deny the need for expertise and knowledge in policy formulation. How then do we bring democracy, politics and knowledge together?"

Indeed, this is a most important question, but one, alas, that will not be answered in Brussels, nor by having artists in residence. But it may be answered by an artist-led research programme that deals with the problems that STEM people have become, and does not shy away from the failing nature of modern institutions, and the insidious commitments to technocracy, scientism, and scientific government.

So why did you not, Jeremy Rifkin, in your book, *The European Dream: How Europe's Vision of the Future is Quietly Eclipsing the American Dream,* deal with the deeply ingrained technocratic nature of the European Commission?

No answer!

Now you know why the European Dream will become the European Nightmare - yet another one! It is already taking shape and the STARTS initiative is a prime example - the product of corruption, ignorance and technocracy, which will fail, but will lead to a conspiracy to hide this fact. The FET Flagships demonstrate the point very clearly. And we have the proof. The European Commission is corrupt and probably also acting illegally - back to STARTS again.

And yes we are going to publish this proof, but in our own special way. It posses a significant artistic challenge for the book about this has already been written, and you have encountered it already. To remind you, this is how it ends: Twelve voices were shouting in anger, and they were all alike. No question, now, what had happened to the faces of the independent experts. Paul, standing outside looked from independent expert to European Commission official, and from European Commission official to independent expert, and from independent expert to European Commission official again; but already it was impossible to say which was which.

**339**

The Soviet Union had a parliament called the Supreme Soviet with deputies elected by the citizens of the Soviet Union, and thus the Soviet Union was a representative democracy based on universal suffrage.

Now we are going to consider, once again, democracy in that most technocratic of unions called the European Union. We do this partly because in the United Kingdom there has been much talk about the sovereignty of Parliament, and a legal case in which a very rich person has sought to use their wealth to obtain a legal ruling that will force the Government to seek Parliamentary approval to invoke Article 50 of the Treaty of the European Union. The people using their wealth in such a manner are doing so in the hope that Parliament will act to override the democratic will of the people, that we should leave the European Union. The people bringing this case claim that they are acting in the national interest. They are without doubt though, acting in their own interest. We believe!

They are however, not politicians; nor do they understand politics. The politicians in Parliament will not go against the wishes of the people - we will be leaving the European Union regardless of what these rich people want. The nice thing about this though is that these rich people will be out of pocket by several millions of pounds in legal fees. We believe!

We thought therefore that it would be interesting to present here the powers of the European Parliament, and yes we have been reading the *Treaty on European Union*. But actually there is another treaty that needs to be read if one wants to discover the powers of the European Parliament; this second treaty is called the *Treaty on the Functioning of the European Union*. Actually both treaties go together to form, what we suppose people are referring to when they talk about, the *Treaty of Lisbon*.

Let us explore why the president of the European Union goes around saying that he has been elected, even though no citizen of the European Union ever voted for him!

Article 14 of the *Treaty on European Union* says that the European Parliament will elect the president of the European Commission. But it does not elect from within its own number because Article 17 says that, taking into account the elections to the European Parliament and after having held the appropriate consultations, the European Council, acting by a qualified majority, shall propose to the European Parliament a candidate for president of the Commission. This candidate shall be elected by the European Parliament by a majority of its component members. If he does not obtain the required majority, the European Council, acting by a qualified majority, shall within one month propose a new candidate who shall be elected by the European Parliament following the same procedure.

The European Council is that body, often in the news, that consists of the heads of state or government of the Member States, together with its president, and the president of the Commission. And who elects all those people who are called European Commissioners? Actually it is the European Council, for Article 17 also states that the Council, by common accord with the president-elect, shall adopt the list of the other persons whom it proposes for appointment as members of the Commission. They shall be selected, on the basis of the suggestions made by Member States, in accordance with the criteria set out in paragraph 3, second subparagraph, and paragraph 5, second subparagraph. And then, the president, the high representative of the Union for Foreign Affairs and Security Policy and the other members of the Commission shall be subject *as a body* to a *vote of consent* by the European Parliament. See it is only a vote of consent. Then on the basis of this consent the Commission shall be appointed by the European Council, acting by a qualified majority.

If you are wondering about who *The Commission* are, they are the European Commission. As for the Council, this is not that body called the European Council, but another one, which represents the governments of the Member States. It is only referred to in the treaty as the *Council*, but elsewhere it is called the *Council of the European Union*.

Are you getting confused? Perhaps you are supposed to be confused? Who can tell?

Now to confuse you more, for there are two types of legislative procedures - ordinary ones and special ones. The European Parliament is only consulted on the latter, but votes on the former. Did you know that?

We refer now to Article 294 of the *Treaty of the Functioning of the European Union* to explain the ordinary legislative procedure.

Where reference is made in the Treaties to the ordinary legislative procedure for the adoption of an act, the following procedure shall apply: The Commission shall submit a proposal to the European Parliament and the Council. There then follows a *First Reading*. The European Parliament then adopts its position at this first reading and communicates this to the Council. If the Council approves the European Parliament's position, the act concerned shall be adopted in the wording which corresponds to the position of the European Parliament. If the Council does not approve the European Parliament's position, it shall adopt its position at first reading and communicate it to the European Parliament. The Council then informs the European Parliament fully of the reasons which led it to adopt its position at first reading. The Commission shall inform the European Parliament fully of its position.

How many of you realised that the Council can override the wishes of the European Parliament? No expensive court cases needed!

Of course Article 294 has more to say, than the above, for when the Council overrides the European Parliament, it can react because there is a *Second Reading* stage, in which the European Parliament can just accept the Council's position and that then becomes law. Or, it can reject, by a majority, the Council's position at first reading, and then the act is deemed to have not been adopted. The European Parliament can at this stage also propose amendments to the Council's position, which are then delivered to the Council and the Commission, who then both deliver their opinions on the amendments.

If then the Council approves all the amendments the act is deemed to have been adopted. But it only has to do this on the basis of a qualified majority. If they are not all approved, then the Council and the European Parliament form a Conciliation Committee. However, to be noted is this point: The Council shall act unanimously on the amendments on which the Commission has delivered a negative opinion.

And so it goes on as follows:

The Conciliation Committee, which shall be composed of the members of the Council or their representatives and an equal number of members representing the European Parliament, shall have the task of reaching agreement on a joint text, by a qualified majority of the members of the Council or their representatives and by a majority of the members representing the European Parliament within six weeks of its being convened, on the basis of the positions of the European Parliament and the Council at second reading. The Commission takes part in the Conciliation Committee's proceedings, taking all necessary initiatives with a view to reconciling the positions of the European Parliament and the Council. Then, if, within six weeks of its being convened, the Conciliation Committee does not approve the joint text, the proposed act shall be deemed not to have been adopted. Or, if there is agreement, then there is a *Third reading.*

In this third reading the Conciliation Committee's approved joint text is put before the European Parliament and the Council. The former, acting by a majority of the votes cast, and the latter, acting by a qualified majority, shall each have a period of six weeks from that approval in which to adopt the act in question in accordance with the joint text. If they fail to do so, the proposed act shall be deemed not to have been adopted.

We continue to explore whether there has been any case where any of the above has ever happened. It is a difficult exercise! It should not be so.

The European Parliament is not sovereign. It has been rendered an irrelevant side show. It is now also being manipulated by the European Commission. So instead of exerting scrutiny it is being drawn into participating in the European Commission's dangerous fantasies. The STARTS initiative illustrates the point very clearly. Actually the European Parliament has no power of scrutiny over the European Commission. Who has? It seems the answer is no one. It is a power unto itself. Sound familiar?

The Soviet Union was a representative democracy. So is the European Union. It says so in the Treaty of the European Union – Article 10. This also states that citizens are directly represented at Union level in the European Parliament, as was the case with citizens of the Soviet Union being directly represented in the Supreme Soviet. So relax you have nothing to worry about.

But you say there are differences! Indeed they are. The Soviet Union was a one party state you say, but the European Union is multi party. Actually you are wrong because the European Union is a zero party state, because the political system of parties is still primarily organised at Member State level, which is why after the elections for the European Parliament, a process takes place in Brussels of forging alliances between different groups. And there is no opposition party in this system. No opposition party – just like the Soviet Union!

So please do relax for you have nothing to worry about. Just keep on pressing those send, share, like, and tweet buttons, for that seems now to represent an increasing number of people's reality. Keep on listening to all that information (disinformation? propaganda?) that everyday flows through the traditional and the social media world into your mind, making you feel well, and loyal, and inclined to wave a blue flag with yellow stars, twelve of them, set in a circle. Blue is a reassuring colour. Ask one of those artists in residence that are coming your way courtesy of the European Union and at your expense – We are talking about that thing called STARTS yet again.

**340**

Normative: Establishing, relating to, or deriving from a standard or norm, especially of behaviour. *Modus ponens*: The rule of logic which states that if a conditional statement ('if p then q') is accepted, and the antecedent (p) holds, then the consequent (q) may be inferred.

Here is Michael Gelb to speak:

"One of Leonardo's achievements was, 40 years before Copernicus, that the sun does not move, and that the earth is not the centre of the circle of the sun, nor the centre of the universe."

Well so what? Such information was – as we would say today – in the public domain in Leonardo's time, in a book ascribed to Aristotle called

DE CAELO (On the Heavens) - his chief work on the cosmos. It was also discussed in other works of the *Middle Ages*. Here is Aristotle.

"It is true that in my book I discuss what you now refer to as a heliocentric model. I wrote that it remains to speak of the earth, of its position, of the question whether it is at rest or in motion, and of its shape. As to its position there is some difference of opinion.

"Most people, all in fact, who regard the whole heaven as finite, say it lies at the centre. But the Italian philosophers known as Pythagoreans take the contrary view. At the centre, they say, is fire, and the earth is one of the stars, creating night and day by its circular motion about the centre. They further construct another earth in opposition to ours to which they give the name counter-earth.

"In all this they are not seeking for theories and causes to account for observed facts, but rather forcing their observations and trying to accommodate them to certain theories and opinions of their own. But there are many others who would agree that it is wrong to give the earth the central position, looking for confirmation rather to theory than to the facts of observation. Their view is that the most precious place befits the most precious thing: but fire, they say, is more precious than earth, and the limit than the intermediate, and the circumference and the centre are limits. Reasoning on this basis they take the view that it is not earth that lies at the centre of the sphere, but rather fire."

Through books, and people talking about the heliocentric model, is most probably how Copernicus came to the issue of heliocentricity. Do you agree, Arthur Koestler?

"I do. It was towards the end of his Italian studies that the heliocentric system began to take shape in Copernicus' mind. It was much discussed in Italy at the time. During the later Middle Ages the heliocentric system was favoured by scholars who took an interest in astronomy. Nicolas of Cusa is an example of (another priest) advocating the earth moving around a centre.

"Copernicus therefore did not conceive the theory of a heliocentric universe."

Thanks Arthur for that confirmation. Here we just add that what Copernicus did was develop a model that fitted the available data - Ptolemy's data. The model of course was still not the final say on the matter, as the Copernicus still stuck to the ancient dogma that the orbits were circular orbits.

And while Gelb is still here we must now tell him that the idea of using a convex lens to study the surface of the moon, did not, as Gelb implies, originate with Galileo, so Leonardo did not, 60 years before Galileo, come up with this idea as Gelb claims. This idea actually was developed in The Netherlands at the beginning of the sixteenth century - the invention was called the telescope.

We will leave you the reader to reflect on what it means to be a New Renaissance man in modern times.

*New Leonardos* - what a lot of nonsense.

C.P. Snow will now like to speak to you about his writing career. He is referring here to characters in his novel *The Search*, which was first published in 1934:

"Arthur Miles is not much like me. I was educated as a scientist, as Miles was: but I never had his single-minded passion, and in fact knew my own ultimate vocation from the time I was about eighteen. Miles' intellectual attitudes are simpler than mine were; I deliberately made him four or five years older than myself, so that he could represent the pure idealist in science, less involved in either politics or applied science than would have been possible for my exact contemporaries or for me.

"But he is like me in one respect. I, like him, have the knack, sometimes inconvenient, of getting immersed in what I am doing, more or less irrespective of what it is. So becoming almost by accident a professional scientist as a young man, I got immersed in the scientific life. The account I give of scientific experience in *The Search* is first-hand. A good many of my scientific friends, including some of the most creative, among them Rutherford, have told me it is true."

Now here is a spokesperson for the Encyclopedia.com:

"Snow's brief career as a research physicist was devoted mainly to extensions of his doctoral work on the infrared spectra of simple diatomic molecules and to crystallographic studies of complex organic molecules.

"A blunder he made in one of these studies convinced him he would never be a successful scientist. Snow and a colleague, Philip Bowden, thought they had hit upon a method for synthesizing vitamin A. However, their publication, which received excited attention, contained an obvious and critical error. The embarrassment determined the direction that Snow's creative energy would take."

Here is Stefan Collini:

"At first Snow seemed marked out for a successful career as a research scientist, but in 1932 he suffered a setback which redirected his life. He and a colleague believed that they had discovered how to produce Vitamin A by artificial methods. The discovery promised to be of immense theoretical and practical importance, and, following the announcement in *Nature*, the president of the Royal Society confirmed to the national press the significance of the findings. But alas, their calculations had been faulty, their discovery had to be recanted amid considerable publicity, and the trauma after all the publicity put Snow off scientific research irrevocably.

"That Snow was a trained scientist was crucial to the authority with which he was later to treat the question of the *two cultures*, but, as those scientists uneasy with this self-appointed champion of the scientific culture were to remark, his credentials were in fact somewhat shaky. By the time he came to give his Rede lecture, it was more than twenty years since he had been engaged in first-hand scientific research, and his achievements as a scientist had been patchy at best."

Yes indeed, but the *two cultures* nonsense was about something else. Snow had an agenda. Have you not guessed what it was? Perhaps we should invite him back to explain what this was?

Now here is a spokesperson for the Royal Society.

"I would like to confirm that C.P. Snow was never elected a Fellow of the Royal Society."

And we can see why! But this begs the question why, in Robert Root-Bernstein's paper, (whose paper we have already discussed) in that table of scientists who publicly exhibited works of art, published works of fiction, etc, Snow has been listed at all, and designated as, what Root-Bernstein calls a member of the Royal Society. Putting this wrong terminology aside, Snow was more a person who gave up science, and followed a literary career while also becoming involved in government. So strictly Snow was not a scientist who had a second career as a fiction writer. His scientific career actually never developed. So in Snow's case, art did not foster scientific success. Is Root-Bernstein offering the world art-science fakes?

**341**

So our attention was momentarily diverted away from the European Union and its Parliament – the European Parliament – but not sidetracked!

Now John A. Armstrong will provide some information about the Parliament of the Soviet Union – the Supreme Soviet:

"According to the Constitution, the Supreme Soviet was the primary legislative body of the USSR. But it was not. It only met twice yearly. The sessions seemed like mere ceremonies of ratification, characterised by set speeches rather than debate, and by unanimous acceptance of legislation introduced by government spokesmen. The elected deputies lacked the familiarity with government business needed to exercise real influence on legislative proposals.

"In spite of its docility, the Supreme Soviet was not always permitted to exercise even the formal power of passing the legislation that the Constitution assigns to its sphere. Many major laws were in fact enacted as decrees of the Presidium of the Supreme Soviet. This body, nominally elected by the Supreme Soviet, could be more easily manipulated than the entire legislative body.

"In view of the circumstances, it is no exaggeration to describe the Supreme Soviet not as a real legislature but as a rubber stamp."

And who among you knows if the European Parliament does any more than, in the end, rubber stamp decrees handed down by the European Commission and the Council of the European Union? And we have not mentioned the special legislative procures - decrees - for which no rubber stamping by the European Parliament is required.

Talking about technocracy, here is Claudio Radaelli:

"Technocracy has been with us for a long time. Looking beyond Plato and his philosopher kings, which was not really technocracy, the first precursor is Francis Bacon, who in 1627 published his *New Atlantis*, a political utopia where, what we would today call scientists, have replaced Plato's philosophers. With Bacon a technocratic theme appeared, that is the idea of using a scientific administration in lieu of politics.

"The idea reappears over many centuries, until in the twentieth century, in 1921, we have Thorstein Veblen, in the United States, advocating engineers running society."

This of course is just a brief overview. Claudio Radaelli one more time:

"Technocrats seek legitimacy by arguing for the government of science. The reality is that government by reason will always mean government of scientists. Governments will always be of persons, never of things. Hence there is little doubt as to who will be the king; if a government of science were ever to exist, it would be a government of scientists."

And if a government of science were to ever exist it would come into being in the European Union, and that government would be the European Commission - and they would do exactly what those before them did - the

Roman Catholic Church, the Marxist-Leninist's, the National Socialists. We leave you to imagine all the horror and suffering that would thus flow, not to mention the mess they will make of all those things that they will apply their religion to. Incompetence, lack of knowledge, corruption, cognitive illusions, and so forth - these are just some of the reasons why they will fail. The European Commission DG CONNECT's STARTS Platform serves to illustrate the point. Then they will resort to make-believe - propaganda - to hide the truth. They are already doing it.

**342**

More C.P. Snow nonsense will now follow. Here is Willem Elias:

"The concept of the third culture was first introduced by C.P. Snow in 1959 as he lamented the gap that had formed since the nineteenth century between humanities and natural science. This gap, according to Snow, negatively impacted both fields. He predicted that a new generation of scientists would close the communication gap between the *two cultures*. In addition I would like to say that during the Renaissance no distinction was made between the artist and the scientist, and this is the proverbial quality of that time."

Can you see how important it is to get you facts right? Scientist is a modern word! We say no more! Here is C.P. Snow:

"I deny all the above. I did not propose or introduce a third culture. What I said in 1964 (the publication date on the second edition of the *Two Cultures* book), was that there is something like a third culture, but only in terms of my very specific reasons for using the term *two cultures*.

"When I spoke about the two-cultures I should have stressed more heavily that I was speaking as an Englishman, from experience drawn mainly from English society. I did in fact say this and I said also that this cultural divide seems at its sharpest in England. I now realise that I did not emphasise it enough.

"In the United States, for example, the divide is nothing like so unbridgeable. There are pockets of the literary culture, influenced by the similar culture in England, which are as extreme in resisting communication and in ceasing to communicate: but that isn't generally true over the literary culture as a whole, much less over the entire intellectual society. And, just because the divide is not so deep, just because the situation is not accepted as a fact of life, far more active steps are being taken to improve it. This is an interesting example of one of the laws of social change: change doesn't happen when things are at their worst, but when they are looking up. So it is at Yale and Princeton and

Michigan and California, that scientists of world standing are talking to non-specialised classes: at MIT and CalTech where students of the sciences are receiving a serious humane education. In the last few years, all over the country, a visitor cannot help being astonished by the resilience and inventiveness of American higher education ruefully so, if he happens to be an Englishman.

"I think also that writing as an Englishman made me insensitive to something which may, within a few years, propel the argument in another direction or which conceivably may already have started to do just that. I have been increasingly impressed by a body of intellectual opinion, forming itself, without organisation, without any kind of lead or conscious direction, under the surface of this debate.

"This body of opinion seems to come from intellectual persons in a variety of fields - social history, sociology, demography, political science, economics, government (in the American academic sense), psychology, medicine, and social arts such as architecture. It seems a mixed bag: but there is an inner consistency. All are concerned with how human beings are living or have lived - and concerned, not in terms of legend, but of fact. I am not implying that they agree with each other, but in their approach to cardinal problems - such as the human effects of the scientific revolution, which is the fighting point of this whole affair - they display, at the least, a family resemblance.

"I ought, I see now, to have expected this. I haven't much excuse for not doing so. I have been in close intellectual contact with social historians most of my life: they have influenced me a good deal: their recent researches were the basis for a good many of my statements. But nevertheless I was slow to observe the development of what, in the terms of our formulae, is becoming something like a third culture. I might have been quicker if I had not been the prisoner of my English upbringing, conditioned to be suspicious of any but the established intellectual disciplines, unreservedly at home only with the *hard* subjects. For this I am sorry."

Note the apology, and the admission that he is only at home with the hard subjects of the natural sciences! Snow continues:

"It is probably too early to speak of a third culture already in existence. But I am now convinced that this is coming. When it comes, some of the difficulties of communication will at last be softened: for such a culture has, just to do its job, to be on speaking terms with the scientific one. Then, as I said, the focus of this argument will be shifted, in a direction which will be more profitable to us all.

"There are signs that this is happening. Some social historians, as well as being on speaking terms with scientists, have felt bound to turn their attention to the literary intellectuals, or more exactly to some manifestations of the literary culture at its extreme.

"Concepts such as the *organic community* or the nature of pre-industrial society or the scientific revolution are being dealt with, under the illumination of the knowledge of the last ten years. These new examinations are of great importance for our intellectual and moral health."

So what was C.P. Snow really doing when he launched his attack on a specific group of people - the literary elite in whose circle he moved? Perhaps we should also hear from the social sciences, for this is what he was referring to as a third culture, as well as humanities, which were far more developed in 1964 than he credits.

Why is so much discussion in the art-science and art-technology love affairs shaped by this failed scientist, by a man with an agenda, by a man who, by his own admission, is unreservedly at home only with the *hard subjects*?

The latter is actually not true!

**343**

It is a truth, not universally understood, that no where in Leonardo da Vinci's famous letter (which may be a fake) to the Duke of Milan does he claim to be an inventor, or to have invented any of the things he says he can make, has knowledge of, or are his secrets. Thus we invite Charles Gibbs-Smith and Sir Joseph Needham to compare notes, so to speak.

Go ahead Charles.

"Leonardo designed a transmission unit with a multiplying rotation for the axle of a wagon. A large horizontal toothed wheel turns the axle many times for each of its own revolutions, thus converting a slow action by a prime mover into a faster rate of movement at the cart wheels."

So Joseph what do you say.

"I have described a road wheel drive, where on one of the axles there is a gear-wheel (in the vertical plane) enmeshing at right angles with a horizontal gearing to effect rapid rotation of hay-cutting knives at the front. This road-wheel drive was a very ancient device in Chinese engineering.

"It is true that there are several machines with transmission mechanisms that were introduced to China from Europe. But it is also true that there were several that were already known in China, and did not need to be introduced. Gear-wheels, including ratchets and right-angle gearing, are one such."

Charles again:

"Leonardo drew chain links. These are an important mechanical device. The fact that he drew these illustrates conclusively that the chain drive developed by de Vaucanson in 1770 was not quite new - Leonardo had depicted it nearly three hundred years earlier, although it is not possible to say from these drawings what practical application he intended for his chain-drive."

Over to you Joseph:

"da Vinci was not the only fifteenth century European to have an interest in chain drives - Taccola designed an endless one - but it was de Vaucanson who developed the first industrially practical chain drive. Chain drives have however been used in practical power transmission applications in China since +eighth century."

So what is the point of this? We do not say that da Vinci did not design, or even invent. He was schooled it would appear in the mechanical arts, which included painting, so one can suppose he was familiar with many mechanical principles. But he was evidently not alone in doing such things - Brunelleschi for example seems to have designed machines to aid the construction of buildings. It is also possible though, that da Vinci may also have received information and knowledge from other sources - China for example - and what he drew are not his inventions or designs at all. We will never know. Sketches and notes on pieces of paper do not prove anything. The lesson here is as before, when you cross disciplinary boundaries, different methods, different types of proof, different modes of thinking, different challenges and values apply. Do not be so arrogant as to assume that your methods, your type of proof, your mode of thinking, your challenges and values, are superior or more important that those of others.

Unfortunately this type of mind-set is all too familiar to those who have to suffer working with people from the world of STEM. And to such people, we say this: You have not been as successful as you think or believe. Your measures of success are mostly material in nature, and while greater comfort and ease of life are welcome, they have been purchased at too high a cost - for humanity and for nature. It is time to change, to walk a different path, before you destroy both humanity and nature. Do you have the courage, the vision, to find such a path, or are you so caught-up in your ideology and dogma, that you are just like those Aristotelians who went before you - the modern version of the scholastics, with minds rigidly closed to anything that does not fit with your cognitive biases.

On the question of *Theory of Knowledge,* of the unity of knowledge, and the belief systems that would underpin this, whether we should, as Descartes would have us do, build from the ground up, a new house, or, what Neurath would suggest, be like the sailors who have to rebuild their ship on the open sea, without ever being able to dismantle it in dry-dock and reconstruct it from its best components. Many belief systems prevail in the world. The purpose in seeking universality is not to render all but one system redundant, to find the one-best system, according to some Abrahamic mind-set, some European cultural colonial agenda, but to recognise the need to live with many belief systems, many ways of seeing the world, many foundations for interpreting the world, while also recognising the matter of context. What applies in one context does not necessary apply in others. Such sophisticated thinking lies beyond the mind-set of *Hawkins Dawkins Wilson Wolpert.* It lies beyond what the *Vainglorious Enlightened Ones,* the technocrats, the believers in scientism, can tolerate. It does not lie beyond that we can accommodate and, more importantly, it does not lie beyond what we are able to imagine and to bring into existence. Now we know what sciptovisualism is for.

All knowledge exists in the context of belief systems. Did you consider this point Edward O. Wilson? Or were you too caught up in your own belief system to recognise this?

**344**

Change of coordinates one more time. Now we are in Brussels. It is the final meeting of the FET Flagship evaluation panel. The report has been significantly modified since its first draft. The conclusions that were reached in the former headquarters of IG Farben have been suppressed. Efforts have been made to cover-up the failure of the FET Flagships. The report has many contradictions. We do not bother to mention these. It will no doubt end-up with more contradictions as once we have completed our tasks, there will be further work on the document - out of sight! What is the right word to describe what will be done to the report? Perhaps we will not say. Oh yes we will - it will be further *fictionalised*! Evidently the world of STEM and the people of Europe have learned nothing from their tragic past. Thus does Prometheus reinvent himself in exactly the same form he was yesterday!

It would appear that we are winding-up this odyssey. So perhaps now is the time for C.P. Snow to come clean and admit to what he was up to when he started writing and speaking about *two cultures*. Here he is.

"Congratulations for identifying that I have an agenda, which is that I want scientists active in all levels of government."

Oh yes, the rule of science and reason! God help us all. Continue please Charles.

"Indeed, I will. By scientists I mean people trained in the natural sciences, not only engineers, though I want them too."

Oh yes, those who, like yourself, with fragmented and reductive minds, with their love of hierarchy, and authority, who are so focused on their narrow disciplines, knowing a lot about very little, do not understand or even recognise the existence of other disciplines like the social sciences. God help us all. Continue please Charles.

"Yes indeed. I did say I was sorry. But I want a special requirement for the scientist proper, because, partly by training, partly by self-selection, they include a number of speculative and socially imaginative minds. While engineers - more uniform in attitude than one would expect a professional class to be - tend to be technically bold and advanced but at the same time to accept totally any society into which they may have happened to be born. The scientist proper are nothing like so homogeneous in attitude, and some of them will provide a quality which it seems to me we need above everything else.

"In my view, and it is one of the points from which I started, it would be a real gain. It is a clear advantage to the Soviet Union that they have, right at the top of the political and administrative trees, a fairly high proportion of men with scientific or technical training."

Of advantage to the Soviet Union you say. But evidently not of advantage to the ordinary people of the Soviet Union. Yes you do indeed by your own words condemn yourself, C.P. Snow. This is what one can expect from the rule of science and reason - people praising oppressive and totalitarian systems of government, because they have scientists and technical people running the State. God help us all.

"But they have made better technical choices than us, and this collective influence of scientist and technical people must have been a help. But though this is a real gain, the main benefit is foresight."

Oh dear, what are we going to do with you C.P. Snow, technocrat supreme? Perhaps more important what are we going to do about those caught-up in art-science and art-technology love affairs. What a nutty lot of people. God help us all.

"Transformation will cut across the art and sciences by treating them as a continuum."

That was a spokesperson for the very short-lived art-science-architecture magazine called *Transformation: arts, communication and environment*, which was published from 1950-1952 (only three issues). C.P. Snow did not mention this. Very few other people have. A continuum! Is that relevant to a new *Theory of Knowledge*?

**345**

Today we stood in the Rauschenberg exhibition in Tate Modern. We will have to make a second visit in 2017. It is evidently important to understand his work if one is to understand E.A.T.

We have invited Bernard Connolly to speak about the 2012 edition of his 1995 book entitled, *The Rotten Heart of Europe: The Dirty War for Europe's Money*. Bernard was once, back in the mid 1990s, an employee of the European Commission, and was Head of the Unit responsible for analysis of the European Monetary System and national and Community monetary policies. Obviously after publishing such a book, he is no longer an employee of the European Commission!

"It is only sixty-odd years in Europe since the restoration of peace, democracy and legitimacy, with the foundations of economic prosperity put in place. Yet Europe is now a continent of widespread economic misery, of financial collapse, of disappearing faith in mainstream political parties and rising support for extremist parties, of loss of sovereignty and thus legitimacy and democratic control, and of the destruction of law, both domestic and international, by the judicially larcenous European Court of Justice. Perhaps worst of all, the relations of amity and cooperation among European countries which could be observed thirty of forty years ago have been replaced by a return to mutual resentment, distrust, jealousy, contempt, ridicule, anger and even hatred. This is what Europe has wrought; and its most destructive instrument, along with the European Court of Justice, has been monetary union.

"The outcome of monetary union has been, as predicted in my 1995 book, the destruction not only of prosperity but of political legitimacy, in every country in Europe, and of friendship and cooperation, an outcome that creates an extremely serious threat to peace. How did this happen? Why was it allowed to come about?

"Back in 1995, propaganda, disinformation and political correctness – the suppression of rational thought – had created worldwide enthusiasm for the monetary union project in Europe. Now, not even its proponents can hide the fact that monetary union has brought anguish.

"I would argue that the European Union deserves better to be known as the New Soviet Union – one with symptoms more sinister than mere painful micturition. Monetary union was and is, not only economically perverse because of the chaos, suffering and despair, which was so predictable, but also politically perverted. Its real purpose was very different from its advertised purpose. The intention of its progenitors was to use it to create an empire defined above all by the Anglo-Saxon model of the relations between the state and the individual."

STARTS is also perverse and sinister, and is more about propaganda and image making – the product of an ignorant and manipulative technocracy. History will note that many were taken in! The art-science lovers for example! Exploitation of artists is evil.

It is of course a truth that a person trained in a particular discipline will naturally perceive and act according to the basic principles of the discipline. Which is why we should not believe what those in STEM are saying about art – art fosters scientific success for example. This is also why we should not believe STARTS – the product of STEM thinking.

Totalitarian art: production of art that has everything to do with a top-down, state-driven project to bring about an aesthetic and spiritual union of government and people according to a prefixed dogma. Examples: The Third Reich; The Soviet Union; The European Union (the European Commission – DG CONNECT's STARTS Platform); some aspects of the art-science and art-technology secular religious cults.

**346**

A representative of the European Commission will now read to you Article 245 of the *Treaty of the Functioning of the European Union*:

"The Members of the Commission shall refrain from any action incompatible with their duties. Member States shall respect their independence and shall not seek to influence them in the performance of their tasks.

"The Members of the Commission may not, during their term of office, engage in any other occupation, whether gainful or not. When entering upon their duties they shall give a solemn undertaking that, both during and after their term of office, they will respect the obligations arising therefrom and in particular their duty to behave with integrity and discretion as regards the acceptance, after they have ceased to hold office, of certain appointments or benefits. In the event of any breach of these obligations, the Court of Justice may, on application by the Council acting by a simple majority or the Commission, rule that the Member concerned be, according to the circumstances, either compulsorily retired in accordance with Article 247 or deprived of his right to a pension or other benefits in its stead."

Now we will address the case of the Neelie Kroes a European Commissioner in Barosso's second Commission. She was one of the vice presidents of the European Commission and was the European Commissioner responsible, from 2010 to 2014 for what was initially called DG INFSO, but became, under her rule, DG CONNECT - the same DG that launched the ICT ART CONNECT initiative that became the STARTS Platform in Horizon 2020. Indeed it can be said that it is because of her that this thing called STARTS came about. Pity she was not told that the predecessor DG to DG CONNECT had already done STARTS, 25 years earlier!

Kroes came into the spotlight after leaving her post. Her name appeared in the so-called Bahamas papers, which were leaked into the public domain in September 2016. It transpired that she held office in an off-shore company and did not declare this fact, and is thus in breach of Article 245 of the *Treaty of the Functioning of the European Union*.

Here we also mention that according to the web site of the company that was the lead partner in the FET-OPEN project called FET ART that ran the ICT ART CONNECT initiative, that they have, in effect, been put out of business by the European Commission, because, the European Commission has stopped all current and future payments to the company because of an alleged fraud, following an investigation by OLAF, the European Commission's anti-fraud office.

Here is a spokesperson from OLAF:

"OLAF investigates fraud against the European Union budget, corruption and serious misconduct within the European institutions, and develops anti-fraud policy for the European Commission."

A murky past has this thing called STARTS. What else is lurking there, at the moment unseen, to add to this? Art-science fakes? Perhaps an investigation by the Court of Auditors is needed? Who can tell?

We have seen a book in the British Library - which is today one of the few places where the book can be seen - that is about the art of the theoretical physicist Richard P. Feynman. Feynman was awarded, jointly with two other people, the Nobel Prize for Physics in 1965. The Prize motivation: for their fundamental work in quantum electrodynamics, with deep-ploughing consequences for the physics of elementary particles. Richard Feynman was actually nominated 48 times for a Nobel Prize over the period 1956 to 1965.

Feynman's first sketch is included in the above mentioned book. The sketch is dated 1962. What do you think is the likelihood that starting to draw in 1962 fostered his scientific success? The answer depends on whether you are caught-up in the art-science love affair and believe its dogma, or are still able to think independently! To think critically!

Now for more nonsense and the rewriting of history by the European Commission's DG CONNECT, in collaboration with the European Parliament. Here is Theodoros Karapiperis and Miklós Györffi, of the European Parliament, who will now tell us all about what life was like in the Renaissance, as told to people at the event called EuroScience Open Forum 2016 which was held in Manchester, United Kingdom. He is speaking about a session called *Arts and Science: The Crossroads of Creativity.*

"Before the Renaissance, the arts and the sciences were united. Then, the Western Renaissance initiated an era of specialisation and science became codified as a segregated set of processes and world views. Meanwhile, art moved in its own direction, largely ignoring the agendas of science. In the Industrial Revolution, science inspired technology and technology inspired science while, in counterpoint, the arts moved towards a focus on nature."

What a load of nonsense you are speaking!

"Motivated by an on-going project within the European Parliament, we seek to demonstrate that the interaction between artists, scientists and citizens is essential to the development of functional societies, and brings added value to both arts and sciences. In a changing world, fostering the capacity for cooperation, critical thinking, curiosity and openness to diversity, are necessary for achieving peace and prosperity. Where the economic crisis reveals the limits of our current development model, rethinking the role of arts and science, especially in a research context, is essential."

But you will struggle to find critical thinking among art-sciences. Too much in love with each other! Those that engage in critical thinking tend not to associate themselves with establishment bodies like the European Parliament and its dangerous fantasies. Likewise DG CONNECT! Critical thinking was notably absent in ICT-ART CONNECT. We STARTed looking at STARTS for evidence of critical thinking and found only sycophancy and tall-tales etc. Or, as we now like to refer to these matters - art-science fakes. You are also talking about totalitarian art!

Back now to the event at which two Members of the European Parliament spoke: Eva Kaili MEP and Julie Ward MEP.

So you both agree with this nonsense about the Renaissance do you?

"Well we must do, as we spoke at the event."

The zoologist Desmond Morris is a long established surrealist artist. So what? Perhaps the art-science lovers, the Root-Bernsteins of that community, will want to attribute his scientific success to his art? Perhaps not though, for he was never elected to a Fellowship of the Royal Society! Perhaps that is because he is, by his own admission, anti-establishment.

Did you take that matter into account in your social-psychological research? Or were you too, engaged in biased reductionism?

That question was directed to Root-Bernstein and his supporters.

There is another paper by Root-Bernstein that we read. It is called *Arts and Crafts: Critical to Economic Innovation*. It is biased. One can tell that it is biased as soon as one starts to read it. The method and results are flawed by researcher bias. Evidently the reviewers of the paper thought so too, as there is a statement in the conclusions section addressing the bias. Put there perhaps as a condition of publication? No reputable social science journal would publish such rubbish. Some people would call this paper scientific fraud. We prefer the phrase art-science fake.

**348**

Time to hear from the social sciences; the sociotechnical school in particular, who can lay a sound claim to being able to influence technology - if only engineers and technologists would let them. No artists involved! They can change organisations as well, in fact both technology and organisation simultaneously - if only engineers and technologists would let them!

Albert Cherns is here to explain the idea of joint technical and organisational design of systems which, quite often, contain both a technical sub-system and a social sub-system:

"I have formulated eleven design principles for sociotechnical systems design. These are based on the sound practices we developed at the Tavistock Institute of Human Relations, starting just after the Second World War. These principles are:

"*Compatibility*: The design process by which we design sociotechnical systems needs to be compatible with the objectives we are pursuing. We cannot, for example, bring about employee participation and empowerment by decree. If we aim to develop some form of participatory organisational structure, we must design it by involving people.

"*Minimum Critical Specification*: This principle states that we should specify no more than is absolutely necessary, and this applies to all aspects of the system: tasks, jobs, roles, etc. We can be specific about what has to be done, but we should not over determine how things should be done. This implies that we create a degree of flexibility and openness in job descriptions, group structures, and technologies.

"*Variance Control*. Those variances which we cannot eliminate should be controlled as near to the point of origin of the variance as possible. We define variances are those events that are unexpected or unprogrammed. Some of these variances may be critical, in that they have an important affect on results. It is important that we should control variances at source because, not to do so, often introduces time delays.

"*The Multifunctional Principle - Organism vs. Mechanism*: Our traditional organisations are based on a high level of specialisation and fragmentation of work. This means that we can more easily replace people. The problem with this approach is that when we require a complex array of responses, it becomes easier for us to achieve this variety of responses when the system elements are capable of undertaking or performing several functions.

"*Boundary Location*: In our traditional organisations we primarily organise around fragmented functions, and one result of this is that barriers arise which impede the sharing of data, information, knowledge and experience. Boundaries should be designed around a complete flow of information, or knowledge or materials, so as to enable the sharing of all relevant data, information, knowledge and experience.

"*Information Flow*: We need to provide information at the place where decisions and actions will be taken based on the information. Many of our information systems are designed on the opposite principle. Information is usually provided to those distant from the point where decisions should be taken, thus removing decision making from those who are best placed to take decisions.

"*Support Congruence*: This principle states that we should design our reward systems, performance measurement systems, etc. to reinforce the behaviours that we seek from people. For example, individual reward for individual effort, is not appropriate if team behaviour is required.

"*Design and Human Values*: One of the objectives of sociotechnical systems design is to provide a high quality of working life. If we do not address the human factor, that is, issues of stress, motivation, personal development, etc, then systems will fail.

"*Incompletion*: When we design something, the design is never finished. As soon we complete implementation, its consequences, and on going changes, indicate the need for redesign.

"*Power and Authority*: When we assign responsibilities to people for tasks, we should make available to them the resources that they need, and we should give them the power and authority to secure these resources. In return we should expect people to accept responsibility for the proper use of the resources which we have provided. We should also make sure that people have the knowledge and skills needed, but having provided people with knowledge, skills and resources, we should avoid interfering, unless we are equally knowledgeable and skilled.

"*Transitional Organisation*: We can be faced with two quite distinct problems when creating new organisations. We could be dealing with the design and start-up of plants on greenfield sites, or on brownfield sites (existing plants). The second is much more difficult than the first. We should, in both situations, view the design team, and the processes it uses, as a tool to support the transitions that we wish to see take place. In other words, the design team is an integral part of the process of managing change. As such it should embody the new values and culture and be diffusing them throughout the enterprise. The actions taken by design teams will speak louder than any words embedded in mission statements. Design teams are the messengers of new values and concepts."

Did C.P. Snow know anything about the developments that had taken place in the social sciences in the 1940s and 1950s? He claims in *A Second Look*, that he was suspicious of any but the established intellectual disciplines, unreservedly at home only with the *hard subjects*, acknowledging then, that a third culture might be emerging. But in the early part of his Rede Lecture, he admits that non-scientists had told him that there was a third culture, and that some of his sociological friends did not identify with the literary elites, or (entirely) with the world of the natural sciences. But he chose to ignore this advice. Why?

Elsewhere, in his first novel, Snow demonstrated that he had understanding of disciplinary differences.

**349**

Time to hear from another physicist! This one goes by the name of Arthur I. Miller, and he is speaking about his book *Colliding Worlds: How Cutting Edge Science is Redefining Contemporary Art.*

"Leonardo made no distinction between art and science."

Oh dear here we go again! We do not have the inclination to go over this one more time and to explain why this was not so. Leonardo was not a scientist. There was no science as we know it today. People also made no distinction between barbers and surgeons. Let us therefore see some surgeons with barbers in residence. They can, while the surgeon operates, give the patient a hair cut or a shave!

Oh look a Nobel Laureate is here. He has an art avocation so he must have achieved his Nobel Prize because of his art avocation. Let us see what he has to say. His name is Eric Kandel:

"In 1963 C.P. Snow described the possibility of a third culture. Thirty years later John Brockman advanced Snow's idea in his book *The Third Culture: Beyond the Scientific Revolution*."

Here is John Brockman. Let us see if he agrees or disagrees.

"I disagree. I did not advance Snow's idea. In my book I said very specifically that I was only borrowing Snow's phrase, and that I was not describing the third culture that he was predicting."

So he disagrees! Why is it that these reductive scientific folk, even the ones who are Nobel Laureates, cannot read other peoples' books and understand what is written there?

Anything else to say Eric Kandel, before we send you on your way back to your fragmented and reductive world?

"Yes."

Well then, do proceed.

"Reductionist approaches are central to science. But reductionism is not limited to the sciences. Artists too are reductionists. Mondrian, among others, was a radical reductionist."

We see that you have, no where in your (very well researched?) book, quoted or made reference to the written works of Piet Mondrian. Why?

We collected in one place the written works of Piet Mondrian. We did this because we wanted to understand what he was doing and why. We note here just a brief point, that Mondrian wrote a lot about harmony and balance, and spoke also about the tragedy of the dominance of verticals in nature. This concern, this interest is reflected in his transition from the figurative to the abstract and is manifested in his famous works of lines and planes, starting with Composition with Line (1917).

We wonder why people, who most definitely would not apply such sloppy research practices in their own field, feel that it is quite acceptable to do so when they the cross disciplinary boundaries and start interpreting art. It is of course all highly subjective and anyone can have an opinion and claim all sorts of things. Better though to at least make some line of argument that relates to the artist's intent. Or perhaps not if you are from the world of science and intent on rewriting art history and claiming, as Miller does, that cutting edge science is redefining contemporary art. It is not - it is artists who are creating what is just a part of contemporary art. That is all! Obviously this is not enough for the conquistadors.

It is not much of a third culture John Brockman, when those whom you elevate to this pedestal, cannot get their facts right. Intellectual celibacy! More than that actually - it is a scientific reduction to ignorance!

"The art of any period tends to serve the ideological interests of the ruling class."

That was John Berger speaking.

Now you know about STARTS, SEAD and STEAM. Now you know why STEM people are interested in art. Now you know why STEM people are rewriting art history. Now you know why the European Commission is in the process of appropriating art. Now you know why the European Parliament is in the process of appropriating art. Now you know we are in trouble - serious trouble!

Was Leonardo da Vinci, Aristotelian or Platonian? Have you every come across anyone asking such a question? We already told you that there is a different Leonardo to be found by those that look - not many it would seem bother to look.

"It is ironic indeed that Leonardo, whose intellectual outlook was predominantly Aristotelian, should be presented or represented in the form of a neoplatonic ideal."

That was Patricia Rubin speaking. Anything else to say Patricia?

"Yes. Vasari's Leonardo is a fabrication, not a fiction."

Thank you. The comment is quite valid when applied to some aspects of art-science, and art-technology. So we return to the matter of art-science fakes. STARTS! We leave the matter to future books. There is plenty of material to work with – a diamond mine in fact!

**350**

Now we stand at chapter 350 and the year is drawing to a close. Orion is back in our skies. Winter is here. Trees are bare. Days grow shorter as we move towards the Winter Solstice. Long are the dark cold nights. Birds feed in the garden. They do not though, linger long. By mid afternoon they are looking to roost for the night. Weary is the year. Its time is nearly done. Age descends. Minds turn to Christmas. And then a new year will begin. The cycle repeats. Everlasting. Eternal. Only 16 chapters to write. 16 days to completion. What shall we write? Is there anything left to cover? Surely our mind now is empty. No more will words pour forth from the subconscious in this surrealist odyssey. Life is an odyssey. So is art. Beyond definition. Beyond subversion. Beyond the *Vainglorious Enlightened Ones'* power to influence. Beyond their power to lay claim to. Hubris. Arrogance. Contempt. Ignorance. These are words of knowing. Chapter 350. We say it again. We are nearly done. Rules are broken. Rules are smashed. There are no rules in art. Only people from science, from the world of STEM, cannot see this. STEM, STARTS, STEAM, and SEAD. STEMming from an interest in art, STEM has STARTed STEAMing and SEADing. But to what avail? The pen is far mightier than STEM, STARTS, STEAM and SEAD. The pen and the brush combined even mightier. The picture, the visual composition that we created in the dual process of scriptovisualism – what became of this? What is scriptovisualism? Now we have a better idea. So we continue to write. What are we doing? We of course means Julia and Paul. Who is Julia? A matter of interpretation? Ambiguity? Have DG CONNECT seen our book yet? Do they understand? Probably not! Does it matter? No. They are the past. They will be blown away. They will be gone with the wind. Times they are a changing. No one yet seems to understand that their day, that of the *Vainglorious Enlightened Ones,* is over. The age of Europe is at an end. What comes next? This is what we seek. Forget about the European Union. Forget about Europe. Forget about the West. This is the time of the coming age. Soon it will be with us. The post-European, post-Western era. Thus do we take up the pen to write *366 – A Scriptovisual Composition Unknown: An Artistic Investigation of the Spaces between Spaces in the Quest to Create a post-European Era, post-Enlightenment Science*. Look we are referencing in the book we are writing, the book we are writing. Make any sense? Is it supposed to? It is surreal.

That is what we said we were doing. We are doing. See the evidence before your eyes. You just read it. Chapter 350.

Now it is time to hear some words about *Taoism*.

"Are Chinese Confucian? Yes, indeed. Every Chinese, if tutored at all, cannot help being Confucian, more or less. And yet, there is necessarily the other side - the counterpoise. For every Chinese is likewise a Taoist, and the definition of a Taoist is, normally, what a Confucianist is not! Most Westerners are at a loss in understanding Chinese personality. Not a little of the difficulty, I suggest, has come from ignoring this dual nature of the Chinese. The fact is: we are socially Confucian and individually Taoist. For all the imposing superstructures of our society, with its intricate family ties and relations and its myriad conventions and mannerisms to which we submit as good Confucianists, we as individuals *per se* are irreducibly Taoists.

"Confucianism, with its command of duty, calls for cooperation and conformity; socially speaking, it unites, conserves and perpetuates. Taoism, as an embodiment of individual free will, ever tends toward dissent and dissociation. The contrast between the two is glaring indeed. Yet, like so many paradoxes of life, opposites dwell closest together. The combination of these two modes of thought in one culture, nay, in one person, is not an unhappy mixture in many ways. They supply, shall we say, the positive and the negative elements in Chinese life - the *Yin* and the *Yang*, which complement each other and operate dialectically to lead life out of its periodic impasse; and thanks to them, China is no nation of monomaniacs and monotones.

"One should be on one's guard against confounding Taoism facilely with Western individualism. True, both advocate individual freedom. But it would be sheer naïveté to assume that they mean the same thing. Taoist freedom is the freedom of a pre-social or an asocial being, while the libertarian freedom of the West is the freedom of a socially conscious man."

That was Lin Tung-Chi speaking.

Two men compared: Aristotle and Leonardo da Vinci. Aristotle, a towering intellectual whose work covers many fields, including science, whatever that meant in Ancient Greece. Yet no one speaks about Aristotle in the same way as they do about Leonardo. Aristotle was an accomplished .... Why is that? Two men compared: Aristotle and Leonardo da Vinci. Who is the greater of the two? What was da Vinci up to?

## 351

We stand now at chapter 351 and we wish to ask Joseph Needham to clarify a point. We have been talking about a *post-European Era, post-Enlightenment Science.* Be careful! Understand. Do not jump to conclusions.

So Joseph, speak your words of knowing.

"In conferences one sometimes hears it said that, for example, 'this would be impossible in Muslim science, Japanese science, American or Russian science.' I cannot agree that in the modern world there is such a thing as Muslim science, Japanese science, American or Russian science, except as an abuse of the word science, unless one means Muslim, Japanese, American or Russian *use* of modern science.

"In my opinion there is no science outside modern international science."

We agree.

But, in this book we have said that we are working to develop a Taoist science!

Indeed. You are confused, perhaps? This, you must see, is more complicated than you realised. We mention here some words and phrases: theory of knowledge; ways of knowing; overcoming technocracy; ridding the world of scientism; equivalence; choice; values; beliefs, intellectual celibacy.

This is, one might say, just a reconnaissance in force. This is just a scriptovisual composition unknown. Not unknown anymore! Next comes the detailed work. Where to begin?

John Brockman is still here. We would like to point out to you, John, something concerning what you say in your book, concerning Snow's third culture. You refer to literary intellectuals being on speaking terms with the scientists. This is not correct - Snow did not say this. He speaks about a third culture, which is not defined by literary intellectuals. Indeed he mentions what this third culture is, and states that 'Social historians as well as being on speaking terms with scientists have felt bound to turn their attention to the literary intellectuals, or more exactly to some manifestations of the literary culture at its extreme.'

It is not much of a third culture John, as you propose, when you cannot get your facts right.

Intellectual celibacy? Rewriting history? Or is it scientific reduction to ignorance?

"There is the weak case for art-science interaction and there is the strong case for art-science interaction. The strong case, I argue, is crucial, not only for setting the scientific agenda, but because a changed science and redirected technology will emerge. We accept that individuals such as Leonardo da Vinci, through art-science synthesis, drove changes in the content and methodology of future science and technology. The strong case argues that we are now in a *new Renaissance* where such cases are beginning to occur again."

That was Roger Malina speaking, expressing his beliefs, which are grounded in the art-science lovers' belief system. It is the standard, unquestioned, da Vinci, Platonic idealism - a fabrication.

Renaissance - meaning rebirth, renewal. And the Oxford English Dictionary definition: the revival of European art and literature under the influence of classical models in the period fourteenth to sixteenth centuries.

Of course, European art was not in need of rebirth - that is another fabrication.

*New Renaissance*: another revival of European art and literature under the influence of classical models in the twentieth and twenty-first centuries! *Return to order*, again - in the artistic sense of the meaning of this term?

Autochthonous: (of an inhabitant of a place) indigenous rather than descended from migrants or colonists. Interesting word, with a clear connection to Leonardo, and, science! And, art-science too, perhaps?

We have been autochthonous at times in this book. Why? Poetic justice perhaps, or rhetoric, or both? Needs more exploration!

To be or not to be that is the question; whether it is better to write in Latin, or in the vernacular, and in choosing the latter, to be seen as autochthonous.

Art-science lovers caught in a toxic embrace, a meeting of *epistêmê* and *technê* from ancient European times.

We return to the matter of beliefs, which we again encounter in the justification for the toxic embrace. Highly subjective are these beliefs! Once again we note that knowledge is grounded in beliefs. Then we ask the question: what if the story that is knowledge, which results from the story that is beliefs, is based on a story that does not correspond to what the written record actually says? The *Notebooks of Leonardo da Vinci* for example?

Leonardo, the man who wrote that "[...] the arch in buildings is made up of two segments of a circle, and each of these two segments being in itself very weak desires to fall [...]" Desires to fall! We have met this already. The desire to change state! Very Aristotelian! It is also known as animism. And of those seas - for example the Sea of Tranquillity - that people used to believe were on the moon, Leonardo, not questioning such, says he has "[...] proved that the part of the moon which shines consists of water [...]" and that the reason the waters of the moon "[...] do not descend to the centre of the universe and to join itself to the earth, [...] is clear sign that the moon is clothed with her own elements, namely water, air and fire [...]". Again, very Aristotelian! And very geocentric as well! And we have Michael Gelb saying that one of Leonardo's achievements was, 40 years before Copernicus, that the sun does not move, and that the earth is not the centre of the circle of the sun, nor the centre of the universe. Yet here we see Leonardo writing the exact opposite. Oh to be a new Renaissance man in the modern world!

Leonardo - the man who is adored and worshiped by the art-science lovers. How many of them have actually read the famous Leonardo da Vinci notebooks?

"The above supports my view that those in the arts have hardly a clue as to what science is about."

There may be some truth in this Lewis, but we would not say that it should be applied to all in the arts. Nevertheless it is a concern that must be taken into account, for not to recognise the deeply Aristotelian nature of Leonardo's observations is a major flaw. Such is the nature of love!

If you have not yet realised, the last speaker was Lewis Wolpert.

What was da Vinci really up to? Obviously the art-science lovers have not discovered. This we find useful, otherwise we would not have been drawn to look and to ask. There is no *new Renaissance*. The claim that da Vinci drove changes in the content and methodology of future science and technology is a belief, nothing more. There is no one working at the forefront of art and science, because there is no such thing as art-science. It is a story. Is it a useful story? Or is it a dangerous story? If it leads to knowledge and understanding, to something useful, then perhaps it is a useful story. But does it lead to something useful? Perhaps it does, but not in the sense that the art-science lovers expect. Perhaps instead what is useful, lies in the critique that we do now engage in, and what this, of itself, can give rise to? Obviously the lovers will not be pleased.

And on the matter of what da Vinci was really doing, one can also wonder whether the art-science lovers have read Sigmund Freud's psychoanalytical essay on da Vinci? It is very revealing, for emerging is the picture of neurotic behaviour, with the neurosis manifesting itself in an obsession for investigation. Freud identifies the neurosis as lying in da Vinci's very early childhood, particularly the (accidental) circumstances of his illegitimate birth and the excessive tenderness of his mother, coupled with a strong tendency towards instinctual repressions and an extraordinary capacity for sublimating primitive instincts.

"I place Leonardo close to the type of neurotic that we describe as *obsessional*; and we may compare his researches to the *obsessive brooding* of neurotics."

That was Freud speaking. He also noted in his essay that the research which took the place of artistic creation seems to have contained some of the features which distinguish the activity of unconscious instincts - insatiability, unyielding rigidity and the lack of an ability to adapt to real circumstances. Which we might add, by way of speculation (of course), may be the reason why there is so much that, to modern eyes, is so strange in the results of these investigations, and which looks to include that which appears elsewhere. Could it be that the obsession with investigation was, as is the case with very young children, an attempt to record the world around and to construct explanations - any explanations - regardless? And people - like art historians, and those with fragmented and reductive minds, and those that are art-science lovers - have just chosen to interpret as ... Evidently we have stumbled upon a flaw. It seems that the overly rational, overly objective European world is full of flaws: the Leonardo cult, the art-science lovers, the art-technology lovers, STARTS - what are these, but flawed understandings resulting from beliefs!

So the matter of *Art-science Fakes* keeps cropping up. Scientific reduction to ignorance! The ultimate fate of the *Vainglorious Enlightened Ones*!

There is only one way to find out if there is anything of substance in this art-science love affair, or whether it is just a tool to enable this critique, and that is to electronically search the whole content of the journal called *Leonardo*, every issue from 1968 to 2016, to interrogate it, to seek answers to pertinent questions. Then we will know. Then we will have to write another book. What will be the title? Will the phrase *Art-science Fake*, or *Art-science Tangled Web*, appear in the title? You will have to wait to find out! Are you listening, all you so called *New Leonardos*?

Our interrogation will START in 2018 - happy 50$^{th}$ birthday *Leonardo*!

Margaret Wheatley now speaks:

"I think that we are coming out of this very inhibiting even paralyzing mental model of the world as machine. That was purely in the West, and I believe that we are at the end of it. That would be hopeful except for historical evidence that whenever an old thought form or paradigm is dying, it becomes vicious, it goes out in a blaze of anger and aggression and that's what we are seeing also."

The European Commission – the viciousness, the blaze of anger, the aggression will from them come forth. This is why we are acting now against them. The Europeans once again! They have not learned. And it STARTS with STARTS and what can be seen in people like Kandel, Miller, and numerous others – *Hawkins Dawkins Wilson Wolpert*. Then will come the reaction to the United Kingdom's departure from the European Union. You will not have to wait long after we have departed, before it STARTS. You will deny it of course, such is the scale of delusion associated with this very dangerous idea called the European Union. The events that will trigger what happens, are already STARTing.

Now it is time to stop them – thus do we take up the pen and the brush, and become scriptovisual.

"The literary work can sometimes say more, even about the social realm, than many writings with scientific pretensions, especially when the difficulties that must be overcome in acceding to knowledge are not so much intellectual obstacles as the resistance of the will.

"The denegation that literary expression performs is what permits limited manifestation of a truth which put otherwise, would be unbearable.

"Literature can reveal the truth, without actually speaking the truth."

That was Pierre Bourdieu speaking.

So what happens when someone simultaneously obeys the rules of art while also breaking the rules of art? In your binary world this is not possible. We however do not live in your binary world.

Now we know too that a work of art can be knowledge. But it is also a licence for *art-science fakes* as well, as indeed is that word research. So be careful! Be very careful!

"I love you," said the artist.

"I love you too," the scientist responded, tenderly. "But still no babies, no New Leonardo. What shall we do?"

"We'll pretend, for no one will know, because everyone is blinded by the love they have for us."

Now a feeling comes upon us, and we START to write:

> STARTS weaving the art-science tangled web,
> STARTS new Leonardos in ignorance dwelling,
> STARTS the new Medici's power rising,
> STARTS ...
>
> STARTS the unraveling.

Is it a poem or is it a painting? Or is it a scriptovisual composition - unknown? Perhaps it is another book about STARTS? Who can tell?

This notion of *New Leonardos*: it is a retreat into the past - a past that never existed! It is also a retreat into a past that did exist - one where humanity must never go again! This retreating into the past is what people do when their civilisation begins to collapse. The past is, people believe, a known place, safe and secure, but it is also artificial. Such pasts are also places of delusion - dangerous delusion!

**352**

Now we stand at chapter 352. Max Ehrmann will now read his poem, *Desiderata*:

> "Go placidly amid the noise and the haste, and remember what peace there may be in silence. As far as possible, without surrender, be on good terms with all persons.
>
> "Speak your truth quietly and clearly; and listen to others, even to the dull and the ignorant; they too have their story.
>
> "Avoid loud and aggressive persons; they are vexatious to the spirit. If you compare yourself with others, you may become vain or bitter, for always there will be greater and lesser persons than yourself.
>
> "Enjoy your achievements as well as your plans. Keep interested in your own career, however humble; it is a real possession in the changing fortunes of time.
>
> "Exercise caution in your business affairs, for the world is full of trickery. But let this not blind you to what virtue there is; many persons strive for high ideals, and everywhere life is full of heroism.
>
> "Be yourself. Especially, do not feign affection. Neither be cynical about love; for in the face of all aridity and disenchantment it is as perennial as the grass.

"Take kindly the counsel of the years, gracefully surrendering the things of youth.

"Nurture strength of spirit to shield you in sudden misfortune. But do not distress yourself with dark imaginings. Many fears are born of fatigue and loneliness.

"Beyond a wholesome discipline, be gentle with yourself. You are a child of the universe no less than the trees and the stars; you have a right to be here.

"And whether or not it is clear to you, no doubt the universe is unfolding as it should. Therefore be at peace with God, whatever you conceive Him to be.

"And whatever your labours and aspirations, in the noisy confusion of life, keep peace in your soul. With all its sham, drudgery and broken dreams, it is still a beautiful world. Be cheerful. Strive to be happy."

*Desiderata*: desired things. What, in this age, at this moment in time, are desired things? To keep spending money on those things that amuse the minds of those who have become knowledgeable about very little? To leave the world's richest one percent holding more than forty-six percent of the world's wealth, and the poorest ninety percent with only fourteen percent? To do nothing about scientific hubris, technocracy, and scientism? To leave science and technology unchanged, still living in an Abrahamic past of sole truth? To …

## 353

Now we stand at chapter 353 and we wish to ask Daniel Kahneman to tell us more about System 1 and System 2, which we believe will lead humanity to a disaster on a scale never before experienced:

"So now I speak about heuristics and biases.

"The availability heuristic is a mental shortcut that occurs when people make judgments about the probability of events on the basis of how easy it is to think of examples. The availability heuristic operates on the notion that, if you can think of it, it must be important.

"Substitution is an automatic process where the mind substitutes for a complex question, a much simpler heuristic question that provides a means of not engaging in the careful reasoning needed to answer the complex question. With luck, the heuristic question may be a good enough approximation to the complex question.

"The optimistic bias is a cognitive bias that leads people to overestimate the likelihood of success or to underestimate the challenges involved in being successful, or both.

"Finally I want to say something about *What You See Is All There Is*. This theory states that when the mind makes decisions, it deals with limited information as if it were all there is to know. It builds the best possible story with the information available, and if it is a good story, the mind believes it. The mind has an almost unlimited ability to ignore its ignorance."

And, as we are considering human behaviour, now is a good time to mention that strange behaviour known as *Information Avoidance*! Here people will actively avoid information (even that which is freely available) which would help them to improve decision making or help them avoid ill-founded actions. One reason for *Information Avoidance* is that information may contradict established beliefs, or may create undesirable emotions.

Now you know, yet again why technocracy does not work, why scientism is extremely dangerous. Both need to be banished from our world; consigned to the place where they belong - the past!

Now you also know why there is no third culture as defined by John Brockman - just a lot of deluded people driven by cognitive biases, suffering from cognitive illusions; spouting forth ideological nonsense (for the most part). Richard Dawkins is our prime example. Not the *God Delusion*, but the *Science Delusion*. More appropriately that particular book could be called *My Struggle against Religion* or just *My Struggle*. He has it bad! So it seems do some artists!

**354**

The artist Robert Rauschenberg is here and wishes to say something about E.A.T. So Robert, what do you want to say?

"This is a non profit benefit. Artists have been struggling with insufficient help to get and understand contemporary materials. This foundation has been established to provoke the interest of science, money, industry and art to work together to make work that could not exist otherwise."

Here is Billy Klüver one more time. He and Robert will speak with one voice, what seems to have been the E.A.T. manifesto.

"Maintain a constructive climate for the recognition of the new technology and the arts by a civilised collaboration between groups unrealistically developing in isolation. Eliminate the separation of the individual from technological change and expand and enrich technology to give to the individual variety, pleasure, and avenues for exploration and involvement in contemporary life. Encourage industrial initiative in generating original forethought instead of a compromise in aftermath, and precipitate a mutual agreement in order to avoid the waste of a cultural revolution."

So what does this mean? Did anything beyond the making of art happen? Was there really a true collaboration, or was it all about engineers servicing the needs of artists?

We will of course never know the truth. Why? Because E.A.T. will always be subject to a romantic image of artists and engineers working together. It is because people, many years afterwards, reminisce about those pioneering days, often forgetting to mention all the bad things that happened. It is because they apply post-event rationalisation. It is because it is in the interests of E.A.T. to continue to paint the picture of the image that they want to present to the world. It is because E.A.T., like many of these art-science, art-technology collaborations, are not subject to independent observation. It is because such things as E.A.T. are always subject to interpretation. Hence what we end up with is what people want to end up with. E.A.T. - the stuff that myths are made on.

One has to dig hard, be very critical, before one can understand something of what E.A.T. was. Some aspects of it will always be unknowable. That is the nature of life and the universe!

And what goes for E.A.T. also goes for Leonardo too. What we know is that he was an artist, and being such, he was also very likely schooled in the mechanical arts. But not the liberal arts! The rest is often speculation, invention, misinterpretation, image making, and so forth. He may well have invented as well. Leonardo - the stuff that myths are made on.

Leonardo, Snow, E.A.T. - any relevance?

**355**

Equivalence: the condition of being equal or equivalent in value, worth, function, etc. Transformation: A marked change in form, nature, or appearance; in mathematical logic, a process by which one figure, expression, or function is converted into another one of similar value; in linguistics, a process by which an element in the underlying logical deep structure of a sentence is converted to an element in the surface structure; in biology, the genetic alteration of a cell by introduction of extraneous

DNA, especially by a plasmid. Transfiguration: A complete change of form or appearance into a more beautiful or spiritual state.

Note how the word transformation has different interpretations in different disciplines.

In mathematics there is a concept called decomposition. This is a process whereby a number is broken down into component numbers. One can also decompose functions as well - functional decomposition. One is taking functions apart. One is not engaging in a reduction in the sense that scientists use the term - like ignoring things, or just looking at something in isolation.

Now we return to the claim made by Eric Kandel, in his book *The Age of Insight*, that artists like Piet Mondrian used reductionism. Oh really! There is more to art and what artists are seeking to do, than you, as a scientist, may think or are even able to comprehend. We just demonstrated that. Or did you not notice!

Let us now see what Piet's friend, the poet and writer Michel Seuphor has to say.

"Piet Mondrian was deeply interested in theosophy. This phase lasted from 1899 to about 1917. This highlights his spiritual interests. In some of his paintings, for example, of the sea, we can see an evolution. He begins with naturalistic drawings, and moves to something more abstract. He spoke about the horizontal horizon representing repose. He spoke of horizontals and verticals crossing, referring to masculine and feminine; the spiritual and the material; forming a unity of the whole. He strove for the reconciliation of opposites.

"Later, when he was in his Neo-Plasticism phase, he wrote down the principles of Neo-Plasticism.

"The General Principles of Neo-Plasticism are as follows.

"The plastic medium should be the flat plane or the rectangular prism in primary colours (red, blue, and yellow) and in non-colour (white, black, and grey). In architecture, empty space counts as non-colour. The material can count as colour.

"There must be an equivalence of plastic means. Different in size and colour, they should nevertheless have equal value. In general, equilibrium involves a large uncoloured surface or an empty space, and a rather small coloured surface or space filled with matter.

"The duality of opposing elements in the plastic medium is also required in the composition.

"Abiding equilibrium is achieved through opposition and is expressed by the straight line (limit of plastic means) in its principal opposition, that is to say, the right angle.

"The equilibrium that neutralises and annihilates the plastic means is achieved through the proportions within which the plastic means are placed, and which creates the living rhythm.

"All symmetry shall be excluded.

"With regard to Neo-Plasticism and Form:

"In nature, relations are veiled by matter appearing as form, colour, or its natural properties. The *morphoplasticism* was consciously followed in the past by all arts. Thus in the past, art was *after nature*. For centuries painting plastically expressed relations through natural from and colour until it came in our day to plasticism of relations alone. For centuries, painters composed by means of natural from and colour; at present the composition itself is the plastic expression, the image.

"Now for Neo-Plasticism and colour:

"Despite its *interiorised* plastic expression, Neo-Plasticism still is painting. Its means of expression is pure and definite colour, in planes which remain equivalent to the surface of the picture; in other words, colour remains flat on a flat surface. It is not weakened by having to follow the modulations of the form; thus it is stronger than in morphoplasticism. Colour finds its equivalent opposition in non-colour, that is in white, black, and grey.

"And Psychology and Social Consequences of Neo-Plasticism:

"Balance through the equivalence of nature and mind, of that which is individual and that which is universal, of the feminine and the masculine – this general principle of Neo-Plasticism can be achieved not only in plastic art, but also in man and society. In society, the equivalence of what relates to matter and of what relates to mind can create a harmony beyond anything hitherto known.

"By the interiorisation of what is known as matter, and by the exteriorisation of what is known as the mind – until now the two have been kept pretty far apart! – mind matter becomes a unity.

"Neo-Plasticism demonstrates the exact order. It stands for equity, because the equivalence of the plastic means in the composition demonstrates that it is possible for each, despite differences, to have the same value as the others.

"Equilibrium, through a contrasting and neutralising opposition, annihilates individuals as particular personalities and thus creates the future society as a real unity."

Now Piet Mondrian has the final word on his art:

"Not to concern oneself with form or colour insofar as it is form, that is Neo-Plasticism in art. Not to be dominated by the physical-natural, that is the new mentality.

"To be concerned exclusively with the relations, while creating them and seeking their equilibrium in art and in life, that is the good work of today, that is to prepare the future.

"In life, sometimes the spirit has been over-emphasised at the expense of the body, sometimes one has been preoccupied with the body and neglected the spirit; similarly in art, content and form have alternatively been over-emphasised or neglected because *their inseparable unity* has not been clearly realised.

"To create this unity in art, *balance of the one and the other must be created.*

"It is an achievement of our time to have approached towards such balance in a field in which disequilibrium still reigns.

"Disequilibrium means conflict, disorder. Conflict is also part of life and of art, but it is not the whole of life or universal beauty. Real life is the mutual interaction of two oppositions of the same value but of a different aspect and nature. Its plastic expression is universal beauty."

So Eric Kandel, Nobel Prize winner as you are, what on earth were you on about? Just another fragmented and reductive mind, making fools of themselves by meddling in art, without taking the time to understand, for reasons of ...

Why do they do it? It is scientific reduction to ignorance. We will write a book about it.

If you want to substitute the very complex explanation we have just provided for Neoplasticism, with the simple (simplistic) statement that Neoplasticism is just reductionism, you are free to do so. Likewise we are free to observe that this is just a reductive explanation for a complexity that you have chosen not to understand, and that as soon as you reduce, which is your way, everything is lost - you have embarked upon a *scientific reduction to ignorance.*

**356**

Today is the day of the Winter Solstice and Robert Rauschenberg speaks again on the topic of E.A.T:

"The beginning of E.A.T. Experiments in Art and Technology to function as a catalyst for the inevitable fusing of specialisations creating a responsible man working in the present."

You will find the above words in Robert's autobiography - it is a painting not a book.

So once again we return to scriptovisualism and the notion of a different type of STEM person, living in the present, not the past, working in a responsible way. A different type of science and technology - *Taoist* science and technology! No more technocracy. No more scientism. No more technocentric thinking. No more conquistadors. No more fragmented and reductive minds. No more sole truths. No more *Hawkins Dawkins Wilson Wolperts*. It seems that Robert foresaw our coming into the world.

Ancient Rome, the first European Union, was originally a republic - a democracy of sorts. Then it became a dictatorship. The history of Europe is a history of European Unions. Prometheus sits on his rock. Narcissus gazes at his reflection.

**357**

Now we discuss dynamics. A linear dynamical system can be described by differential equations. For example a mass-spring-damper system is a second-order system because it is described by a second-order differential equation. A circuit with a resistor, capacitor and inductor is also a second-order system for it too is described by a second-order differential equation. In fact the two systems are equivalent.

An alternative way of describing a second-order linear system, is to use a state-space description. This is just a set of two first-order differential equations, appropriately related through matrices. So the two descriptions are equivalent. An *nth* order linear differential equation can be represented by a set of *n* first-order differential equations.

So we have two different ways of describing the world. Both are related to each other, through appropriate numbers. One can be transformed into the other, and vice versa.

We now refer you to Piet Mondrian's work. We now refer you to causal descriptions and purposive descriptions. We now refer you to Enlightenment science and post-Enlightenment science. We now refer you to European science and post-European science. We now refer you to Abrahamic science and Taoist science.

Different ways of looking at the same things, but with radically different implications for humanity, for society, for nature, for the planet, for our future! How are you feeling?

**358**

Rauschenberg and E.A.T. are still on our minds. Here is Michelle Kuo:

"Art and technology were divided in the 1960s, infamously split between the *two cultures*. Many warned that this rift would lead to the dystopian nightmare of machines run amok, of technocratic domination, nuclear winter, ecological ruin. E.A.T. echoed this alarm. As the organisation's inaugural manifesto proclaimed, culture and technology were dangerously developing in isolation. Yet their response was wholly unorthodox. Unlike Snow's technocratic call to arms, which argued that culture abdicate to modern science and which Klüver and Rauschenberg ultimately opposed, E.A.T. posed a way to bring disparate fields together. If artists and engineers actually talked to one another, shared information, broke out of professional specialisations, they might change one another, change their disciplines, change the structure of invention itself."

At last someone who understands that Snow had a technocratic agenda! But is what has just been said about E.A.T., Klüver and Rauschenberg all true? There is no reference to culture and technology dangerously developing in isolation in the E.A.T. manifesto! Actually it says that they are *unrealistically* developing in isolation. But culture and technology were not developing in isolation from each other. One only has to look at the world of television and the film industry to understand this. Also, before E.A.T. there were collaborations between artists and technologists. Walt Disney is a case in point. That of course is commercial art – so what!

The artist Alexander Calder used small electric motors in his kinetic sculptures in the 1930s. Calder is also an example that shows that art-technology collaboration is not new, for he collaborated with engineers to realise his larger sculptures. According to the material exhibited in Tate Modern as part of the *Alexander Calder: Performing Sculptures* Exhibition of 2015/16, in 1939 Calder designed for the New York World's Fair, a spectacular ballet to be performed by jets of water, but although the jets were installed they never worked as, according to Calder, the engineer he worked with was not interested in making it possible.

There were also conflicts and misunderstanding among the artists and engineers involved with *Nine Evenings*. Few people though, ever mention this, which is a pity for these conflicts are the places where real learning concerning the challenges of different disciplines working together lie.

Jasia Reichardt also mentions conflicts when artists collaborate with engineers. So too does the artist Lillian Schwartz. Conflicts also arise when STEM people collaborate with social scientists. These too tend to be covered-up. This is likely to be the case with STARTS, about which the truth is unlikely ever to be told. No doubt they will hold a meeting in Frankfurt in the former headquarters of IG Farben and conspire with DG CONNECT to fabricate a story that they will present to the world, authored by a modern day Vasari. We may do the job ahead of this to save the need to spend public money on such propaganda. Yet another STARTS book is already taking shape in our heads it seems! Perhaps we might call it *Lives of the Most Excellent European STARTS Artists*. We are going to be very busy in 2017 and 2018, writing books about STARTS. More paintings too!

"Rauschenberg's early enthusiasm for working collaboratively was infused with the utopian spirit of the historical avant-gardes as digested through the optimism of the United States and its emerging youth culture after the Second World War."

That was Achim Borchardt-Hume speaking.

Indeed the 1960s was a time of optimism; the growth of consumerism, the moon-shot era. It seemed that much was possible and made so by the power of science and technology. But by the end of the 1960s this had changed and Rauschenberg also moved on.

"The assassination of Senator Robert Kennedy, race riots, anti-Vietnam War protests cooled Rauschenberg's initial faith in the redemptive powers of working collectively. It did not however completely dissipate."

That was Achim Borchardt-Hume again.

One of the keys to understanding Rauschenberg is his inclusiveness. The other is his endless experimentation. Another is his working in the space between art and life. E.A.T. is a natural outcome of these traits.

Dialogue: a conversation between two or more people as a feature of a book, play, or film. Discussion: the action or process of talking about something in order to reach a decision or to exchange ideas.

Was E.A.T. a dialogue or a discussion? The words have slightly different meanings which are reflected in their origins. The Oxford English Dictionary says of the origins of the word dialogue: it comes from Middle

English: from Old French dialoge, via Latin from Greek dialogos, converse with. The origin of the word discussion is different: it comes from Middle English (denoting judicial examination): via Old French from late Latin discussio(n-).

The physicist David Bohm makes a clear distinction between the two:

"Dialogue comes from the Greek *dialogos*. Logos means *the word*, or in this case, *the meaning of the word*, and *dia* means *through*. A dialogue can be among any number of people, not just two. Even one person can have a sense of dialogue with himself. Dialogue will make possible a flow of meaning out of which may emerge some new understanding which may not have been in the starting point at all.

"Discussion is a process of analysing and breaking up and will not get us far beyond our various points of view. Rather, the object of the game is to win or gain points for yourself."

Dialogue, of course, makes more demands of our ability to participate effectively in a game in which the aim is for everyone to win together. In particular, we must be prepared to question our assumptions and make them explicit, which takes effort when many of them, including the most powerful among them, are held unconsciously.

So was E.A.T. dialogue or discussion? We will probably never know, such is the nature of these art-science and art-technology collaborations - love affairs.

**359**

It is Christmas Eve and the work on 366 continues. Why would it stop?

Something now about *not knowing*! Obviously very relevant to what you have been reading! Here is Elizabeth Fisher and Rebecca Fortnum:

"Artists often begin something without knowing how it will turn out. In practice, this translates as *thinking through doing*. Some of the methods used by artists to drive the creative process or work out problems can seem counter-intuitive or irrational: distraction or relinquishing control, embracing chance and collaboration; following a hunch rather than a rationale; privileging the senses over the intellect. Artists think in ways that stray beyond the methods of other disciplines.

"Knowledge is positive; the unknown is often simply seen as the opposite: it is uncertain, invisible, and incomprehensible. *Not* knowing represents a lack or absence, an inadequacy to be overcome. However, for artists not knowing is not only something not to overcome, but is actually sought, explored and savoured; where failure, boredom, frustration and getting lost are constructively deployed alongside wonder, secrets and play."

And what you are now reading is the result of the process just descried - composition unknown. Scriptovisualism embraces *not knowing* as a process and in a double dose, so to speak - in the descriptive and in the visual.

Here is Johan Huizinga:

"Play is older than culture, for culture, however, inadequately defined, always presupposes human society, and animals have not waited for man to teach them their playing. We can safely assert, even, that human civilisation has added no essential feature to the general idea of play."

Playfulness is found in the poetic. Playfulness is found in art.

There are differences everywhere; and commonalities too. Art and science (and art and technology), may have greater differences, than those that to art-science and art-technology ideologies do subscribe, and vain slaves do become.

Elizabeth Fisher and Rebecca Fortnum say this:

"It is difficult to apply the language and logic of other disciplines to art. Yet it is also true that artistic practices have moved closer to other disciplines, which signals a move to align artistic practice with the knowledge and cultural economies of late capitalism. This is perhaps most evident in recent discourses around art and research."

**360**

It is Christmas Day and the work on 366 most definitely does not stop. But what should we write about today? This:

> On Yasmin, they do discuss, if STEM to STEAM, is but hot air, all moist, and probably cloudy too! Thus to this rattling about in non-viscous cyberspace, we now turn, to add a very humble creation of poetic kind. A very delightful title too, we formed, for this strange way of knowing, through performing; a process, you may note this time, is founded in adapting. We call our strange rendition: *Desiderata for People STARTing to STEAM.*
>
> Go placidly amid the hissing noise of STEAM, and remember what peace there may be in silence for it is hard to find in STARTS and STEAM. As far as possible, without surrender, be on good terms

with all persons, even those hissing STEAM and those STARTing to talk nonsense.

Speak your truth quietly and clearly; and listen to others, even those caught up in the ideologies of STARTS and STEAM; they too have their story, even though it is often quite empty. But pick your acronyms carefully, lest poetics be used to expose shallow graves, where lay foul phantoms of inaccuracies, myths and disingenuous claims, ready to lead astray unwary travellers lost in art's romantic mist.

Avoid loud noisy artists and STEM people STARTing to STEAM; they are vexatious to the spirit. If you compare yourself with others, you may find yourself wondering what lesser god STARTers and STEAMers worship.

Enjoy your achievements as well as your plans. Keep interested in your own career, however humble; it is a real possession in the changing fortunes of time, especially now that STARTS and STEAM are in fashion; these are but a passing interest of the grazing herd, so your humble achievements may in the end outshine all that STARTS and STEAMS.

Exercise caution in your business affairs, for the world is full of trickery; especially when STEAM clouds the true nature of STARTS intentions. But let this not blind you to what virtue there is; many persons strive for high ideals, and everywhere life is full of heroism. Even in STEAM there may be a few - a very few - golden threads.

Be yourself, especially, as others most often are not. Do not feign affection for STEM, especially now its destructive nature is revealed. Neither be totally cynical about art-science and art-technology lovers; for in the face of all aridity and disenchantment love is as perennial as the grass, even the naïve sort that is most of art-science and art-technology. Neither of these actually exists though, for there is only art, which is always made with whatever is at hand - paint, data, glass, code, stone, electronic junk, toys, cables, books, photographs ...

START by learning about DG CONNECT's former European Commissioner - the one that could be said to be responsible for ICT ART CONNECT - who now finds herself investigated for not declaring the holding, while in office, of a directorship in an offshore company; a clear contravention of conditions laid down in the Treaty on the Functioning of the European Union. START too by learning of OLAF's investigation into alleged fraud by the company that led the ICT ART CONNECT project. More revelations will no doubt in time come, when the world learns of the artistic voices that DG CONNECT silenced; so reflect before STARTing to STEAM, that DG CONNECT's woes might, one day, be yours too.

Take kindly the counsel of the years, gracefully accepting that inexperience and the pursuit of fortune and glory means that much that is proclaimed as new by those STARTing to STEAM will be found in old reports, papers and books. Beware of bodies whose job it is to scrutinise public expenditure – they may ask for a refund of public money spent just reinventing what is in these aging publications.

Nurture strength of spirit to shield you in the sudden misfortune of having to put up with empty-headed people who START talking nonsense. But do not distress yourself with dark imaginings. Many fears are born of fatigue and loneliness, and soon the empty-headed ones will be gone, as no sooner will they START to STEAM, than they will run out of STEAM.

Beyond a wholesome discipline, be gentle with yourself and ignore most of this STEAMing. You are a child of the universe no less than the trees and the stars; you have a right to be here, and pursue a less noisy, more thoughtful course away from the hissing sounds of STEAM.

And whether or not it is clear to you, no doubt the universe is unfolding as it should. Therefore be at peace with God, whatever you conceive Him to be, for probably God has inflicted STARTS and STEAM on those artists and STEM people who do now so much noise make, as way of punishment for sins committed in the past – poetic justice so to speak!

And whatever your labours and aspirations, in the noisy confusion of STEAM, keep peace in your soul. With all the sham, drudgery and broken dreams of STARTS and STEAM, it is still a beautiful world. Be cheerful. Strive to be happy, and stay away from those noisy people STARTing to hiss STEAM.

Thanks to Max Ehrmann who wrote the original Desiderata. And what aesthetic in the above *Desiderata for People STARTing to STEAM* is to be found? Most certainly playfulness is woven into the words, but what else, you may wonder. Is there more? Can you not see? We leave you to wonder and to discover for yourself, for we are not in the business of being totally unambiguous!

**361**

Boxing Day! Do you know what this is? It is a very British holiday of uncertain origins. Many stories exist. It is like the myths that have grown up around Leonardo, C.P. Snow, and E.A.T. Many stories exist! Seems to be the way of things, especially when it comes to matters of art - people say the strangest of things! Not just artists, but also STEM people as well.

Now we invite Simon Sheikh to speak:

"Thinking is not equivalent to knowledge. Whereas knowledge is circulated and maintained through a number of normative practices - disciplines as it were - thinking is here meant to imply networks of indiscipline, lines of flight and utopian questionings. Naturally, knowledge has great emancipatory potentials, but knowledge, in the sense of being what you know, what you have learned, is also a limitation: something that holds you back, that inscribes you within tradition, within certain parameters of the possible. And thus with certain eliminations of what it is possible to think, possible to imagine - artistically, politically, sexually and socially. Secondly, the notion of knowledge production implies a certain placement of thinking, of ideas, within the present knowledge economy, that is to say, the dematerialized production of current post-Fordist capitalism. And here we can see the interest of capital become visible in the current push for standardization of (art) education and its measurability, and for the moulding of artistic work into the formats of learning and research. There is a direct corollary between the dematerialization of the art object, and thus its potential (if only partial) exodus from the commodity form and thus disappearance from the market system, and the institutional re-inscription and validation of such practices as artistic research and thus a knowledge economy commodity."

What is research? Imagine this: you plan to go on a holiday. But where to go is the question? To answer this you collect and compare information about destinations, flights, airlines, hotels, attractions, cuisine, prices, climate, and so forth. You are doing research. You are engaging in knowledge production. It is however personal research, creating knowledge for your own use, even though you may share the results with others. You will not get a research grant from a government research agency to support your personal research.

What is research? Imagine this: a company employs an organisational development consultant to help managers deal with human and organisational problems. The consultant starts by establishing the *as-is* situation. She interviews people, looks at data (absentee rates, employee turnover figures for each department, etc.) and observes meeting, etc. She is doing research. She is engaging in knowledge production. It is company specific knowledge, but with some elements that are generic, in the sense that the consultant has encountered some things before. The company will not get a research grant from a government research agency to support this research, unless the work is part of some programme specifically directed at researching problems about which there is insufficient knowledge.

What is research? Imagine this: a patient in a physician's consulting room. The doctor is asking questions and undertaking tests to try to establish the symptoms so that she can diagnose the disease. She is doing research. She is engaging in knowledge production. It is patient specific knowledge, but with some elements that are generic, in the sense that the doctor has encountered some things before. The physician will not get a research grant from a government research agency to support this diagnostic work, unless there is a programme specifically directed at undertaking medical research into the disease and symptoms that the patient is presenting.

What is research? Imagine this: an artist is working for a PhD in an academic institution. She is doing research. She is engaging in knowledge production. It may not necessarily be generic research. It could be very specific. She will write a thesis. She may also offer an art object or performance to the examiners, but she will still write a thesis. She might get a research grant from a government research agency to support this research.

What type of research do artists do? It is an important question. Being able to recognise when to fund artists' research is important. Being able to recognise when they are not engaging in advocacy is important. Being able to recognise when they are not engaging in perpetuating the culturally created European myth of the artist as a special person, is important. Being able to recognise that, most probably, no one from the world of STEM is able to do any of the above, is important. Being able to use this knowledge is important.

**362**

It is Tuesday – the last one of 2016. Many things are now the last of 2016.

Helga Nowotny wants to speak, so here she is speaking:

"The rise of modern science and technology has led to the vast improvement of material living conditions that provide the potential to lift the remaining parts of the world population out of poverty."

Oh dear another one! Science and technology will not solve the problems of world poverty. Nor did they solve ours. It was capitalism, to a certain extent, through its contribution to the common good, making some use of science and technology that raised the material standard of living, while at the same time making lots of problems and at an extraordinary cost to people and the environment. But for the most part our problems have been solved by governments through policies and laws that reform society. And policies are based on beliefs and values – politics (not science and reason).

We have heard before reductive statement like the one made by Helga Nowotny. Is that not so Charles?

"Yes indeed, for I said the same in 1956."

Tell us what you said Charles.

"Indeed. I said that the scientific industrial revolution would solve the problems of the developing world and lift those people out of poverty."

And did it happen?

"No."

And why is that?

"Because, I have been told, the problems are far more complicated than any scientist or technologist is able to understand."

That is right Charles, they are. But it does not stop those with fragmented and reductive minds, saying these silly things, does it?

"No."

So why did you say what you said Charles?

"Because I am a technocrat."

Yes indeed, and we know that technocracy does not work. Neither will science and technology save humanity. It will help, but it will not of itself save humanity. Only it seems that there are still people who have not understood this, even those who are not STEM people – Helga Nowotny, for example!

So when it comes to the cunning of uncertainty, one thing is not uncertain and that is the way people think – or not!

So what to say next? This: System 1 and System 2; discontinuities; Prometheus; cognitive biases; morphostatic change; cognitive illusions; morphogenic change; we cannot go on as before … We could also be talking about STARTS! Rooted in the past and a product of …

When will people start listening? When they are standing on the edge of doom? Or before?

**363**

On the subject of Rauschenberg and E.A.T., here is Billy Klüver:

"During the *Nine Evenings*, there was an enormous amount of energy and enthusiasm and we had lots of meetings in the bar, in the basement of the Armoury, and around Rauschenberg's kitchen table and out of these discussion, Rauschenberg, Robert Whitman, Fred Waldhauer and I decided that we needed a foundation to continue the kinds of artist-engineer collaborations that *Nine Evenings* had developed. Because of the tax structure in the United States, we needed a non-profit, tax-exempt foundation, thus Experiments in Art and Technology was founded. The name was invented by our lawyer. None of us liked the name because the word *experiments* doesn't denote something that was a finished work. Artists don't experiment. But there it was."

But artists do experiment! Robert Rauschenberg is known as an experimental artist which is why he worked with you!

Let us see what the Oxford English dictionary has to say about this word experiment. Experiment: a scientific procedure undertaken to make a discovery, test a hypothesis, or demonstrate a known fact. Could easily apply to what some artists do. Just drop the word scientific. Many people experiment. This does not imply that it is an unfinished act!

"Vasari's *Lives* promoted a certain understanding of the figurative arts endorsed by the ruler of Florence to whom the *Lives* are dedicated. Cosimo I placed Florentine artists under his protection and control. It is clear that a complex intersection of individuals with different aims and values produced the *Lives* and made it appear in print as if Vasari were the sole author. There may never be a scholarly consensus on the significance and mechanisms of this fabrication, but the point is that there is no possibility of taking at face value the *Life of Leonardo*, cast in the *Lives* as the first modern hero-artist. Biographies of Leonardo, spanning some 400 years, then became entangled in many different contexts involving many different agents. Embroidered anecdotes that may have once been rightly perceived as praise to properly embellish a moralising life story, sank with leaden literalness into the soft sediment of nineteenth and twentieth century popular culture."

That was Claire Fargo speaking. No doubt Leonardo himself contributed to the fabrication. The journal known by the name *Leonardo* has been one of the entangling contexts, where agents have added further to the fabrication.

So those who criticise the religious for their literal interpretation of sacred texts, are found to be guilty of the same act – literal interpretation of the biographical Leonardo scared text.

How will we use this?

**364**

Is rationality normative? This is a philosophical question. One philosophical answer to the philosophical question is this: no! Another answer is: yes!

John Broome's answer is: no. He has examined the arguments for the answer, yes, and rejected them all.

"I have examined all the arguments and I can find no grounds for thinking that rationality is normative. The matter is of course complex, so I will just give you a flavour. Some people think that rationality consists of responding correctly to reasons. Yet on some occasions, there might be a reason for you to achieve something but, without any irrationality on your part, you might well not believe that this reason exists. If you do not believe it exists, then you might well not respond correctly to it, and your failure will not imply any failure of your rationality. Therefore, rationality cannot consist in responding correctly to reasons.

"For example, suppose the fish in front of you contains salmonella. This is a reason for you not to eat it. But there may be no obvious evidence that it contains salmonella. So you might not believe it contains salmonella, and you might eat it, and nevertheless you might be rational. So you are rational even though you do not respond correctly to the reason.

"Many philosophers find this quick objection convincing. As a result, few support precisely the idea that rationality consists in responding correctly to reasons. Instead they support a modified idea that connects rationality with the reasons you *believe* to exist rather than with actual reasons. They say that rationality consists in responding correctly to the reasons that you *believe* to exist, or something of that sort."

Poor Richard Dawkins – lost in ideology and dogma. The rule of science and reason – what nonsense! The fragmented and reductive mind!

Note that the word *believe* appeared again!

"We artists are losing our ability to distinguish between good art and bad art, between high art and low art, between art that is of the spirit and art that just has to do with the profane."

That was Martin Best speaking.

He is of course right, and the place where this is most noticeable is in the domain of the lovers: art-science and art-technology. Here we have people who do not know what is good and what is bad, prepared to state that just about anything is good, and to write books about it.

Scientists, engineers, technologists and mathematicians have become blinded by their science, engineering, technology and mathematics. Some artists too have become blinded by science, engineering, technology and mathematics. Wolpert would question whether they understand science, engineering, technology and mathematics. Some probably do. One suspects that others do not or only have superficial understandings. We are all ignorant of other peoples' knowledge. Some scientists, engineers, technologists and mathematicians are probably ignorant of art, others are not, while some only have superficial understandings. How many people have knowledge of the social sciences?

We have not yet come to a place where we are ready to write about what differentiates good art that uses science and technology, from bad art that uses science and technology. One day we might get there. The lovers will not be happy. All we will say for the moment is that we are dealing with something that is both subjective and complex, but nevertheless important, especially to agencies that fund art, and those STEM agencies that seem to want to venture into an area that they do not understand. It is important to artists as well!

Assessment starts at the individual level of likes and dislikes, which is entirely valid. Yet, when one sees art-science lovers telling other art-science lovers that their work is amazing, and your intuition is telling you that it is not amazing, then what is opening up is an opportunity for ... We will in due course pursue the opportunity. We have actually been pursuing this for several years. We started to do so when we walked away from ICT-ART CONNECT and started a journey of discovery that leads to ... Art-science, art-technology, already looking tired and overly familiar! Time to move on! We did.

This walking away might have been our moment when we, the tortoise, left the hare, ICT-ART CONNECT (STARTS) behind, as it slept by the side of the road waiting for public handouts from the new patron of the arts - the new Medici that is DG CONNECT - along with all the restrictions on artistic freedoms that it is imposing on artists, but which all those involved will deny. Denial, recall, is always the end stage. Another book on STARTS arises in our imagination. The history of Europe is a history of European Unions ...

"I would like to say, on the matter that you raised about art-science, art-technology, or indeed, any art, that I am against interpretation."

That was Susan Sontag making an interesting point. She may have just said something very profound, which might help in our quest.

"The fact is all Western consciousness of and reflection upon art have remained within the confines staked out by the Greek theory of art as mimesis or representation. It is as though this theory that art as such - above and beyond given works of art - becomes problematic, in need of defence. And it is the defence of art which gives birth to the odd vision by which something we have learned to call form is separated off from something we have learned to call content, and to the well-intentioned move which makes content essential and form accessory."

That was Susan again.

And this art that has emerged from the world of STEM - the art-science lovers - that too is also very Ancient Greek, being also mimetic, and sometimes also *methectic*. The Catholic Mass is also *methectic*. The fact that art-science lovers are producing the *methectic* is not surprising - science is the new religion and like all religions it needs *methectic rituals*.

Adolf Hitler was an artist. It shows - in the nature of National Socialism with its use of colour, form, symbols, performances, methectic rituals, etc. And its use of totalitarian art - that which STEM people and DG CONNECT are interested in. The history of Europe is a history of European Unions.

Back to the matter of interpretation: how do we evade the interpreters? This is a question that Susan once asked.

But do we want to evade them? Or do we want to evade them, while at the same time not evade them, so to speak? Contradictory, perhaps, to those who live in the binary Western world? The issue of interpretation is evidently very important, but in what way? A matter for future books, we think.

"I would like to say that I am in favour of interpretation. What I am against are dualities. In particular I am against the dualism of the absolute truth of an interpretation, and complete relativism in interpretation. There is not a single correct interpretation of a text, but neither are there endless interpretations according to *any interpretation is valid* type of thinking."

That was Louise Rosenblatt speaking.

"I too would like to say something, specifically in relation to quality. Let me point you to relational aesthetics which consists in judging artworks on the basis of the inter-human relations they represent, produce, or prompt. Let me also remind you about totalitarian art which stresses symmetry. Let me too remind you that the relationships between art and technology are much more complicated than they were in the 1960s."

That was Nicolas Bourriaud speaking.

The Japanese concept of Ma: the pure and essential void between things, the frame in which objects and events exist, which provides the framing for these. The void in time, the void in space: the spaces between spaces. It is not the same as the Western concept of negative space.

**365**

In a normal year - what ever that may be - the year has 365 days. So today would have been New Year's Eve. This has not been a normal year.

So we turn at the end, to other matters that might be seen as being irrational. At least to some! The matter is: Ludic Systems. Here is Bill Gaver on the matter of Ludic Systems:

"Ludic is an obscure word meaning playful, but in a very wide sense, including learning, exploration, etc. The Ludic perspective does not just characterise people by thinking or achievements, but also by their Ludic engagement with the world: their curiosity, their love of diversion, their explorations, inventions and wonder. Play is therefore not just perceived as mindless entertainment, but an essential way of engaging with and learning about the world and the people in it. Underestimating the importance of this is a key element creating the pervasive distortions that technology makes to culture."

Here then is a new sociotechncial design principle, suited to the age when technology is no longer just the preserve of the professional user in workplace contexts such as government, research, and business. We call it the *non-utility* principle:

*Non-utility Principle*: Information and Communication Technologies (ICTs) in non-workplace contexts serve purposes beyond mere utility, and ICTs should be designed to enable users to achieve emotional fulfilment through play, exploration, and several other dimensions, that are not traditionally associated with workplace environments.

We can add that this same *non-utility principle* can be also applied in workplace contexts. How irrational is that?

Now John Berger speaks:

"In the first essay on my book *Ways of Seeing*, I say that each of the images of paintings I have reproduced in that essay have become part of my argument which has nothing to do with the painting's original independent meaning. The words have quoted the paintings to confirm their own verbal authority. Other essays in the book that are without words, which are just reproductions of paintings, make this distinction more clear."

John Berger also said that the (art) images that were once the preserve of the ruling class, have now become, because of reproduction, ephemeral, ubiquitous, insubstantial, available, valueless, free. They now surround us in the same way language surrounds us. They have entered the mainstream of life over which they no longer, in themselves have power.

He also said that the new language of images could be used differently, conferring a new power. What matters is who uses this language and for what purpose.

Now we know something more about scriptovisualism. We also now know more about Eric Kandel's book, to which we have already referred. We know more about all those people who worship da Vinci. John Berger used the word, mystification, to describe the processes at work when, people like Kandel, use reproduced works of art, to ascribe meanings to these works, which the originals never had. It can also be said that they do so to as Berger noted, for a reason. That reason is that they, the scientists, who are a new privileged minority, are assuming meanings of art and inventing histories, to justify their role as a new ruling class - who are, as all tyrannies have been - the sole arbiters of the truth.

The white frocked finders and keepers of the secrets of the universe are now becoming the keepers of the secrets they say can be found in art; more than that in fact, for they are also becoming the keepers of a new set of - to use John Berger's term - sacred relicts. These new sacred relics are those created by leading edge science - not, you will note by artists - by leading edge science. They even have a journal - they call it *Leonardo* - where the new sacred relics are worshiped.

"It is a mistake to think that when the Constructivist artist uses exact elementary shapes, preferring to draw them with a precision which makes them look like geometry, he is by that fact alone trying to make a scientific communication. He is not and cannot."

That was the Russian constructivist artist Naum Gabo speaking.

Thus do STARTS and STEAM begin to rewrite history, in support of the new ruling class, which is now in the process of establishing the legitimacy of its rule, which cannot be done without art and the changing of the meaning of art to suit the image of STEM and its agenda which is grounded in technocracy and scientism. And when institutions such as STEM begin to appropriate art in such a way - well - you just know you are in trouble! We are in deep trouble! The fascists are back!

Epistolary: (of a literary work) in the form of letters. That is the Oxford Dictionary's definition. An epistolary novel is one written as a series of documents, such as letters. Other documents are also used these days, for example, blog postings. Jane Austin's Lady Susan is an example of the epistolary form.

The epistolary form can add greater realism to a story, because it mimics the workings of real life. It is thus able to demonstrate differing points of view without recourse to the device of an omniscient narrator. Or it can be used to present to the world, that which it now needs to know.

Our book *STARTS: Science, Technology and the Arts: The Artistic Voices that DG CONNECT Silenced* is epistolary, being a series of documents, some of which were originally published as blogs, with others being written in the same style but never published as blogs. There is something of the epistolary about the book you have just read.

Intertextuality: this is the shaping of a text's meaning by another text.

Now to the sum of things: what have we learned from the 365 chapter we wrote and you have just read? We answer by way of a short exegesis:

Humans make up stories. Humans need stories. Humans cannot live without stories. Stories are created and communicated using languages, of which there are many, some of which involve no words. There are many different stories. Stories beget new stories. Stories can be useful as they are a way of knowing. The same story can be told in different ways, but with different consequences. Stories reflect the beliefs, in both a positive and a negative sense, of the age in which they were created. Some stories are timeless. Some stories have silent narratives. Some stories have content that their authors keep hidden in order to deceive. Some stories require interpretation. Some people do not know that there are several interpretations to stories. Some people prefer simple stories. Some stories that seem simple are complex. Some stories that seem complex are simple. Some people have a preference for one type of story, others for a completely different type of story. Most people think their preferred story is better than other peoples' stories. People think that they believe in only one story, but often do not realise that they also believe in other stories. Some people are able to consciously embrace several often contradictory stories, and wonder why other people cannot do this. Some stories can be dangerous. Some stories become monsters that consume everyone and everything. Some stories are dangerous fantasies, which people follow to the edge of doom and beyond, and yet still - they cannot see. Some stories seek the creation of a paradise, which comes in different forms - material or spiritual, rarely both. People become enslaved by their stories. Many people think that only one story can be true. Some people believe that other peoples' stories are false superstitions, and that conversion is necessary, so that all who have different stories, come to believe in one story - the one perfect true story. All stories are bound together in ways that are beyond comprehension. People usually do not want to know how their stories are entwined with other stories. Thus does time become a singularity! The serpent beguiles. It is time to set people free. Time to break those chains! You will not like having your chains broken.

So was Lewis Wolpert right about science and art being strange bedfellows? To some extent he was.

All knowledge is grounded in beliefs. A much longer exegesis will follow in time.

Now it is time for the final words …

**366**

In the end there was the image and the word …

This is *366 – A Scriptovisual Composition Unknown*. Now it is known. It was an artistic investigation of the spaces between spaces in the quest to create a *post-European Era, post-Enlightenment Science*. This is how we will rid the world of technocracy and scientism. When we began to write this book, at about 6am on January 1$^{st}$ 2016, we did not know this. This is how artists think - well some at least!

But the story we have just written is more than that which is in the title stated. It is also a place to begin building a new path – the path of harmony, balance, tolerance and love for ones fellow human beings.

Marcel Duchamp does now speak:

"Art has been turned into religion by some moderns. What matters is life, not art."

We agree. You obviously foresaw the religion of the art-science and art-technology lovers. And DG CONNECT!

Scriptovisualism reveals complementarity, not comparability. In life that is ... Scriptovisualism is not retinal art. It is also anti-art.

DG CONNECT's flawed use of art is founded on much ignorance, but not understanding the difference between retinal and non-retinal art is certainly a significant ignorance.

Have you ever heard of anyone speaking about the convergence of art and the social sciences? There is no such thing as the convergence of art and science. Have you ever heard of anyone speaking about the nexus of social science, technology and the arts? There is no such thing as the nexus of science, technology and the arts.

Social scientists do research. They have had much to say about science, technology and engineering. They have published much on these topics, as have other disciplines from the area known as the humanities. No one waves their arms about when this happens. When artists make claims that they are doing research in science and technology, suddenly there is a lot of arm waving, noise, and general jumping around. But where is the beef? Now that we know why this is happening, we intend to demonstrate, to prove the point, that the beef is quite rare! We will start with STARTS and art-science fakes.

What is left to say? Only this: all conventions are boundaries waiting to be transcended. One can transcend a boundary only if one can first conceive of doing so. Separation is an illusion.

We did transcend boundaries. We did so because we are able to conceive of doing so: scriptovisualism.

And thus it ends as it began; in the darkness.

The light shines in the darkness, and the darkness has not understood it. Did you understand?

The End

**Bibliography**

We gratefully acknowledge the various contributions (some ambiguous) to the performance known as *366 – A Scriptovisual Composition Unknown*, of the following people and organisations:


Allen, P.B., (1995). *Art is a Way of Knowing*. Boston: Shambhala Publications.

Alternattiva Demokratika (2014). *Manifesto European Parliament Election 2014 – We are Europe!* Retrieved from: http://www.alternattiva.org/manifesto-european-parliament-election-2014-we-are-europe/

Andalo, D. (2014). Row over Calls to Abolish the Role of Chief Scientific Adviser for the European Commission. *The Pharmaceutical Journal: A Royal Pharmaceutical Society of Great Britain Publication*. Retrieved from http://www.pharmaceutical-journal.com/news-and-analysis/news/row-over-calls-to-abolish-the-role-of-chief-scientific-adviser-for-the-european-commission/20066028.fullarticle.

Anderson, H,C. (2008). *The Emperor's New Clothes (M.R. James, Trans.)*. Retrieved from https://www.gutenberg.ca/ebooks/andersen-emperor/andersen-emperor-00-h.html. (*Originally published in 1837*).

Andrew, E. (1975). A Note on the Unity of Theory and Practice in Marx and Nietzsche. *Political Theory, 3*(3), pp. 305-316.

Apostle Didymos Judas Thomas (n.d.). *The Gospel of Thomas: Scholars Version*. Retrieved from https://selfdefinition.org/christian/Gospel-of-Thomas-Scholars-Version-15-pages-1961.pdf

Aristotle (n.d.). *De Caelo (J. L. Stocks, Trans.)*. London: Clarendon Press.

Armstrong, J.A. (1973). *Ideology, Politics and Government in the Soviet Union: An Introduction (3rd ed.)*. London: Nelson.

Arts Catalyst (2016a). *Transdidsciplinary Collaboration: Exhibiting Archives and Re-staging Exhibitions*. London: Arts Catalyst.

Arts Catalyst (2016b). *Descriptive Panels Selected from The Story of E.A.T., Written and Compiled by Billy Klüver, Together with Julie Martin, Director of E.A.T.* Arts Catalyst Experiments in Art and Technology Exhibition, 22 September – 29 October. London: Arts Catalyst.

Ascott, R. (2003). *Telematic Embrace: Visionary Theories of Art, Technology, and Consciousness (E.A. Shanken, Ed.)*. Berkeley: University of California Press.

Ascott, R. and Girão, L.M. (Eds.) (2011). *Presence in the Mindfield: Art, Identity and the Technology of Transformation*. Planetary Collegium's 12th Annual Research Conference, Consciousness Reframed: Art and Consciousness in the Post-biological Era. Portugal: Universidade de Averio.

Ascott, R. Facebook Page (n.d). *Quotes taken from Roy Ascott's Facebook Page*. https://en-gb.facebook.com/roy.ascott

Ashby, E. (1958). *Technology and the Academics: An Essay on Universities and the Scientific Revolution*. London: MacMillan & Co.

Atkin, A. (2013). Peirce's Theory of Signs. In: E.N. Zalta (Ed.), *The Stanford Encyclopaedia of Philosophy (Summer Edition)*, Retrieved from https://plato.stanford.edu/archives/sum2013/entries/peirce-semiotics.

Austin, J.L. (1962). *How to do Things with Words: The William James Lectures Delivered at Harvard University in 1955 (J.O. Urmson, Ed.)*. London: Clarendon Press.

Bacon, F. (1660). *New Atlantis*. In Lord Verulam, *New Atlantis: A Platform for Monarchial Government*. London: John Crooke. (*Originally published in 1627*).

Barrie, L. (1991). Oswald de Andrade's "Cannibalist Manifesto". *Latin American Literary Review, 19*(38), pp. 35-47.

Bacon, F. (1902). *Novum Organum (J. Devey, Ed.)*. New York: P.F. Collier. (*Originally published in 1620*).

Barthelme, D. (1997). *Not-knowing: The Essays and Interviews of Donald Barthelme (K. Herzinger, Ed.)*. New York: Vintage International.

Baum, L.F. (1981). *The Wizard of Oz*. London: Fontana Paperbacks. (*Originally published in 1900*).

Beck, U. and Cohn-Bendit, D. (n.d.). *Manifesto for Re-building Europe from the Bottom Up*. Retrieved from http://evs4all.eu/manifesto/

Berger, J. (1972). *Ways of Seeing*. London: Penguin Books.

Berlin, I. (1999). *The Roots of Romanticism: The A.W. Mellon Lectures in the Fine Arts 1965, the National Gallery of Art, Washington, DC (H. Hardy, Ed.)*. London: Chatto & Windus.

Berthoin Antal, A. and Strauβ, A. (2016). Multistakeholder perspectives on searching for evidence of values-added in artistic interventions in organisation. In: U. Johansson Sköldberg, J. Woodilla and A. Berthoin Antal, *Artistic Interventions in Organizations* (pp. 37-59). Abingdon: Routledge.

Bignall, V. and Fortune, J. (1984). *Understanding System Failures*. Manchester: Manchester University Press.

Bingham, J. (2014). *Richard Dawkins: 'immoral' to allow Down's syndrome babies to be born*. Retrieved from: http://www.telegraph.co.uk/news/health/news/11047072/Richard-Dawkins-immoral-to-allow-Downs-syndrome-babies-to-be-born.html

Birkett, D. (2016). Leonardo: Artist and Engineer. *IET E&T Magazine 11*(1), pp. 68-71.

Biswas, A. (2008). Managing Art-technology Research Collaborations. *Int. J. Art & Tech. 1*(1), pp. 66-89.

Blake, W. (1904). *Jerusalem: The Emanation of the Giant Albion (E.R.D. Maclagan and A.G.B. Russell, Eds.)*. London: A.H. Bullen. (*Originally published in 1804*).

Blincoe, N. ( 2010). *Why Does Stephen Hawking Think Science has Overtaken Philosophy?* Retrieved from: https://www.theguardian.com/commentisfree/belief/2010/sep/08/stephen-hawking-philosophy-maths

Bloomberg BusinessWeek (2008). *IG Farben and Hitler: A Fateful Chemistry. How a Company whose Nobel-winning Scientists Discovered Vital Medicines became a Nazi Collaborator*. Retrieved from http://www.bloomberg.com/news/articles/2008-08-13/ig-farben-and-hitler-a-fateful-chemistry.

Bolter J.D. and Grusin, R. (1999). *Remediation: Understanding New Media*. London: MIT Press.

Bohm, D. (1996). *On Dialogue*. London: Routledge.

Bond, H.L. (1997). Introduction. In: *Nicholas of Cusa: Selected Spiritual Writings (H.L. Bond, Trans.)*. New York: Paulist Press.

Borchardt-Hume, A. (2016). Robert Rauschenberg: Five Propositions. In: L. Dickerman and A. Borchardt-Hume (Eds.), *Rauschenberg* (pp. 15-25). London: Tate Publishing.

Bourdieu, P. (1996). *The Rules of Art: Genesis and Structure of the Literary Field (S. Emanuel, Trans.)*. Cambridge: Polity Press.

Bourriaud, N. (2002). *Relational Aesthetics (S. Pleasance & F. Woods, Trans.)*. Dijon, France: Les presses du réel.

Bowden, M.A. (2010). *Creativity and Art*. Oxford: Oxford University Press.

Brady, R.A. (1937). *The Spirit and Structure of German Fascism*. London: Victor Gollancz Ltd.

Brattström, V. (2016). Playing the 'Magic If': A Theatre Director's Perspective on Intervening. In: U. Johansson Sköldberg, J. Woodilla and A. Berthoin Antal, *Artistic Interventions in Organizations* (pp. 164-182). Abingdon: Routledge.

Breitenbach, A. (2009). Teleology in Biology: A Kantian Perspective. In: D.H. Heidemann (Ed.), *Kant Yearbook 1/2009, Teleology* (pp. 31-56). Berlin: Walter de Gruyter.

Breitenbach, A. (2014). Biological Purposiveness and Analogical Reflection. In: I. Goy & E. Watkins (Eds.), *Kant's Theory of Biology* (pp. 131-148). Berlin: De Gruyter.

Breton, A. (1972). *Manifestoes of Surrealism (R. Seaver and H.R. Lane, Trans.)*. Ann Arbor: University of Michigan Press, Ann Arbor Paperbacks.

Brier, S. (2013). Cybersemiotics: A New Foundation for Transdisciplinary Theory of Information, Cognition, Meaningful Communication and the Interaction between Nature and Culture. *Integral Review* 9(2), pp. 220-263.

Brockman, J. (1995). *The Third Culture*. New York: Simon & Schuster.

Broome, J. (n.d.). *Is Rationality Normative?* Retrieved from http://users.ox.ac.uk/~sfop0060/pdf/Is%20rationality%20normative.pdf

Bruijnis, M., Blok, V., Stassen E., and Gremmen, B. (2014). *The Lesser of Two Evils: Killing Day-old Chicks or the use of Genetically Modified Chickens?* Paper presented at the MVI conference, May 22-24, The Hague (The Netherlands). Retrieved from https://www.academia.edu/6954650/_The_lesser_of_two_evils_Killing_day-old_chicks_or_the_use_of_genitically_modified_chickens_Bruijnis_M._Blok_V._Stassen_E._Gremmen_B._._Paper_presentation_at_the_MVI_conference_May_22-24_2014_The_Hague_The_Netherlands_

Burley, D.M. (1974). *Studies in Optimisation*. Leighton Buzzard: Intertext Books.

Burnet, J. (1908). *Early Greek Philosophy (2nd ed.)*. London: Adam and Charles Black.

Butcher, D. (2016). *A World without Down's Syndrome*. Retrieved from: http://www.radiotimes.com/tv-programme/e/fdg4yq/a-world-without-downs-syndrome/

Candy, L. (2006). *Practice Based Research: A Guide*. Retrieved from https://www.creativityandcognition.com/resources/PBR%20Guide-1.1-2006.pdf.

Carlson, S. (2007). *INTERACT Artists in Industry Placements Evaluation Report*. Retrieved from http://www.artsactive.net/descargas/11-3-2008-1710342_INTERACT%20REPORT_SCarlson_jan2007.PDF.

Carrier, D. (2008). *Leonardo* and Leonardo da Vinci. *Leonardo 41*(1), pp. 36-38.

Carson, R. (2000). *Silent Spring*. London: Penguin Modern Classics. (*Originally published in 1962*).

Cennini, C. (1945). Book of the Art. In: R. Goldwater & M. Treves (Eds.) *Artists on Art: From the XIV to the XX century* (pp. 21-26). New York: Pantheon Books.

Chaplin, E. (1994). *Sociology and Visual Representation*. London: Routledge.

Checkland, P. (1981). *Systems Thinking, Systems Practice*. Chichester: John Wiley.

Checkland, P. (1983). OR and the Systems Movement: Mappings and Conflicts. *J. Operational Research Society, 34*(8), 661-75.

Cherns, A. (1976). The Principles of Sociotechnical Design. *Human Relations 29*(8), pp. 783-792.

Cherns, A. (1987). The Principles of Sociotechnical Design Revisited. *Human Relations 40*(3), pp.153-162.

Christie, A. (1999). *And Then There Were None*. London: Harper Collins. (*Originally published in 1939*).

Clarke, D.M. (1982). *Descartes' Philosophy of Science*. Manchester: Manchester University Press.

Collini, S. (1993). *Introduction to Canto Edition of C.P. Snow The Two Cultures*. Cambridge: Cambridge University Press.

Committee for Information Technology and Creativity (2003). *Beyond Productivity: Information Technology, Innovation, and Creativity*. Washington DC: The National Academies Press.

Confederation of British Industry (n.d.). *10 facts about EU Trade Deals*. Retrieved from http://www.cbi.org.uk/business-issues/brexit-and-eu-negotiations/eu-business-facts/10-facts-about-eu-trade-deals-pdf/

Connolly, B. (2012). *The Rotten Heart of Europe: The Dirty War for Europe's Money*. London: Faber and Faber.

Contemporary Culture Index (n.d.) *Transformation: Arts, Communication, Environment*. New York: Wittenborn, Schultz. Retrieved from: http://www.ccindex.info/iw/transformation/

Cooley, M. (1987). *Architect or Bee? The Human Price of Technology*. London: The Hogarth Press.

Copeland, J. B. (2015). The Church-Turing Thesis, In: E.N. Zalta (Ed.), *The Stanford Encyclopaedia of Philosophy (Summer Edition)*, Retrieved from https://plato.stanford.edu/archives/sum2015/entries/church-turing.

Corporate Europe Observatory (2014). *Looking for Real Transparency – An Open Letter to the European Commission's Chief Scientist: A Corporate Attack on the Precautionary Principle, and an Appeal to Mrs Anne Glover's Transparency Commitments*. Retrieved from https://corporateeurope.org/lobbycracy/2013/10/looking-real-transparency-open-letter-european-commission-chief-scientist.

Cox, V. (2016). Creating Human-Centered Cultures with Human-Centered Technologies. In: National Academies of Sciences, Engineering, and Medicine, *Art, Design and Science, Engineering and Medicine Frontier Collaborations: Ideation, Translation, and Realization: Seed Idea Group Summaries* (pp 59-60).Washington, DC: The National Academies Press. doi: 10.17226/23528.

Coxeter, H.S.M. (1979). The Non-Euclidean Symmetry of Escher's Picture 'Circle Limit III'. *Leonardo 12*(1) pp. 19-25.

Culler, J. (2011). *Literary Theory: A Very Short Introduction*. Oxford: Oxford University Press.

Cybersemiotics Facebook Group (n.d.) https://www.facebook.com/groups/cybersemiotics/

Dagstuhl Aesthetic Computing Workshop (2003). *Aesthetic Computing Manifesto. Leonardo 36*(4), p. 255.

Daniel W.G. (2015). Heraclitus. In: E.N. Zalta (Ed.), *The Stanford Encyclopaedia of Philosophy (Fall Edition)*, Retrieved from https://plato.stanford.edu/archives/fall2015/entries/heraclitus.

Darsø, L. (2004). *Artful Creation: Learning-tales of Arts-in-Business*. Frederiksberg: Sanfundslitteratur.

Darsø, L. (2008). Prologue: Interview with Margaret Wheatley. *J. of Management and Organisation, Special Issue on Re-conceiving the Artful in Management Development and Education, 14*(5), pp. 482-85.

Darwin, C. (1968). *The Origin of Species by Means of Natural Selection, or the Preservation of Favoured Races in the Struggle for Life (J.M. Burrow, Ed.)*. Harmondsworth: Penguin Books. (*Originally published in 1859*).

Davis, S.M. (1987). *Future Perfect*. Reading, Massachusetts: Addison-Wesley.

Davis, E. (Presenter) (2016). EU Special: What Would Out Look Like? In: *BBC Newsnight EU Referendum Specials*. London: BBC Two. Available at http://www.bbc.co.uk/programmes/b07crvy7 (YouTube: Search for Oxford Professor on UK Brexit Trade Deals: e.g. https://www.youtube.com/watch?v=3kPOdSFKD-Q)

Dawkins, R. (2006). *The Selfish Gene (30th anniversary ed.)*. Oxford: Oxford University Press. (*Originally published in 1976*).

de Andrade, O., (1928). *Manifesto Antropofago (L. Barrie, Trans.)*. In L. Barrie, (1991). Oswald de Andrade's "Cannibalist Manifesto". *Latin American Literary Review, 19*(38), pp. 35-47.

Denscombe, M. (2008). Communities of Practice: A Research Paradigm for the Mixed Methods Approach. *J. of Mixed Methods Research* 2(3), pp. 270-283.

Deuze. M. (2012). *Media Life.* Cambridge: Polity Books.

Dewey, J. (1920). *Reconstruction in Philosophy.* New York: Henry Holt and Company.

Dewey, J. (2005). *Art as Experience (Perigee ed.).* NewYork: Penguin Group. (*Originally published in 1934*).

Dickens, C., (1995). *A Christmas Carol in Prose being a Story of Christmas.* London: Wordsworth Classics. (*Originally published in 1843*).

Ding, Z., Xi, Z. and Wang, H. (2008). Quantum Mechanics, Control Theory and Quantum Control. *Trans. Inst. Measurement & Control 30*(1), pp. 17-32.

Drake, S. (1960). *Introduction to the Controversy of the Comets of 1618.* Philadelphia: University of Pennsylvania Press.

Earth Sky Lab's Facebook Page (2016, November 16). *Yes, the Future is Shared Ownership, but not of Things Delivered by Magic Anonymous Zero Energy Drones to your Isolated Castle, but in Community.* Retrieved from https://www.facebook.com/EarthSkyLab/. (See also *Delivery drones will mean the end of ownership.* Retrieved from https://www.theverge.com/a/verge-2021/google-x-astro-teller-interview-drones-innovation).

Editors of Encyclopædia Britannica (n.d.). Nature Worship: Celestial Phenomena as Objects of Worship or Veneration. In: *Encyclopaedia Britannica.* Retrieved from https://www.britannica.com/topic/nature-worship/Celestial-phenomena-as-objects-of-worship-or-veneration#toc38294.

Editors Encyclopedia.com (n.d.). Snow, Charles Percy. In *Encyclopedia.com.* Retrieved from http://www.encyclopedia.com/people/history/historians-miscellaneous-biographies/cp-snow.

Ehrmann, M. (n.d.). *Desiderata.* Retrieved from: https://allpoetry.com/Desiderata---Words-for-Life.

Elias, W. (2008). Reflection on Art and Science: Creative Fusion. In: European Commission Directorate General for Research, *Art & Science – Creative Fusion* (pp. 7-8). Luxembourg: Office for Official Publications of the European Commission.

Emets, E. (2015). *KIMA. Clot Magazine.* Retrieved from http://www.clotmag.com/kima.

Encyclopaedia of Mathematics (n.d.). *Variational Problem.* Retrieved from http://www.encyclopediaofmath.org/index.php?title=Variational_problem&oldid=28278.

Epstein, R. (2016). *The Empty Brain: Your Brain does not Process Information, Retrieve Knowledge or Store Memories. In Short: Your Brain is not a Computer.* Retrieved from https://aeon.co/essays/your-brain-does-not-process-information-and-it-is-not-a-computer.

Europe of Freedom and Direct Democracy Group in the European Parliament (n.d.). *Group's Pamphlet.* Retrieved from http://www.efddgroup.eu/images/Brochure/efd_brochure.pdf.

European Commission (2012). *A Blueprint for a Deep and Genuine Economic and Monetary Union: Launching a European Debate.* Retrieved from http://ec.europa.eu/archives/commission_2010-2014/president/news/archives/2012/11/pdf/blueprint_en.pdf

European Commission (2014). *Europe in a Changing World – Inclusive, Innovative and Reflective Societies: Horizon 2020 Workprogramme 2014-2015.* Retrieved from http://ec.europa.eu/research/participants/data/ref/h2020/wp/2014_2015/main/h2020-wp1415-societies_en.pdf.

European Commission (n.d.). *Commission Expert group for the Interim Evaluation of FET Flagships (E03383)* http://ec.europa.eu/transparency/regexpert/index.cfm?do=groupDetail.groupDetail&groupID=3383

European United Left/Nordic Green Left Alliance Group in the European Parliament (n.d.). *Group's Pamphlet.* Retrieved from http://www.guengl.eu/policy/publication/who-we-are-what-we-work-for pamphlet.

EuroScience Open Forum (2016). Arts and Science: The Crossroads of Creativity. In: *ESOF 2016 Programme,* Manchester, UK, July 24-27 (p. 74). Retrieved from http://manchester2016.esof.eu/en/full-programme/downloadable-programme.html

Experiments in Art and Technology (1967). *E.A.T. Manifesto.* Reproduced in J. Scott (Ed.) *Artists in Labs: Processes of Inquiry* (p. 8). Wien: Springer. Also reproduced in document retrieved from http://www.vasulka.org/archive/Writings/EAT.pdf.

Experiments in Art and Technology (1998). *Experiments in Art and Technology: A Brief History and Summary of Major Projects.* Retrieved from http://www.vasulka.org/archive/Writings/EAT.pdf

Fargo, C. (2015). The Absolute Leonardo. *J. of Art Historiography 13* CF1. Retrieved from https://arthistoriography.wordpress.com/13-dec15/.

FEAT Project (2016). *FEAT @ Ars Electronica,* Workshop Result Paper, June 8th, Linz, Austria. Retrieved from http://www.featart.eu/fileadmin/user_upload/FEAT_3rdWS_result_paper.pdf.

Fell, J. (2016). Am I Making a Monster? *IET E&T Magazine 11*(6), pp. 24-28.

Feyerabend, P. (1993). *Against Method (3rd ed.).* London: Verso.

Feynman R.P. and Leighton, R. (1985). *Surely You're Joking Mr Feynman!: Adventures of a Curious Character as told to Ralph Leighton, (E. Hutchings, Ed.).* London: W.W. Norton.

Feynman, R.P. (1995). *The Art of Richard P. Feynman: Images by a Curious Character (compiled by Michelle Feynman).* Basel: G+B Science.

Fisher, E. and Fortnum, R. (2013). *On Not Knowing: How Artists Think.* London: Black Dog Publishing.

Ford, H. (1924). *My Life and Work (6th Australian ed.) (in assoc. with Samuel Crowther)*. Sydney: Cornstalk Publishing Co.

Foucault, M. (1983). *This is Not a Pipe (J. Harkness, Trans.)*. Berkeley: University of California Press.

Francis Crick Institute (2013). *Strategy Document 2013*. Retrieved from https://www.crick.ac.uk/media/131115/tfc_full_document_for_web_single_pages.pdf.

Frayling, C. (1993). Research in Art and Design. *Royal College of Art Research Papers Series* 1(1). London: Royal College of Arts.

Freud, S. (1957). Leonardo da Vinci and a Memory of his Childhood. In: J. Strachey (Ed.), *The Standard Edition of the Complete Psychological Works of Sigmund Freud, Vol. XI (1910), Five Lectures on Psycho-Analysis of Leonardo da Vinci and other Works* (pp. 59-137). London: The Hogarth Press and the Institution of Psycho-Analysis. (*Original essay published in German in 1910*).

Frost, R. (n.d.). *The Road Not Taken*. Retrieved from https://www.poetryfoundation.org/resources/learning/core-poems/detail/44272.

Fukushima on the Globe (n.d.). *Evacuation Orders and Restricted Areas*. Retrieved from http://fukushimaontheglobe.com/the-earthquake-and-the-nuclear-accident/evacuation-orders-and-restricted-areas.

Gabel, R.A. and Roberts R.A. (1987). *Signals and Linear Systems (3rd ed.)*. New York: John Wiley & Sons.

Gabo, N. (1956). Art and Science. In: G. Kepes (Ed.), *The New Landscape in Art and Science* (pp. 61-3). Chicago: Paul Theobald and Co.

Galileo Galilei (1960). *The Assayer (S. Drake, Trans.)*. In: S. Drake and C.D. O'Malley, Eds. *The Controversy of the Comets of 1618*. Philadelphia: University of Pennsylvania Press. (*Originally published in 1623*).

Gambetta, D. and Hertog, S. (2009). Tinker, Tailor, Engineer, Jihadi ... *New Scientist*, 13 June, pp. 26-27.

Gambetta, D. and Hertog, S. (2016). *Engineers of Jihad: The Curious Connection between Violent Extremism and Education*. Princeton: Princeton University Press.

Gaver, W. (2008). *Curious Things for Curious People*. Originally Retrieved from http://www.goldsmiths.ac.uk/interaction/pdfs/36.gaver.curiousThings. inPress.pdf (Now available at https://www.researchgate.net/publication/242534029_CURIOUS_THINGS_FOR_CURIOUS_PEOPLE).

Gaze, T. and Jacobson, M. (n.d.). *An Anthology of Asemic Handwriting*. Publishing house: Adelaide & Minneapolis.

Gee, K. (2000). *Visual Arts as a Way of Knowing*. Los Angeles: The Galef Institute.

Gelb, M. (2004). *How to Think like Leonardo da Vinci: Seven Steps to Genius Everyday*. London: Thorsons.

Genocide Watch (n.d.). *The Eight Stages of Genocide*. Retrieved from http://www.genocidewatch.org/aboutgenocide/8stagesofgenocide.html.

Gent, E. (2016). Big Data's Dark Side. *IET E&T Magazine 11*(9), pp. 32-35.

Gibbs-Smith, C. (1978). *The Inventions of Leonardo da Vinci*. Oxford: Phaidon Press.

Giddens, A. (1971). *Capitalism and Modern Social Theory: An Analysis of the Writings of Marx, Durkheim and Max Weber*. Cambridge: Cambridge University Press.

Glinkowski, P. and Bamford A. (2009). *Insight and Exchange: An Evaluation of the Wellcome Trust's Sciart Programme*. London: Wellcome Trust. Retrieved from https://wellcome.ac.uk/sites/default/files/wtx057228_0.pdf

Glover, A. (2013). *The Role of Science in Future EU Policy Making*. First Spinoza Lecture, November, European Academy of Sciences and the Arts. Retrieved from http://www.euro-acad.eu/downloads/events/Glover_Lecture_Agenda_19_Nov_2013.pdf

Glover, A. and Müller, J.M. (2013). Finding an Element of Trust. *Financial Times*. Retrieved from https://www.ft.com/content/0105b82e-2b67-11e3-bfe2-00144feab7de

GM Free Cymru (2014). *EC's Chief Scientist Anne Glover must Resign*. Retrieved from http://www.gmfreecymru.org.uk/documents/anne-glover-resign.html.

Goldstein, S. (2013). Bohmian Mechanics. In: E.N. Zalta (Ed.), *The Stanford Encyclopaedia of Philosophy (Summer Edition)*, Retrieved from https://plato.stanford.edu/archives/spring2013/entries/ qm-bohm.

Goldwater. R. and Treves, M. (1945). *Artists on Art*. New York: Pantheon Books.

Golomstock, I. (2011). *Totalitarian Art: In the Soviet Union, the Third Reich, Fascist Italy and the People's Republic of China (R. Chandler, Trans.)*. London: Duckworth

Gorbachev. M. (n.d.). BrainyQuote.com. Retrieved from https://www.brainyquote.com/quotes/quotes/m/mikhailgor402972.html

Griffin, A. (2015). *Stephen Hawking: Space Travel will save Mankind and we should Colonise other Planets*.
Retrieved from http://www.independent.co.uk/news/science/stephen-hawking-space-travel-will-save-mankind-and-we-should-colonise-other-planets-10058811.html.

Grene, M. (Ed.) (1969). *Towards a Unity of Knowledge*. Psychological Issues, Vol. VI, No. 2, Monograph 22. New York: International Universities Press.

Grudin, R. (2015). Humanism. In: *Encyclopaedia Britannica*. Retrieved from https://www.britannica.com/topic/humanism.

Habermas, J. (2015). *The Lure of Technocracy*. Cambridge: Polity Press.

Hacking, I. (2012). Introductory Essay. In: *T. Kuhn, The Structure of Scientific Revolutions (4th edition – 50th anniversary)*. Chicago: The University of Chicago Press.

Hamilton, J. (2001). Artists, Scientists and Events. In: J. Hamilton (Ed.), *Fields of Influence: Conjunctions of Artists and Scientists 1815-1860* (pp. 1-30). Birmingham: University of Birmingham Press.

Handy, C. (1996). *Beyond Certainty: The Changing Worlds of Organisations*. London: Arrow Business Books.

Handy, C. (2002). *The Age of Unreason*. London: Arrow Business Books. (*Originally published in 1986*).

Haraway, D. (1988). Situated Knowledges: The Science Question in Feminism and the Privilege of Partial Perspective. *Feminist Studies* 14(3), pp. 575-599.

Harris, C. (Ed.) (1999). *Art and Innovation: The Xerox PARC Artist in Residence Programme*. Cambridge Massachusetts: The MIT Press.

Hartel. P. (1988). *The Brush and the Compass: The Interface Dynamics of Art and Science*. Lanham Maryland: University Press of America.

Haseman, B. (2006). A Manifesto for Performative Research. *Media International Australia incorporating Culture and Policy, theme issue "Practice-ledResearch" 118*, pp. 98-106.

Hassler, S. (2016). STEM Crisis? What about the STS Crisis? *IEEE Spectrum*, October, p. 8.

Hawkins, S. (2001). *The Universe in a Nutshell*. London: Bantam Press.

Heller, P.G. (1995). *Drama as a Way of Knowing*. Los Angeles: The Galef Institute.

Henderson, D. (2010). The Coincidence of Opposites: C.G. Jung's Reception of Nicholas of Cusa. *Studies in Spirituality 20*, pp. 101-113.

Heydenreich, L.H. (2016). Leonardo da Vinci: Italian Artist, Engineer, and Scientist. In: *Encyclopaedia Britannica*. Retrieved from https://www.britannica.com/biography/Leonardo-da-Vinci

Higgins, D. (2001). Intermedia. *Leonardo 34*(1), pp. 49-54.

Holocaust Education and Archive Research Team (n.d.). *I.G. Farben: I.G. Farbenindustrie AG German Industry and the Holocaust*. Retrieved from http://www.holocaustresearchproject.org/economics/igfarben.html.

Huizinga, J. (1955). *Homo Lunden: A Study of the Play-element in Culture*. Boston: The Beacon Press. (*Originally published in 1938*).

Hyman, I. (2009). Filippo Brunelleschi: Italian Architect. In: *Encyclopaedia Britannica*. Retrieved from https://www.britannica.com/biography/Filippo-Brunelleschi.

IET (2016). Was Leonardo Really an Engineer? *IET E&T Magazine* 11(3), Letters Section, p. 26.

Iggulden, A. (2007). Silent Speech. In: E. Barrett and B. Bolt (Eds.), *Practice as Research: Approaches to Creative Arts Enquiry* (pp. 65-79). London: I.B. Tauris.

International Atomic Energy Agency (n.d.). *The International Nuclear and Radiological Event Scale*. Retrieved from https://www.iaea.org/sites/default/files/ines.pdf.

Jaspers, K. (1953). *The Origin and Goal of History (M. Bullock, Trans.).* London: Routledge & Kegan Paul Ltd.

Jefferies, J. (2010). Artist as Researchers in a Computer Mediated Culture. In: H. Gardiner and C. Gere (Eds.), *Art Practice in a Digital Culture* (pp. 27-42). Aldershot: Ashgate Publishing. Copy retrieved from https://cutonthebiasworkshop.files.wordpress.com/2011/05/jj-gere-final-1st-june.pdf

Jeffreys, D. (2008). *Hell's Cartel: IG Farben and the Making of Hitler's War Machine.* London: Bloomsbury Publishing.

Johnson, G. (1988). Process of Managing Strategic Change. *Management Research News 11*(4/5), pp. 43-46.

Jones, T. (1980). A Discussion Paper on Research in the Visual Fine Arts Prepared for the Birmingham Polytechnic, England, in 1978. *Leonardo 13*, pp. 89-93.

Jones, C. (1981). *Design Methods: Seeds of Human Futures.* Chichester: John Wiley.

Jordan, N. (1963). Allocation of Functions between Man and Machine in Automated Systems. *J. Applied Psychology 47*(3), pp. 161-165.

Jung, C.G. (1967). *Memories, Dreams, Reflections. (Aniela Jaffé, Recorder and Editor; Richard and Clara Winston (Trans.).* London: Collins.

Kahnenman, D. (2011). *Thinking, Fast and Slow.* London: Penguin Books.

Kandel, E.R. (2012). *The Age of Insight: The Quest to Understand the Unconscious in Art, Mind and Brain from Vienna 1900 to the Present.* New York: Random House.

Kandinsky, W. (1914). *The Art of Spiritual Harmony (M.T.H. Sadler, Trans.).* New York: Houghton Mifflin Company.

Kant, I. (1892). *Critique of Judgement (J.H. Bernard, Trans.).* London: Macmillan & Co. (*Originally published in 1790*).

Katz, V. (2016). Let's Do It Together. *Tate etc. 38,* Autumn, pp. 42-51.

Keller, E. F. (1982). Feminism and Science. *Signs 7*(3), Feminist Theory, pp. 589-602.

Kepes, G. (Ed.) (1956). *The New Landscape in Art and Science.* Chicago: Paul Theobald & Co.

Kidd, P.T. (1994). *Agile Manufacturing: Forging New Frontiers.* Woking: Addison Wesley.

Kidd, P.T. (1997). *Revolutionising New Product Development: A Blueprint for Success in the Global Automotive Industry.* London: FT Automotive.

Kidd, P.T. (2008a). Agile Holonic Network Organizations. In: G.D. Putnik & M.M. Cunha (Eds.) *Encyclopaedia of Networked and Virtual Organizations* (pp. 35-42). Hershey, Pennsylvania: IGI Global Publications.

Kidd, P.T. (2008b). Towards New Theoretical and Conceptual Frameworks for the Interdisciplinary Design of Information Society Technologies. *Int. J. of Interdisciplinary Social Sciences 3*(5), pp. 39-46.

Kidd, P.T. (2009). On the Social Shaping of the Semantic Web. In: M.M. Cunha, E.F. Oliveira, A.J. Tavares, & L.G. Ferreira (Eds.) *Handbook of Research on Social Dimensions of Semantic Technologies and Web Services* (pp. 341-356). Hershey, Pennsylvania: IGI Global Publications.

Kidd, P.T. (2013a). *Encounter with a Wise Man*. Macclesfield: Cheshire Henbury.

Kidd, P.T. (2013b). *A Tale of Two Deserts*. Macclesfield: Cheshire Henbury.

Kidd, P.T. (2014a). *Enigma*. Macclesfield: Cheshire Henbury.

Kidd, P.T. (2014b). *Moments in Time*. Macclesfield: Cheshire Henbury.

Kidd, P.T. (2016a). Why do we need to Adopt da Vinci as an Engineer? *IET E&T Magazine, 11*(2), Letters Section, p. 24.

Kidd, P.T. (2016b). *STARTS – Science, Technology and the Arts: The Artistic Voices that DG CONNECT Silenced.* Macclesfield: Cheshire Henbury.

Klüver, B. (1998). *Archives: Bell Labs and the Origins of the Multimedia Artist*. Video available at Engineering and Technology History Wiki, http://ethw.org/Archives:Bell_Labs_%26_The_Origins_of_the_Multimedia_Artist#Video

Klüver, B. (2001). The Great Northeastern Power Failure (1966). In: R. Packer and K. Jordan (Eds.) *Multi Media from Wagner to Virtual Reality* (pp. 33-38). New York: W.W. Norton & Company.

Knowlton, K. (1972). Collaborations with Artists: A Programmer's Reflections. In: F. Nake (Ed.), *IFIP working conference on graphic languages*, Vancouver, May 22-26, (pp. 399-418). Amsterdam: North-Holland Publishing.

Koestler, A. (1966). *The Act of Creation*. London: Pan Books.

Koestler, A. (1967). *The Ghost in the Machine*. Chicago: Henry Regnery Company.

Koestler, A. (1989). *The Sleepwalkers: A History of Man's Changing Vision of the Universe*. London: Arkana, Penguin Books. (*Originally published in 1959*).

Kris, E. (1953). *Psychoanalytic Explorations in Art*. London: George Allen & Unwin Ltd.

Kris, E. and Kurz, O. (1979). *Legend, Myth and Magic in the Image of the Artist: A Historical Experiment (A. Laing, Trans.)*. New Haven: Yale Univeristy Press. (*Originally published in German in 1934*).

Kuhn, T. (2012). *The Structure of Scientific Revolutions (4th edition – 50th Anniversary)*. Chicago: The University of Chicago Press. (*Originally published in 1962*).

Kuo, M. (2016). Inevitable Fusing of Specialisations: Experiments in Art and Technology. In: L. Dickerman and A. Borchardt-Hume (Eds.), *Rauschenberg* (pp. 260-271). London: Tate Publishing.

Kwastek, K. (2013). *Aesthetics of Interaction in Digital Art*. Cambridge Massachusetts: The MIT Press.

Lanteigne, M. (2014). The Sino-Swiss Free Trade Agreement. *CSS Analyses in Security Policy.* No. 147, February. Retrieved from http://www.css.ethz.ch/content/dam/ethz/special-interest/gess/cis/center-for-securities-studies/pdfs/CSSAnalyses147-EN.pdf.

Lanteigne, M. (2015). What the EU could learn from Switzerland's Free Trade Agreement with China? *The London School of Economics and Political Science EUROPP European Politics and Policy Blog, May 5th.* Retrieved from http://blogs.lse.ac.uk/europpblog/2015/05/05/what-the-eu-could-learn-from-switzerlands-free-trade-agreement-with-china/

Lao Tsu (n.d.). *Tao Te Ching (Kindle, 1st ed.),* (14 Sept. 2013). Prague: e-artnow.

Lapointe, F-J. (2014). Hybrids are Hubs: Transdisciplinarity, the Two Cultures and the Special status of Artscientists. *Leonardo* 47(3), p. 277.

Lehrer, J. (2008). *The Future of Science… is Art? Fourth Culture.* Retrieved from: http://seedmagazine.com/content/article/the_future_of_science_is_art/

Leslie, C.R. (1951). *Memoirs of the Life of John Constable: Composed Chiefly of his Letters (J. Mayne, Ed.).* Oxford: Phaidon Press. (*Originally published in 1843*).

Li, Yanzu (2014). Ecological Outlook in Chinese Classical Design: From 'Union of Heaven and Man' to 'Following Natural Rules of Heaven'. *Leonardo* 47(2), pp. 170-172.

Lin, Tung-Chi (1947). The Chinese Mind: Its Taoist Substratum. *J. of the History of Ideas* 8(3), pp. 259-272.

Lindgren, N. (1969a). Art and Technology I. Steps toward a New Synergism. *IEEE Spectrum Magazine,* April, pp. 59-68.

Lindgren, N. (1969b). Art and Technology II. A Call for Collaboration. *IEEE Spectrum Magazine,* May, pp. 46-56.

Living Memory Project (n.d.) *Lessons Learnt from LIVING MEMORY @ 1:3 – Listening to and Developing Technology for Ordinary People.* Retrieved from: http://www.i3net.org/ser_pub/services/magazine/february2003/i3originals.pdf.

Los Angeles County Art Museum (1971). *A Report on the Art and Technology Program of the Los Angeles County Museum of Art 1967-1971.* Los Angeles: Los Angeles County Art Museum.

Luyckx, M. (2008). *The Knowledge Society: A Breakthrough Toward Genuine Sustainability.* Cochin, India: Arunachala Press.

MacCurdy, E. (1955). *The Notebooks of Leonardo da Vinci.* New York: George Braziller.

Mackenzie, W.J.M. (1979). *Biological Ideas in Politics: An Essay on Political Adaptivity.* Manchester: University of Manchester Press.

MacKenzie D. and Wajcman, J. (Eds.) (1985). *The Social Shaping of Technology: How the Refrigerator got its Hum.* Milton Keynes: Open University Press.

Maclagan E.R.D. and Russell A.G.B. (1904). *The Prophetic Books of William Blake: Jerusalem.* London: A.H. Bullen.

Majchrzak, A. and Gasser, L. (1992). HITOP-A: A Tool to Facilitate Interdisciplinary Manufacturing Systems Design. *Int. J. Human Factors in Manufacturing* 2(3), pp. 255-276.

Mao, Tse-Tung (1967). *Quotations from Chairman Mao Tse-Tung*. Peking: Foreign Language Press.

Marquis of Lothian (1939). *The Ending of Armageddon or the Federal Principle as the Only Basis for International Peace, Liberty and Prosperity*. London: Federal Union.

Marx, K. (1954). *Capital: A Critique of Political Economy Vol. I. (S. Moore, E. Aveling, Trans., F. Engels, Ed.)*. Moscow: Progress Publishers. (*Originally published in 1867*).

Marx, K. and Engels, F. (1967). *The Communist Manifesto*. London: Penguin Books. (*Originally published in 1848*).

Matarasso, F. (2005). *How the Light Gets In: The Value of Imperfect Systems of Cultural Evaluation*. Retrieved from https://parliamentofdreams.files.wordpress.com/2012/05/2005-how-the-light-gets-in-matarasso.pdf.

Malina, R. (2006). Welcoming Uncertainty: The Strong Case for Coupling the Contemporary Arts to Science and Technology. In: J. Scott (Ed.) *Artists in Labs: Processes of Inquiry* (pp. 15-23). SpringerWeinNewYork.

Malina, R. (2008). A Call for New Leonardos. *Leonardo 41*(1), p. 2.

McGill, V.J. and Parry, W.T. (1948). The Unity of Opposites: A Dialectical Principle. *Science & Society* 12(4), pp. 418-444.

McLuhan, M. (2001). *Understanding Media: The Extensions of Man*. London: Routledge Classics. (*Originally published in 1964*).

McManus, I.C. (2006). Measuring the Culture of C.P. Snow's *Two Cultures*. *Empirical Studies of the Arts* 24(2), pp. 219-227.

McNiff, S. (1998). *Art-based Research*. London: Jessica Kingsley Publishers.

Miles, M. (2014). *Eco-Aesthetics: Art, Literature, and Architecture in a Period of Climate Change*. London: Bloomsbury Publishing.

Miller, P. (1998a). The Engineer as Catalyst: Billy Klüver on Working with Artists. *IEEE Spectrum Magazine*, July, pp. 20-29.

Miller, P. (1998b). Technology for Art's Sake. *IEEE Spectrum Magazine*, July, pp. 30-37.

Miller, A.I. (2014a). *Colliding Worlds: How Cutting Edge Science is Redefining Contemporary Art*. London: W.W. Norton & Co.

Miller, A.I. (2014b). *Colliding Worlds: How Cutting Edge Science is Redefining Contemporary Art. A Book Reading and Signing with Arthur I. Miller.* DASAR - DC Art Science Evening Rendezvous Video Recording. Available at https://www.youtube.com/watch?v=xW5-kyZ8LUM

Mirand, C. (2016). *Fully Integrating Arts in the S&T Research and Innovation Agenda: The Role of Imagining and Making in the Creation of Knowledge for Innovation.* Retrieved from: https://ec.europa.eu/futurium/en/content/fully-integrating-arts-st-research-and-innovation-agenda-role-imagining-and-making-creation.

Moedas, C. (2015). *European Commissioner's Speech on Open Innovation, Open Science and Open to the World.* Retrieved from http://europa.eu/rapid/press-release_SPEECH-15-5243_en.htm.

Mondrian, P. (1945). *Plastic Art and Pure Plastic Art.* New York: Wittenborn & Co.

Monod, J. (1972). *Chance and Necessity: An Essay on the Natural Philosophy of Modern Biology (A. Wainhouse, Trans.).* London: Collins.

Mortensen, B.M.E. (1998). August Strindberg: Swedish Dramatist. In: *Encyclopaedia Britannica.* Retrieved from https://www.britannica.com/biography/August-Strindberg.

Needham, J. (1927). *Man a Machine. In Answer to a Romantical and Unscientific Treatise Written by Sig. Eugenio Rignano and Entitled 'Man not a Machine'.* London: Kegan Paul, Psyche Miniatures.

Needham, J. (1954). *Science and Civilisation in China: Vol. 1 Introductory Orientations.* Cambridge: Cambridge University Press.

Needham, J. (1956). *Science and Civilisation in China: Vol. 2 History of Scientific Thought.* Cambridge: Cambridge University Press.

Needham, J. (1959). *Science and Civilisation in China: Vol. 3 Mathematics and the Sciences of the Heavens and the Earth.* Cambridge: Cambridge University Press.

Needham, J. (1962). *Science and Civilisation in China: Vol. 4 Physics and Physical Technology: Part I Physics.* Cambridge: Cambridge University Press.

Needham, J. (1965). *Science and Civilisation in China: Vol. 4 Physics and Physical Technology: Part II Mechanical Engineering.* Cambridge: Cambridge University Press.

Needham, J. (2004). *Science and Civilisation in China: Vol. 7 Part II General Conclusions and Reflections (K.G. Robinson, Ed.).* Cambridge: Cambridge University Press.

Nelson, E. (1966). Derivation of the Schrödinger Equation from Newtonian Mechanics. *Physical Review 150*(4), pp.1079-1085, 28 October.

Nelson, R. (Ed.) (2013). *Practice as Research in the Arts: Principles, Protocols, Pedagogies, Resistances.* London: Palgrave Macmillan.

Nicholas of Cusa (1997). *Nicholas of Cusa: Selected Spiritual Writings (H.L. Bond, Trans.).* New York: Paulist Press.

Nobel Prize Official Web Site (n.d.). *The Nobel Prize in Physics 1965.* Retrieved from http://www.nobelprize.org/nobel_prizes/physics/laureates/1965/.

Nobel Prize Official Web Site (n.d.). *Richard P. Feynman – Nominations.* Retrieved from http://www.nobelprize.org/nobel_prizes/physics/laureates/1965/feynman-nomination.html.

Normile, D. (2016). Five Years after the Meltdown, is it Safe to Live Near Fukushima? *Science*. doi: 10.1126/science.aaf4136. Retrieved from http://www.sciencemag.org/news/2016/03/five-years-after-meltdown-it-safe-live-near-fukushima.

Nowotny, H. (2016). *The Cunning of Uncertainty*. Cambridge: Polity Press.

OECD (2015). *Frascati Manual*. Retrieved from http://www.oecd-ilibrary.org/docserver/download/9215001e.pdf?expires=1497369061&id=id&accname=guest&checksum=143D3D942691DBAC9D5D28E6E89E00AB.

OECD Centre for Educational Research and Innovation (1972). *Interdisciplinarity: Problems of Teaching and Research in Universities*. Paris: OECD

Oettinger, G. (2016). *European Commissioner, in Charge of the Digital Economy and Society, Words about Artistic Freedom Forming part of the STARTS Consultation*. Retrieved from https://ec.europa.eu/futurium/en/content/open-consultation-starts-science-technology-arts.

Official Journal of the European Union (2016). *Consolidated Versions of the Treaty on European Union and the Treaty on the Functioning of the European Union*. C202, 59, 7 June. Retrieved from http://eur-lex.europa.eu/legal-content/EN/TXT/PDF/?uri=OJ:C:2016:202:FULL&from=EN.

Oltermann, P. (2015). *Jürgen Habermas's Verdict on the EU/Greece Debt Deal – Full Transcript*. Retrieved from https://www.theguardian.com/commentisfree/2015/jul/16/jurgen-habermas-eu-greece-debt-deal

Olynyk, P. (2012). Minding the Gap: Risk Capital and the Myth of the Two Cultures. *Leonardo 45*(1), p. 2.

Orwell, G. (1951). *Animal Farm*. Harmonsworth: Penguin Books. (*Originally published in 1945*).

Owen, W. (1918). *The Next War*. Retrieved from https://allpoetry.com/The-Next-War

Oxford Online English Dictionary. https://en.oxforddictionaries.com/

Paalen, W. (1943). Art and Science, *DYN 3*, pp. 4-9. In: Wolfgang Paalen's DYN: The Complete Reprint, Christian Kloybe (Ed.), 2000, Springer-Verlag View and NY.

Parker, P. and Pollock, G. (Eds.) (1987). *Framing Feminism: Art and the Women's Movement 1970-85*. London: Pandora.

Parkinson, G. (2008). *Surrealism, Art and Modern Science: Relativity, Quantum Mechanics, Epistemology*. New Haven: Yale University Press.

Parry, R. (2014). *Epistêmê* and *Technê*. In: E.N. Zalta (Ed.), *The Stanford Encyclopaedia of Philosophy (Fall Edition)*. Retrieved from https://plato.stanford.edu/archives/fall2014/entries/episteme-techne/

Picasso, P. (1945). Pablo Picasso: An Interview. In: R. Goldwater & M. Treves (Eds.) *Artists on Art: From the XIV to the XX century* (pp. 416-421). New York: Pantheon Books.

Piore, M.J. and Sabel, C.F. (1984). *The Second Industrial Divide: Possibilities for Prosperity*. New York: Basic Books.

Plato (n.d.). *The Republic*. Retrieved from http://classics.mit.edu/Plato/republic.mb.txt

Pledge, H.T. (1939). *Science Since 1500: A Short History of Mathematics, Physics, Chemistry, Biology*. London: HMSO.

Poincaré, H. (1905). *Science and Hypothesis*. London: The Walter Scott Publishing Co.

Polanyi, M. (2009). *The Tacit Dimension*. Chicago: University of Chicago Press. (*Originally published in 1966*).

Popper, K.R. (1968). *The Logic of Scientific Discovery*. London: Hutchinson & Co.

Popper, K.R. (1972). *Conjectures and Refutations: The Growth of Scientific Knowledge*. London: Routledge and Kegan Paul.

Pretty, J. (2008). Agricultural Sustainability: Concepts, Principles and Evidence. *Phil. Trans. R. Soc. B 363*, pp. 447-465.

Quinn, D. (1999). *Beyond Civilization: Humanity's Next Great Adventure*. London: Harmony Books.

Rabbi Akiba Ben Joseph (attributed author) (n.d.). *The Book of Formation (or Sepher Yetzirah) (K. Stenring, Trans.)*. Berwick, ME: Ibis Press.

Radaelli, C.M. (1999). *Technocracy in the European Union*. London: Longman.

Ray, P.H. and Anderson, S.H. (2000). *The Cultural Creatives: How 50 Million People are Changing the World*. New York: Three Rivers Press.

Rees, M.J. (2009). Foreword. In: Royal Society, *Reaping the Benefits: Science and the Sustainable Intensification of Global Agriculture* (p. v). London: Royal Society.

Reichardt, J. (1971). *The Computer in Art*. London: Studio Vista.

Reutlinger, A., Schurz, G. and Hüttemann, A. (2014). *Ceteris Paribus* Laws. In: E.N. Zalta (Ed.), *The Stanford Encyclopaedia of Philosophy (Spring Edition)*. Retrieved from https://plato.stanford.edu/archives/spr2016/entries/ceteris-paribus/

Reynolds, F. and Capps, D. (Eds.) (1976). *The Biographical Process: Studies in the History and Psychology of Religion*. The Hague: Mouton.

Richard Dawkins Foundation for Science and Reason (n.d.). *What We Do*. Retrieved from https://richarddawkins.net/aboutus/

Rifkin, J. (2004). *The European Dream: How Europe's Vision of the Future is Quietly Eclipsing the American Dream*. Cambridge: Polity Press

Rifkin, J. (2005). *The European Dream: How Europe's Vision of the Future is Quietly Eclipsing the American Dream*. Vortrag an der Humboldt-Universität zu Berlin am 18 April (FCE3/05). Retrieved from http://www.whi-berlin.eu/documents/rifkin.pdf

Rinpoche, S. (1993). *The Tibetan Book of Living and Dying*. San Francisco: HarperSanFrancisco

Ritchie, D.G. (1900). Nature and Mind. *Philosophical Review IX*, p. 241.

Robertson, A. (1977). Conrad Hal Waddington. 8 November 1905 – 26 September 1975. *Biographical Memoir of Fellows of the Royal Society 23*, pp. 575-622.

Ronan, C.A. (1978). *The Shorter Science and Civilisation in China: An Abridgement of Joseph Needham's Original Text – Vol. I: Vols. I & II of the Major Series*. Cambridge: Cambridge University Press.

Root-Bernstein, R., *et al.* (2008). Art Fosters Scientific Success: Avocations of Nobel, National Academy, Royal Society and Sigma Xi Members. *J. of Psychology of Science and Technology 1*(2), pp. 51-63.

Root-Bernstein, R., *et al.* (2013). Arts and Crafts: Critical to Economic Innovation. *Economic Development Quarterly 27(3)*, pp. 221-229.

Rosenblatt, L.M. (1978). *The Reader, the Text, the Poem: The Transactional Theory of the Literary Work*. Carbondale and Edwardsville: Southern Illinois University Press.

Rosenblatt, L.M. (1983). *Literature as Exploration (4th ed.)*. New York: The Modern Language Association of America.

Rosenblatt, L.M. (1993). The Transactional Theory: Against Dualisms. *College English 55(4)*, pp. 377-386.

Rosenbrock, H.H. (1985). A Variational Principle for Quantum Mechanics. *Physics Letters 110A*(7,8), pp. 343-346, 12 August.

Rosenbrock, H.H. (1986a). *On Wave/Particle Duality*. *Physics Letters 114A*(1), pp. 1-2, 27 January.

Rosenbrock, H.H. (1986b). Three Approaches to Quantum Mechanics. *Physics Letters 114A*(2), pp. 63-64, 3 February.

Rosenbrock, H.H. (1986c). The Quantum-mechanical Probability Density. *Physics Letters A 116*(9), pp. 410-412, 14 July.

Rosenbrock, H.H. (Ed.) (1989). *Designing Human-centred Technology: A Cross-disciplinary Project in Computer-aided Manufacturing*. London: Springer-Verlag.

Rosenbrock, H.H. (1990). *Machines with a Purpose: Challenging the View that 'In Science Man is a Machine'*. Oxford: Oxford University Press.

Rosenbrock, H.H. (1995). A Stochastic Variational Treatment of Quantum Mechanics. *Proc. R. Soc. Lond. A 450*, pp. 417-437.

Rosenbrock, H.H. (1997). A Correction to a Stochastic Variational Treatment of Quantum Mechanics. *Proc. R. Soc. Lond. A 453*, pp. 983-986.

Rosenbrock, H.H. (1999). The Definition of State in the Stochastic Variational Treatment of Quantum Mechanics. *Physics Letters A 254*, pp. 307-313, 26 April.

Rosenbrock, H.H. (2000). Doing Quantum Mechanics with Control Theory. *IEEE Trans. Automatic Control 45*(1), pp. 73-77.

Rosenbrock, H.H. and Ding, Z. (2008). Quantum Mechanics and Dynamic Programming. *Trans. Inst. Measurement & Control 30*(1), pp. 33-46.

Rotblat, J. (1995). *Nobel Lecture - Remember your Humanity*. Retrieved from http://www.nobelprize.org/nobel_prizes/peace/laureates/1995/rotblat-lecture.html

Rotblat, J. (1999). A Hippocratic Oath for Scientists. *Science 286*, Issue 5444, Nov. 19, p. 1475. doi: 10.1126/science.286.5444.1475

Royal Society (2007). *List of Fellows of the Royal Society 1660 - 2007*. Retrieved from https://royalsociety.org/~/media/Royal_Society_Content/about-us/fellowship/Fellows1660-2007.pdf

Royal Society (2009). *Reaping the Benefits: Science and the Sustainable Intensification of Global Agriculture*. London: Royal Society.

Royal Society/Royal Academy of Engineering (2012). *Shale Gas Extraction in the UK: A Review of Hydraulic Fracturing*. London: Royal Society.

Rubin, P. (1992). 'What Men Saw': Vasari's Life of Leonardo da Vinci and the Image of the Renaissance Artist. In: F. Ames-Lewis (Ed.); A. Bednarek (Asst.), *Nine Lectures on Leonardo da Vinci* (pp. 96-108). London: Birkbeck College, University of London.

Sacks, O. (1991). *Seeing Voices: A Journey into the World of the Deaf*. London: Pan Books.

Sacks, J. (2011). *The Great Partnership: God, Science and the Search for Meaning*. London: Hodder & Stoughton.

Sadler, M.T.H. (1914). Translator's Introduction. In: W. Kandinsky, *The Art of Spiritual Harmony* (pp. ix-xxv). New York: Houghton Mifflin Company

Saini, A. (2012). Social Engineering. *IET E&T Magazine 7*(2).

Salinger, J.D. (2010). *The Catcher in the Rye*. London: Penguin Books. (*Originally published in 1953.*)

Sargent, E. (2016). *States of Mind: Tracing the Edge of Consciousness Exhibition Pamphlet*. London: Wellcome Collection.

Sarukkai, S. (2009). The Missing link in Art-science Discourse or Art and the Social Sciences. *Leonardo* 42(2), p. 106.

Schwartz, L.F. (1992). *The Computer Artist's Handbook: Concepts, Techniques, and Applications*. New York: W.W. Norton & Co.

Science Museum (2016). *Leonardo da Vinci: The Mechanics of Genius Exhibition Pamphlet*. London: Science Museum.

Scudder, H.E. (1899). *The Complete Poetical Works and Letters of John Keats (Cambridge Edition)*. New York: Houghton, Mifflin & Co.

Seuphor, M. (1956). *Piet Mondrian: Life and Work*. New York: Harry N. Abrams, Inc.

Short, K.G. (1997). *Literature as a Way of Knowing*. Los Angeles: The Galef Institute.

Shakespeare, W. (n.d.). *The Merchant of Venice*. Retrieved from http://shakespeare.mit.edu/merchant/full.html

Shakespeare, W. (1983). *Romeo and Juliet (R.E.S Youngs and P. de Ste. Croix, Eds.)*. London: Harper Collins.

Shakespeare, W. (n.d.). *Sonnet 116: Let Me Not to the Marriage of True Minds*. Retrieved from https://www.poets.org/poetsorg/poem/let-me-not-marriage-true-minds-sonnet-116

Shanken, E. (2005). Artists in Industry and the Academy: Collaborative Research, Interdisciplinary Scholarship and the Creation and Interpretation of Hybrid Forms. *Leonardo 38*(5), pp. 415-418.

Shaw, G.B. (1903). *Man and Superman: A Comedy and a Philosophy*. Westminster: Arcibald Constable & Co. Ltd.

Sheikh, S. (2009). Objects of Study or Commodification of Knowledge? Remarks on Artistic Research. *Art and Research* 2(2). Retrieved from http://www.artandresearch.org.uk/v2n2/sheikh.html.

Shelly, M. W. (2003). *Frankenstein or the Modern Prometheus*. London: Penguin Books. (*Originally published in 1818.*)

Shih, H. (1953). The Natural Law in the Chinese Tradition. In: E.F. Barrett (Ed.), *University of Notre Dame Natural Law Institute Proceedings Vol. V*, (pp. 117-153). Notre Dame, Indiana: University of Notre Dame Press.

Shklovsky, V. (1917). *Art as Technique*. Reprinted In: D.H. Richter (Ed.) *The Critical Tradition: Classic Texts and Contemporary Trends, 3rd ed.*, 2006. New York: Bedford Books.

Simanowski, R. (2010). Digital Anthrophagy: Refashioning Words as Image, Sound and Action. *Leonardo 43*(2), pp. 149-163.

Simanowski, R. (2011). *Digital Art and Meaning: Reading Kinetic Poetry, Text Machines, Mapping Art and Interactive Installations*. Minneapolis: University of Minnesota Press.

Simon, F. (2014). *EU Twisting Facts to fit Political Agenda, Chief Scientist Says*. Retrieved from http://www.euractiv.com/section/science-policymaking/news/eu-twisting-facts-to-fit-political-agenda-chief-scientist-says/

Slingerland, E. and Collard, M. (Eds.) (2012). *Creating Consilience: Integrating the Sciences and the Humanities*. Oxford: Oxford University Press.

Smith, A. (1761). *The Theory of Moral Sentiments (2nd ed.)*. London: A. Millar.

Smith, K.K. (1982). Philosophical Problems in Thinking about Organisational Change. In: P.S. Goodman and Associate (Eds.), *Change in Organizations: New Perspectives on Theory, Research, and Practice* (pp. 316-374). San Francisco: Jossey-Bass Inc.

Smith, M.T. (1993). *'All Nature is but Art': The Coincidence of Opposites in English Romantic Literature.* West Cornwall, CT: Locust Hill Press

Smith, R. (2010). British Thought on the Relations between the Natural Sciences and the Humanities, c. 1870-1910. In: U. Feest (Ed.) *Historical Perspectives on Erklären and Verstehen,* (pp. 161-185). Archimedes Series: Vol. 21. Dordrecht: Springer Netherlands.

Smith, N. (2011). Keeping the Lights On. *IET Members News,* Oct. pp. 16-18.

Snow, C. P. (1956). The Two Cultures. *New Statesman,* 6 October.

Snow, C.P. (1961). *Science and Government.* London: Oxford University Press.

Snow, C.P. (1964). *The Two Cultures and a Second Look: An Expanded Version of the Two Cultures and the Scientific Revolution.* Cambridge: Cambridge University Press.

Snow, C.P. (1965). *The Search.* Harmondsworth: Penguin Books. (*Originally published in 1934*).

Sokal, A. and Bricmont, J. (1999). *Intellectual Impostures: Postmodern Philosophers' Abuse of Science.* London: Profile Books.

Somerville, M.A. and Rapport, D.J. (Eds.) (2000). *Transdisciplinarity: Recreating Integrated Knowledge.* Oxford: EOLSS Publishers.

Sontag, S. (2009). *Against Interpretation and Other Essays.* London: Penguin Modern Classics. (*Originally published in 1961*).

Soussloff, C.M. (1997). *The Absolute Artist: The Historiography of a Concept.* Minneapolis: University of Minnesota Press.

Steinbeck, J. (2000). *The Grapes of Wrath.* London: Penguin Modern Classics. (*Originally published in 1939*).

Sypher, W. (1971). *Literature and Technology: The Alien Vision.* New York: Vintage Books.

Swiss Confederation Government (n.d.). *Free Trade Agreement between the Swiss Confederation and the People's Republic of China: Main Text, Appendices and Annexes.* Retrieved from https://www.seco.admin.ch/seco/en/home/Aussenwirtschaftspolitik_Wirtschaftliche_Zusammenarbeit/Wirtschaftsbeziehungen/Freihandelsabkommen/Partner_weltweit/china/Abkommenstexte.html

Swiss Confederation Government (2014). *Fact Sheet. Free Trade Agreement between Switzerland and China.* Retrieved from https://www.seco.admin.ch/seco/en/home/Aussenwirtschaftspolitik_Wirtschaftliche_Zusammenarbeit/Wirtschaftsbeziehungen/Freihandelsabkommen/Partner_weltweit/china.html

Tate Modern (2015). *Alexander Calder Performing Sculpture Exhibition Pamphlet.* London: Tate Modern.

Tate Britain (2016). *Conceptual Art in Britain 1964-1979 Exhibition Pamphlet.* London: Tate Britain.

Taylor, F.W. (1907). On the Art of Cutting Metals. *Trans. ASME 28,* pp. 31-350.

Tharumarajah, A., Wells, A.J., and Nemes, L. (1996). Comparison of the Bionic and Holonic Manufacturing System Concepts. *Int. J. of Computer Integrated Manufacturing* 9(3), pp. 217-226.

The Economist (2011, November 19). *Minds like Machines: Government by Experts Sounds Tempting, Especially in a Crisis. It Can Work. But Brief Stints have the best Chances.* Retrieved from http://www.economist.com/node/21538698.

Toffler, A. (1984). Foreword: Science and Change. In: I. Prigogine and I. Stengers, *Order out of Chaos: Man's New Dialogue with Nature.* London: Fontana Paperbacks.

Tomkins, C. (1981). *Off the Wall: Robert Rauschenberg and the Art World of Our Time.* Harmonsworth: Penguin Books.

Turing, A.M. (1952). The Chemical Basis of Morphogenesis. *Philosophical Trans. of the Royal Society of London. Series B, Biological Sciences,* 237(641), pp. 37-72.

University of Colorado (n.d.). *Variational Calculus Overview.* Retrieved from http://www.colorado.edu/engineering/CAS/courses.d/AVMM.d/AVMM.Ch01.d/AVMM.Ch01.pdf.

University of Lyon (n.d.). *Quantum Mechanics.* Retrieved from http://math.univ-lyon1.fr/~attal/Mescours/QM.pdf

Unknown Authorship (n.d.). *I Ching – The Book of Change (T. Cleary, Trans.).* Boston: Shambhala.

Vasari, G. (1898). *Lives of the Most Eminent Painters, Sculptors, and Architects Vol. II (J. Foster, Trans.).* London. George Bell & Co. (*Originally published 1568*).

Vasari, G. (1945). From his *Lives.* In: R. Goldwater & M. Treves (Eds.) *Artists on Art: From the XIV to the XX century* (pp. 94-99). New York: Pantheon Books.

Visible Language (n.d.). *The Journal of Visual Communication Research: Submissions – Writing Style.* Retrieved from http://visiblelanguagejournal.com/submission.

Voltaire (2005). *Candide, or Optimism (T. Ciffe, Trans.).* London: Penguin Books. (*Originally published in 1759*).

von Hayek, F.A. (1942). Scientism and the Study of Society: Part I. *Economica* 9(35) pp. 267-291.

von Hayek, F.A. (1943). Scientism and the Study of Society: Part II. *Economica 10*(37) pp. 34-63.

von Hayek, F.A. (1944a). Scientism and the Study of Society: Part III. *Economica 11*(41) pp. 27-39.

von Hayek, F.A. (1944b). *The Road to Serfdom.* London: Routledge.

von Hayek, F.A. (1955). *The Counter-Revolution of Science: Studies on the Abuse of Reason.* New York: The Free Press.

Ward, J. (1906). *Naturalism and Agnosticism: The Gifford Lectures Delivered before The University of Aberdeen in the Years 1896-1898 (3rd ed.) Vol. II.* London: Adam and Charles Black.

Waddington, C.H. (1948). *The Scientific Attitude (2nd ed.)*. West Drayton: Penguin Books.

Waddington, C.H. (1969). *Behind Appearance: A Study of the Relations between Painting and the Natural Sciences in this Century*. Edinburgh: Edinburgh University Press.

Waddington, C.H. (1972). *Biology and the History of the Future*. Edinburgh: Edinburgh University Press.

Weitz, M. (1956). The Role of Theory in Aesthetics. *J. of Aesthetics and Art Criticism*, XV, pp. 27-35.

Weismann, D.L. (1968). *Language and Visual Form: The Personal Record of a Dual Creative Process*. Austin: University of Texas Press.

Weizenbaum, J. (1984). *Computer Power and Human Reason: From Judgement to Calculation*. Harmondsworth: Penguin Books.

Wejchert, J. (2004). Foreword. The Human Touch: Reflections on i3. In: D.N. Snowdon, E.F. Churchill and E. Frécon (Eds.), *Inhabited Information Spaces: Living with your Data*. London: Springer.

Whewell, W. (1840). *The Philosophy of the Inductive Sciences Founded upon their History. Vol. I.* London: John W. Parker.

Whewell, W. (1848). *The Philosophy of the Inductive Sciences Founded upon their History. Vol. II.* London: John W. Parker.

Whitaker, E. (1946). *Space and Spirit: Theories of the Universe and the Arguments for the Existence of God*. London: Thomas Nelson and Sons Ltd.

White, H. (1978). *Tropics of Discourse: Essays in Cultural Criticism*. Baltimore: The Johns Hopkins University Press.

Whitehead, A.N. (1927). *Science and the Modern World*. Cambridge: Cambridge University Press.

Whitney, J. (1980). *Digital Harmony: On the Complementarity of Music and Visual Art*. Peterborough, NH: Byte Books.

Wilczek, F. (2015). *Why Physics needs Art to help Picture the Universe*. Retrieved from https://aeon.co/ideas/perspective-is-a-lesson-in-how-science-collaborates-with-art.

Wilde, O. (n.d.). *De Profundis*. Retrieved from http://www.gutenberg.org/files/921/921-h/921-h.htm.

Williams, E. (Ed.) (2013). *An Anthology of Concrete Poetry*. New York: Primary Information.

Wilsdon, J. and Rebecca Willis, R. (2004). *See-through Science: Why Public Engagement needs to Move Upstream*. London: DEMOS.

Wilson, E.O. (1999). *Consilience: The Unity of Knowledge*. London: Abacus.

Wilson, S. (2000). *Myths and Confusions in Thinking about Art/Science/Technology*. Retrieved from http://userwww.sfsu.edu/swilson/papers/wilson.caapaper.html.

Wilson, S. (2012). *Art + Science Now*. London: Thames & Hudson.

Wokoma, I. Shum, L.L., Sacks, L. and Marshall, I. (2005). *A Biologically-inspired Clustering Algorithm Dependent on Spatial Data in Sensor Networks.* Proceedings of the 2nd European Workshop on Wireless Sensor Networks. Piscataway, NJ: IEEE Publications. Retrieved from http://eprints.lancs.ac.uk/27279/1/27279.pdf.

Wolpert, L. (1992). *The Unnatural Nature of Science*. London: Faber and Faber.

Wolpert, L. (2002). *Oath*. Retrieved from http://www.nobelprize.org/nobel_prizes/themes/medicine/wolpert/

Wolpert, L. (2002). *Is Science Dangerous?* Retrieved from https://www.nobelprize.org/nobel_organizations/nobelfoundation/symposia/interdisciplinary/ns120/lectures/wolpert.pdf

Wolpert, L. (2005). *Strange Bedfellows: Superimposing Art and Science as Kindred Concepts may be Fashionable, but is it Justified?* Retrieved from http://www.lablit.com/article/39

Wyndham, J. (1951). *The Day of the Triffids*. London: Michael Joseph.

Yetisen, A.K. *et al.* (2016). Art on the Nanoscale and Beyond. *Advanced Materials 28,* pp. 1724-1742.

Yu Xiaodong (2016). Party Power. *China Report (English edition)*, August, pp. 10-11.

Zilsel, E. (1942). The Sociological Roots of Science. *American J. of Sociology 47*(4), pp. 544-562.

Zuboff, S. (1988). *In the Age of the Smart Machine: The Future of Work and Power*. New York: Basic Books.

**Bibliography - Sacred Religious Texts**

*The Torah*. Translated by the Jewish Society of America. London: Kuperard.

*The Holy Bible*. New International Version. Lutterworth: Gideons International.

*The Holy Qur'an*. Translated by A. Yusuf 'Ali, 1934. New Delhi: Kitab Bhavan.

*The Holy Qur'an*. Translated by Dr Rashad Khalifa, 2010. Smashwords Edition.

*The Holy Qur'an*. Translated by M.A.S. Abdel Haleem. 2010. Oxford: Oxford University Press.

**Bibliography - Dangerous Secular Fantasy Texts**

We are writing in the year 2099 about warnings from history - secular texts that brought misery, torment and death to tens upon tens of millions of people, of all races, religions, and political beliefs, in ways that defy comprehension:

*Mein Kampf* by Adolf Hitler, 1939 ed., J. Murphy (Trans.). London: Hurst and Blackett Ltd. Publishers.

*My Struggle against Religion (aka The God Delusion)* by Richard Dawkins, 2006. London: Transworld Publishers.

Made in the USA
Monee, IL
04 November 2026

67382998R00312